Gallium Arsenide

The Wiley Series in Solid State Devices and Circuits

Edited by

M. J. Howes* and D. V. Morgan†

*Department of Electrical and Electronic Engineering, University of Leeds

† Department of Physics, Electronics and Electrical Engineering, University of Wales, Institute of Science and Technology, Cardiff

Microwave Devices

Variable Impedance Devices

Charge-coupled Devices and Systems

Optical Fibre Communications

Large Scale Integration

Reliability and Degradation

Gallium Arsenide

Gallium Arsenide

Materials, Devices, and Circuits

Edited by

M. J. Howes
Department of Electrical and Electronic Engineering, University of Leeds

and

D. V. Morgan
Department of Physics, Electronics and Electrical Engineering, University of Wales, Institute of Science and Technology, Cardiff

A Wiley–Interscience Publication

JOHN WILEY & SONS
Chichester · New York · Brisbane · Toronto · Singapore

Copyright © 1985 by John Wiley & Sons Ltd.

All rights reserved.

No part of this book may be reproduced by any means, nor transmitted, nor translated into a machine language without the written permission of the publisher.

Library of Congress Cataloging in Publication Data:

Main entry under title:

Gallium arsenide.

   (The Wiley series in solid state devices and circuits)
   Includes index.
   1. Gallium arsenide semiconductors. 2. Metal semi-conductor field-effect transistors. I. Howes, M. J.
II. Morgan, D. V. III. Series.
TK7871.15.G3G35   1985   621.3815′2   84-17351

ISBN 0 471 90048 6

British Library Cataloguing in Publication Data:

Howes, M. J.
   Gallium arsenide.—(Wiley series in solid
   state devices and circuits)
   1. Gallium arsenide semiconductors
   I. Title II. Morgan, D. V.
   537.6′22   TK7871.15.G3.

ISBN 0 471 90048 6

Typeset by Macmillan India Ltd., Bangalore 25

Printed by Page Bros. (Norwich) Ltd.

Contributors

R. T. Chen — Rockwell International, California, USA

I. A. Dorrity — GEC, Hirst Research Centre, Wembley, UK

R. C. Eden — GigaBit Logic Inc., California, USA

F. H. Eisen — Rockwell International, California, USA

K. R. Elliott — Rockwell International, California, USA

J. D. Grange — VG Semiconductors Ltd., Sussex, UK

D. E. Holmes — Rockwell International, California, USA

M. J. Howes — University of Leeds, UK

C. G. Kirkpatrick — Rockwell International, California, USA

F. S. Lee — GigaBit Logic Inc., California, USA

D. L. Lile — Naval Ocean Systems Center, San Diego, California, USA

D. Maki — General Electric Company, Electronics Park, Syracuse, USA

D. V. Morgan — University of Wales, Institute of Science & Technology, Cardiff

S. D. Mukherjee — Cornell University, Ithaca, New York, USA

D. H. NEWMAN   British Telecom Research Laboratory, Ipswich, UK

C. J. PALMSTRØM   Cornell University, Ithaca, New York, USA

J. SMITH   STC Components Ltd., Paignton, Devon, UK

B. TURNER   RSRE, Malvern, UK

B. M. WELCH   GigaBit Logic Inc., California, USA

D. K. WICKENDEN   GEC, Hirst Research Centre, Wembley, UK

D. WIGHT   RSRE, Malvern, UK

D. WOODARD   Cornell University, Ithaca, New York, USA

Series Preface

The Oxford Dictionary defines the word revolution as 'a fundamental reconstruction'; these words fittingly describe the state of affairs in the electronic industry following the advent of solid state devices. This 'revolution', which has taken place during the past 25 years, was initiated by the discovery of the bipolar junction transistor in 1948. Since this first discovery there has been a worldwide effort in the search for new solid state devices and, although there have been many notable successes in this search, none have had the commercial impact which the transistor has had. Possibly no other device will have such an impact; but the commercial side of the electronics industry stands poised, awaiting the discovery of new devices as significant perhaps as the transistor.

Research and development in the field of solid state devices has concerned itself with two important problems. On the one hand we have device physics, where the aim is to understand in terms of basic *physical concepts* the mode of operation of the various devices. In this way one seeks to optimize the technology in order to achieve the best performance from each device. The second aspect of this work is to consider the important contribution of the circuit to the operation of a device. This problem has been called *device–circuit interaction*. It is a great pity that in the past these two major aspects of the one problem have been tackled by separate groups of scientists with little exchange of ideas. In recent years, however, this situation has been somewhat remedied, the improvement being due directly to the very rigorous system specifications demanded by industry. Such demands constantly require greater performance from devices, which can only be brought about by coordinated team work.

The objective of this new series of books is to bring together the two aspects of this problem: device physics and device–circuit interactions. We hope to achieve, by coordinated co-authorship of leading experts in the respective fields, a varied and balanced review of past and current work. The books in this series will cover many aspects of device research and will deal with both the commercially successful and the more speculative devices. Each volume will be an in-depth account of one or more devices centred on some common theme. The level of the text is designed to be suitable for the graduate student or research worker wishing to enter the field of research concerned. Basic physical concepts in semiconductors and elementary ideas in passive and active circuit theory will be assumed as a starting point.

University of Leeds   M. J. HOWES
University of Wales   D. V. MORGAN
December 1983

Preface

Gallium arsenide has for some years been called 'the material of the future' and many engineers and scientists have patiently been waiting to see if the many promises it holds will be realized in practical device and integrated circuit products. At this point in time GaAs has begun to make that important transition from a novel material into the important area of engineering products. There can be little doubt that this step forwards owes much to the development of the MESFET transistor, which has opened the way to both discrete device applications in the microwave area together with the development of an integrated circuit technology to parallel that of silicon. GaAs is not thought of as a replacement for silicon but more as a specialist material which will find a role in high-frequency active devices, for use in both analogue (microwave) integrated circuits (with the potential large market products such as the complete integrated front end receivers for direct home reception in satellite television) and for digital integrated circuits to operate at speeds beyond the capability of silicon systems.

In this book it is sought to bring together both the fundamental and circuit aspects of current GaAs devices. Chapter 1 provides an introduction to the basic physical and electrical properties of GaAs, emphasizing some of its key and novel advantages over silicon: the transferred electron effect, a high electron mobility and, because of its direct band gap, useful optoelectronics properties. Chapters 2 and 3 deal with the growth of device-grade material; Chapter 2 concentrates on bulk ingot growth whilst Chapter 3 considers epitaxial growth.

Technology is also pursued in Chapters 4 and 5 (etching of GaAs; and ion implantation and damage in GaAs), Chapter 6 (ohmic and Schottky contacts), and Chapter 7 (metal–insulator–GaAs structures). At this juncture the book moves into the realm of devices and circuits. Discrete devices are treated in Chapters 8–11: transferred electron devices (Chapter 8), IMPATTS (Chapter 9), MESFETS (Chapter 10), and optical devices (Chapter 11). The final two chapters present and discuss recent results on GaAs integrated circuits:

monolithic microwave integrated circuits (Chapter 12) and digital integrated circuits (Chapter 13).

The editors wish to thank most warmly the authors who have contributed to this volume and are particularly indebted to Professor L. F. Eastman (Cornell University), Dr Ken Gray (Thorn EMI UK), Dr Fred Eisen (Rockwell), and Mr Doug Maki (GE Syracuse USA).

Contents

1. **The Physical and Electronic Properties of GaAs** 1
   D. R. Wight
   1.1 Introduction 1
   1.2 Lattice properties 2
   1.3 Electrical properties 7
   1.4 Optical properties 23

2. **Growth of Bulk GaAs.** 39
   C. G. Kirpatrick, R. T. Chen, D. E. Holmes, and K. R. Elliott
   2.1 Introduction 39
   2.2 Basics of LEC growth technique 40
   2.3 Progress in LEC technology 43
   2.4 Application of LEC GaAs to device fabrication 88
   2.5 Conclusions 90

3. **Epitaxial Growth of GaAs** 95
   I. A. Dorrity, J. D. Grange, and D. K. Wickenden
   3.1 Introduction 95
   3.2 Liquid phase epitaxy 95
   3.3 Vapour phase epitaxy 99
   3.4 Molecular beam epitaxy 105
   3.5 Intercomparison of techniques 112

4. **Etching and Surface Preparation of GaAs for Device Fabrication** . 119
   S. D. Mukherjee and D. Woodard
   4.1 Introduction 119
   4.2 Wet etching 123
   4.3 Anodic etching 143
   4.4 Surface preparation 147

4.5 Dry etching 151
4.6 Conclusions 157

5. Ion Implantation and Damage in GaAs 161
D. V. Morgan and F. H. Eisen
5.1 Introduction 161
5.2 Ion–crystal interaction 162
5.3 Radiation-induced damage and semi-insulating layers 166
5.4 Ion implantation doping of GaAs. 171
5.5 Ion-implanted semiconductor devices and integrated circuits. . 180
5.6 Conclusions 187

6. Metallizations for GaAs Devices and Circuits 195
C. J. Palmstrøm and D. V. Morgan
6.1 Introduction 195
6.2 Electrical properties of metal/GaAs systems 195
6.3 Schottky barrier metals 206
6.4 Ohmic contacts 219
6.5 Conclusions 253

7. Metal–Insulator–GaAs Structures 263
D. L. Lile
7.1 Introduction 263
7.2 The MIS system 265
7.3 Device technology 276
7.4 Device performance 283
7.5 Status 288

8. Transferred Electron Devices 299
M. J. Howes
8.1 The transferred electron device 299
8.2 Electronically tuned oscillators: YIG tuning 307
8.3 Varactor tuned TEOs 308
8.4 Operation of pulsed TEOs 315
8.5 Transferred electron device amplifiers 321

9. GaAs IMPATT Diodes 331
J. Smith
9.1 Introduction 331
9.2 Basic model 331
9.3 Technology 350
9.4 Present performance of GaAs IMPATTS 354
9.5 Future trends 357

Contents

10. GaAs MESFETs **361**
B. Turner
10.1 Introduction 361
10.2 Fabrication technology, device parameters, and equivalent circuit. 370
10.3 GaAs MESFET theory—analytical methods. 383
10.4 GaAs MESFET theory—short gates and computer modelling. 394
10.5 Low-noise GaAs MESFETS 400
10.6 GaAs power MESFETS 408
10.7 Special MESFET configurations. 417

11. GaAs Optoelectronic Devices **429**
D. H. Newman
11.1 Introduction 429
11.2 Principles of device operation. 430
11.3 Light-emitting diodes 431
11.4 Lasers 436
11.5 Other components 445

12. GaAs Microwave Monolithic Circuits **453**
D. Maki
12.1 Introduction 453
12.2 Transmission lines 454
12.3 Lumped elements. 459
12.4 Active devices 473
12.5 Integrated circuit processing 477
12.6 Monolithic circuit design 485
12.7 Conclusions. 512

13. GaAs Digital Integrated Circuit Technology **517**
B. M. Welch, R. C. Eden, and F. S. Lee
13.1 Introduction 517
13.2 Why GaAs is superior for high-speed digital ICs 518
13.3 Comparison of GaAs device approaches for IC applications . 524
13.4 Logic approaches for digital GaAs ICs 534
13.5 GaAs IC fabrication technologies 545
13.6 Performance of GaAs digital ICs 556

Index. **575**

Gallium Arsenide
Edited by M. J. Howes and D. V. Morgan
© 1985 John Wiley & Sons Ltd

CHAPTER 1

The Physical and Electronic Properties of GaAs

D. R. WIGHT

1.1 INTRODUCTION

The proprties of gallium arsenide have been the subject of intensive study for over 20 years and a very large literature has evolved. The reasons which gave rise to such intense interest are clear:

(a) GaAs is a compound semiconductor combining group III and group V elements from the same row in the periodic table as the archetypal group IV semiconductor, germanium. The charge exchange inherent in the bonding of the lattice adds an ionic component to the predominantly covalent bond found for Ge and the band gap is expected to increase, offering among other things a wider range of electrical resistivity and the maintenance of extrinsic semiconductor behaviour to high temperatures—features of value to the device physicist and engineer.

(b) Because substitutional impurity atoms from groups II, IV, and VI will act as shallow donors and acceptors in the GaAs lattice, extra degrees of freedom are available when compared to the situation in germanium.

(c) When the whole family of III–V compounds is envisaged, from InSb to AlP including interperiodic compounds such as InP, AlSb, and alloys such as GaAsP (the first commercially significant III–V semiconductor) or GaAlAs (which allows the band gap to be changed without significantly changing the interatomic distance), the potential advantages over the group IV sequence Sn, Ge, Si, C (diamond) are even more obvious. The study of GaAs as a prototype is clearly justified in both an academic and an applied sense.

The true significance of GaAs as a compound semiconductor thus reaches fields outside those covered by this book, and reviewing the present state of knowledge would be a truly mammoth task. Because this book is oriented towards device engineers and scientists, we will restrict ourselves to the properties of the binary compound only, and dwell only on those properties that offer a significant potential advantage in known device functions when compared to the market leader, silicon. The room-temperature properties will be emphasized because of their relevance to the operation of the vast majority of devices, and thus a large amount of low-temperature data have been omitted.

In making this restriction of the field, the author is aware of neglecting the most powerful feature of GaAs technology, which when viewed with hindsight at the turn of the century may have made the most significant impact on the world of silicon electronics. This feature is the ability to form lattice-perfect heterojunction structures with GaAlAs either abrupt to a single atomic plane or graded over any arbitrary distance. Not only does this offer major new degrees of freedom in conventional device design and operation (FETs, bipolar transistors, LEDs, and detectors) but it also opens up a whole new field of semiconductor device physics involving the properties of quantum wells and superlattices, where the periodic potential structure of the semiconductor crystal is under direct control in one dimension.

An attempt has been made to organize the material in a consistent manner: Sections 1.2, 1.3, and 1.4 cover lattice properties, electrical properties, and optical properties, respectively. The major points of advantage over Si are highlighted by the inclusion of Si data in some of the figures and further comments in the text.

Simple discussions of the underlying physical principles are included in some areas and it is hoped that readers who are not semiconductor physicists will find this interesting and helpful.

The references cited seldom refer to the most significant original publications and in the main refer to books and review articles that give a much more comprehensive treatment of the individual topics and their literatures.

1.2 LATTICE PROPERTIES

1.2.1 Phase equilibria

The provision of single-crystal material is central to semiconductor technology and a number of growth techniques have been established for GaAs that have allowed the current knowledge of its properties to be established. Bulk ingots are most usually produced using the liquid-encapsulated Czochralski (LEC) technique[1] in which the crystal is pulled from a melt through a molten boric oxide encapsulant, at the melting temperature of GaAs (1513 K). While this is a lower temperature than that used for Si pulling (1690 K), the equilibrium overpressure of As of 1.0 atm ($As_2 + As_4$ species) and the problems of stoichiometry control must be regarded as negative features in comparison with Si. The Bridgeman technique[1] is also used to produce ingot GaAs especially when highly conducting n-type material is needed, but this imposes size restrictions and tends to produce material with more variable electrical properties. LEC crystals with constant 2 inch diameter are the current industry standard, with a (100) pulling axis so that square dice can be cleaved when chips are processed. A 3 inch diameter capability is also available and will become more usual as volume requirements are established.

Material of the highest electronic quality is produced using epitaxial techniques at temperatures between 900 and 1100 K. These low temperatures enable both impurity and stoichiometric defects to be controlled at concentrations below 10^{15} cm^{-3}, which cannot be achieved in ingots. Liquid phase epitaxy[2] from gallium solution and hydride or chloride vapour phase epitaxy were initially successful and have been developed into small-scale production for a variety of electrical and optical devices. More recently, molecular beam epitaxy (MBE) and metal-organic chemical vapour deposition (MOCVD) processes have emerged,[2] which offer superior control of dopants and alloy constituents as well as excellent surface morphology. Of these epitaxial techniques, MBE and MOCVD offer significant advantages over the standard silane process used to produce epitaxial Si, largely because of their lower growth temperatures. It is interesting to note that the growth of silicon using MBE is only now beginning to be studied in depth, and that this represents a reversal of the usual situation where GaAs workers strive to exceed the power of silicon-based technology.

In all growth processes the crystal is exposed to either a vapour or liquid (Ga + As) phase, and the free concentrations of arsenic and gallium species are parameters controlling the growth rates and stoichiometric or impurity defect concentrations. Knowledge of the equilibrium concentrations is therefore very important and these are shown in Figure 1.1 for both the liquid and vapour phases.

Note that the equilibrium arsenic species in the vapour phase are molecular As$_2$ and As$_4$ and that the equilibrium pressure of free arsenic atoms is insignificant. Below 950 K the gallium and arsenic vapour species have approximately equal equilibrium pressures, which implies that GaAs evaporates congruently below this temperature. Above 950 K the excess pressure of the arsenic species over the gallium species indicates that arsenic is lost preferentially during (arsenic-free) heat treatments, giving rise to a gallium-rich liquid on the surface of the crystal.

1.2.2 Crystal structure

GaAs has a density of 5.3174 g cm^{-3} and crystallizes into the zinc blende structure, which consists of two equivalent, interpenetrating face-centred cubic lattices, one containing Ga atoms and the other As atoms. The two sublattices are separated by 2.447 93 Å along the body diagonal of the unit cube whose side length is 5.653 25 Å at 300 K.[5] The lattice is broadly equivalent to that of silicon, which has a diamond structure, but the presence of dissimilar atoms gives rise to

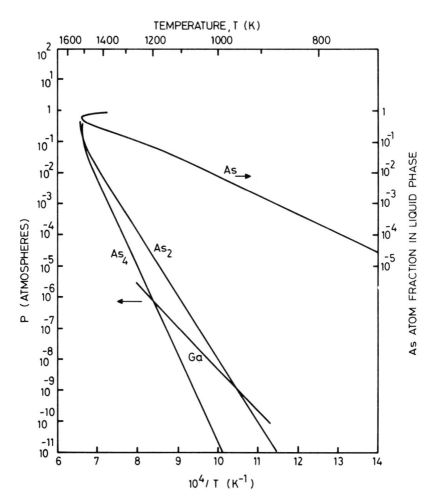

Figure 1.1 Experimental data on the equilibrium vapour pressures above solid (stoichiometric) GaAs and the arsenic concentration in the (Ga + As) liquid phase in equilibrium with GaAs. The vapour phase data are due to Arthur[3] and Richman[4] as compiled by Blakemore[5] and the liquid phase data are due to Panish[6]

some significant differences in the crystal properties. A shift of valence charge from gallium to arsenic atoms produces a mixed (ionic/covalent) bond compared to the covalent bond in germanium and silicon. This has very significant effects on the electronic band structure and increases the bond strength—the melting point of GaAs is higher than that of Ge. However, the hardness of GaAs is between 4 and 5 and that of Ge is 6 on Mohs' scale. The degree of charge transfer of the bond is about 0.3 e,[5] compared with the substantially complete transfer (1.0 e) of the alkali halides (NaCl, etc.). The symmetry of the GaAs crystal is lower than that of silicon and only the latter possesses a centre of symmetry, which makes the absorption of infrared radiation by optical phonons a forbidden process in silicon but allowed in GaAs (see Section 1.4).

A further consequence of the presence of dissimilar atoms is a marked difference in the chemical activity of opposite (111) crystal faces in GaAs. If one considers the (111) planes in the crystal, it can be seen (Figure 1.2) that all gallium atoms are coplanar and that all As atoms are coplanar and separated by only 0.8 Å from the adjacent gallium plane. The next nearest gallium plane is 2.4 Å away. The chemical activity of the (111) face in etching experiments is determined by the ease with which these closely spaced pairs of atom planes can be removed by chemical attack, which in turn depends on whether the pair of planes presents its gallium or arsenic face. The (111)A or gallium face (vertically uppermost in Figure 1.2) is resistant to many etches that attack the (111)B face quite readily, and it is noticed that stacking faults and dislocations promote local acceleration of the etches on (111)A surfaces, producing a rough surface finish covered with etch pits. Various etches show further orientation selectivities, which are used in device fabrication technology to produce sloping or undercut mesa structures.[7]

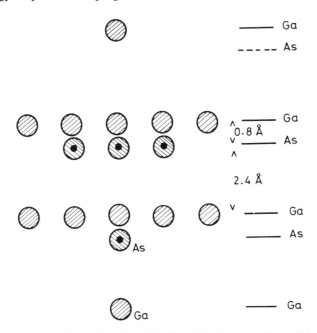

Figure 1.2 A unit cube of the Ga sublattice viewed parallel with the (111) planes along the [11̄2] direction. Only those As atoms located within the unit cube are shown and when the extended lattice is included the atoms are seen to lie in the planes shown

When mechanically stressed at 300 K, a crystal wafer cleaves most readily on the (110) planes. If stresses less than those required to fracture the crystal are applied at elevated temperatures (say 1100 K), plastic deformation occurs. Slip on (110) planes is preferred, with large densities of dislocations being generated along the slip plane. Similar slip phenomena are also seen when thermal expansion stresses are generated by transient local heating using, for example, laser or electron beams.

1.2.3 Elastic and vibrational properties

The elastic properties of GaAs are not isotropic and three elastic moduli are required to describe its elastic response to small stresses using tensor analysis. The degree of anisotropy exhibited is typically less than a factor of 2 for properties such as Young's modulus and the speed of sound; Table 1.1 lists the major parameters of GaAs[5] for the principal [100] directions (where appropriate) to give an overview of its properties compared to silicon.[8]

Table 1.1 Elastic properties (300 K)

	GaAs	Si	Units
Bulk modulus	75.5	98	10^{10} dyne cm^{-2}
Young's modulus (100)	85.5	131	10^{10} dyne cm^{-2}
Shear modulus (100)	32.6	51	10^{10} dyne cm^{-2}
Poisson's ratio (100)	0.55	0.28	
Speed of sound (100)			
longitudinal waves	4.73	8.4	10^5 cm s^{-1}
transverse waves	3.35	5.8	10^5 cm s^{-1}

The vibrational properties of the GaAs crystal are very important in determining the fundamental limits to some optical and electrical properties of immediate practical significance. They also give rise to more extensive phenomena that have attracted much academic research but are perhaps best viewed as details which can aid the analysis of defect properties or physical mechanisms by the applied scientist. The electronic conduction properties of GaAs are fundamentally limited by electron–phonon scattering, which is clearly of paramount significance in this sense, whereas magneto–phonon effects and phonon cooperation in impurity absorption and luminescence phenomena are less significant and will not be covered in this chapter.

The lattice dynamics of GaAs have been studied using neutron scattering, which enables the full phonon dispersion curves to be established. Plots of the quantum energy (or frequency v) vs wave vector q for the four phonon species (longitudinal optical, transverse optical, longitudinal acoustic, and transverse acoustic) are shown in Figure 1.3. As with the more familiar electron dispersion curves, which describe the electronic band structure of materials, the vibrational lattice waves are analysed along the principal crystal axes in the first Brillouin zone of reciprocal space. The optical modes are restricted to a relatively narrow band of energies above the acoustic modes, giving rise to a large peak in the density of phonon states just below the maximum single-phonon energy, which is the long-wavelength ($q \to 0$) LO phonon with a frequency of 8.6×10^{12} Hz (energy = 36 meV). This feature plays a significant role in the infrared optical properties of GaAs (see Section 1.4). The slope of the acoustic phonon curves near $q = 0$ determines the velocity of sound (v) in the crystal ($v/q = v\lambda = v$), and the anisotropy mentioned above can be deduced with both the LA and TA branches having slopes which depend on the propagation direction in the crystal.

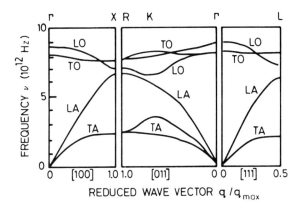

Figure 1.3 The phonon dispersion curves of GaAs (300 K).[9] The quantum energy is obtained by multiplying the frequency by Planck's constant (h)

1.3 ELECTRICAL PROPERTIES

1.3.1 Band structure

The electronic band structure of a semiconductor is its most significant attribute. It determines the major electrical and optical properties from which device functions are generated. A complete and accurate description of the band structure is difficult to obtain because experimental data tend to yield accurate but only limited information at particular points in the Brillouin zone. The elucidation of the band structure in GaAs has therefore involved a sequence of theoretical attempts to describe the real crystal completely under a range of approximations and assumptions, with considerable uncertainty being generated in the absolute and differential electron energies predicted. On the other hand, improved understanding of the experimentally discernible parameters has been achieved progressively so that the various theoretical treatments have improved as time has elapsed. The current state of play is shown in Figure 1.4, which is the result of a pseudopotential calculation. A similar silicon band structure is included in the figure for comparison. In the perfect intrinsic crystal at low temperature, the electron states below the zero energy reference level (set at the top of the valence band states in the figure) are filled and those above this energy are empty conduction band states.

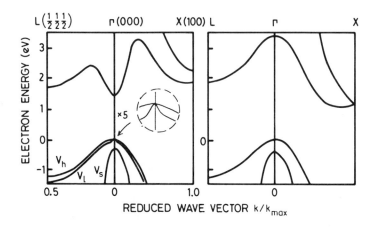

Figure 1.4 The partial band structures of GaAs[5, 10] and Si[11] at 300 K, showing the uppermost valence bands and the lower conduction bands along two of principal directions in the crystals. The magnified ($\times 5$) insert shows the detail at Γ of the light and heavy hole bands in GaAs (V_l and V_h)

When the temperature is raised or impurity dopants are added, vacant states (holes) are generated at the highest-lying valence band states and the lowest conduction band states become occupied.

In GaAs and Si this valence band maximum occurs at the zero wave vector point (Γ) and the nature of the bands at this point controls simple hole conduction processes. The minimum allowed energy for conduction band states also occurs at Γ in GaAs but is located near the X point in silicon, so the electron wave vector has a finite magnitude and direction [100] for conduction electrons in Si. This difference has profound effects.

First, the energy gap separating the valence and conduction band extrema is larger in GaAs (1.42 eV) than in Si (1.08 eV). Thermal generation of carriers (electrons and holes) across this gap at 300 K thus produces lower steady-state carrier densities in GaAs (10^6 cm^{-3}) compared to Si (10^{10} cm^{-3}), so more highly insulating materials and a superior device isolation capability are potentially available in GaAs. Further, the temperature at which thermal generation across the band gap overrides any particular carrier density achieved by doping in devices is higher in GaAs and a better temperature range of device functions ensues.

Secondly, and in a similar manner to that observed in the phonon dispersion curves, the shape of the conduction energy bands near this minimum determines the motion of the electrons in diffusion processes or under applied electric fields. Under the plane wave approximation for free electrons, the mass of the particle can be determined by the second differential of energy with respect to momentum

$$\mathrm{d}^2 E / \mathrm{d}k^2 = \hbar^2 / m^*$$

(where m^* = 'curvature' effective mass) and a parabola is generated in the E–k dispersion curve. If we approximate the conduction band shape at the minimum value to a parabola, an effective mass for conduction electrons is determined. Since the conduction band varies much faster with wave vector (momentum) at the Γ point in GaAs than at the minimum point in Si, the effective electron mass is much lower in GaAs than in Si. The force on an electron in an electric field therefore accelerates electrons in GaAs much more quickly than electrons in Si and, since the scattering processes are of similar magnitude in the two materials, the mean drift velocity of low-energy carriers is much higher in GaAs than in Si. We can thus see that the low-field electron mobility will be much higher in GaAs, and faster electronic devices should be achievable.

Thirdly, the difference in wave vector for low-energy holes and electrons in silicon means that intrinsic optical absorption and luminescence processes across the band gap require the participation of momentum-conserving phonons. This makes these processes much less efficient in silicon than in GaAs, and implies that higher-performance optical devices should be achievable.

These major differences will be elucidated further in subsequent sections.

1.3.2 Effective mass

Neither the valence band nor the conduction band extrema are accurately parabolic or isotropic. The conduction band minimum departs from the idealized paraboloid of revolution progressively as the energy increases. At about 0.2 eV above the bottom of the band, the band is showing small anisotropy (the (100) energy differs from the (111) energy by about 10 meV, but both values are about 50 meV below the true paraboloid). This non-ideality is sometimes of little significance but it leads to complications that can be confusing. For example, the curvature effective mass changes with carrier concentration when the Fermi level penetrates the conduction band (degeneracy condition), and divergence in the value of effective mass is seen depending on how the mass is measured, which would not appear in the ideal case. The situation is analogous to the problems of defining resistance in a component that does not obey Ohm's law. Care has to be taken to determine the correct definition of mass appropriate to the measurements made, or to use the correct average values of mass in theoretical calculations involving the summation of conduction band states to derive matrix elements or defect wavefunctions. In this chapter we will leave these problems behind and quote only GaAs 'curvature' masses at the zone centre (Γ).

The situation in the valence band extrema is worse in that three bands attain a maximum value at Γ, two of which are degenerate and one of which (the 'split-off' band) is displaced 0.34 eV to lower energy. Each of these bands is less parabolic and isotropic than the conduction band, and the descriptions of hole transport and other phenomena require very detailed study to obtain satisfactory (but hardly rigorous) theoretical treatments. 'Averaged' values for the valence bands are the simplest approach to a broad appraisal of the basic phenomena, which are usually dominated by the upper two bands. Figure 1.4 shows that these two degenerate bands have very different curvatures at Γ, so they are usually differentiated by the terms 'light' and 'heavy' hole bands. Table 1.2 lists these averaged 'curvature' effective mass values.

Table 1.2 Effective masses for the primary band extrema at 300 K

		GaAs	Si
Conduction band	m_e^*	$0.063m_0$ (Γ)	$0.3m_0$ (near X)[a]
Valence bands			
Heavy hole band V_h	m_{hh}^*	$0.50m_0$ (Γ)	$0.5m_0$ (Γ)
Light hole band V_l	m_{lh}^*	$0.076m_0$ (Γ)	$0.16m_0$ (Γ)
'Split-off' band V_s	m_{so}^*	$0.155m_0$ (Γ)	

[a] Averaged value, large band anisotropy $m_e^* = m_0 \to 0.2m_0$.

1.3.3 Carrier concentration

1.3.3.1 'Intrinsic' bulk properties

The carrier concentrations in a non-degenerate semiconductor are given by the Boltzmann approximation of the Fermi–Dirac treatment:

$$n = N_c \exp\left(-\frac{E_c - E_F}{kT}\right) \tag{1.1}$$

$$p = N_v \exp\left(-\frac{E_F - E_v}{kT}\right) \tag{1.2}$$

where N_c and N_v are the effective volume densities of electron and hole states in the conduction and valence bands, and E_F is the Fermi energy, which in a non-degenerate sample lies between the lowest conduction band energy (E_c) and the highest valence band energy (E_v).

The effective densities of states are again linked to the shapes of the relevant band extrema used to define the effective masses:

$$N_c, N_v = 2\left(\frac{2\pi m^* kT}{h^2}\right)^{3/2} \quad (1.3)$$

where m^* refers to the averaged 'curvature' mass of the appropriate band or bands.

In the intrinsic case, thermal generation across the gap (and recombination to balance it) creates equal numbers of electrons and holes so the intrinsic carrier density n_i is obtained by equating equations (1.1) and (1.2) and multiplying them to obtain:

$$n_i p_i = n_i^2 = N_c N_v \exp(-E_g/kT) \quad (1.4)$$

where E_g is the gap energy ($E_c - E_v$).

Since in GaAs $m_e^* = 0.063 m_0$ and $m_h^* = [(m_{lh}^*)^{3/2} + (m_{hh}^*)^{3/2}]^{2/3} = 0.53 m_0$, the effective density of states in the conduction band (4.2×10^{17} cm^{-3}) is much less than that in the valence band (9.6×10^{18} cm^{-3}), so both the density of states terms and the band gap difference suppress the intrinsic carrier concentration in GaAs (2×10^6 cm^{-3}) with respect to that in Si (2×10^{10} cm^{-3}).

The foregoing analysis is somewhat immaterial when one considers that particular impurity or defect species would need their concentrations to be controlled to better than one part in 10^{16} to achieve GaAs crystals of genuine intrinsic quality, whereas control to about one part in 10^{10} is the best ever achieved in any semiconductor.

The practical realization of the isolation advantages of GaAs was (and still is?) achieved by the *accidental* incorporation of defects that create deep electronic states near the centre of the band gap and lock the position of the Fermi level close to its intrinsic position. (This intrinsic position has to lie within a few tens of millivolts from the middle of the gap because of the nature of the Fermi function and the requirement $n_i = p_i$.) The mechanism that achieves this locking of the Fermi level is easily understood if one recognizes that the Fermi level denotes that energy position where the probability of a state being occupied is 0.5. Defects can produce electronic states at any position in the band structure and many produce states near the middle of the band gap. These deep states may be occupied or empty, and when they are in a dominant concentration and in thermal equilibrium they can completely fill or empty all other defect states in the crystal. The electron occupancy of the crystal is thereby controlled by the deep state and the Fermi level is pinned to the neighbourhood of the defect state energy. Some single defects such as transition metal impurity atoms produce multiple deep defect states which can both accept electrons from higher-lying states and donate electrons to fill lower-lying states, so whatever the nature of the other defect states the deep-level defect will control the position of the Fermi level.

The proper description of these deep states in GaAs and other semiconductors is one of the major areas of ignorance that continues to attract a great deal of scientific study. Some progress is being made on the defects that control the resistivity of bulk (pulled crystal) GaAs. Substitutional chromium atoms on the gallium sites are known to produce a deep state close to the centre of the gap and to control resistivity.[12] In the absence of chromium impurities, a further defect level near the centre of the band gap (designated EL2) controls the resistivity, and this defect level is thought to be generated by an imperfection in the intrinsic host lattice and not to be associated with impurities.[13] Lattice damage caused by irradiation with fast particles (electrons, protons, ions) produces a great variety of deep defect states[14] and local high-resistivity material is achieved. The local generation of isolating regions in devices and circuits is usually achieved with proton irradiation because the irradiation depth and spreading characteristics achieved with these particles offer the best spatial definition for device purposes.

No further comment on the deep-level behaviour of bulk material is appropriate in a review of this nature, but the electrical properties of GaAs surfaces deserve mention.

1.3.3.2 Surface properties

In the same way that point deep-level defects affect the bulk electrical properties, extended lattice defects such as dislocations and grain boundaries generate deep-level states and can cause a local pinning of the Fermi level. Local depletion regions can be formed in conducting materials which both scatter the free charge carriers and locally enhance electron–hole recombination rates. The termination of the lattice at the crystal surface is the major manifestation of these effects, and the situation in GaAs is significantly different to that in Si. Since present semiconductor technology is mainly based on planar (surface) devices whose active dimensions are very small (0.1–1 μm), the role of the surface can be critical. When metal contacts are applied to the surface, depletion barriers can be generated which may reflect the adjustment of the system to the Fermi level (work function) difference between the isolated metal and the semiconductor. If this barrier height is very small or if steps are taken to ensure a locally high carrier concentration in the semiconductor, and the barrier becomes thin enough to allow electrons to tunnel through it, ohmic contacts are formed. If these conditions are not met, the barrier produces a rectifying Schottky contact. Figure 1.5 shows the barrier height for various metals on n-type Si and GaAs, and an interesting difference is seen. The barrier height depends systematically on the metal work function in Si but not in GaAs. Although the full understanding of this phenomenon has not yet been obtained, it is clear that the silicon interface produces fewer deep-level states so that Fermi level pinning can be overcome as electrons adjust themselves across the interface. The strong pinning seen in GaAs, which fixes the barrier height in Figure 1.5, is independent of the nature of the

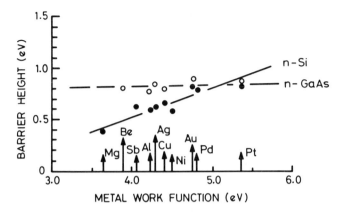

Figure 1.5 Barrier height data for metals on n-GaAs and n-Si (after Cowley and Sze[15])

interface, being similar for typical free surfaces and metal contacts. It has therefore been suggested[16] that the accumulation of intrinsic point defects near the surface is the controlling process and that there would be an interesting link between bulk and surface properties. It is certainly true that the EL2 pinning level at about 0.8 eV below the conduction band is consistent with the barrier height results in Figure 1.5.

This difference in behaviour sets a major limitation to the device potential in GaAs, in that inversion mode devices cannot be simply manufactured as in the ubiquitous and powerful silicon CMOS technology. However, the power of the GaAs system does allow inversion devices to be fabricated if the active GaAs material is terminated by a GaAlAs heterojunction. This new device (the HEMT) offers further potential speed and/or power advantages and is receiving considerable attention.[17]

The Fermi level pinning energy in the band gap of GaAs is independent of the type of crystal, so that the barrier height on p-type samples is significantly lower, and metal Schottky barrier devices are generally only feasible at hole concentrations below 10^{16} cm^{-3}.

1.3.3.3 *Extrinsic carrier concentration*

In GaAs, simple substitutional impurity atoms from groups II or IV on the Ga site and groups IV or VI on the As site generally produce electronic states close to the band extrema. The reason for this is well established by analogy with the hydrogen atom because the major binding mechanism for free carriers in the band states is the (net) unit coulombic charge on the impurity atom. Since the carriers can be treated as free under the effective mass approximation, the binding energy or depth of the defect ground state from the band edge is given by:

$$E_b = \frac{m^* q^4}{2\varepsilon^2 \hbar^2} \quad (1.5)$$

where m^* is the appropriate effective mass, ε is the dielectric constant, and the other symbols take their usual definitions.

The correct values to insert for m^* in GaAs and Si require some discussion. For the GaAs conduction band edge (Γ), 'curvature mass' will suffice because the band is sufficiently close to the ideal paraboloid. In the Si conduction band there are six equivalent minima along the six [100] axes and each minimum is strongly anisotropic. The average 'conductivity' mass ($m^* = 0.26m$) is required in this case,[18] which is formed by averaging the reciprocals of the three principal mass components that define the anisotropy. In the valence band the position is complex but similar in both materials. The split-off band can be ignored because of its separation, but the conductivity masses for the light and heavy hole bands need averaging by weighting the reciprocal masses with the relative hole populations of the two bands. The densities of states are very different in the two bands at Γ, with the heavy hole band containing over 10 times the hole density of the light hole band. If we assume that this conductivity effective mass for the valence band is valid for insertion into equation (1.5), $m^* \sim m_{hh}^* = 0.5m$. Setting $\varepsilon = 12$ in both materials, the following 'hydrogenic' ground state binding energies are obtained:

	GaAs	Si
Acceptor	50 meV	50 meV
Donor	7 meV	26 meV

These values are crudely in agreement with experiment where the range of values found for well studied impurities are shown below:

	GaAs		Si	
Acceptor	(Zn, C, Mg, Ge, Cd)	25–45 meV[19]	(B, Ar, Ga)	45–65 meV[18]
Donor	(Si, S, Se, Te)	5–7 meV[19]	(P, As, Sb)	40–55 meV[18]

There are clearly however some problems with the overall agreement for GaAs acceptors and silicon donors, which have been the subject of further detailed study. From our point of view the major difference between GaAs and silicon is the very low value of E_b for donors, which gives GaAs the capability of maintaining its conducting property to very low temperatures ($\lesssim 20$ K) so that low-temperature FET devices can be achieved, which offer remarkable improvements in the power × speed product of switching circuits.

These 'hydrogenic' levels are used to control the type and conductivity of materials in devices, because at room temperature they all become thermally ionized and release their electrons or holes from their ground states to the free band states. One can easily see how this process controls the position of the Fermi level in the band gap. For non-degenerate GaAs materials, the Fermi level is within the band gap and

$$pn = n_i^2 \sim 10^{13} \text{ cm}^{-6}$$

where p and n are the equilibrium hole and electron concentrations in the bands. So with, say, 10^{16} cm^{-3} hydrogenic donors that are virtually fully ionized at room temperature ($kT > E_b$), there are only 10^{-3} holes/cm^3 in the valence band. The Fermi level must move up from the intrinsic position to lie below the donor ground state, its exact position being that required to ensure the correct fractional occupancy of the donor levels and the conduction band states.

The 'hydrogenic' impurities are inserted into the crystal during growth, by diffusion or by ion implantation techniques. Although Zn diffuses readily into GaAs at temperatures near 750°C, the other acceptors and all donors require much higher temperatures to achieve any realistic penetration depths, and thus sample decomposition becomes a severe problem (see Section 1.2). This is a drawback in GaAs technology when compared to silicon, where p- and n-type diffusion processes are an industry standard. This has had the effect of increasing the research and development of GaAs doping control by ion implantation and epitaxial techniques, the latter being now at a superior state of development than for silicon and offering unique degrees of control for fine-geometry devices (e.g. high-performance quantum-well lasers with 60 Å active layer widths have been achieved). A further limitation (which has not been turned to advantage) is the limited solubility of donor species in GaAs. Figure 1.6 shows how the carrier concentration depends on impurity concentration in the gas phase in vapour phase epitaxial (MOCVD) GaAs.[20]

While hole concentrations up to those best described as Zn alloys can be achieved, the electron concentration limit is at 1×10^{19} carriers/cm^3. This kind of behaviour is universal, regardless of the growth technique, and is more pronounced for donors other than Se. It is thought to be caused by a fundamental solubility limit for the substitutional species so that, although the impurity concentration can exceed 10^{19} cm^{-3}, the donor atoms start to occupy different sites or precipitate as a second phase when this solubility limit is exceeded.

The onset of degeneracy occurs at hole concentrations of about 1×10^{19} cm^{-3} and electron concentrations near 5×10^{17} cm^{-3}.[19] As the impurity concentrations approach and exceed these values, a number of effects occur. (1) For example, the hydrogenic bound state (ground and excited state) wavefunctions

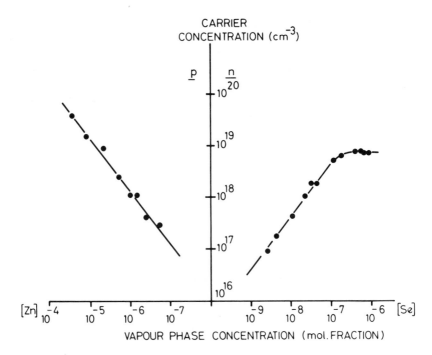

Figure 1.6 Doping calibration curves for MOCVD-grown GaAs, measured by Hall effect experiments for zinc and selenium doping

overlap and form continuous bands so that the ionization energies of the donors and acceptors tend to zero. The impurity states are then wholly merged with the band states and the band gap energy decreases. (2) A similar reduction in the band gap energy is caused by the random distribution of the increasing concentration of ionized (charged) donor or acceptor centres. Tails of states are generated by the random electric fields generated, which give rise to exponential degradations of the band edges. (3) An apparently opposite effect is caused by conduction band filling (degeneracy) where the absorption and luminescence processes move to energies well above the band gap. These optical effects will be quantified in Section 1.4, but we will not pursue the details of the electrical behaviour at these high doping levels because they do not bear heavily on device performance.

1.3.4 Carrier transport

1.3.4.1 Low-field conditions

In uniform extrinsic crystals where the carrier concentrations are predominantly of one type, the electrical conductivity is determined by the acceleration of the carriers due to the applied electric field and by those processes which arrest their motion through the lattice. Impurities and phonons create local electric fields which scatter the carriers, whereby the carrier trajectories are randomized and their excess kinetic energy is lost to the lattice. It should be remembered that the carriers are moving at random with high average velocity ($\sim 10^7$ cm s^{-1} at 300 K) because of their thermal energy, so the field generates a small distortion of this velocity distribution which imparts a small 'drift' velocity to the carrier 'gas'. As would be expected this drift velocity is determined by the mean free time of the carriers between collisions with the lattice (τ_f), this being the time for which they are accelerated in the direction of the field. Using Newton's laws of motion, the drift velocity is given by

$$v_D = \frac{qE\tau_f}{2m^*} \tag{1.6}$$

with m^* being the conductivity effective mass, and E the field in the crystal. In a more rigorous treatment which accounts for the thermal velocity distribution of the carriers:[21]

$$v_D = \frac{qE\bar{\tau}}{m^*} \tag{1.7}$$

where $\bar{\tau}$ is the exponential relaxation time for the carrier distribution to reach equilibrium on removal of the field ($\bar{\tau} \simeq \tau_f$), and the carrier drift mobility is

$$\mu = \frac{v_D}{E} = \frac{q\bar{\tau}}{m^*} \tag{1.8}$$

The significance of the effective mass on transport properties is now obvious. Ohm's law is obeyed at low fields ($< 10^3$ V cm^{-1}) because m^* and $\bar{\tau}$ do not vary with drift velocity ($v_D \ll 10^7$ cm s^{-1}), and the conductivity is given by

$$\sigma = n(\text{or } p)q\mu \tag{1.9}$$

Carrier mobilities are universally measured by a combination of Hall effect and conductivity measurements.[21] The Hall voltage generated by the Lorentz force in a magnetic field is used to measure the type and concentration of the carriers, so that mobility is extracted from equation (1.9). When the thermal velocity distribution of the carriers and the nature of the scattering processes are taken into account, the carrier concentrations deduced by the simplest analysis are seen to be inaccurate, and a correction factor (usually lying between 1 and 2) can be applied. Ignorance of the true value for this correction factor at the particular magnetic field used leads to its omission and the mobilities so obtained are then termed 'Hall mobilities'.

The Hall mobilities of GaAs and Si doped with hydrogenic donors and acceptors are shown in Figure 1.7.

The mechanisms controlling the GaAs mobilities in Figure 1.7 will now be addressed.

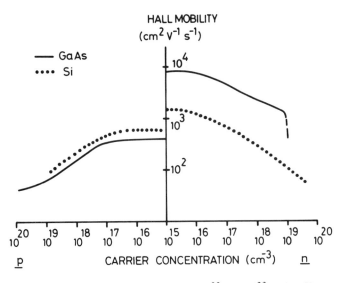

Figure 1.7 Hall mobilities for GaAs[22] and Si[23] at 300 K

At low carrier concentrations (10^{13}–10^{16} cm^{-3}), the mobility is controlled by phonon scattering and does not vary significantly with doping. At higher carrier concentrations, impurity scattering from the coulombic potentials of the ionized impurity centres increasingly dominates by reducing the mean free time. Many attempts have been made to analyse the detail of this behaviour by fitting the temperature dependences to various theories.[22] But when one considers that polar and non-polar mode scattering from optical phonons, acoustic phonon scattering, the effects of free carrier screening, degeneracy, and the non-ideality of the band structure need to be established independently as functions of temperature, the results cannot be viewed as rigorous.

Polar phonon scattering is due to the dipole fields generated by the opposing motions of the adjacent atoms in optical phonons, and the coupling to LO modes is much stronger than to TO modes. Ionic or partially ionic materials can generate such fields, but the symmetrical Si and Ge lattices obviously will not. Deformation potential scattering is in some circumstances significant for all types of phonons. The scattering field is generated by the local changes in lattice constant set up by the phonon, which produce local changes in the absolute energy of the relevant band edge (since in all semiconductors the band energies change with lattice compression, all lattice vibrations produce some effect). The terms non-polar optical scattering and acoustic phonon scattering refer to this process.

Acoustic phonon scattering[22,24] has often been identified as being present in semiconductors by its $T^{3/2}$ temperature dependence but the deformation potential constant for the requisite band is seldom well established. The current thoughts on the 300 K mobility-limiting processes are that polar LO phonon scattering controls the limit for electrons in GaAs, but that deformation potential scattering may dominate for holes.[22] At higher carrier densities, the Brooks–Herring formula[25] for coulombic scattering is broadly effective in describing the experimental results and gives a very useful estimate of the compensation levels in samples, because the density of coulombic scattering centres is the sum of the donor and acceptor densities $(N_D + N_A)$, whereas Hall measurements give the carrier concentration which for shallow states at room temperature is $|N_D - N_A|$.

The limit of about 8000 cm^2 V^{-1} s^{-1} for electrons at 300 K is the major electrical advantage of GaAs over Si and it is worth noting that values up to 10^6 cm^2 V^{-1} s^{-1} can be obtained at low temperatures so that 'COLDFETS' may be of commercial significance for very fast circuits, offering similar performance to that of superconducting Josephson junctions.

Before leaving the topic of low-field transport, a comment on ambipolar effects is worthwhile.

When near-intrinsic conditions are approached or high *uniform* densities of carriers are excited in the crystal, the carrier densities in the two bands are similar and an ambipolar drift mobility (μ_i^*) is appropriate. By calculating and adding the conductivities in the two bands, it is trivial to show that

$$\mu_i^* = \frac{n\mu_n + p\mu_p}{n+p} \quad (1.10)$$

However, an ambipolar drift mobility often quoted is that for excess carriers in a non-uniform distribution, say carriers excited at a localized point or plane in the crystal:[21]

$$\mu^* = \frac{(n_0 - p_0)\mu_n \mu_p}{n\mu_n + p\mu_p} \quad (1.11)$$

where n_0 and p_0 denote the uniform thermal equilibrium carrier concentrations. This value is strikingly different from μ_i^* and is zero for $n_0 = p_0$. The reason for this is that the non-uniform excess carrier density populations cannot migrate in opposite directions under the applied field because the very strong fields produced by such charge separation lock the populations together in space and no drift of the excess carrier distributions is produced. The niceties of how non-uniform the distribution has to be to determine how equations (1.10) and (1.11) are applied, and the role of contact structures, are intriguing but will not concern us here.

1.3.4.2 *High-field conditions*

Electrons in the Γ conduction band minimum at low fields can reach energies at high fields that are above that of the L minimum (see Figure 1.4). The carriers can then transfer to the L minimum by phonon scattering processes which accommodate the crystal momentum (wave vector) change under momentum conservation laws. The effective mass of the L minimum is $0.56m_0$ so the electron mobility is much lower than that in the Γ minimum. The average drift velocity of electrons thus reduces to a value that reflects the steady-state population of the two bands.

The drift velocity reduction and current continuity considerations cause charge accumulation domains to form which generate high-field dipoles. These high-field domains travel through the crystal and produce high-frequency voltage oscillations in two terminal Gunn oscillator devices.[23]

Figure 1.8 shows the experimentally determined variation of drift velocity with field, and theoretical Monte Carlo calculations have given good agreement with these data.[5] The Si data in Figure 1.8 show no negative differential resistance region and thus Si cannot be exploited in Gunn oscillator devices. The figure also shows the much reduced low-field mobility as the low-field intercept on this plot.

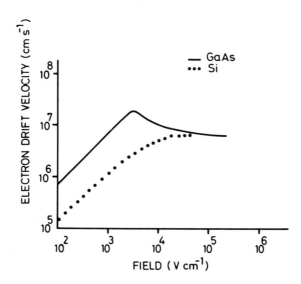

Figure 1.8 The velocity–field characteristics of GaAs and Si conduction electrons

1.3.4.3 Avalanche breakdown

High-energy carriers can excite valence electrons across the band gap (E_g) by impact excitation and cause carrier multiplication. The simplest classical energy and momentum equations are, assuming all particles have mass m^*:

$$u^2 = \frac{2E_g}{m^*} + v^2 + v_1^2 + v_2^2 \tag{1.12}$$

$$u = v + v_1 + v_2 \tag{1.13}$$

where u and v are the initial and final velocities of the primary particle and v_1 and v_2 are the velocities of the impact-excited electron and hole. If s is the sum of three positive numbers (n_1, n_2, n_3), it is easy to show that $n_1^2 + n_2^2 + n_3^2 \geq 3(s/3)^2$, so a minimum value for u is obtained when $v = v_1 = v_2$ and equation (1.12) then gives

$$\frac{2u^2}{3} = \frac{2E_g}{m^*}$$

or

$$\tfrac{1}{2} m^* u^2 = \tfrac{3}{2} E_g \qquad (1.14)$$

The threshold energy for the initial particle is thus $3E_g/2$, well above the band gap energy.

In GaAs the calculation is complicated by the difference in the electron and hole masses, but it is clear that very high electron or hole energies are needed to cause avalanche multiplication. The larger band gap of GaAs thus enables higher-field operation of appropriate devices (FETs, p–i–n and Schottky barrier diodes, etc.) than for Si.

Devices that use avalanche effects (e.g. IMPATTs) are conditioned by the rates at which electron and hole impact excitations occur, which depend on the nature of the bands. Figure 1.9 shows the relevant ionization coefficients for electrons and holes in the two materials.

The difference in the ratio of electron (α) and hole (β) ionization coefficients is a further factor affecting device performance.[26]

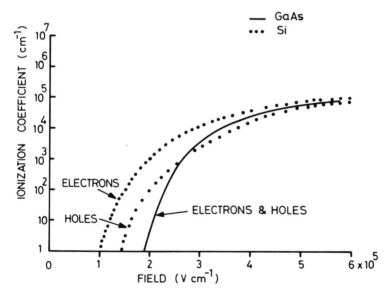

Figure 1.9 Approximate values for the electron and hole ionization coefficients in GaAs and Si at 300 K[26]

1.3.4.4 Non-radiative carrier recombination

The description of 300 K carrier recombination kinetics is dealt with later (Section 1.4.2) but we will now briefly discuss the major differences in non-radiative recombination properties of GaAs and Si.

In typical extrinsic GaAs materials at 300 K, carrier recombination is dominated by the bulk radiative recombination mechanism across the direct band gap. In Si this process is much less efficient because of its indirect gap, and non-radiative recombination through deep-level impurity or defect states often dominates. The Shockley–Read model[27] for such recombination processes has given some insight into the macroscopic kinetics and capture cross sections for charged defects.[28,29] Very limited knowledge of the mechanisms that occur after the capture of the particles has been obtained, and it revolves around the same problems as the description of deep-level defect states, although a generalized treatment of the likely phenomena[30] has been useful. In high-quality (long bulk lifetime) silicon the kinetics are usually controlled by non-radiative surface recombination, which can be avoided in GaAs by terminating the material with GaAlAs heterojunction barriers using epitaxial growth techniques. Bulk non-radiative recombination at dislocations is also very significant in Si, but in heavily doped GaAs very large defect concentrations are needed before it dominates the radiative process. Figure 1.10 shows the bulk minority carrier lifetimes (300 K) in GaAs and Si materials, which are controlled by recombination at dislocations.[31] The hatched line shows the predictions of a simple theoretical model which assumes that the carrier recombination rate is limited by the diffusion of carriers to the defect, so that, providing the recombination at the defect is strong enough, the nature of the defect is immaterial and this represents the shortest possible lifetime at each defect density. It can be seen that dislocations are much stronger recombination centres in GaAs than in Si.

A similar behaviour is exhibited by free surface recombination in the two materials, which is probably controlled by the non-radiative recombination rates for surface states. The surface recombination velocity, which describes the rate of loss of carriers at the surface, is

$$S = \delta p_s D \frac{\delta p}{\delta x}\bigg|_{x=0}$$

where δp_s is the surface concentration of minority carriers, D is the diffusion constant, and $(\delta p/\delta x)|_{x=0}$ is the gradient of the minority carrier concentration at the surface.

In GaAs, S is usually so large ($> 10^6$ cm s^{-1}) that the recombination rate is diffusion limited and its value cannot be determined in simple experiments. In Si its value is readily measurable[21] ($\sim 10^3$ cm s^{-1}) and an interesting parallel with dislocations and surface pinning phenomena is seen.

There are of course other intrinsic and extrinsic non-radiative recombination

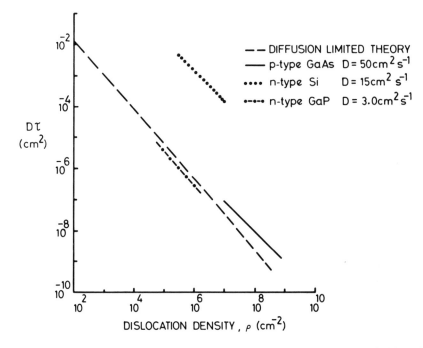

Figure 1.10 An analysis of dislocation-controlled minority carrier lifetime (τ) in Si, GaAs, and GaP. The theory is that for the diffusion-limited recombination rate. The experimental data are plotted as $D\tau$ vs ρ, where D is the diffusion coefficient of the minority carrier and ρ is the dislocation density, so that the data from the three materials can be directly compared. The fact that dislocations in GaAs and GaP show recombination strengths at the limiting value associates this behaviour with the lower symmetry of the III–V compound lattices rather than the indirect gap/direct gap difference in the Si/GaAs band structure (Section 1.3.1)

mechanisms such as free carrier or defect Auger recombination which are the inverse of impact ionization processes. The free carrier Auger effect is usually dominant only at very high carrier concentrations because the recombination rate depends on the square of the carrier density and it does not have great direct device significance.

1.4 OPTICAL PROPERTIES

The major optical properties of a material are interrelated as a result of fundamental electromagnetic theory.[32] At normal incidence the reflectivity (R) and relative phase (ϕ) of the reflected ray are given by

$$R = \frac{(n-1)^2 + k^2}{(n+1)^2 + k^2} \tag{1.15}$$

$$\tan \phi = \frac{2k}{1-n^2-k^2} \qquad (1.16)$$

where $n - \mathrm{i}k$ is the complex refractive index, n being the refractive index and k the extinction coefficient, related to the absorption coefficient ($\alpha = 4\pi k/\lambda$) at the particular wavelength in question, λ.

The complex and real dielectric constants are given by

$$\epsilon = \epsilon_1 + \mathrm{i}\epsilon_2 = (n^2 + k^2) + \mathrm{i}2nk \qquad (1.17)$$

and

$$\varepsilon = \sqrt{\epsilon \cdot \epsilon^*} = n^2 - k^2 \qquad (1.18)$$

The Kramers–Kronig relation[33] shows that ϕ can be found by integrating a function of R and λ with respect to wavelength, and refractive index and absorption coefficient data are often obtained using this powerful technique. However, the physical mechanisms behind the observed data are not determined, so the theoretical analysis of the various features is in no way aided.

1.4.1 Optical absorption

A compilation of absorption coefficient data for 'intrinsic', n- and p-type GaAs is shown in Figure 1.11. 'Intrinsic' Si data are also shown (dotted lines). The major differences between the two materials are seen in the absence of strong absorption at the longest wavelengths in Si and the less abrupt absorption edge in Si at the fundamental band gap wavelengths (Si, 1.1 μm; GaAs, 0.85 μm). We will now briefly describe the dominant mechanisms, starting at the long-wevelength features.

1.4.1.1 *Phonon absorption*

As was mentioned previously (Section 1.2), the partially ionic nature of the GaAs lattice bond produces large electric dipoles for the displacement generated by optical phonons. This necessarily implies that strong optical absorption will occur at the optical phonon frequencies that will generate optical phonons. Photons have small momenta so the long-wavelength (Γ) phonons will be strongly favoured and strong resonance effects for TO(Γ) and LO(Γ) phonons, which have slightly different energies, can be deduced from simple harmonic oscillator theory.[5] These effects produce the 'Restrahlen' single-phonon resonance peak in the figure. The fact that the LO(Γ) and TO(Γ) phonon energies have to be equal in silicon[33] is interesting when considering phonon effects, but immaterial here because the symmetry of the crystal does not allow the production of the necessary dipole, optical absorption is forbidden, and the Restrahlen peak is absent.

When more than one phonon is generated by optical absorption, the process becomes allowed in both materials and 'multiphonon' structure is seen in the

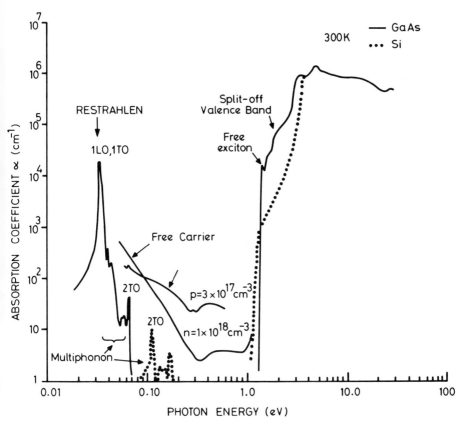

Figure 1.11 Optical absorption in GaAs and Si (from refs 5, 23, and 32)

absorption spectra. This structure is complex and knowledge of the phonon dispersion curves and phonon selection rules is needed to interpret it successfully.[33] Studies of the multiphonon spectra are useful, however, in determining the siting of impurities in the lattice. Simple theories predict the energies and strengths of extrinsic peaks in the spectrum caused by the mass difference of the foreign atom, so impurity concentrations can be measured and their lattice siting verified.[34] This technique usually requires thick samples and relatively high ($> 10^{16}$ cm^{-3}) impurity concentrations to achieve measurable absorption strengths, and light atoms are favoured because they produce one-phonon features at energies above those of the strong, intrinsic one-phonon features in GaAs.

1.4.1.2 *Free carrier absorption*

In conducting materials, the phonon absorption is swamped by absorption processes caused by the free carriers, so phonon absorption studies have to be carried out on 'intrinsic' samples (which entails the irradiation or deep-level impurity diffusion of heavily conducting samples to detect hydrogenic impurities). In Figure 1.11 the major mechanisms are displayed for GaAs.

In n-type materials, the electrons absorb energy so that they are excited to higher levels within the Γ minimum. The balancing energy-loss process is of course phonon generation (heat production) as the electrons relax to lower energies. This process is called interband absorption and requires that the electrons interact with the lattice. The mechanism is most simply understood as the inverse process of Bremsstrahlung—where electron deceleration by lattice scattering produces photons. So incident photons will equally produce the acceleration of scattering electrons. Experimentally it is observed to follow the form

$$\alpha \propto N\lambda^b \qquad (1.19)$$

where N is the impurity concentration and b lies between 1.5 and 3.5 in various materials.[35] N is chosen as an appropriate parameter because ionized impurity scattering as well as electron density can be important in deciding the strength of this process. In n-GaAs (Figure 1.11) $b = 3$ and the process becomes significant at wavelengths longer than about 3 μm.

In p-type material, the other free carrier process, intraband absorption is dominant. This process is the excitation of free holes from one of the three highest valence bands to another. The threshold for transitions to the split-off band is thus seen near 3.6 μm (~ 0.34 eV; Section 1.3). At longer wavelengths the transitions are limited to those between the light and heavy hole bands (Section 1.3). The analysis of intraband hole absorption is fraught with the same problems as the description of hole mobility, in that a proper account of intraband scattering is very hard to prove unambiguously.

1.4.1.3 Band gap absorption

Figure 1.11 shows that the steepness and strength of the edge in GaAs is much greater than that in Si because strong phonon participation is mandatory in the indirect gap material to conserve momentum for the transition. This is one of the key factors of GaAs (cf. Si) which allows efficient light-emitting devices to be produced (see Section 1.4.2) and also allows faster light-detecting devices to be achieved. For direct gap semiconductors, the shape of this edge can be predicted from simple theory[19] by

$$\alpha(E) = AE \int_{-\infty}^{\infty} p_c(E')p_v(E'') |M|^2 [f(E'') - f(E')] \, dE \qquad (1.20)$$

where A is a constant, $p_c(E')$ and $p_v(E'')$ are the (energy) densities of states in the conduction and valence bands at energies E' and E'' respectively, $E' - E'' = E$ being the required (absorbed) photon energy, M is the matrix element for the transition probability for absorption between these single states at E'' and E', and $f(E')$ and $f(E'')$ are the occupancy probabilities for the two states using Fermi–Dirac statistics. The matrix element is found by performing an integration of the overlap of the two electron state wavefunctions in real space multiplied by the interaction Hamiltonian determined from the solution of the time-dependent Schrödinger equation.

Again we can see that the band structure controls this absorption process, and it can be shown that equation (1.20) predicts that the absorption coefficient increases with the square root of the incremental energy above the band gap, i.e.

$$\alpha \propto \sqrt{(h\nu - E_g)} \qquad (1.21)$$

This does not, however, agree with experiment because among other things it ignores the fact that the free electron and free hole can associate to form free excitons. This phenomenon permits further absorption processes to occur just below and around the band gap energy and greatly enhances the absorption coefficient.[38] If these effects are taken into account, theory can be made to predict the experimental data at and above the edge with acceptable precision. The small feature in Figure 1.11 (see also Figure 1.12) is evidence of this exciton process in GaAs at 300 K.

A more detailed plot of the absorption edge for intrinsic and extrinsic material is shown in Figure 1.12. Below the edge on intrinsic data at low temperatures the curve is less steep than the exciton plus band-to-band theory and the presence of

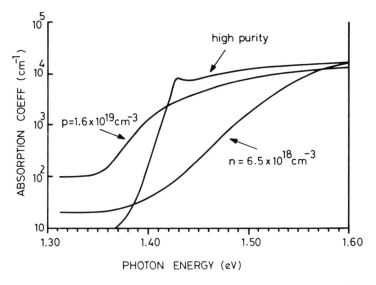

Figure 1.12 The band gap absorption of GaAs (after Casey and Stern[36]). The heavily doped samples show residual absorption[36] for $h\nu \sim 1.35$ eV, which is the result of free carrier absorption (see Figure 1.11)

an exponential tail is deduced.[5] At 300 K the theoretically ideal edge would be expected to degrade as a result of the thermal population of phonon modes, and again exponential tails can be generated.[33] Further, the presence of (unoccupied) shallow impurity states, lattice microstrains, and non-uniform distributions of charged centres also may cause tails. It would seem that statements on the origin of these tails in insulating samples are rather speculative and very sample dependent. However, by using very high-quality undoped materials, the experimental data can be fitted to produce the band gap energy with good precision. The energy gap has been found[5] to follow the equation

$$E_g = 1.519 - 5.405 \times 10^{-4} T^2/(T+204) \quad \text{eV} \quad (1.22)$$

for absolute temperatures (T) in the range 0 K $< T <$ 1000 K. This expression is accurate to two millivolts and gives a 300 K value of 1.424 eV. The situation is better when we consider the effects of doping. In p- and n-type samples, the edge degrades due to band tails caused by the progressive merging of the hydrogenic impurity states with the band, and, as degeneracy is reached and exceeded, band filling occurs and shifts the absorption edge to high energy. In p-type samples, the band tails result in a shift to lower photon energy which is reversed for $p > 1 \times 10^{19}$ cm^{-3}, when band filling starts to move the edge in the opposite sense. In n-type samples, degeneracy sets in at much lower doping, so the effects of the band tailing are less obvious. Figure 1.12 shows two examples of the edge for high p- and n-type carrier concentrations, and the shift in n-type material due to band filling is very pronounced.

1.4.1.4 *Absorption above the edge*

Reflectance data are used to produce data on the absorption coefficient above the edge and features can be identified with transitions at other critical points in the band structure. As an example, the transition from the split-off valence band to the Γ conduction band minimum is indicated in Figure 1.11. This has also been observed by direct absorption measurements.[5] These features have little direct device significance.

1.4.2 Luminescence

A considerable amount of work has been done on low-temperature luminescence studies of intrinsic and extrinsic semiconductors for its own sake, to elucidate recombination mechanisms or as an aid to impurity identification, but the advent of very powerful analysis techniques such as secondary ion mass spectrometry has reduced its value to the applied scientist in this latter area. The details of mechanisms occurring at 300 K are usually very hard to prove from luminescence studies at this temperature and the approach of van Roosbroeck and Shockley[39] to define the radiative band-to-band processes is the most useful procedure. Here the recombination of minority carriers at low excitation densities is viewed as the inverse of the optical absorption process across the band gap. Since most doped GaAs luminesces quite strongly via this mechanism at 300 K, we will restrict ourselves to a brief description of these results. Lack of space will also mean that high excitation density conditions which lead to absorption saturation, optical gain, and laser action in solid state GaAs lasers[19] will also be ignored.

1.4.2.1 Luminescence spectra

At thermal equilibrium, the carrier generation and recombination processes across the band gap are in balance. So by considering the absorption of blackbody radiation by the energy gap van Roosbroeck and Shockley[39] showed that the radiative spontaneous emission process could be fully described. The internally generated emission spectrum is a band about kT wide near the absorption edge, given by

$$I(hv) = \frac{8\pi n^2 h^2 v^2 \alpha(hv)\delta}{h^3 c^2 [\exp(hv/kT) - 1]} \qquad (1.23)$$

where $I(hv)$ is the photon intensity within the energy interval $hv \to hv + \delta$, $\alpha(hv)$ is the (band-to-band) absorption coefficient at that photon energy, and n is the refractive index. In most luminescence experiments the non-radiative surface recombination (Section 1.3) is severe, so the steady-state emission intensity under low excitation conditions increases with depth in the sample. This means that the majority of the luminescence signal leaving the crystal has passed through a narrow region of relatively unexcited crystal, so the spectrum of equation (1.23) is distorted by absorption in this material. If this and other self-absorption distortions are taken into account, the agreement with experiment is good.[40] Using electron beam excitation, Cusano[41] studied the 300 K spectra for n- and p-type GaAs. While the proper corrections were not made in this work, the broad behaviour expected from the absorption edge data (Figure 1.12) was observed. The spectra broadened with doping, shifting to high energy for n-type and to low energy for p-type samples.

1.4.2.2 Radiative lifetime and minority carrier diffusion length

By integrating equation (1.23) with respect to photon energy, the total photon flux (I_{tot}) can be obtained, which represents the minority carrier recombination rate in extrinsic samples at thermal equilibrium. Using equation (1.4), for p-type samples

$$p\delta n = n_i^2 \qquad (1.24)$$

where δn is the equilibrium electron concentration.

If we define a mean radiative lifetime τ for the minority carrier we then have

$$I_{tot} = \frac{\delta n}{\tau} = \frac{n_i^2}{p\tau}$$

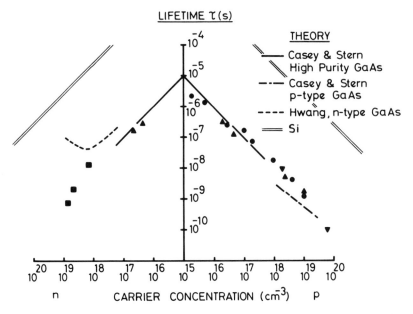

Figure 1.13 The radiative lifetime of minority carriers in GaAs and Si (after Hall[42]). The solid lines are values calculated from high-purity GaAs absorption edge data using the detailed balance principle.[36] The experimental results are due to Nelson and Sobers (o),[43] Wight et al. (▼),[45] t'Hooft et al. (▲),[44] and Hwang (■).[37] Theoretical estimates of the lifetime at high carrier concentrations are included, which are due to Casey and Stern[36] and Hwang[37]

or

$$\tau = \frac{n_i^2}{pI_{tot}} \qquad (1.25)$$

When the minority density is increased by carrier injection or generation processes, this time constant would also determine the response of carriers and the luminescence signal. If the materials are not compromised by recombination at bulk defects* or surfaces, these lifetime values can be approached in doped materials. The radiative lifetime in GaAs and Si predicted from equation (1.25) and those measured for high-quality GaAs are shown in Figure 1.13. At high carrier concentrations ($n \gtrsim 2 \times 10^{18}$ cm^{-3}, $p \gtrsim 1 \times 10^{19}$ cm^{-3}), the observed lifetimes are reduced below the radiative limits, supposedly by non-radiative free carrier Auger or defect recombination. It is interesting to note that the high doping predictions in Figure 1.13 show the effects of the shifts in the absorption edge in the exponential term of equation (1.23), lying above the extrapolated high-

---

* Shockley–Read kinetics are used to analyse these processes.[27]

purity prediction for n-type GaAs and below it in p-type material, although the experimental results hardly confirm these effects.

The migration of excess minority carriers by thermal diffusion is central to the operation of bipolar transistors and is characterized by the minority carrier diffusion length, L. Following simple diffusion theory,

$$L = \sqrt{(D\tau)} \tag{1.26}$$

where D is the diffusion constant for the minority carrier. The minority carrier diffusion constant is related to the minority carrier drift mobility μ_m by the Einstein relationship:

$$D = \mu_m \frac{kT}{q} \tag{1.27}$$

μ_m takes the ambipolar form (equation (1.11)) when $n \simeq p$ but in extrinsic materials at low excitation levels it can be roughly approximated by the mobility of the (majority) carrier in material of the opposite type and the same impurity concentration. Measured values of L compared with the simple theory above (Figure 1.13 and equations (1.26) and (1.27)) are shown in Figure 1.14. When allowance is made for the sample thickness and surface recombination, the agreement is good, even at the longest values of L observed.

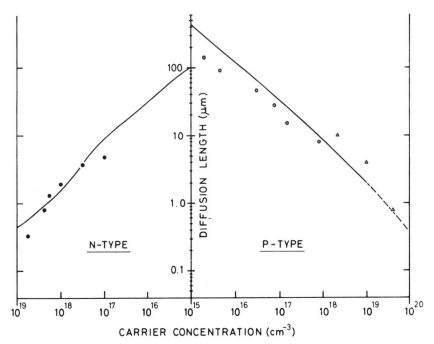

Figure 1.14 Diffusion lengths in n- and p-type GaAs. Data from refs 45–47 are compared with values predicted from the lifetime data in Figure 1.13

It is instructive to note at this point that all of the materials parameters required to predict τ and L can be estimated from a knowledge of the E–k diagrams for electrons and phonons in GaAs, which shows the fundamental significance of this information.

Before leaving this topic, it is important to mention some relatively new results which have yet to be fully assimilated by the semiconductor world at large. When the recombination process is *entirely* dominated by the radiative processes in Figure 1.13, the internal luminescence efficiency becomes 100%. The photons produced can then be reabsorbed in regions of lower excitation density to recreate the minority carrier. Non-uniform minority carrier distributions, for example in forward biased p–n junctions, cannot then be modelled in the standard way by ascribing a diffusion length to the carrier since the recombination event does not terminate its existence, as would be the case in typical silicon materials controlled by Shockley–Read recombination at bulk defects. The effects of GaAs surface recombination can be avoided by confining the crystal with heterojunctions so the carrier lifetime can be controlled by the photon loss from the sample, thus being geometry controlled in a rather new way. These processes are more familiar to laser physicists but Nelson[46] has shown that they can be very significant at minority carrier densities very much lower than that required to produce stimulated emission in lasers. When this photon-recycling process is dominant, the simple theories behind the band gap absorption and luminescence processes in Section 1.4.1 need modification and their experimental verification becomes more complex. There is as yet no comprehensive body of published work which treats and verifies this new situation in GaAs, so it is clear that discussions of the detailed agreement of data and theory in this section are redundant. However, the departure of the τ and L data above the theory lines in Figures 1.13 and 1.14 probably reflect this phenomenon.

1.4.3 Reflectivity

The normal-incidence reflectivity of 'intrinsic' and n-type GaAs is shown in Figure 1.15. The intrinsic data show the relatively small effects of particular absorption processes occurring in the band structure above the energy gap, with a severe reduction in R at the highest energies associated with 'plasma type' (see below) oscillations of the valence band electrons. The major features near 35 meV are the result of the strong resonance absorption of single LO and TO phonons (Section 1.4.1) and the concomitant refractive index changes. This is simply modelled using classical harmonic oscillator theory.[32] Although the fundamental gap absorption is stronger than the phonon absorption, the wavelength dependence of the extinction coefficient means that the refractive index changes are less and hence the band edge feature is relatively weak (equation (1.15), Section 1.4).

In the n-type sample, a new feature is seen near 70 meV which is the result of

The physical and electronic properties of GaAs

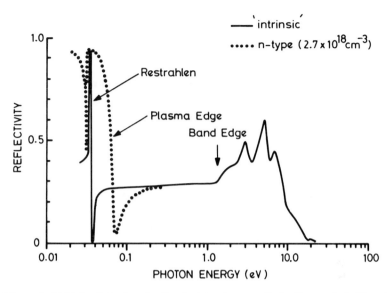

Figure 1.15 Reflectivity data for 'intrinsic' and n-type GaAs (from refs 5, 47, and 48)

by plasma oscillations. Consider the electron gas in an n-type sample and the fixed positive charges of the ionized hydrogenic donors in the lattice. Strong electric fields are generated if the electron density is locally disturbed and plasma oscillations occur for which the quantum particles are called plasmons. For long-wavelength vibrations the frequency tends to that generated by the bodily displacement of the electrons with respect to the fixed ion charges, which is easily shown to be:[32]

$$\omega_p = \left(\frac{4\pi n q^2}{m^*}\right)^{1/2} \quad (1.28)$$

where n is the carrier concentration, and m^* is the curvature effective mass for the case of non-degenerate GaAs, but

$$m^* = \frac{h^2 k}{\delta E/\delta k}$$

evaluated at the Fermi level in the usual degenerate condition under which this effect is prominent. The direct displacement of charge involved gives strong optical coupling to light at this plasma frequency and the plasma reflection edge results. In this case, the extinction coefficient (and absorption coefficient) is relatively small and the reflectivity is controlled by the refractive index until the critical frequency for the metallic state (ω_p) is reached. For $h\nu < \hbar\omega_p$ the refractive index has no real meaning because such optical waves cannot be supported in a

34 Gallium arsenide

metallic conductor. The extinction coefficient, however, can still be defined and this increases as $h\nu$ decreases below $\hbar\omega_p$. These optical absorption features are not strong enough to be detected in the presence of the very strong free carrier absorption in n-GaAs (Figure 1.11).

The Restrahlen feature for the n-type sample in Figure 1.15 is modified by the coupling of phonon and plasma modes and could be thought of as a resonance loss process where the coupling of the plasma to the lattice (usually weak) becomes strong enough to affect the optical properties.

1.4.4 Refractive index

The refractive index has some significance in the light-guiding properties of laser cavities and could become of growing significance in the fields of passive waveguides and active integrated optics devices in GaAlAs materials. Figure 1.16 shows the refractive index for intrinsic and n-type samples. Apart from the major excursions due to the optical phonon resonance in the intrinsic data and the effects of the plasma resonance in the n-type sample, a fairly featureless curve is measured,[48] with the reduction at high energies being related to the excitation of electron oscillations in the valence band states. In the n-type sample, the refractive index goes to zero at the plasma resonance frequency; and, as was pointed out in Section 1.4.3, optical waves cannot propagate at lower frequencies, so the refractive index ceases to have meaning.

The fine detail of the refractive index data near the absorption edge is shown for

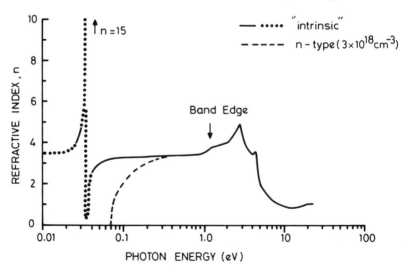

Figure 1.16 Refractive index data for intrinsic and n-type GaAs (from refs 5, 48). The dotted portion of the 'intrinsic' curve was obtained by fitting measured data to simple theory[5]

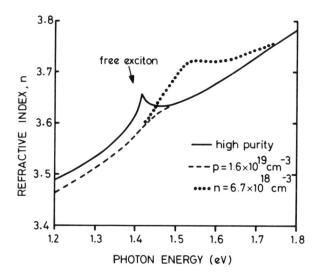

Figure 1.17 Detailed refractive index data for high-purity, n- and p-type GaAs (from ref. 19)

'intrinsic', n- and p-type samples in Figure 1.17 and the major features are related directly to the absorption edge structures seen in Figure 1.12.

At energies near 1.2 eV (below the band gap), where the materials are relatively transparent, the refractive index change is given by:

$$\delta n \approx -0.006 \, n/10^{18}$$

and

$$\delta n \approx -0.002 \, p/10^{18}$$

for electrons and holes, respectively. This behaviour is caused by the band-filling shift of the absorption edge (Section 4.1.3) rather than the direct effect of the free carrier absorption.

REFERENCES

1. C. G. Kirkpatrick, R. T. Chen, D. E. Holmes, and K. R. Elliott, Chapter 2 in this book.
2. I. A. Dority, J. D. Grange, and D. K. Wickenden, Chapter 3 in this book.
3. J. R. Arthur, *J. Phys. Chem. Solids,* **28,** 2257 (1967).
4. D. Richman, *J. Phys. Chem. Solids,* **24,** 1131 (1963).
5. J. S. Blakemore, *J. Appl. Phys.,* **53,** 10 (1982).
6. M. B. Panish, *J. Cryst. Growth,* **27,** 6 (1974).
7. D. W. Shaw, *J. Electrochem. Soc.,* **128,** 874 (1981).

8. M. Neuberger, *Handbook of Electronic Materials*, vol. 7, New York, IFI/Plenum (1972).
9. J. L. T. Waugh and G. Dolling, *Phys. Rev.*, **132**, 2410 (1963).
10. J. R. Chelikovsky and H. Cohen, *Phys. Rev. B*, **14**, 556 (1976).
11. C. Kittel, *Introduction to Solid State Physics*, New York, John Wiley (1968).
12. G. M. Martin, *Semi-Insulating III–V Materials* (Nottingham 1980), Nantwich, Cheshire, Shiva (1980).
13. J. Lagowski, J. Parsey, M. Kamimsea, K. Wada, and H. Gatos, *Semi-Insulating III–V Materials* (Evian 1982), Nantwich, Cheshire, Shiva (1982).
14. L. W. Aukerman, *Semiconductors and Semimetals*, vol. 4, New York, Academic Press (1968).
15. A. M. Cowley and S. M. Sze, *J. Appl. Phys.*, **36**, 3212 (1965).
16. W. E. Spicer, S. Eglash, I. Lindau, C. Yu, and P. Skeath, *Thin Solid Films*, **89**, 447 (1982).
17. S. Hiyamizu, T. Mimura, and T. Ishikawa, *Jap. J. Appl. Phys.*, **21** (Suppl. 21-1), 161 (1981).
18. R. A. Smith, *Semiconductors*, London and Cambridge, Cambridge University Press (1964).
19. H. C. Casey and M. B. Panish, *Heterostructure Lasers*, New York, Academic Press (1978).
20. S. J. Bass and P. E. Oliver, Unpublished Data, RSRE Malvern (1983).
21. J. P. McKelvey, *Solid State and Semiconductor Physics*, New York and London, Harper and Row (1966).
22. D. L. Rode, *Semiconductors and Semimetals*, vol. 10, New York, Academic Press (1973).
23. S. M. Sze, *The Physics of Semiconductor Devices*, New York, John Wiley (1969).
24. W. Shockley, *Electrons and Holes in Semiconductors*, Princeton NJ, Van Nostrand (1950).
25. H. Brooks and C. Herring, as reported in: L. M. Falicou and M. Cuevas, *Phys. Rev.*, **164**, 1025 (1967).
26. J. G. Smith, Chapter 9 in this book.
27. W. Shockley and W. T. Read, Jr *Phys. Rev.*, **87**, 835 (1952).
28. M. Lax, *Phys. Rev.*, **119**, 1502 (1960).
29. V. N. Abakumov and I. N. Yassievich, *Sov. Phys. JETP*, **44**, 345 (1976).
30. C. Henry and D. Lang, *Phys. Rev. B*, **15**, 989 (1977).
31. D. R. Wight, *J. Phys. D: Appl. Phys.*, **10**, 431 (1977).
32. T. Moss, G. Burrel, and B. Ellis, *Semiconductors and Opto-Electronics*, London, Butterworths (1973).
33. E. J. Johnson, *Semiconductors and Semimetals*, vol. 3, New York, Academic Press (1967).
34. W. G. Spitzer, *Semiconductors and Semimetals*, vol. 3, New York, Academic Press (1967).
35. H. Y. Fan, *Semiconductors and Semimetals*, vol. 3, New York, Academic Press (1967).
36. H. C. Casey and F. Stern, *J. Appl. Phys.*, **47**, 631 (1976).
37. C. J. Hwang, *J. Appl. Phys.*, **40**, 9 (1969); *Phys. Rev. B*, **6**, 1355 (1972).
38. J. O. Dimmock, *Semiconductors and Semimetals*, vol. 3, New York, Academic Press (1967).
39. W. Van Roosbroeck and W. Shockley, *Phys. Rev.*, **94**, 1558 (1954).
40. D. J. Day, Private Communication, RSRE Malvern.
41. D. A. Cusano, *Solid State Commun.*, **2**, 353 (1964).
42. R. N. Hall, *Proc. IEE*, **106**, 923 (1960).

43. R. J. Nelson and R. G. Sobers, *J. Appl. Phys.*, **49**, 6103 (1978).
44. G. W. t'Hooft, C. Van Opdorp, H. Veevliet, and A. T. Vink, *J. Cryst. Growth*, **55**, 173 (1981).
45. Values derived from D. W. Wight, P. E. Oliver, T. Prentice, and V. Steward, *J. Cryst. Growth*, **55**, 183 (1981).
46. J. R. Nelson, *GaAs and Related Compounds 1978* (St Louis), *Inst. Phys. Conf. Ser.* **45**, Bristol and London, Institute of Physics, p. 256 (1979).
47. H. C. Casey, B. I. Miller, and E. Pinkas, *J. Appl. Phys.*, **44**, 1281 (1973).
48. B. D. Seraphin and H. E. Bennett, *Semiconductors and Semimetals*, vol. 3, New York, Academic Press (1967).

Gallium Arsenide
Edited by M. J. Howes and D. V. Morgan
© 1985 John Wiley & Sons Ltd

CHAPTER 2

Growth of Bulk GaAs

C. G. Kirkpatrick, R. T. Chen, D. E. Holmes, and K. R. Elliott

2.1 INTRODUCTION

Progress in developing liquid-encapsulated Czochralski (LEC) techniques for the growth of bulk semi-insulating GaAs has resulted in readily available sources of large-diameter round wafers suitable for use in standard semiconductor processing equipment. This achievement has significantly improved the manufacturability of GaAs integrated circuit technology. In this chapter, recent progress in our laboratory in the growth, characterization, and application of LEC GaAs for integrated circuit fabrication is presented. Approaches for improving the yield of semi-insulating substrates, reducing dislocation, densities, and qualifying GaAs materials for digital IC processing are described. The successful application of these materials to ion implantation device fabrication is discussed, as well as the use of LEC substrates for epitaxial processing.

The development of GaAs integrated circuit technologies offering superior performance including high speed, low power, and radiation hardness, has spurred demand for high-quality, standard-size wafers with reliable electrical behaviour. The new technologies for digital integrated circuits (IC's), monolithic microwave integrated circuits (MMICs), and charge-coupled devices (CCDs) have generally utilized ion implantation to form active layers for planar device structures. The uniformity, reproducibility, high yield, and low cost of ion implantation processing make this approach particularly suitable for IC fabrication. However, the use of the substrate as the medium in which active layers are formed and as the isolation between devices places requirements on the electrical characteristics of the substrates.

A second generation of higher-performance GaAs technologies is also under development, which utilize molecular beam epitaxy (MBE) and metal-organic chemical vapour deposition (MOCVD) for the formation of active device layers. Materials for these new devices, based on high electron mobility transistors (HEMTs) and heterojunction bipolar transistors (HBTs), undergo essentially the same screening as materials destined for ion implantation processing in our laboratory. Undoped LEC materials are used as substrates in epilayer growth to minimize the possibility of impurity outdiffusion into epitaxial regions.

Substrates for integrated circuit fabrication must meet specifications for large-diameter round wafers: uniform, reproducible electrical characteristics, including high resistivity, thermal stability, and good activation of implanted dopants; low background impurities; and high degrees of crystalline perfection. To meet these substrate requirements, we have investigated the growth, characterization, and selection of LEC GaAs materials for integrated circuit processing. In this chapter, major findings of this research effort are described. In Section 2.2, the basics of the LEC growth technique are described. Section 2.3 describes detailed recent findings from our research into improving LEC growth technology and characterizing the resulting materials. This discussion includes the results of studies directed towards reduction of twins and dislocations, the determination of the compensation mechanism in undoped materials, and the effects of growth parameters on electrical characteristics. In Section 2.4, the requirements of device fabrication on substrates are described. Conclusions of this work and implications for future directions of bulk GaAs research are discussed in Section 2.5.

2.2 BASICS OF LEC GROWTH TECHNIQUE

All GaAs crystals used in this study were grown in a Melbourn high-pressure LEC puller manufactured by Metals Research Ltd. The LEC technique was first applied to the growth of PbTe by Metz,[1] applied to III–V materials by Mullin et al.,[2] and adapted for use with pyrolytic boron nitride (PBN) crucibles by Swiggard[3] and AuCoin.[4] The puller consists of a resistance-heated 6 inch diameter crucible system capable of containing charges up to 10 kg. The growth process is monitored through a closed-circuit vidicon television camera. This type of high-pressure puller features *in situ* synthesis of GaAs from elemental Ga and As. The technique eliminates the need for a separate high-temperature synthesis step before crystal growth, reducing the potential for contamination.

A schematic of the LEC crucible configuration is shown in Figure 2.1. The high-purity (6–9s) Ga and As components are weighed and loaded into a 6 inch diameter high-purity pyrolytic boron nitride crucible with the Ga lying on top of the As, topped by a preformed disc of boric oxide (B_2O_3) with known moisture content as specified by the manufacturer. Charges with a total weight of approximately 3–4 kg were used in these studies. B_2O_3 discs typically weighed 500 g. The crystals were grown in the $\langle 100 \rangle$ direction with the weight varying from 2.4 to 3.6 kg. The crystal diameters were typically slightly greater than 3 inch in order to allow for grinding to exact 3 inch diameters after growth. Both quartz and PBN crucibles have been successfully utilized in the growth of semi-insulating GaAs. Although the initial cost of PBN crucibles is high (generally $4000–6000, depending on quantity and manufacturer), the crucible can be

Growth of bulk GaAs

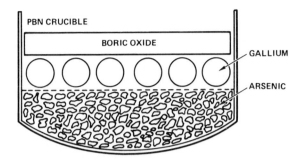

Figure 2.1 LEC crucible configuration before growth, with charge of elemental Ga and As and preformed B_2O_3 disc

cleaned and re-used about a dozen times. Furthermore, owing to the considerably higher yield in growing high-resistivity, single-crystal GaAs, the use of PBN crucibles has become a standard in the growth of undoped, semi-insulating LEC GaAs crystals.

The stoichiometry of the GaAs melt can be changed by varying the composition of the charge. In order to study the effect of melt stoichiometry on the properties of undoped LEC GaAs crystals, the initial melt composition of each growth run has to be accurately determined. It is therefore necessary to take into account the loss of As from the charge during the heat-up cycle resulting from incomplete wetting of the B_2O_3 to the PBN crucible before synthesis. The weight loss is calculated by comparing the weight of the initial charge with the weight of the crystal and charge remaining in the crucible after growth. We have effectively varied the As concentration of the melt from 0.46 to 0.51. The melt composition for samples along the crystal's length was calculated by adjusting the initial melt composition value for the crystal weight at that point of the growth. Crystal weight and length as a function of time are routinely recorded during a growth run.

The Ga and As compounding process is initiated by heating the crucible under an initial chamber pressure of 600 psi. The Ga, which is solid to just above room temperature, melts first and encapsulates the As. As the temperature reaches 450–500°C, the boric oxide melts, flows over the Ga and As charge, and seals at the crucible wall. The boric oxide melts before significant arsenic sublimation occurs. Compound synthesis ($Ga_{liquid} + As_{solid} = GaAs_{solid}$) occurs rapidly and exothermically at about 800°C under high argon overpressures (approximately 1000 psi). The presence of the boric oxide and the use of high argon overpressures prevent significant loss of As due to sublimation and evaporation during and subsequent to the synthesis. The boric oxide also shields the melt against contamination from the crucible and growth chamber.

The crystal growth procedure begins as soon as the melt temperature is stabilized. It is initiated by dipping the seed, which is held on the pull shaft, through the transparent B_2O_3 and into the melt. The crystal is pulled by slowly withdrawing the seed, generally sliced from low dislocation material, from the melt. The growth configuration is illustrated in Figure 2.2. Gradually the crystal diameter is allowed to increase until full diameter is reached. The crystals were grown with iso-rotation (the crystal and crucible rotating in the same direction at 6 and 15 rpm, respectively) and counter-rotation (the crystal and crucible rotating in opposite directions at 5 and 15 rpm, respectively). Typical pull rate for this investigation was 7 mm h^{-1} and the crucible lift rate was 1.4 mm h^{-1}. At the end of the growth cycle, the crystal was situated over the boric oxide encapsulant, and the system was slowly cooled at a constant rate between 30 and 80°C h^{-1}.

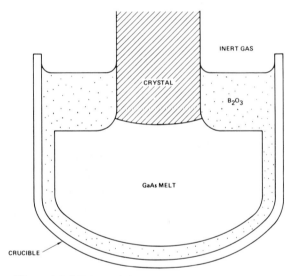

Figure 2.2 LEC crucible configuration during growth

The diameter of the growing crystal is monitored manually with the differential weight signal. This signal emanates from the load cell, a weighing device on which the crystal and pull shaft are mounted in the LEC puller. Changes in the differential weight indicate variations in diameter; an increase or decrease in differential weight signifies an increase or decrease, respectively, in diameter. In addition, the crucible configuration is viewed at all times on a TV monitor to assist in parameter control.

The crystal is grown in three main stages, as shown in Figure 2.3. Once the seed contacts the melt and pulling has begun, a neck is formed by reducing the crystal diameter below the seed diameter (approximately 4 mm) to 1–3 mm. The crystal diameter is then gradually allowed to increase, forming the cone. After the desired diameter is reached, the body of the crystal is pulled maintaining this dimension.

Growth of bulk GaAs

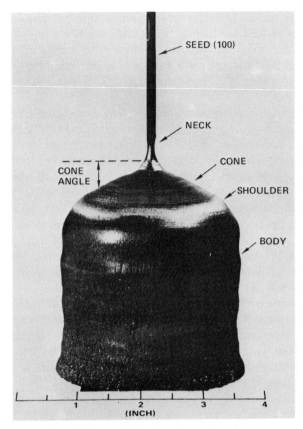

Figure 2.3 LEC GaAs (100) crystal and seed

2.3 PROGRESS IN LEC TECHNOLOGY

The demand for superior quality, large-diameter substrates has been the impetus for studies directed towards the optimization of GaAs substrates for device fabrication. In this section, recent findings at our laboratory concerning the development of LEC growth techniques and the characterization of the resulting materials are detailed. Three main areas of crystal growth and characterization research are discussed; these include the growth of single crystals, the reduction of the dislocation density, and control of the semi-insulating behaviour.

2.3.1 Growth of single crystals

The formation of twins can be a serious problem affecting the yield of GaAs material suitable for device processing. Twinning can cause alteration of the crystallographic orientation of the material and can also result in the formation of polycrystalline material or large grain boundaries. For a high yield of single-crystal wafers for device applications, twinning must be avoided. In our studies, control over the melt stoichiometry was the most important factor in preventing twin formation in undoped, large-diameter, $\langle 100 \rangle$ GaAs crystals grown by the LEC technique.

In experiments to investigate twinning phenomena, 20 GaAs crystals were grown from stoichiometric and non-stoichiometric melts. The results of these investigations, summarized in Table 2.1, showed that twinning incidence is significantly reduced when the crystals are pulled from As-rich melts. For crystals grown from Ga-rich melts, four of 12 (33%) were single. In contrast, seven of eight (88%) crystals grown from As-rich melts were single. Other growth parameters, for example, the wetness of the boric oxide,[4] the cone angle, or fluctuations in the crystal diameter, did not exhibit correlations with twinning incidence. In additional studies involving the growth of more than 30 other undoped $\langle 100 \rangle$ crystals from As-rich melts, the positive impact of melt stoichiometry has been confirmed, with a single-crystal yield of over 90% consistently obtained.

Previous research[5,6] has shown that the incidence of twinning in small-diameter GaAs crystals can be reduced through the use of steep cone angles. In our work involving large-diameter crystals, however, no correlation was observed between twinning and the cone angle. In fact, the high single-crystal yield from As-rich melts reported here was demonstrated with small cone angles ranging from 0 to 35°.

The results described above were obtained in growths from pyrolytic boron nitride crucibles. Specific experiments were not undertaken with As-rich melts in quartz crucibles to compare with the results from Ga-rich melts. However, in a separate series of investigations focusing on the growth of Se, Si, and Zn-doped crystals from As-rich melts in quartz crucibles, the incidence of twin formation was also very low. Of the nine ingots grown, eight were single crystal.

Crystals with twins were examined to determine the twin morphology, and separated into two categories. One group of crystals exhibited only one longitudinal twin, which nucleated at the crystal surface and intersected the crystal obliquely on a (111) plane. X-ray analysis[7] revealed that the twinned region of the crystal under examination was oriented with the $\langle 122 \rangle$ direction parallel to the growth direction. A second category of crystals exhibited multiple twins. In all the crystals studied, the twins invariably nucleated at one of the four peripheral facets running axially through the crystals. These peripheral facets form at the intersection of (111) As and (111) Ga facet planes with the crystal edge along $\langle 110 \rangle$ directions perpendicular to the $\langle 100 \rangle$ growth axis. Neither the As nor Ga peripheral facets appeared to dominate as twin nucleation sites.

The importance of As-rich melts in the growth of single-crystal GaAs has been reported for other growth techniques as well. Reduced twin formation through

Growth of bulk GaAs 45

Table 2.1 Incidence of twinning in large-diameter (100) LEC GaAs crystals

Crystal[c] number	Crucible material	Melt stoichiometry	Initial melt composition	Final As atom fraction[a]	Result	Cone angle (deg)[b]	Twin morphology
1	PBN	Ga-rich	0.462	0.445	twin	65	multiple twins
2	PBN	Ga-rich	0.477	0.459	twin	30	multiple twins
3	PBN	Ga-rich	0.486	0.439	twin	60	multiple twins
4	PBN	Ga-rich	0.488	0.434	single	60	
5	PBN	Ga-rich	0.489	0.439	single	50	
6	PBN	Ga-rich	0.492	0.457	twin	30	one longitudinal twin
7	PBN	Ga-rich	—	—	twin	40	multiple twins
8	quartz	Ga-rich	—	—	twin	20	multiple twins
9	quartz	Ga-rich	—	—	single	70	
10	quartz	Ga-rich	—	—	twin	10	multiple twins
11	quartz	Ga-rich	—	—	twin	50	multiple twins
12	quartz	Ga-rich	—	—	single	60	
13	PBN	As-rich	0.500	0.500	single	30	
14	PBN	As-rich	0.500	0.500	single	25	
15	PBN	As-rich	0.500	0.501	single	30	
16	PBN	As-rich	0.501	0.508	single	30	
17	PBN	As-rich	0.502	0.512	single	30	
18	PBN	As-rich	0.502	0.502	single	0	
19	PBN	As-rich	0.504	0.534	twin	35	one longitudinal twin
20	PBN	As-rich	0.506	0.536	single	30	

[a] matter to come
[b] matter to come
[c] not in chronological order of growth.

46   *Gallium arsenide*

the use of As-rich melts has been demonstrated in Bridgman[8] and modified Gremmelmaier[5] growth experiments. The impact of melt stoichiometry on twin formation may be fundamental to GaAs crystal growth, rather than to a particular growth technique. The significant shift in twinning incidence over a relatively limited range of melt compositions in our studies indicates that local stoichiometry of the solid at the growth interface may be a dominant effect.

2.3.2 Reduction of dislocation density

Since much of the current interest in the LEC technique is due to the demand for high-quality undoped GaAs semi-insulating substrates for integrated circuit applications, the dislocation density in these materials is of great concern. The dislocation densities observed in GaAs LEC materials are high compared to those in silicon materials used for integrated circuits, and the distribution of dislocations is non-uniform. These dislocations may negatively impact circuit yield and performance in several ways. Dislocations located near or under the channel of a field-effect transistor may cause variable channel conductivity and non-uniformity of the threshold voltage. Dislocations may affect Schottky diode leakage or ohmic contact reliability. The lifetime and reliability of heterojunction bipolar devices may be affected by the dislocation density. Dislocation climb into epitaxial layers may also affect the quality of modulation-doped layers for high electron mobility transistors.

While there is much speculation regarding the role of dislocations on device yield and performance, there are, as yet, few pieces of direct experimental evidence to reveal the role of dislocations. Recent reports[9,10] have indicated that variations in threshold voltage may be attributable to the dislocation density distribution on LEC GaAs wafers.

As first steps in understanding the role of dislocations on devices, characterization of the density and distribution of dislocations on large-diameter LEC GaAs wafers and growth experiments exploring means of reducing and controlling dislocation densities have been undertaken. The results of these studies will be described in terms of radial and longitudinal distributions, and the parameters affecting dislocation reduction.

The major cause of dislocation formation in bulk elemental and compound semiconductors is stress induced by temperature gradients during growth and postgrowth cooling. While large-diameter (4–5 inch) Si crystals can be grown essentially dislocation-free, only small-diameter (less than 0.5 inch) GaAs crystals have been reported with comparable structural perfection.[11,16] Although the thermal conditions for GaAs growth can be comparable to those for Si, GaAs is fundamentally more susceptible to dislocation formation under the same growth conditions for two reasons. First, the strength of GaAs compared to Si is much lower, so dislocations form more easily. The critical resolved shear stress σ_{CRSS} ranges from 4 to 40 g mm^{-2} for GaAs, increasing with increasing doping, whereas

σ_{CRSS} for Si ranges from 60 to 150 g mm^{-2}.[17] Secondly, the thermal conductivity of GaAs (0.54 W cm^{-1} K^{-1}) is about a factor of 3 lower than for Si (1.4 W cm^{-1} K^{-1}). Therefore, the reduction of thermal gradients in the GaAs crystal is more challenging. As a result, it is more difficult to control the thermal gradient-induced stress below σ_{CRSS} is GaAs crystal growth.

High temperature gradients are of particular concern in large-diameter Czochralski growth configurations (LEC, Gremmelmaier). Most of the published work on dislocation studies concentrates on small-diameter (less than 0.5 inch diameter) crystals grown by the LEC,[11,13] Bridgman,[14,15] and modified Gremmelmaier[16] techniques. Since the thermal gradients generally decrease with decreasing crystal diameter, effectively dislocation-free small crystals have been demonstrated.[11,13,15,16] For small-diameter LEC crystals, several parameters have been reported to reduce radial gradients, including the boric oxide thickness[11,18] and the cone angle.[19] Other material properties observed to suppress dislocations include impurity concentration[13,20,21] and melt stoichiometry.[12,14,15] Relatively little information has been published regarding the growth of large-diameter LEC crystals.

In our studies, over 50 undoped LEC crystals were grown and analysed. Samples were obtained by slicing crystals according to the diagram shown in Figure 2.4. After lapping and polishing wafers on both sides, wafer dislocation densities and distributions were evaluated by preferential etching in potassium hydroxide (KOH) for 25 min at 400°C. The KOH preferentially etches dislocations intersecting the sample surface, forming hexogonal etch pits. X-ray topography has demonstrated in this work and other investigations[22,23] that etch pit density (EPD) corresponds directly to the dislocation density. The EPD data in these studies were derived by counting etch pits over 1.3 × 1.0 mm regions utilizing low-magnification (70 ×) micrographs. For EPDs greater than 1×10^5 cm^{-2}, higher magnifications (140 × or 280 ×) were required. The error in etch pit counting was estimated to be less than 5%.

Our investigations focused on exploring the effects of seven growth parameters on the dislocation density and distribution. These parameters included: (1) cone angle, (2) seed quality, (3) seed necking, (4) diameter control, (5) melt stoichiometry,

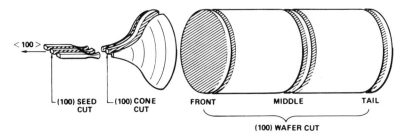

Figure 2.4 Diagram indicating (100) crystal slicing pattern for LEC studies[22]

(6) thickness of the boric oxide layer, and (7) ambient pressure. In these studies, the cone angle was varied from 0 to 65°. The seed crystals used to initiate growth exhibited EPD ranging from 1.5×10^3 to 5×10^5 cm^{-2}. Both low and high EPD seeds were utilized, with and without Dash-type seed necking.[24] The diameter of the crystal necks ranged from 1.2 to 3 mm. The best diameter control (deviations from the average value of ± 1.1 mm) resulted from controlling the diameter through the cooling rate with minimal direct temperature adjustment. Initial melt stoichiometries varied from 0.462 to 0.506 As atom fraction. In most of the growth experiments discussed here, the thickness of the boric oxide encapsulant over the melt was approximately 17 mm, resulting from 500 g of boric oxide. Additional experiments utilized 9 and 13 mm boric oxide thicknesses (170 and 390 g). The typical ambient pressure during growth was 300 psi. Low-pressure (50 psi) growth was also investigated.

2.3.2.1 Radial dislocation distribution

As shown in Figure 2.5, the radial dislocation distribution forms a four-fold symmetry pattern indicative of the $\langle 100 \rangle$ wafer orientation. The considerable variations in EPD which may appear across a (100) wafer are illustrated in the microscopic view in Figure 2.6. The radial dislocation distribution exhibits several main features: (1) minimum EPD is observed within a wide annular ring between the centre and edge of a wafer (region 1, or 'ring' region); (2) intermediate EPD values appear at the wafer centre (region 2, or 'centre' region); (3) maximum

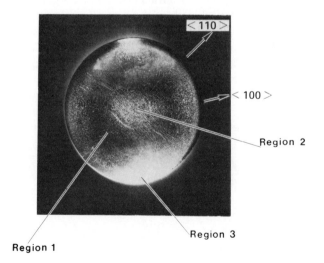

Figure 2.5 Photograph of 3 inch (100) LEC GaAs wafer after KOH etch, revealing four-fold symmetry in dislocation density and distinct (1) ring, (2) centre, and (3) edge regions[22]

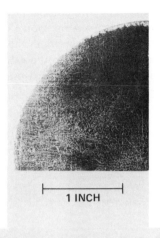
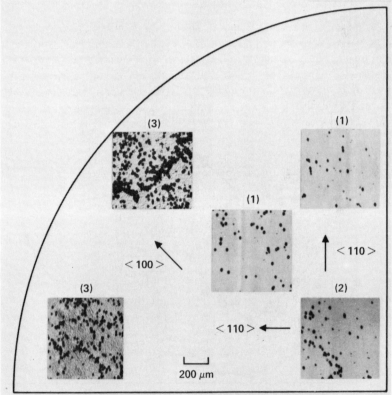

Figure 2.6 Photomicrographs of selected regions on (100) LEC GaAs wafer after KOH etch, showing radial dislocation distribution[23]

EPD is observed at the wafer edge (region 3, or 'edge' region). The dislocation density in the ring and edge regions are sensitive to crystallographic direction; EPD along the $\langle 100 \rangle$ direction is higher than the $\langle 110 \rangle$ direction, for a (100) wafer (Figure 2.6). Typically, measurements of dislocation densities across the full diameter of LEC wafers exhibits a 'W'-shaped profile, as shown in Figure 2.7.

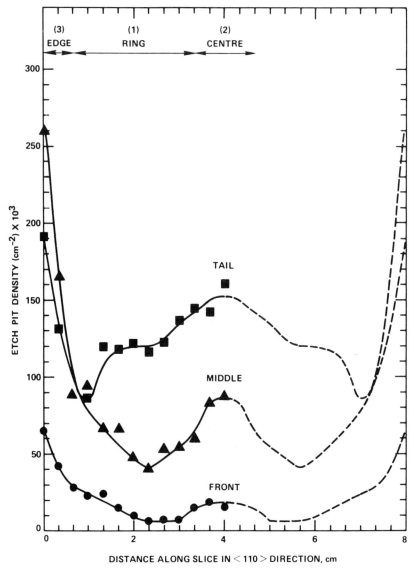

Figure 2.7 Radial and longitudinal etch pit density measurements for 3 inch LEC (100) GaAs wafer[22]

Such experimentally determined radial dislocation distributions correlate with predictions based on theoretical thermoelastic analyses of Czochralski crystals by Penning[25] and Jordan et al.[26,27] In Jordan's work, the total stress in the crystal was calculated in terms of twelve (111) ⟨110⟩ slip systems, while assuming that the dislocation density is proportional to the total stress, within an additive constant. The periphery and centre of the crystal are under tension and compression, respectively, because the crystal edge is cooler than the centre as the crystal emerges from the melt.

The calculated stress is greatest at the crystal periphery, consistent with the measured maximum in EPD at the wafer edge; the lowest calculated stress is in the transition area between regions of tension and compression, where the lowest EPDs ('ring' region) are measured. The relatively higher EPD along the ⟨100⟩ compared to ⟨110⟩ direction may have theoretical basis in the greater density of slip systems along the ⟨100⟩ direction contributing to the total stress. The results of both theoretical and experimental studies indicate that radial gradient-induced stress is the main source of dislocations in these crystals, as substantiated by other investigations of (100) GaAs[27] as well.

Dislocation networks A prominent feature in morphology studies of the microscopic distribution of dislocations is the presence of etch pit networks. Several types of structures are observed. Cellular networks, shown in Figure 2.5, exhibit interconnected networks of cells formed by dislocations, with only a few dislocations inside each cell. In such a structure, the cell diameters are approximately 500 μm, with an EPD of 2×10^4 cm^{-2}. As the EPD increases, the cell diameter decreases (100 μm cell diameter for an EPD of approximately 1×10^5 cm^{-2}). For EPDs less than approximately 2×10^4 cm^{-2}, the network morphology appears as lineage, with the etch pits forming observable wavy lines. These lines, as shown in Figure 2.5, can extend for a few millimetres to more than one centimetre, and appear oriented along the ⟨110⟩ direction.

Such networks of dislocations may result from a polygonization process,[28] in which the dislocations orient themselves after solidification to minimize the crystal strain energy. Both glide and climb processes may contribute towards this realignment. Dislocations in zinc blende structures generally can realign to form walls defined by (110) planes perpendicular to (111) slip planes. Such walls would intersect (100) planes in ⟨110⟩ directions, as observed experimentally. The cellular network thus consists of a high density of lineage-type structures, where dislocations may intersect.

2.3.2.2 Longitudinal dislocation distribution

Variation in dislocation density along the longitudinal (growth) direction was studied by comparing radial distributions from the front, middle, and tail of crystals, as indicated in Figure 2.7. Although the radial profiles remain 'W'-shaped, the EPD always increased from front to tail (Figure 2.7). This result may be indicative of increasing stress along the length of the crystal, or multiplication of the dislocations after growth, or both. Comparatively, the average EPD (see Table 2.2) increased by a factor of 8, 7, and 1.5, in the 'ring', 'centre', and 'edge' regions, from the front to the tail. The radial EPD distribution became more uniform towards tails of the crystal.

2.3.2.3 Parameters affecting dislocation density

The observed dislocation density in GaAs materials varies widely, depending on the growth approach and techniques utilized. Commercial 2 inch Bridgman materials typically exhibit dislocation densities ranging from 2×10^3 to 3×10^4 cm^{-2} and commercial 3 inch LEC GaAs from 1×10^4 to 2×10^5 cm^{-2}. We have grown undoped semi-insulating 3 inch GaAs LEC material with dislocation densities as low as 6×10^3 cm^{-2}, with EPD below 2×10^4 cm^{-2} over 75% of the wafer area. Investigations including the growth of approximately fifty 3 inch LEC crystals resulted in the development of improved techniques and optimized parameters for reducing the dislocation density. Quantitative dependence of EPD on the cone angle, boric oxide thickness, ambient pressure, seed quality and necking, diameter control, and melt stoichiometry was investigated. The impact of each growth parameter was determined by assessing the variation in EPD as that parameter was varied independently. The changes in EPD in the three regions, ring, centre, and edge, were used to define the impact of parameter changes; reduction of EPD in all three areas simultaneously was used to define the effect of an individual growth parameter. Sufficient spatial resolution (averaged over 1.3×1.0 mm areas) was utilized to observe real variations in the average EPD on wafers, while minimizing effects due to polygonization-associated microscopic density fluctuations. Higher magnifications were used to measure EPD in the edge regions, and these data are accurate within approximately 50%. Data from the centre and ring regions were more accurate indications of actual EPD variations from crystal to crystal, and represent approximately 80% of the crystal area.

Cone angle We have evaluated the effect of the cone angle on the dislocation density by studying the EPD of full-diameter wafers from the front of LEC crystals. These results, which appear in Figure 2.8, indicate no correlation between the cone angle and dislocation density for cone angles greater than 25°. For example, crystals grown under very similar conditions except for the cone angle, 62° and 30°, displayed essentially no difference in EPD values in the centre and ring regions.

In contrast, the flat-top crystal (0° cone angle), exhibited EPD in the low 10^5 cm^{-2} range (see Figure 2.8), which is comparatively high. The longitudinal distribution in this crystal is inverted in the front half of the crystal, decreasing

Growth of bulk GaAs

Table 2.2 Summary of EPD measurements on LEC ingots

Ingot no.	Wafer location[a]	EPD (cm^{-2})			EPD ratio (centre to ring)
		Ring	Centre	Edge	
1	F	7.6×10^4	4.6×10^4	3.0×10^5	0.6
	T	6.1×10^5	6.1×10^5	1.1×10^5	1.0
2	F	2.2×10^4	5.0×10^4	2.3×10^5	2.3
	M	3.2×10^4	7.3×10^4	2.5×10^5	2.3
3	F	4.0×10^4	8.0×10^4	2.9×10^5	2.0
	T	n.a.	n.a.	n.a.	–
4[b]	F	4.0×10^4	1.4×10^5	4.0×10^5	3.5
	M	3.0×10^4	1.0×10^5	n.a.	3.3
5	F	1.5×10^4	3.4×10^4	1.7×10^5	2.3
	T	1.2×10^5	1.4×10^5	2.1×10^5	1.2
6	F	1.8×10^4	2.6×10^4	8.0×10^4	1.4
	T	8.6×10^4	7.7×10^4	2.0×10^5	0.9
7	F	1.0×10^4	2.5×10^4	5.6×10^4	2.5
	M	2.5×10^4	3.9×10^4	7.8×10^4	1.6
8	F	1.4×10^4	3.7×10^4	1.0×10^5	2.6
	T	n.a.	n.a.	n.a.	–
9	F	1.4×10^4	2.0×10^4	2.5×10^5	1.4
	T	1.0×10^5	1.0×10^5	2.4×10^5	1.0
10	F	1.0×10^4	2.1×10^4	1.1×10^5	1.9
	T	n.a.	n.a.	n.a.	–
11	F	7.5×10^3	1.3×10^4	1.9×10^5	1.7
	T	8.1×10^4	1.8×10^5	1.8×10^5	2.2
12	F	1.2×10^4	1.7×10^4	2.5×10^5	1.4
	T	9.0×10^4	1.0×10^5	2.2×10^5	1.1
13	F	3.5×10^4	1.0×10^5	1.5×10^5	2.9
	T[d]	8.0×10^4	1.1×10^5	2.0×10^5	1.1
14	F	6.0×10^3	1.8×10^4	9.6×10^4	3.0
	T	n.a.	n.a.	n.a.	–
15[c]	F	1.1×10^5	2.4×10^5	2.7×10^5	2.2
	T	1.3×10^5	2.3×10^5	1.6×10^5	1.8
16	F	1.3×10^4	2.8×10^4	1.7×10^5	2.2
	T	1.4×10^5	2.2×10^5	2.4×10^5	1.6
17	F	1.1×10^4	2.0×10^4	1.1×10^5	1.8
	T	1.5×10^5	2.5×10^5	1.7×10^5	1.7
18	F	8.5×10^3	1.6×10^4	1.2×10^5	1.9
	T	9.7×10^4	1.3×10^5	2.4×10^5	1.3

[a] F, front; M, middle; T, tail.
[b] High EPD seed used.
[c] Flat-top growth.
[d] ~ 3/4 ingot length area.

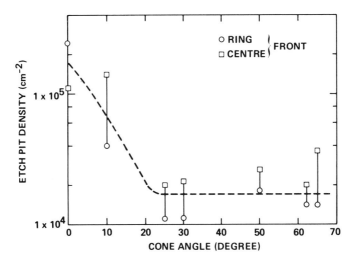

Figure 2.8 Dependence of etch pit density on (100) LEC GaAs wafers as a function of crystal cone angle.

from the front to the tail, then increasing again as is more commonly observed. This particular crystal expanded rapidly as the top of the crystal was pulled from the boric oxide encapsulating layer. A bulge resulted at the front of the crystal equivalent in height to the boric oxide layer thickness. Evidently the crystal underwent significant cooling when emerging from the boric oxide; the convective heat transfer from the crystal to the surrounding ambient was considerable relative to heat transfer to the liquid encapsulant. This increase in cooling may be responsible for increased stress at the top of the crystal, resulting in the very high dislocation density.

Longitudinal cross sections of cone regions were mapped (see Figure 2.9) to determine the dislocation density distribution in the growth direction for various cone angles. These maps clearly reveal the 'W'-shaped radial profile (see Figure 2.7). The longitudinal EPD increased below the neck, attained a maximum value, and then decreased prior to reaching full crystal diameter. As the cone angle increased as shown in Figure 2.8, the maximum observed value of EPD decreased. Crystals produced with shallow cones also displayed a high density of slip traces. The highest value of longitudinal dislocation distribution appeared directly below the neck for shallow cones, and narer the centre for high-angle cones.

The measured location of the EPD maximum and variation of EPD with cone angle are consistent with the observed behaviour for the flat-top crystal. It appears that the same mechanism controls the EPD value and distribution at the seed end of all of the crystals, with the flat-top sample representing the limiting case of 0° cone angle. The dislocations at the maximum form a secondary

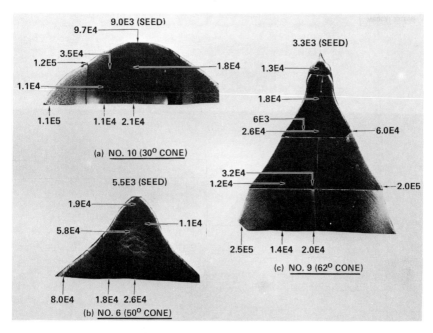

Figure 2.9 Photographs of KOH-etched longitudinal cross sections of crystal cones with various cone angles[22]

distribution, in addition to the primary grown-in distribution forming at the solidification front.

Our investigations of the secondary dislocation distribution support a heat flow model in which the growing crystal is located at the top surface of the liquid encapsulant. The shape of the isotherm is regulated by the relative radial and vertical heat flow components. Vertical heat flow is predominant when the crystal is thin, as the neck is pulled through the boric oxide, and the isotherm is relatively flat. As the cone is pulled through the liquid encapsulant, radial heat flow increases, and the isotherm shifts to a concave shape with the increase of the radial gradient. For shallow cone angles, the radial gradient increases, resulting in more noticeable maxima in EPD. After the shoulder of the crystal, decreased isotherm curvature results in reduced gradients during the growth of the vertical crystal body.

Boric oxide thickness The impact of thickness of the boric oxide encapsulating layer on the dislocation density was assessed by varying the B_2O_3 height from 9 to 17 mm. The resulting data, shown in Figure 2.10, show that the EPD increased as the layer height increased. This trend is especially pronounced at the seed end of the crystals. The secondary cone dislocation distribution appeared to be

Gallium arsenide

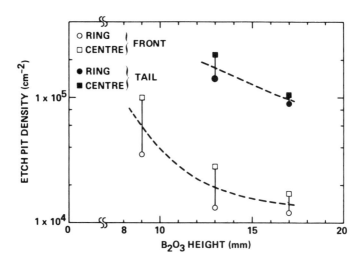

Figure 2.10 Dependence of etch pit density on (100) LEC GaAs wafers as a function of B_2O_3 thickness

independent of the boric oxide height. This indicates that the thicker boric oxide layer results in reduced radial gradients near the crystal–melt interface due to more effective thermal isolation between the crystal and the Ar ambient. This experimental observation conflicts with the theory of Jordan et al.,[27] which predicts that the radial gradients decrease as the boric oxide thickness decreases.

Based on these data, we attempted to use a very thick 35 mm encapsulating layer for LEC GaAs growth. The reduction in thermal gradients was obvious during the growth process as the diameter was extremely difficult to control, as apparent in Figure 2.11. Isolated regions of this crystal display dislocation densities as low as 5000 cm^{-2}; however, these data did not demonstrate the significant EPD reduction expected. We believe that the significant transient thermal gradients applied to control the diameter of the crystal negated the impact of using thicker boric oxide. The application of automatic diameter control in LEC growth may allow the thicker encapsulating layers to be used to better advantage.

Ambient pressure Typical pressure for high-pressure LEC growth is approximately 300 psi. We have grown LEC GaAs at 50 psi to study the effect of ambient pressure on the dislocation density. A problem with this approach is the thermal degradation of the crystal surface as it is pulled from the liquid encapsulant. This can be a significant problem in low-pressure LEC growth. Large Ga droplets formed near the front of the crystal, and thermally migrated through the crystal resulting in a polycrystalline tail. The dislocation density in

Figure 2.11 LEC crystal grown with thick B_2O_3, displaying large diameter fluctuations due to reduced radial gradients

the ring region was 5000–6000 cm^{-2}, a 50% reduction, and the front of the crystal was semi-insulating. The considerable thermal degradation made it evident that surface protection was needed to make this approach viable for high-resistivity materials. However, the experiment did highlight the impact of convective heat transport through the ambient on dislocation density. The theoretical work of Jordan[26] predicts that the heat transfer coefficient at the crystal–ambient interface should increase as the square root of the pressure. This indicates a maximum reduction in the heat transfer coefficient in this experiment of 2.5, by reducing the pressure from 300 to 50 psi. Our data revealing a 50% EPD reduction are consistent with this prediction.

Seed quality and necking The impact of seed quality and Dash-type seed necking were evaluated in a series of growth experiments utilizing high and low dislocation seeds for crystals with and without thin necks. The dislocation densities at the front of each crystal at full diameter were measured and

Gallium arsenide

Table 2.3 Effect of seed quality and necking in dislocation density

Ingot no.	Necking[a]	Seed	EPD (cm^{-2})	Tail
1	no	high (5 × 10^4)	ring centre edge	7.6 × 10^4 4.6 × 10^4 3.0 × 10^5
5	yes	high (5 × 10^5)	ring centre edge	1.5 × 10^4 3.0 × 10^4 1.7 × 10^5
9	yes	low (3.3 × 10^3)	ring centre edge	1.4 × 10^4 2.0 × 10^4 2.5 × 10^5
16	no	low (4.5 × 10^3)	ring centre edge	1.3 × 10^4 2.8 × 10^4 1.7 × 10^5

[a] All cone angles > 25° and other growth parameters are similar.

compared. The resulting data, presented in Table 2.3, show that low-dislocation crystals (EPD less than 2.5×10^3 cm^{-2}) can be produced using low EPD seeds with and without seed necking, as well as by using high EPD seeds with necking.

Longitudinal crystal cross sections of the neck regions were studied to evaluate the effect of seed necking. Owing to deformation of neck regions with diameters less than 2.5 mm, the grown-in dislocations could not be determined (see Figure 2.12). For neck diameters between 2.5 and 3.5 mm, however, significant reduc-

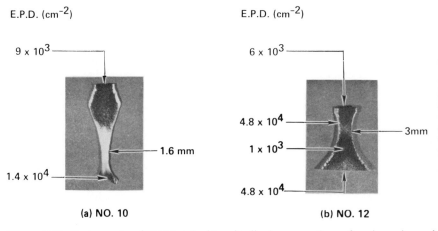

Figure 2.12 Photographs of KOH-etched longitudinal cross sections of seeds, necks, and cones for (a) 1.6 mm neck diameter and (b) 3 mm neck diameter[22]

Growth of bulk GaAs

tions in dislocation density were observed, as shown in Figure 2.12. These data reveal that Dash-type seed necking does reduce the dislocation density, independent of the seed EPD. Apparently, the dislocations can be transmitted from the seed to the crystal, an effect diminished by seed necking. However, the impact of seed necking is limited because dislocations will be formed in the crystal even for a dislocation-free seed.

Diameter control Diameter control effects were studied by evaluating the dislocation density differences in the tails of LEC crystals. The data of Table 2.4 show that the dislocation density is reduced by improving the diameter control. Crystals 6 and 9, for example, were pulled under essentially similar conditions except for the degree of diameter control. The dislocation density at the front of crystal 6, the more highly controlled diameter, is higher than that at the front of

Table 2.4 Effects of melt stoichiometry and diameter control on dislocation density

Ingot no.	Initial melt composition[a]	Diameter variation (mm)		EPD (cm^{-2}) Front	EPD (cm^{-2}) Tail
8	53.0% Ga	±4.0	ring centre edge	1.4×10^4 3.7×10^4 1.0×10^5	n.a.
6	51.5% Ga	±3.0	ring centre edge	1.8×10^4 2.6×10^4 8.0×10^4	8.6×10^4 7.7×10^4 2.2×10^5
9	51.5% Ga	±7.1	ring centre edge	1.4×10^4 2.0×10^4 2.5×10^5	1.0×10^5 1.0×10^5 2.4×10^5
10	50.7% Ga	±4.5	ring centre edge	1.1×10^4 2.1×10^4 1.1×10^5	n.a.
5	stoichiometric	±8.5	ring centre edge	1.5×10^4 3.0×10^4 1.7×10^5	1.2×10^5 1.4×10^5 2.1×10^5
12	50.1% As	±1.6	ring centre edge	1.2×10^4 1.7×10^4 2.5×10^5	9.0×10^4 1.0×10^5 2.2×10^5
16	50.3% As	±1.5	ring centre edge	1.3×10^4 2.8×10^4 1.7×10^5	1.4×10^5 2.2×10^5 2.4×10^5
11	50.6% As	±1.5	ring centre edge	7.5×10^3 1.3×10^4 1.9×10^5	8.1×10^4 1.8×10^5 1.8×10^5

[a] All cone angles > 25° and other growth parameters are similar.

crystal 9; however, the tail dislocation density of crystal 6 is lower. The apparent impact of diameter control is less than that of other parameters, such as cone angle, seed quality, and seed necking. However, crystals produced under less stable diameter control conditions appear to be subject to greater transient gradient-induced stress, resulting in higher dislocation densities.

Melt stoichiometry Stoichiometric and non-stoichiometric melts were utilized in the growth of LEC GaAs crystals to study the effect of melt stoichiometry. For compositions less than 0.503 As atom fraction, no correlation between EPD and the melt stoichiometry was observed, whether the melt was Ga- or As-rich (Table 2.4). However, crystals 11 and 12, pulled under essentially the same conditions except for melt composition, displayed significant differences. Crystal 11 exhibited lower EPD values in the seed end compared to crystal 12. This reduction in dislocation density at the crystal front indicates that an As-rich condition is advantageous provided the melt composition is greater than about 0.505 As atom fraction. The tail of crystal 11 did not exhibit noticeable improvement, which may support the hypothesis for the existence of a window of melt compositions between 0.505 and 0.535 for minimizing EPD.

2.3.3 Semi-insulating behaviour

The semi-insulating behaviour of GaAs substrates utilized in IC device fabrication with selective ion implantation is of critical importance. The insulating property is essential for providing device isolation, and consistent and uniform activation of the implanted dopants is vital to control the threshold voltages of FETs. A principal issue in the investigation of semi-insulating LEC GaAs has been the development of growth techniques to control impurity and trap levels. Characterization studies have been successful in providing data for modelling the semi-insulating behaviour in terms of background impurity levels, intrinsic defects, and growth parameters.[29,32]

Chemical analyses of undoped semi-insulating GaAs material grown by the LEC technique typically indicate low concentrations of residual impurities common to GaAs, such as Si, S, Se, Mg, Cr, Mn, and Fe (see Table 2.5). The principal impurity species, as determined by local vibrational mode spectroscopy or secondary ion mass spectroscopy (SIMS), are boron, which is present in a concentration range from 1×10^{16} to 6×10^{17} cm^{-3}, and carbon, which ranges from 2×10^{15} to 1.3×10^{16} cm^{-3}. Si contamination can be eliminated as a problem by growing crystals in pyrolytic boron nitride crucibles. The main source of boron contamination is the B_2O_3 encapsulant. The incorporation of boron has been shown to depend on the water content of the encapsulant, with the boron concentration decreasing as the water content increases[33] (see Figure 2.13).

Table 2.5 Analysis of chemical impurities in LEC GaAs

Technique	Crucible	Impurities (cm^{-3})									
		S	Se	Te	Mg	Cr	Mn	Fe	C	Si	B
LEC	quartz	2E15	<1E14	<1E14	<5E14	<5E14	<1E15	<9E15	n.d.–9E15	5E14–3E16	1E14–2E17
LEC	PBN	1.5E15	5E14	5E13	2E14	<5E14	1E15	3E15	2E15–1E16	2E15	1E14–2E17
Bridgman	quartz	3E15	3E14	4E13	4.5E14	3.1E16	4.7E14	4.7E15	n.d.	2E16	2E14

n.d. = not detected.

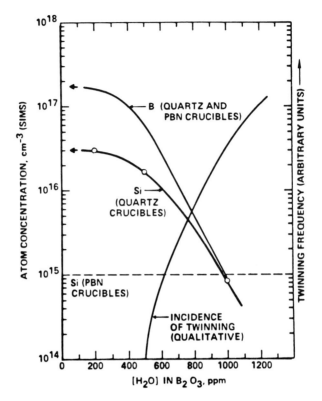

Figure 2.13 Dependence of Si and B impurity levels and twinning in undoped LEC GaAs on B_2O_3 water content

2.3.3.1 *Intrinsic defects, and C and B impurities*

Evaluation of the electrical properties of undoped LEC GaAs reveals a strong dependence on melt stoichiometry, as shown in Figures 2.14 to 2.17. As indicated in Figure 2.14, the material is semi-insulating above and p-type below a critical As concentration in the melt—approximately 0.475 As atom fraction. The variation in resistivity is explained in terms of the corresponding free carrier concentration and Hall mobility, as shown in Figures 2.15 and 2.16. The semi-insulating material grown above the critical composition is n-type with carrier concentrations ranging between 1×10^7 and 1×10^8 cm^{-3}. The material grown below the critical composition is p-type with hole concentrations in the range from 1×10^{15} to 3×10^{16} cm^{-3}. The mobilities in the n- and p-type material are typically 4000–5000 and 200–300 cm^2 V^{-1} s^{-1}, respectively, and are relatively independent of composition.

Growth of bulk GaAs

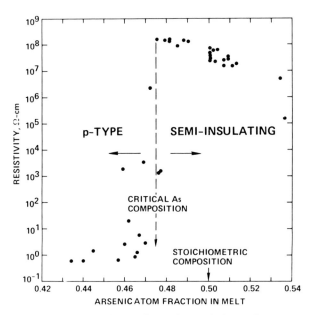

Figure 2.14 Dependence of electrical resistivity of undoped LEC GaAs on melt stoichiometry[30]

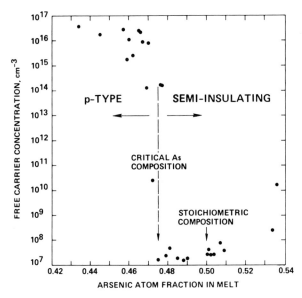

Figure 2.15 Dependence of free carrier concentration of undoped LEC GaAs on melt stoichiometry[30]

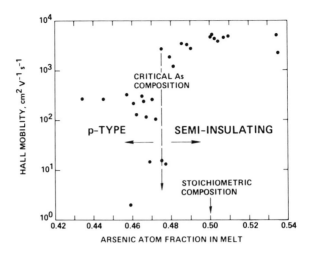

Figure 2.16 Dependence of Hall mobility of undoped LEC GaAs on melt stoichiometry[29]

The evaluation of the electrical and optical properties of the semi-insulating material indicates that the donor, commonly referred to as EL2, is the predominant deep centre.[34] An optical absorption band between 1 and 1.4 μm previously identified with the EL2 centre was observed in all the semi-insulating material.[31,35] In addition, the activation energy of the electron concentration, obtained from plots of the temperature-corrected free electron concentration as a function of the reciprocal of temperature, was 0.75 ± 0.02 eV. This energy is consistent with published values for the activation energy of EL2.[34] The behaviour of the photoconductivity thresholds above and below 120 K was also found to be consistent with the presence of EL2.

The concentration of EL2 was determined by optical absorption using the cross section reported by Martin et al.[35] Absorption due to unoccupied EL2 centres was not observed, and the Hall measurements indicated that the centre was more than 90% occupied. Consequently, the absorption was taken to be proportional to the total EL2 concentration. The concentration at EL2 was found to depend on the melt stoichiometry, as shown in Figure 2.17, increasing from about 5×10^{15} to 1.7×10^{16} cm^{-3} as the As atom fraction increased from about 0.48 to 0.51. The concentration remained constant as the As fraction increased further to about 0.535.

The concentration of a residual double acceptor observed in Hall measurements and also in the far-infrared absorption spectra of the p-type material (Figure 2.18) is also shown in Figure 2.19. It should be noted that the concentrations shown in Figure 2.19 represent absolute concentrations of the two defects as opposed to a measurement of the compensation of one defect by the other. The carrier concentrations obtained from the Hall measurements

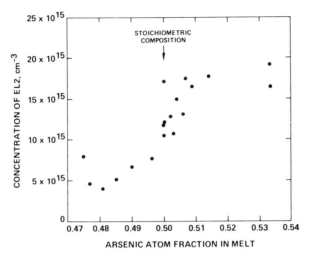

Figure 2.17 Dependence of EL2 concentration in undoped LEC GaAs, as determined by optical absorption, on melt stoichiometry[30]

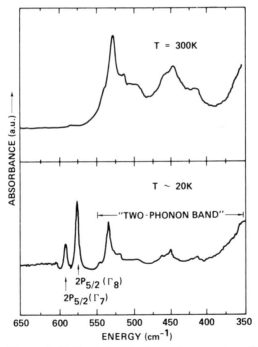

Figure 2.18 Far-infrared absorption spectra of 78 meV acceptor in undoped LEC GaAs. At room temperature, only phonon absorption is observed; at lower temperature, spectrum associated with the acceptor is observed[31]

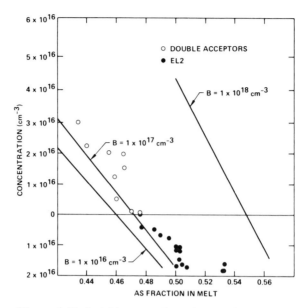

Figure 2.19 Stoichiometry dependence of EL2 deep donors and residual double acceptors in undoped LEC GaAs. Concentration of double acceptors is plotted on positive axis; concentration of EL2 deep donors is plotted on negative axis. Straight lines indicate predictions of model for various boron concentrations[40]

indicate compensation ratios for As_{Ga} (EL2) in the semi-insulating material of between 10 and 100 typically, indicating negligible compensation by residual acceptors. In the p-type Ga-rich material, Hall measurements indicate the presence of residual shallow carbon acceptors which have not been compensated by EL2, placing an upper limit on the EL2 content at 2×10^{15} cm^{-3} in this material. The presence of the uncompensated carbon also guarantees that the deeper double acceptor is uncompensated, and that the infrared absorption measurements give an uncompensated measure of the concentration.

Such behaviour explains in part the dependence of the electrical properties on stoichiometry. Semi-insulating material is obtained only when EL2 is present, whereas p-type material is obtained in the presence of the double acceptor. The correlation of EL2 with the semi-insulating behaviour suggests that the electrical activity of the material is determined by the deep donor EL2 and additional residual compensating acceptors. Hall measurements in the p-type Ga-rich material suggest that these acceptors can be identified with carbon. Carbon is the only acceptor identified in these measurements other than the deep double acceptor. This double acceptor is not observed in material simultaneously with

EL2. Thus, the indicated mechanism for semi-insulating behaviour is the compensation of EL2 by carbon acceptors. The following analysis supports this interpretation.

The ionization of EL2 produces an ionized centre plus an electron in the conduction band:

$$\text{neutral EL2} \rightarrow \text{ionized EL2} + e^- \tag{2.1}$$

According to the law of mass action, the concentration of ionized centres N_I, the concentration of electrons n, and the concentration of neutral centres N_U are related by the following equation:

$$\frac{N_I n}{N_U} = K \tag{2.2}$$

where K is a constant determined by the thermodynamics of the system. N_I is equal to the net acceptor concentration, given as the difference in concentration between shallow acceptors N_A and shallow donors N_D:

$$N_I = N_A - N_D \tag{2.3}$$

The concentration of acceptors is given as the sum of the concentrations of carbon and other residual acceptors N_A^R:

$$N_A = [\text{carbon}] + N_A^R \tag{2.4}$$

The concentration of neutral centres is equal to the EL2 concentration as determined by optical absorption. That is, only EL2 centres that are occupied by electrons contribute to the optical absorption process:

$$N_U = [\text{EL2}] \tag{2.5}$$

By substituting (2.3)–(2.5) into (2.2), the following expression for the free electron concentration is obtained in terms of the predominant centres in the material:

$$n = K \frac{[\text{EL2}]}{[\text{carbon}] + N_A^R - N_D} \tag{2.6}$$

This expression can be rewritten in the following form:

$$[\text{carbon}] = K \frac{[\text{EL2}]}{n} + N_D - N_A^R \tag{2.7}$$

Therefore the carbon concentration is proportional to the ratio of the EL2 concentration to the electron concentration.

Semi-insulating material grown from melts ranging from 0.475 to 0.535 As atom fraction was evaluated according to equation (2.7) from measurements of the carbon concentration (by local vibration mode studies), the EL2 concentration (by optical absorption), and the electron concentration (by Hall effect measurements) for each sample. A plot of the carbon concentration as a function of the ratio of the EL2 concentration to the electron concentration, shown in Figure 2.20, follows linear behaviour, indicating that the electron concentration is indeed controlled by the balance between EL2 and carbon. This result is independent of possible errors in the published values of the optical cross sections for carbon and EL2. It is important to note that if some other impurity were the predominant acceptor, such as Mn, Fe, Cu, or Zn, the linearity predicted on the basis of equation (2.7) would still necessarily hold. However, the linearity would not be distinguishable because the term [carbon] would be small compared to N_A^R; i.e. the figure would be a scatter plot. In fact, the scatter in these data probably reflects actual fluctuations in the concentration of other background impurities rather than random error in the experimental measurements. The small value of the intercept $(N_D - N_A^R)$ of the least-squares fit to the data also indicates the predominance of carbon acceptors. Thus, EL2 deep donors and carbon acceptors control the electrical compensation in semi-insulating LEC GaAs grown from melts ranging from 0.475 to 0.535 As atom fraction.

The identification of EL2 has been a controversial subject. However, evidence correlating EL2 with the defect As_{Ga} has recently been obtained. Electron

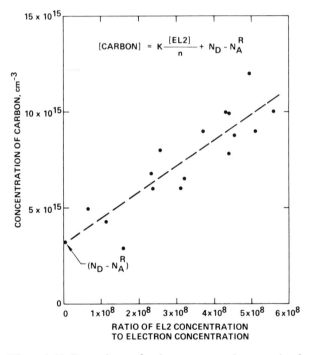

Figure 2.20 Dependence of carbon concentration on ratio of EL2 concentration to electron concentration in undoped LEC GaAs[30]

paramagnetic resonance (EPR) experiments have identified the presence of As_{Ga} in the samples used to obtain the data of Figure 2.17.[36] In addition, as shown in Figure 2.21, the EPR signal varies linearly with carbon content in identical fashion to the quantity [EL2]/n shown in Figure 2.20. Since the Fermi level is pinned at EL2 for the temperatures at which the EPR is measured, the change in EPR signal with carbon content can only occur if As_{Ga} and EL2 as measured by optical absorption are the same defect. Other experiments supporting this view include measurements of the optically induced change of the As_{Ga} EPR signal, which is identical to that of EL2.[37] It should be noted that the As_{Ga} EPR signal is not observed in the Ga-rich p-type material.

The 78 meV double acceptor is most likely due to either boron impurities or gallium atoms substitutionally incorporated on As lattice sites (Ga_{As} or B_{As}). It is possible to rule out other impurity related defects on the basis of SIMS measurements.[31] It is also possible to rule out defects which have symmetry lower than tetrahedral (T_d) on the basis of the transitions and splittings observed in the far-infrared spectra.[31] Both B_{As} and Ga_{As} defects are expected to have properties that can explain the behaviour observed for the double acceptor.

A comparison of the expected behaviour for B_{As} and Ga_{As} with experiment supports the identification of the residual double acceptor as being associated

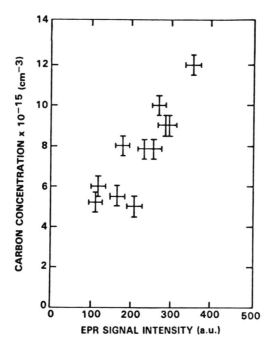

Figure 2.21 Relative EPR signal as a function of carbon concentration for undoped LEC GaAs[36]

with one or both of these defects. We can model these defects by using known energies for the double acceptor Zn in Ge. To obtain the appropriate energies by using this model for the antisite Ga_{As} or B_{As} in GaAs, one simply scales the values for the acceptor Zn in Ge by the ratio of effective rydbergs, 2.6, for the two materials. The results of this procedure are tabulated in Table 2.6, showing excellent agreement with the experimental data.[38]

Table 2.6 Estimated and experimental energies for the residual double acceptor in GaAs. E_{Acc} refers to the hole binding energy and $C-D$ is the energy difference between the excited states observed in the far-infrared absorption

	$C-D$ (cm^{-1})	E_{Acc} (meV)
Experimental Ga_{As}	16	78
Estimated Ga_{As} (Zn_{Ge} scaled by 2.6)	15.9	83
Experimental Ga_{As}	70	200
Estimated Ga_{As} (Zn_{Ge} scaled by 2.6)	70	221

B_{As} defects have been observed using local vibrational mode spectroscopy in some of the Ga-rich crystals. B_{As} centres are routinely observed in similarly grown p-type Ga-rich crystals which have been electron irradiated.[39] In addition, crystals which have anomalously high (2×10^{16} cm^{-3}) or low (1×10^{16} cm^{-3}) boron content show behaviour which deviates from the curve shown in Figure 2.17. Other researchers have found a correlation between the boron content and presence of the 78 meV acceptor in their material.[30]

The following model explains the results shown in Figure 2.17, supports the identification of antisites as the principal defects introduced during growth of bulk GaAs, and helps explain the role of boron in the GaAs defect chemistry.

We assume that the following reactions determine the defect chemistry at the melt temperature:

$$As_{Ga} + Ga_{As} \to As_{As} + Ga_{Ga} \qquad (2.8)$$

$$Ga_{As} + Ga_{Ga} + 2As_{liq} \to 2Ga_{liq} + As_{Ga} + As_{As} \qquad (2.9)$$

$$Ga_{As} + B_{Ga} \to Ga_{Ga} + B_{As} \qquad (2.10)$$

The first of these reactions is an annihilation reaction between As_{Ga} and Ga_{As} as suggested by Figure 2.14. The second reaction determines equilibrium between the native defects Ga_{As} and As_{Ga} for a particular melt composition. These two reactions are sufficient to determine the equilibrium intrinsic antisite (As_{Ga} and Ga_{As}) concentration in the bulk crystal near the growth interface. The third equation describes the equilibrium between boron impurities and the intrinsic antisite defects. Following growth, reactions (2.8) and (2.10) remain important since during cooldown the defects can continue to react. Since reaction (2.9) is the only reaction directly influenced by the melt composition, the net stoichiometry of the crystal cannot change during this cooldown period.

A number of simplifications have been made to facilitate development of this model. First, we have ignored the presence of defects and defect reactions involving vacancies, interstitials, etc. These defects are in equilibrium with the melt at the growth temperature and can be treated similarly. Provided these defects do not complex with the antisite defects described here, their presence will not affect the results predicted by the model. Condensation of vacancies into dislocation loops and other extended defects may, in fact, be important in determining macroscopic defect densities. Secondly, we have assumed that the defects do not change charge state while reacting. At the growth temperature, virtually all defects are ionized and this assumption is a good approximation. We have also assumed thermodynamic equilibrium during growth. The reaction dynamics may be important, however, in determining defect densities during cooldown.

These assumptions simplify modelling of the defect incorporation during growth by reducing the number of equations one must deal with. Using the law of mass action, (2.8), (2.9), and (2.10) yield the following equations:

$$K_1 = N^2 \exp(-\Delta_1/kT) = [As_{Ga}][Ga_{As}] \quad (2.11)$$

$$K_2 = \exp(-\Delta_2/kT) = [x/(1-x)]^2 [Ga_{As}]/[As_{Ga}] \quad (2.12)$$

$$K_3 = N \exp(-\Delta_3/kT) = [Ga_{As}][B_{Ga}]/[B_{As}] \quad (2.13)$$

The total boron content can be expressed as the sum of the boron on As and Ga sites:

$$[B_{As}] + [B_{Ga}] = [B] \quad (2.14)$$

N refers to the density of lattice sites in GaAs and x is the As atom fraction in the melt at the growth temperature. Δ_1, Δ_2, and Δ_3 are the free energies associated with reactions (2.8), (2.9), and (2.10).

For a given melt composition and boron content in the solid, these equations yield values for the defect densities of As_{Ga}, Ga_{As}, B_{As}, and B_{Ga}. Following growth, reaction (2.9) becomes inactive and the net stoichiometry of the crystal becomes invariant. Hence the net difference in deep donor and deep acceptor concentrations, $N_{DD} - N_{DA}$ remains unchanged and can be expressed as

$$N_{DD} - N_{DA} = [As_{Ga}] - [Ga_{As}] - [B_{As}] \quad (2.15)$$

If the cooling rate is sufficiently slow, reactions (2.8) and (2.10) proceed to completion. Thus, one is left with one of the three defects, As_{Ga}, Ga_{As} or B_{As} or a combination of B_{As} and Ga_{As}.

Equations (2.7)–(2.10) have been solved as functions of As atom fraction and boron concentration to obtain $N_{DD} - N_{DA}$. The effect of changes in Δ_1, Δ_2, and Δ_3 on the predictions of the model can be easily established by inspecting (2.7)–(2.10). For $\Delta_2 = 0$ and zero boron content, the concentration of defects is determined simply by the melt composition and Δ_1 with a zero intercept occurring at $x = 0.50$. For finite values of Δ_2 and non-zero boron concentrations, changes in Δ_1 will primarily affect the slope of the $N_{DD} - N_{DA}$ prediction. Changes in Δ_2 affect the zero intercept with the melt composition axis for low boron concentrations; changes in Δ_3 will primarily affect the slope of the $N_{DD} - N_{DA}$ prediction. Changes in Δ_2 affect the zero intercept with the melt composition axis for low boron content, whereas changes in Δ_3 affect both the intercept and the slope of the prediction for high boron content ($> 5 \times 10^{17}$ cm^{-3}).

The predictions of the model are shown in Figure 2.19 for various boron concentrations;[42] the quantity $N_{DD} - N_{DA}$ is plotted versus melt composition. To obtain Δ_1 and Δ_2, the predicted curves were fitted to the data in Figure 2.19 from low boron concentration crystals ($< 2 \times 10^{17}$ cm^{-3}). To obtain Δ_3, the data for high boron concentration crystals were used along with a comparison to the results published previously by Ta et al.[41] This crystal had double acceptor densities $\sim 6 \times 10^{16}$ cm^{-3} for melt compositions near 0.50 As atom fraction with $B \sim 1 \times 10^{18}$ cm^{-3}. The fit agrees with the data published for high boron concentrations crystals by Ta et al.[41]

The best values obtained for the energies Δ_1, Δ_2, and Δ_3 are $\Delta_1 = 3.3$ eV, $\Delta_2 = 0.072$ eV, and $\Delta_3 = 1.2$ eV. The values for Δ_1 and Δ_2 compare favourably with the estimates made by Van Vechten for antisite formation.[42] These estimates yield virtual enthalpy values of ~ 3.7 eV for reaction (2.4) and 0 eV for reaction (2.5).

This modelling helps explain the role of boron in determining the electrical properties of LEC GaAs. Inspection of (2.9) shows that the ratio of B_{As} to Ga_{As} depends linearly on boron concentration. Solution of the equations for low boron content (10^{16}–10^{17} cm^{-3}) indicates that in the p-type Ga-rich material, Ga_{As} defects are formed in greater amounts than B_{As} during growth. For high boron content, the opposite is true and B_{As} defects are formed in greater numbers. Hence, for low boron content, the p-type character of the material is largely independent of boron and is determined principally by reactions (2.4) and (2.5). In the high boron concentration material, $N_{DD} - N_{DA}$ is almost linearly dependent on the boron content, converging to the Ga_{As} value for low boron concentration. In both cases, the sign and magnitude of Δ_3 suggest that many of the Ga_{As} defects anneal into B_{As} defects during cooldown. Hence, one can obtain B_{As} defects, but the density of these defects is reasonably independent of boron content for moderate boron doping ($< 2 \times 10^{17}$ cm^{-2}). As can be seen from Figure 2.19, the scatter in the measured data can be large enough to obscure any dependence of $N_{DD} - N_{DA}$ on boron concentration in this doping range.

Other studies utilizing deep-level transient spectroscopy (DLTS) spectra of

Growth of bulk GaAs

As-rich samples grown with Se doping indicate low levels of electron traps other than EL2. DLTS spectra of p-type Ga-rich material show low levels of residual hole traps other than the 78 meV level (Figure 2.22a). One can only speculate on the concentration of hole traps in the As-rich material or electron traps in the Ga-rich p-type material. The previous analysis indicates that levels of these traps tend to be low.

Residual electron traps are particularly important in low carbon material. In such material the concentration of these defects may be sufficiently high to render the material conducting by compensating carbon in the material. We can identify the levels in addition to EL2 in Figure 2.22b as EL5 (0.42 eV) and EL3 (0.57 eV).[34] EL5 has the same energy as the defect E3 observed in electron-irradiated material.[43] It has been established that E3 exhibits recombination-enhanced defect motion and is associated with a primary defect formed by displacements on the As sublattice.[43,44] In view of the As-rich nature of this material, and assuming that E3 is the same trap as observed here, identification of the defect as an As interstitial would seem reasonable. The large diffusion coefficient of this defect is consistent with such an interpretation.

Residual hole traps are important because they affect the implant activation for direct implant processes. The DLTS spectra in Figure 2.22 show a broad band near room temperature. Residual iron shows a level at the correct energy to cause a peak at this temperature, although other levels such as the 'B' level observed in LPE material may also contribute to this peak. Iron has been detected in some of our samples in low concentration in EPR and also by SIMS. The concentration of iron ranges from 2×10^{14} to 2×10^{15} cm^{-3} under standard growth conditions.

2.3.3.2 Distribution of EL2 deep donors in LEC crystals

The EL2 deep donor plays a dominant role[30] in controlling the electrical compensation of the undoped semi-insulating material. The free electron concentration, for example, is determined by the balance between EL2 deep donors and carbon acceptors. As a result, the electron concentration increases, and the resistivity decreases, as the EL2 concentration increases. Since the yield of GaAs ICs fabricated by ion implantation depends in part on the uniformity and reproducibility of the substrate electrical properties, we have begun to evaluate the distribution of EL2 through our LEC GaAs crystals.

It has been shown that the average EL2 concentration along GaAs crystals (from the seed end to the tail) is controlled by the melt stoichiometry.[30,45,46] The average EL2 concentration either increases or decreases or remains constant depending on whether the melt is As-rich, Ga-rich, or near-stoichiometric, respectively. Uniform longitudinal distributions are obtained by growing from near-stoichiometric melts. It is preferable that the melt be slightly As-rich because the EL2 concentration can decrease under Ga-rich conditions to such as extent that the material becomes semi-insulating p-type.[30] The use of As-rich near-

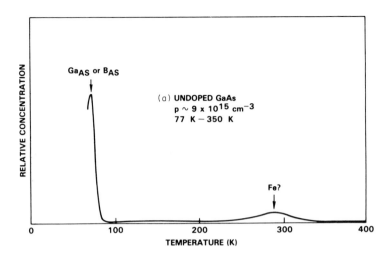

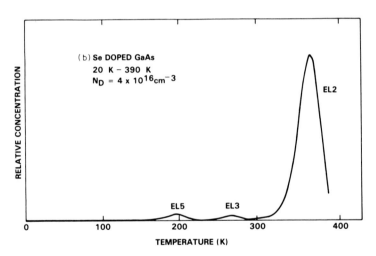

Figure 2.22 DLTS spectra for (a) undoped LEC GaAs and (b) Se-doped LEC GaAs

stoichiometric melts is now a routine procedure in our laboratory and elsewhere, and the yield of semi-insulating material is over 90% in our crystal growth process.

Of more immediate concern is the origin of radial variations in EL2 (variations across the sliced (100) crystal). A limited number of studies have been published concerning radial EL2 variations. In these studies, linear EL2 profiles across 2 and 3 inch LEC GaAs crystals were obtained by optical scanning techniques.[47,48]

Growth of bulk GaAs 75

It was observed that the EL2 profiles followed a characteristic 'W' pattern, and that the dislocation density followed this same pattern.

To study the possible correlation between EL2 and dislocations, and to evaluate the role of melt stoichiometry in controlling radial EL2 variations, we have determined contour maps of the EL2 deep-level concentration across 3 inch diameter semi-insulating ⟨100⟩ GaAs crystals grown in our laboratory. The EL2 contour maps were obtained utilizing 4 mm thick slices cut perpendicular to the ⟨100⟩ growth axis at the seed and tail end of each crystal. Thick slices were also obtained from the middle of selected crystals. The slices were cut into bars, and the EL2 profile along each bar was determined by optical absorption[47] at 1.1 μm. Measurements were made every 3 mm through a 3 × 3 mm square window. The data from each bar were collected, as shown in Figure 2.23, and the contour lines were drawn consistent with the data.

Contour maps of the EL2 concentration across the seed end of the crystals invariably had a four-fold symmetrical pattern, as shown in Figure 2.23. The EL2 concentration is highest at the edge, decreases towards the centre along any radial direction, and then increases slightly at the centre. Local minima always occur along ⟨110⟩ directions approximately midway between the centre and the edge, imparting four-fold symmetry to the pattern.

It is important to emphasize that the standard deviation of the EL2 distributions was typically about 10%, which is comparable to the percentage ionization of EL2 in our material. Since optical absorption measures only the concentration of un-ionized (occupied) EL2, the observed variations in EL2 could have reflected concentration variations in carbon, the predominant

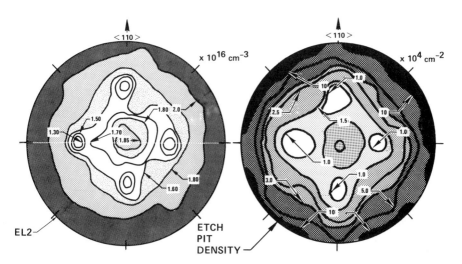

Figure 2.23 Mappings of EL2 and dislocation density for seed end of (100) undoped LEC GaAs crystal showing correlation[49]

compensating centre, rather than the total EL2 concentration. However, our analysis shows that the EL2 profiles are uncorrelated with the carbon profiles for two reasons, as summarized in Figure 2.24. First, carbon profiles across the crystal are uniform to within the sensitivity of the measurements rather than M-shaped. (An M-shaped profile would lead to a W-shaped EL2 profile.) Secondly, the shape of the EL2 profile is independent of the absolute carbon concentration, and therefore independent of the occupancy of the EL2 level, as indicated in the figure. These results show that the EL2 patterns obtained from our crystals indeed reflect the behaviour of the total EL2 concentration rather than variations in compensating centres.

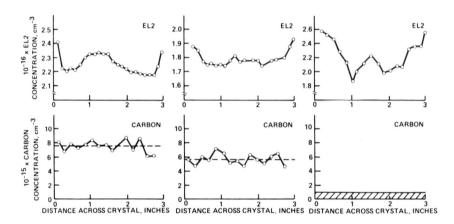

Figure 2.24 Radial profiles of EL2 and carbon concentration for (100) undoped LEC GaAs showing non-correlation

The basic, four-fold symmetric EL2 pattern was found to be independent of crystal and crucible rotation rates and relative sense of rotation. The crystal depicted in Figure 2.23 was grown with iso-rotation. For comparison, the basic EL2 pattern in a crystal grown with counter-rotation, shown in Figure 2.25, is also four-fold symmetric. Two additional features are observed in the crystal grown with counter-rotation. First, the EL2 concentration is highest at the edge along the $\langle 100 \rangle$ direction compared to the $\langle 110 \rangle$ direction. Secondly, there is a secondary set of maxima and minima along the $\langle 100 \rangle$ directions. However, these additional features were not unique to this particular crystal, but were also observed in other crystals grown with iso-rotation.

The basic EL2 pattern was also found to be independent of the melt stoichiometry. The crystal depicted in Figure 2.23 was grown from an As-rich near-stoichiometric melt. For comparison, the contour map of the seed end of a Ga-rich grown crystal, shown in Figure 2.26, is also four-fold symmetric. Some

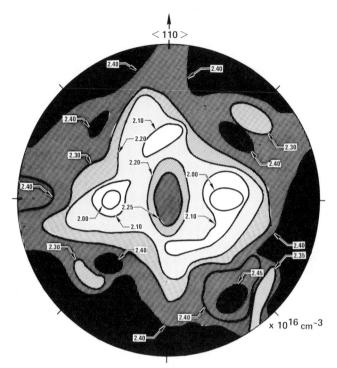

Figure 2.25 Mapping of EL2 at seed end of (100) undoped LEC GaAs crystal grown with counter-rotation

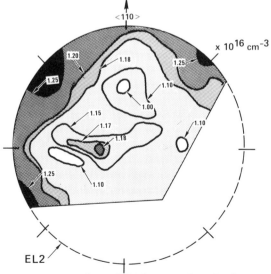

Figure 2.26 Mapping of EL2 at seed end of (100) undoped Ga-rich LEC GaAs crystal

shifting of the central EL2 maximum from the geometrical centre of the sample is evident. However, the EL2 patterns were never perfectly symmetrical is any of our crystals. In addition, the standard deviation of the EL2 concentration in about 20 crystals studied showed no correlation with the deviation of the melt composition from the stoichiometric value.

It has been shown[30,31,45,46] that the average EL2 concentration along GaAs crystals is controlled by the melt stoichiometry, either increasing, decreasing, or remaining constant when the melt is As-rich, Ga-rich, or near-stoichiometric. The possible relationship between variations of the melt stoichiometry across the solidification front during crystal growth and radial EL2 profiles has also been discussed.[48] This possible mechanism involves convection-induced variations of the thickness of the compositional boundary layer, as briefly discussed below.

A compositional diffusion boundary layer forms in a GaAs melt ahead of the growth interface when the melt is non-stoichiometric. For example, an As-rich layer builds up as a crystal is pulled from an As-rich melt. Variations in the fluid flow velocity and direction in the melt near the interface lead to variations in the thickness (δ) of the boundary layer. The As/Ga ratio in the melt at the interface increases as δ increases. Since the EL2 concentration increases[30] as the As fraction in the melt increases, the EL2 concentration would increases as δ increases.

The present experimental results were evaluated in terms of the model discussed above, and four inconsistencies were noted. First, according to the model, the EL2 pattern would have radial rather than crystallographic symmetry. It would be difficult to explain the observed EL2 patterns in terms of corresponding variations of δ considering the fact that the crystal and crucible are not stationary.

Secondly, the use of counter-rotation would be expected to cause changes in the EL2 distribution at a sufficiently high crystal rotation rate. This is because the mode of convection would change from free towards forced, accompanied by a change in the fluid flow pattern. Although we cannot be sure that the rotation rates used in these experiments were sufficiently high to promote the free-to-forced transition, the apparent independence of the symmetrical EL2 pattern on these growth parameters is inconsistent with the stoichiometry model.

Thirdly, EL2 distributions of Ga-rich grown crystals would be inverted compared to distributions in As-rich grown crystals on the basis of the stoichiometry model for radial EL2 variations. This is because the As fraction in the melt decreases as δ increases under Ga-rich conditions. However, the results show the same pattern independent of melt stoichiometry.

Finally, the standard deviation of the EL2 concentration would increase as the deviation of the melt composition from the stoichiometric level increases on the basis of the model, yet no dependence was observed experimentally. These results indicate strongly that variations in the melt stoichiometry across the growth interface have no significant effect on the radial distribution of EL2 in LEC GaAs crystals, even though the average EL2 concentration along the crystal is in fact controlled by the melt stoichiometry.

A good correlation is observed between the spatial distribution in EL2 and dislocations at the seed end of the crystals, as shown in Figure 2.23. All the important features of the EL2 patterns, such as the localized maxima at the edge along $\langle 100 \rangle$ (shown in Figures 2.25 and 2.26) and the secondary set of maxima and minima along $\langle 100 \rangle$ (shown in Figure 2.25), were found to correlate with the dislocation patterns in the respective crystals.

At the tail of the crystals, on the other hand, the contour maps of EL2 and dislocations correlated in only about 50% of the crystals. Whereas the dislocation patterns always remained four-fold symmetric, the tendency of the EL2 pattern was to become less symmetrical (i.e. two-fold or mirror-plane symmetry), as shown in Figure 2.27. Profiles (line scans) across the tail were often U-shaped or inverted U-shaped as well as W-shaped and M-shaped. In all crystals, both the EL2 and dislocation distributions were more uniform at the tail compared to the seed end as evidenced by the standard deviation of the respective concentration maps. EL2 patterns in the middle of a limited number of crystals were also investigated. These patterns were symmetric, and the uniformity, as indicated by standard deviation determinations, was intermediate between the seed end and the tail, indicating that the EL2 profiles become progressively more uniform towards the tail.

Although there can be a strong correlation between the distribution of EL2 and dislocations across LEC crystals, the relationship is not quantitative. For example, it has already been discussed how the average EL2 concentration

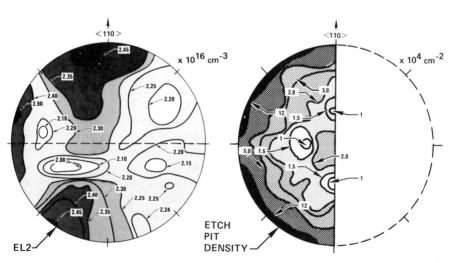

Figure 2.27 Mappings of EL2 and dislocation density at tail end of (100) undoped LEC crystal showing non-correlation

increases from the seed end to the tail under As-rich conditions or decreases under Ga-rich conditions. The average dislocation density, on the other hand, always increased in our crystals.[22] It has also been shown[15] that the EL2 concentration in very low dislocation density (about $500\,\text{cm}^{-2}$) GaAs is comparable to that in our material, indicating that EL2 formation can be independent of the dislocation density. Furthermore, whereas the melt stoichiometry controls the average EL2 concentration along the crystals, the current results shown that radial EL2 variations are practically independent of the melt composition. This behaviour indicates that EL2 could form by two distinct mechanisms. Thus, referring to the typical W profile (see Figure 2.28), the minimum EL2 concentration between the centre and edge of the crystal could be determined by the melt stoichiometry. This distribution is uniform across the crystal. Additional EL2 forms at the centre and edge of the crystal as a result of the second mechanism. The superposition of the 'second' distribution imparts four-fold symmetry to the overall pattern.

Martin et al.[47] first proposed that the correlation between EL2 and dislocations across LEC crystals could result from stress. It is known that the radial

	EL2 VARIATIONS	
	ALONG CRYSTAL	ACROSS CRYSTAL
STOICHIOMETRY	YES	NO
STRESS	NO	YES

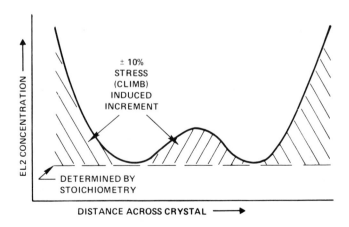

Figure 2.28 Model for EL2 formation

distribution of dislocations depends in part on stress arising from radial temperature gradients in the crystal during growth and postgrowth cooling. The stress pattern is crystallographic[27] because of the thermoelastic properties of GaAs. The radial stress patterns in (100) GaAs are four-fold symmetric. Martin proposed that EL2 is a native defect or native defect complex forming in the presence of stress. The detailed relationship between EL2, and stress and dislocations is not understood at this time. However, there seem to be three possible models: (1) the stress-enhanced equilibrium model, (2) the dislocation climb model, and (3) the dislocation gettering model. These models are discussed below in light of our experimental results.

Stress-enhanced equilibrium model The first model involves stress-enhanced formation of EL2. Assume that EL2 forms by a defect reaction such as the reaction for antisite formation proposed by Lagowski *et al.*,[45] given by

$$V_{Ga} + As_{As} \rightarrow As_{Ga} \text{ (EL2)} + V_{As} \qquad (2.16)$$

where V_{Ga} is a gallium vacancy, As_{As} is an arsenic atom on its normal site, and V_{As} is an arsenic vacancy. The EL2 concentration is determined by V_{Ga}, which is a grown-in defect controlled by the melt stoichiometry. V_{Ga} incorporates uniformly on a macroscopic scale. Reaction (2.16) (or some other relevant defect reaction) could be sensitive to stress owing to thermodynamic considerations, and the formation of EL2 is enhanced by stress during postgrowth cooling. As a result, the stress pattern in the crystal would impart a corresponding symmetrical pattern of EL2 superimposed on the stoichiometry-controlled (V_{Ga}) distribution.

Dislocation climb model The second model is based on the dislocation climb mechanism proposed by Weber *et al.*[37] Weber applied electron paramagnetic resonance (EPR), photoelectron paramagnetic resonance, and photoluminescence to plastically deformed GaAs and showed that the antisite defect (As_{Ga}) forms under deformation. They showed that the assignment of the antisite to EL2 can be explained on the basis of similarities between photoquenching of the EPR of the antisite and of the photocapacitance of EL2. In addition, Weber noted the results of deep-level transient spectroscopy and other studies, which showed the enhancement of EL2 after plastic deformation. Weber proposed that dislocation climb, which occurs during plastic deformation, is accompanied by the formation of EL2. One possible mechanism is the reaction of a gallium vacancy and an arsenic interstitial during climb to form the antisite. Presumably, the climb-induced EL2 distribution would be superimposed on the uniform, stoichiometry-controlled background. The fraction of EL2 resulting from dislocation climb would be highest in more heavily dislocated regions of the crystal and in regions subjected to the highest stress during postgrowth cooling.

Dislocation gettering model Brozel and coworkers[50,51] recently applied high spatial resolution infrared imaging to study the distribution of EL2. Fine structure superimposed on the macroscopic EL2 pattern was revealed to consist of cells and bands ('sheets' and 'streamers'). The fine structure was found to correlate with localized networks and bands of dislocations. Stirland discussed how either gettering of EL2 to dislocation cores through strain–field interactions or dislocation climb could possibly account for the high optical absorption associated with the dislocation cores. He discounted the gettering mechanism because EL2 does not behave like a simple substitutional impurity (i.e. the EL2 concentration can increase, decrease, or remain constant along the crystal). However, we have already addressed why EL2 does not behave like a simple substitutional impurity, and that the relationship between the total EL2 concentration and the dislocation density is not quantitative. Therefore, we believe that the gettering effect suggested by Stirland is a feasible explanation for the EL2–dislocation correlation. The infrared imaging studies are also consistent with the dislocation climb model. However, in demonstrating a correlation between dislocation and EL2 inhomogeneties on the microscale, these infrared imaging studies would seem to refute the simple stress-enhancement model (the first mechanism discussed in this section) for radial EL2 variations. This is because the EL2 concentration across the crystal would vary smoothly according to the model corresponding to variations in stress.

The lack of correlation between EL2 and dislocation patterns at the tail of the crystals, and the progressive improvement in EL2 uniformity from the seed end to the tail, is probably related to differences in thermal history. For example, the effective cooling rate of the material at the seed end of the crystal is determined by the pulling rate and the vertical temperature gradient as the crystal moves into a cooler region of the growth chamber. The cooling rate at the tail of the crystal, on the other hand, is determined by the programmed cooling rate of the heater because the crystal is not withdrawn from the hot zone of the furnace during postgrowth cooling. We estimate that the seed end typically cools to the 400–500°C range about twice as fast as the tail. The crystals had not typically been rotated during postgrowth cooling as well. Differences in cooling rate, thermal gradients, or other growth parameters between the seed end and tail of the crystals could have led to variations in the formation of EL2. For example, the progressively more uniform distribution of dislocations towards the tail could result in more uniform EL2 distributions in accordance with the dislocation climb model. On the other hand, the initial EL2 distribution towards the tail of the crystal could become altered through a time-dependent process such as bulk or pipe diffusion towards the periphery of the crystal. Further work is needed to explain the variations in the EL2 patterns along the LEC GaAs crystals.

In summary, macroscopic variations in the EL2 concentration across 3 inch ⟨100⟩ LEC GaAs crystals were characterized by determining quantitative, two-dimensional contour maps. The effects of melt stoichiometry, dislocations, and stress on the EL2 distribution were evaluated. EL2 patterns correlate with dislocation patterns across the crystals towards the seed end, although the relationship is not quantitative along the crystals. While it has been shown that the melt stoichiometry controls the average EL2 concentration along the crystal, the present results show that variations in EL2 across the crystals are practically independent of the melt composition. This behaviour indicates that two separate mechanisms could be involved in controlling EL2 formation. The primary stoichiometry-controlled EL2 distribution is probably uniform on the macroscale. The 10–20% variations across the crystal could be induced by dislocation climb, dislocation gettering, or stress-enhanced EL2 formation during postgrowth cooling. The climb or gettering mechanisms appear to be more consistent with our results and previously published work of Brozel et al. Changes in the EL2 patterns towards the tail of the crystals were discussed in terms of the differences in thermal history along the crystal. It is believed that an understanding of changes in the EL2 patterns along the crystal will be helpful in determining the mechanism controlling radial EL2 variations.

Finally, it is important to note two implications of our findings with respect to both the electrical properties of LEC GaAs and to correlating the variations in device parameters and material properties across processed wafers. First, it has been shown that the compensation mechanism[30] of undoped LEC GaAs is controlled by the balance between EL2 deep donors and carbon acceptors. As the EL2 concentration decreases, the free electron concentration decreases and the resistivity increases. Therefore, it can be expected on the basis of our EL2 contour maps that the resistivity varies inversely with EL2. In fact, Bonnet et al.[52] recently published resistivity contour maps for 3 inch LEC GaAs showing four-fold symmetry, consistent with our EL2 maps. Improved uniformity in EL2 should therefore improve the uniformity of the electrical properties.

Secondly, there have been additional reports[9,10] of correlations between variations in device parameters, such as source-to-drain current and threshold voltage, and the distribution of dislocations in semi-insulating GaAs wafers. Although EL2 may not necessarily directly or indirectly affect these or other device parameters, the relationship between EL2 and dislocations is of interest when analysing device performance uniformity parameters.

2.3.3.3 Carbon in LEC GaAs

Owing to the important role of carbon in the compensation mechanism in undoped semi-insulating GaAs, we have investigated the impact and control of carbon in greater detail. In particular, the effect of carbon in the substrate on device parameters is of interest. We have found that the carbon concentration in LEC material strongly affects the depletion or threshold voltage for implanted

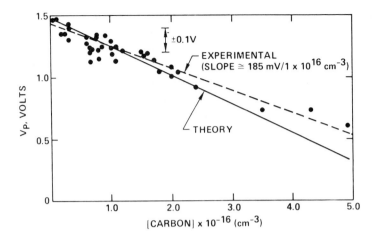

Figure 2.29 Dependence of depletion voltage for ion-implanted layers in undoped LEC GaAs on background carbon concentration

layers such as used for FETs (Figure 2.29). As the carbon concentration increases, the threshold voltage decreases.

This finding is supportive of the compensation mechanism model discussed earlier in this chapter. In this model, in which the electron concentration is controlled by the balance between EL2 deep donors and carbon acceptors, EL2 is typically 10% ionized by carbon. However, in n-type implanted layers, EL2 is virtually 100% occupied. The compensation of the implanted donors by carbon increases as the carbon concentration increases, decreasing the effective implant activation. Higher compensation reduces the threshold voltage. The slope of the experimental curve of $185 \text{ mV}/1 \times 10^{16} \text{ cm}^{-3}$ is in good agreement with theory, as indicated in Figure 2.29. The data in this figure represent variations in threshold voltage corresponding to variations in carbon on a crystal-to-crystal basis.

With these data, the need to study the distribution of carbon across crystals, representing radial variations on wafers, and along the growth axis, representing variations from wafer to wafer, came into sharp focus. In response to this need, we developed a room-temperature profiling technique to facilitate carbon profiling studies.

The carbon concentration in GaAs is determined by localized vibrational mode (LVM) far-infrared absorption. The conventional method[53] involves the measurement of the absorption coefficient at 582 cm^{-1} at 77 K. Such low-temperature measurements are time consuming and not amenable to profiling studies. Therefore, a room-temperature measurement suitable for profiling was developed and applied. Variations in the background absorption in the infrared spectrometer were determined to be an important source of scatter, which could be significantly reduced by calibrating the spectrometer with a reference sample of GaAs containing no measurable carbon. By subtracting the sample absorption, the absorption due to carbon is determined more precisely. Measurements made with the modified room-temperature method were shown to be in excellent agreement with conventional 77 K measurements. The background detection limit is about 1×10^{15} cm^{-3}, slightly higher than the level of about 5×10^{14} cm^{-3} achievable by the conventional method.

Carbon profiles across the seed end, middle, and tail of several crystals were determined by the modified measurement technique. A series of profiles along one crystal, which is representative of the behaviour of all crystals analysed, is shown in Figure 2.30. These results highlight two important issues with respect to threshold voltage uniformity. First, the carbon distribution across the crystals is uniform to within $\pm 6 \times 10^{14}$ cm^{-3}, the apparent precision of the measurements. According to the results in Figure 2.29, the contribution to the non-uniformity in threshold voltage corresponding to these compositional fluctuations is only ± 11 mV or less. Radial carbon variations would not, therefore, lead to significant variations in threshold voltage across a single wafer. On the other hand, the carbon concentration invariably decreased towards the tail of LEC crystals by 40% or more. The corresponding effect on variations in threshold voltage would depend on the absolute carbon concentration. For example, the threshold voltage would increase by about 100 mV from seed end to tail along a crystal doped at 1×10^{16} cm^{-3} carbon at the seed end. In contrast, a crystal doped to the 1×10^{15} cm^{-3} level would be expected to show only a 10 mV change. The

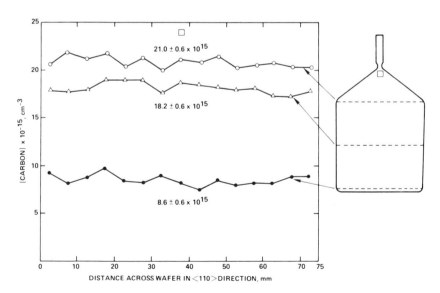

Figure 2.30 Radial profiles of carbon in undoped LEC GaAs

implication of these results is, of course, that variations in carbon levels along the crystal may lead to variations in threshold voltage over its length, requiring adjustment of the implant dose.

Analysis of carbon segregation along (100) undoped LEC GaAs crystals was made. A classical 'normal segregation' analysis of a limited number of crystals indicated that the distribution coefficient k_0 of carbon in GaAs is about 2 (k_0 is defined as the impurity concentration in the crystal at the growth interface divided by the impurity concentration in the liquid at the growth interface). A value of $k_0 > 1$ indicates that the impurity segregates into the crystal, depleting the melt during growth. As a result, the impurity concentration decreases towards the tail.

Further analysis of carbon segregation indicated that, although the value of k_0 obtained from the segregation analysis is consistently greater than 1, there is a measurable spread of k_0 values in the 1.4 to 2.4 range. This variation is shown by the different slopes of the curves obtained in the normal segregation analysis, as illustrated in Figure 2.31. Further analysis showed that the measured distribution

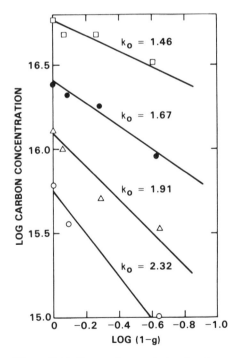

Figure 2.31 Segretation analysis of carbon. Slope of log [carbon] as a function of log (1 −g), where g is fraction of melt solidifed, gives distribution coefficient k

Growth of bulk GaAs

coefficient of carbon is a function of carbon concentration, decreasing as the carbon concentration increases, as shown in Figure 2.32.

The significant reduction of the carbon concentration enabled study of seed-to-tail variation of the threshold voltage in low ($< 1 \times 10^{15}$ cm^{-3}), medium ($(1-10) \times 10^{15}$ cm^{-3}) and high ($> 1 \times 10^{16}$ cm^{-3}) carbon crystals. The results, summarized in Figure 2.33, highlight two important issues with respect to wafer-to-wafer control of the threshold voltage.

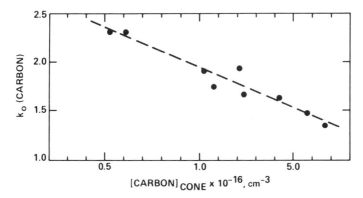

Figure 2.32 Concentration dependence of distribution coefficient indicating non-ideal carbon segregation behaviour

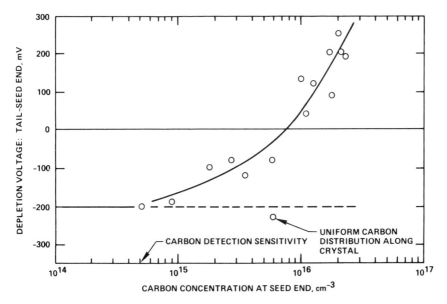

Figure 2.33 Dependence of depletion voltage as a function of carbon concentration at crystal seed end

First, the threshold voltage increases towards the tail of the high carbon crystals, as determined by carbon segregation. However, the opposite behaviour is observed in the low and medium carbon crystals. That is, the threshold voltage decreases towards the tail, although improved uniformity was expected. The results show the existence of a second mechanism in the low carbon crystals, and possibly in all of the crystals, that tends to reduce the threshold voltage towards the tail. This reduction in threshold voltage at the tail of crystals with uniform carbon profiles and medium carbon levels further emphasizes the effect of a second mechanism.

The second important result is that uniform threshold voltage can be achieved by intentionally doping with carbon in the mid 10^{15} cm^{-3} range. Since the carbon concentration can be reproducibly controlled in the 10^{14} cm^{-3} range, the crystals can be controllably back-doped to the desired range. At present, back-doping with carbon represents the most practicable near-term solution for the growth of material with reproducible threshold voltages for ion-implanted ICs.

2.4 APPLICATION OF LEC GaAs TO DEVICE FABRICATION

The development of LEC GaAs materials is already having significant impact on device fabrication activities, making manufacturing of GaAs ICs feasible. In this section, the selection and application of bulk GaAs materials to digital device fabrication in our laboratory will be discussed.

Variability in the electrical and crystalline characteristics of GaAs substrates available in the past, in particular those grown by the Bridgman technique and doped with substantial amounts of chromium, mandated the application of material qualification or selection techniques before application in device fabrication. The specifications we utilize to select GaAs substrates for digital IC processing include high resistivity ($> 10^7 \, \Omega$ cm), good activation of implanted dopants ($> 80\%$ for n-type dopants), dislocation density ($< 50\,000$ cm^{-2}) and crystalline perfection (single crystal) requirements, and dimensional (3.00 ± 0.015 inch) and flatness ($\pm 1.5 \, \mu$m cm^{-1} variation) criteria.

At present, we utilize 3 inch (100) undoped, semi-insulating substrates in our digital research and development activities as well as our pilot line work. This includes devices based on MESFETs (metal–semiconductor field-effect transistors), HEMTs (high electron mobility transistors), and HBTs (heterojunction bipolar transistors). The 3 inch round wafer format was selected to facilitate use of standard semiconductor processing equipment. A major and minor flat are utilized, to orient each wafer uniquely with respect to the two non-identical planes in each (100) wafer, which can effect the threshold voltages in FETs.[54] Relatively thick (25 mil; 0.025 inch) wafers with edge bevelling are used in device processing to minimize breakage. A featureless, flat wafer surface is sought, to maximize yields in lithography.

Evaluation of the electrical characteristics of semi-insulating substrates for device fabrication applications is particularly important for high yields and performance. This is particularly the case for structures which utilize the semi-insulating substrate directly in the formation of active device layers, such as GaAs MESFETs (Figure 2.34). In these devices, the high-resistivity substrate functions as the isolation between devices, and the active layers are formed by individual ion implantation steps directly in the substrate. The thermal stability of substrates used in ion implantation processing is particularly critical, since the annealing steps which follow to activate dopants and remove radiation damage can result in the formation of conducting skins on the wafer surface for some materials. GaAs materials made semi-insulating through the addition of compensating impurities such as chromium are particular susceptible to such changes. Thermal processing can cause redistribution of the Cr.[55,56] If the background of donor impurities (usually Si) is sufficiently high, an uncompensated surface conducting layer is formed, making the material unsuitable for device fabrication utilizing the substrate for isolation. In addition, this effect can manifest itself in the carrier profile of the lightly doped channel layer of the FET as shown in Figure 2.35.

While the application of undoped, semi-insulating LEC substrates has largely eliminated these problems, careful evaluation of the implant activation and profile, as well as the substrate resistivity before and after annealing, is carried out to ensure that only materials which display standard properties are utilized in device fabrication. Materials for HEMT and HBT processing undergo essentially the same screening process. Undoped LEC materials are used as substrates for molecular beam epitaxy (MBE) layer growth for these devices,[57,58] as well as for metal-organic chemical vapour deposition (MOCVD) to minimize the possibility of impurity outdiffusion into epitaxial layers.

As indicated in Figure 2.36, the uniformity of undoped semi-insulating GaAs is excellent for ion implantation processing. The resulting uniformity in device parameters has allowed successful fabrication of low-power and high-speed (1 ns

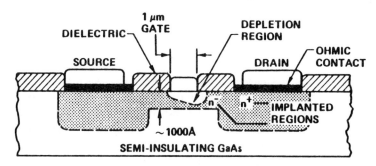

Figure 2.34 Cross section of ion-implanted MESFET device formed in semi-insulating substrate

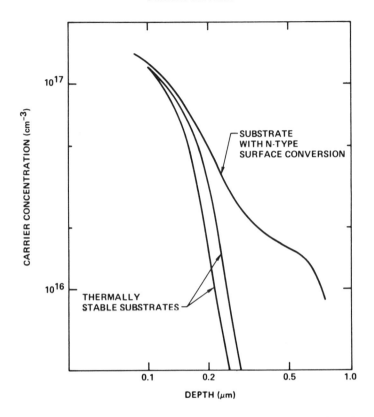

Figure 2.35 Carrier concentration profiles for ion-implanted FET channel layers

access time) 256-bit RAM devices and 8-bit by 8-bit (1008 gate) multiplier circuits. In addition, these same undoped LEC GaAs materials have been utilized to fabricate heterojunction bipolar devices with cutoff frequencies as high as 16 GHz,[59] and HEMT inverters with 12.2 ps propagation delay.[58]

2.5 CONCLUSIONS

Significant improvements in the size, electrical characteristics, crystalline features, and supply of semi-insulating GaAs materials for IC applications have resulted from the development of the high-pressure LEC growth technique. Understanding of the compensation mechanism in undoped, semi-insulating material, together with the utilization of techniques to reduce dislocation densities and twins, has markedly improved material quality and availability. The existence of a reliable supply of GaAs substrates for IC fabrication is facilitating the move of high-performance GaAs device technology from the laboratory into

Growth of bulk GaAs 91

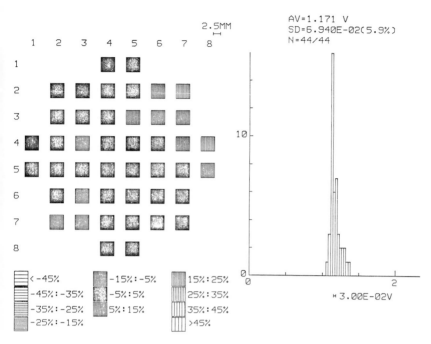

Figure 2.36 Uniformity of threshold voltage for ion-implanted FETs over 3 inch undoped LEC GaAs wafer

manufacturing and systems. LEC materials, with uniform sizes conforming to standard semiconductor processing equipment, as well as thermal stability and high purity, are excellent substrates for ion implantation based processing, as well as epitaxial growth.

In the future, continued improvements will be sought to control substrate parameters which affect device characteristics and reduce and homogenize dislocation densities. The identification of those material parameters impacting device performance or radiation hardness will continue to be subjects for investigation. In addition, to provide cost-effective production of larger quantities of wafers, increased melt sizes produced with automatic diameter control will be needed in the near term.

REFERENCES

1. E. P. A. Metz, R. C. Miller, and R. Mazelsky, A technique for pulling single crystals of volatile materials, *J. Appl. Phys.*, **33**, 2016–17 (1962).
2. J. B. Mullin, R. J. Heritage, C. H. Holliday, and B. W. Straughan, Liquid encapsulation crystal pulling at high pressures, *J. Cryst. Growth*, **34**, 281–5 (1968).
3. E. M. Swiggard, S. H. Lee, and F. W. Von Batchelder, GaAs synthesized in pyrolytic

boron nitride (PBN), *Gallium Arsenide and Related Compounds 1976* (St Louis), Inst. Phys. Conf. Ser. 33b, Bristol and London, Institute of Physics, pp. 23–7 (1977).
4. T. R. AuCoin, R. L. Ross, M. J. Wade, and R. O. Savage, Liquid encapsulated compounding and Czochralski growth of semi-insulating GaAs, *Solid State Tech.*, **22**, 59–67 (1979).
5. A. Steinemann and V. Zimmerli, Growth peculiarities of gallium arsenide single crystals, *Solid State Electronics*, **6**, 597–604 (1963).
6. W. A. Bonner, Reproducible preparation of twin-free InP crystals using the LEC technique, *Mater, Res. Bull.*, **15**, 63–72 (1980).
7. M. D. Lind, Personal Communication (1981).
8. L. R. Weisberg, J. Blanc, and E. J. Stofko, On the crystallinity of GaAs grown horizontally in quartz boats, *J. Electrochem. Soc.*, **109**, 642–3 (1962).
9. S. Miyazawa, T. Mizutani, and H. Yamazaki, Leakage current I_L variation correlated with dislocation density in undoped, semi-insulating GaAs, *Jap. J. Appl. Phys.*, **21**, L542–4 (1982).
10. Y. Nanishi, S. Ishida, T. Honda, H. Yamazaki, and S. Miyazawa, Inhomogeneous GaAs FET threshold voltages related to dislocation distribution, *Jap. J. Appl. Phys.*, **21**, L335–7 (1982).
11. B. C. Grabmaier and J. G. Grabmaier, Dislocation-free GaAs by the liquid encapsulation technique, *J. Cryst. Growth*, **13/14**, 635–9 (1972).
12. J. C. Brice, An analysis of factors affecting dislocation densities in pulled crystals of gallium arsenide, *J. Cryst. Growth*, **7**, 9–12 (1970).
13. Y. Seki, H. Watanabe, and J. Matsui, Impurity effect on grown-in dislocation density of InP and GaAs crystals, *J. Appl. Phys.*, **49**, 822–8 (1978).
14. J. C. Brice, An analysis of factors affecting dislocation densities in pulled crystals of gallium arsenide, *J. Cryst. Growth*, **7**, 9–12 (1970).
15. J. Parsey, Y. Naminski, J. Lagowski, and H. C. Gatos, Electron trap-free low dislocation melt down GaAs, *J. Electrochem. Soc.*, **128**, 936–7 (1981).
16. A. Steinemann and U. Zimmerli, Dislocation-free GaAs single crystals, *Proc. Int. Crystal Growth Conf.* (Boston, MA, 1966), ed. H. S. Peiser, Oxford, Pergamon, pp. 81–7 (1967).
17. M. G. Mil'vidski and E. P. Bochkarev, Creation of defects during the growth of semiconductor single crystals and films, *J. Cryst. Growth*, **44**, 61–74 (1978).
18. S. Shinoyama, C. Uemura, A. Yamamoto, and S. Tokno, Growth of dislocation-free undoped InP crystals, *Jap. J. Appl. Phys.*, **19**, L331–4 (1980).
19. P. J. Roksnoer, J. M. P. L. Huijbregts, W. M. van de Wijgert, and A. J. R. de Kock, Growth of dislocation-free gallium phosphide crystals from a stoichiometric melt, *J. Cryst. Growth*, **40**, 6–12 (1977).
20. T. Suzuki, S. Akai, K. Kohe, Y. Nishida, K. Fujita, and N. Kito, Development of large size 'dislocation-free' GaAs single crystal, *Sumitomo Electronics Tech. Rev.*, **18**, 105–11 (1978).
21. M. G. Mil'vidsky, V. B. Osvensky, and S. S. Shifrin, Effect of doping on formation of dislocation structure in semiconductor crystals, *J. Cryst. Growth*, **52**, 396–403 (1982).
22. R. T. Chen and D. E. Holmes, Dislocation studies in 3-inch diameter liquid encapsulated Czochralski GaAs, *J. Cryst. Growth*, **61**, 111–24 (1983).
23. J. Angilello, R. M. Potenski, and G. R. Woolhouse, Etch pits and dislocations in (100) GaAs wafers, *J. Appl. Phys.*, **46**, 2315–16 (1975).
24. W. D. Dash, Growth of silicon crystals free of dislocations, *Growth and Perfection of Crystals*, eds R. H. Doremus, B. W. Roberts, and D. Turnbull, New York, Wiley, p. 361 (1957).
25. P. Penning, Generation of imperfections in germanium crystals by thermal strain, *Philips Res. Rep.*, **13**, 79–97 (1958).

26. A. S. Jordan, An evaluation of the thermal and elastic constants affecting GaAs crystal growth, *J. Cryst. Growth*, **49**, 631–42 (1980).
27. A. S. Jordan, R. Caruso, and A. R. Van Neida, A thermoelastic analysis of dislocation generation in pulled GaAs crystals, *Bell System Tech. J.*, **59**, 593–637 (1980).
28. R. E. Reed-Hill, *Physical Metallurgy Principles*, 2nd edn, Princeton, NJ, Van Nostrand, p. 274 (1973).
29. D. E. Holmes, R. T. Chen, K. R. Elliott, and C. G. Kirkpatrick, Compensation mechanism in liquid encapsulated Czochralski GaAs, *IEEE Trans. Microwave Devices*, **MTT-30**, 949–55 (1982).
30. D. E. Holmes, R. T. Chen, K. R. Elliott, and C. G. Kirkpatrick, Stoichiometry controlled compensation in liquid encapsulated Czochralski GaAs, *Appl. Phys. Lett.*, **40**, 46–8 (1982).
31. K. R. Elliott, D. E. Holmes, R. T. Chen, and C. G. Kirkpatrick, Infrared absorption of the 78 meV acceptor in GaAs, *Appl. Phys. Lett.*, **40**, 898–901 (1982).
32. K. R. Elliott, Residual double acceptors in bulk GaAs, *Appl. Phys. Lett.*, **42**, 274–6 (1983).
33. J. R. Oliver, R. D. Fairman, and R. T. Chen, Undoped, semi-insulating GaAs: a model and a mechanism, *Electronics Lett.*, **17**, 839–41 (1981).
34. G. M. Martin, A. Mitonneau, and A. Mircea, Electron traps in bulk and epitaxial GaAs crystals, *Electronics Lett.*, **13**, 191–3 (1977).
35. G. M. Martin, Optical assessment of the main electron trap in bulk semi-insulating GaAs, *Appl. Phys. Lett.*, **39**, 747–8 (1981).
36. K. R. Elliott, R. T. Chen, S. G. Greenbaum, and R. J. Wagner, Identification of As_{ga} antisite defects in liquid encapsulated Czochralski GaAs, *Appl. Phys. Lett.*, **39**, 747 (1983).
37. E. R. Weber, H. Ennen, U. Kaufmann, J. Windschief, J. Schneider, and T. Wasinki, Identification of As_{ga} antisites of plastically deformed GaAs, *Appl. Phys. Lett.*, **53**, 6140–3 (1982).
38. K. R. Elliott, Residual double acceptors in bulk GaAs, *Appl. Phys. Lett.*, **42**, 274–6 (1983).
39. J. Woodhead, R. C. Newman, I. Grant, D. Rumsby, and R. M. Ware, Boron impurity antisite defects in p-type gallium-rich GaAs, *J. Phys. C: Solid State Phys.*, **16**, 5523–33 (1983).
40. K. R. Elliott, *J. Appl. Phys.*, to be published.
41. L. B. Ta, H. M. Hobgood, and R. N. Thomas, Evidence of the role of boron in undoped GaAs grown by liquid encapsulated Czochralski, *Appl. Phys. Lett.*, **41**, 1091–3 (1982).
42. J. A. Van Vechten, Simple theoretical estimates of the enthalpy of antistructure pair formation and virtual enthalpies of isolated antisite defects in zincblende and wurtzite type semiconductors, *J. Electrochem. Soc.*, **122**, 423–9 (1975).
43. D. V. Lang, Review of radiation induced defects in III–V compounds, *Radiation Effects in Semiconductors 1976*, Inst. Phys. Conf. Ser. 31, Bristol and London, Institute of Physics, pp. 70–94 (1977).
44. S. Loualiche, G. Giullot, A. Nouhailhat, and J. Bourgoin, Defect identification in electron irradiated GaAs, *Phys. Rev. B*, **26**, 7090–2 (1982).
45. J. Lagowski, H. C. Gatos, J. M. Parsey, K. Woda, M. Kaminski, and W. Walakeiwicz, Origin of the 0.82 eV electron trap in GaAs, *Appl. Phys. Lett.*, **40**, 342–4 (1982).
46. L. B. Ta, H. M. Hobgood, A. Rohatgi, and R. N. Thomas, Effects of stoichiometry on thermal stability of undoped semi-insulating GaAs, *J. Appl. Phys.*, **53**, 5771–5 (1982).
47. G. M. Martin, G. Jacob, G. Poiblaud, A. Goltzene, and C. Schwab, Identification and

analysis of near-infrared absorption bands in undoped and Cr-doped semi-insulating GaAs crystals, *Defects and Radiation Effects in Semiconductors 1980* (Oiso), Inst. Phys. Conf. Ser. 59, Bristol and London, Institute of Physics (1981).
48. D. E. Holmes, K. R. Elliott, R. T. Chen, and C. G. Kirkpatrick, Stoichiometry-related centers in LEC GaAs, *Semi-Insulating III–V Compounds*, eds S. Makram-Ebeid and B. Tuck, Nantwich, Cheshire, Shiva, pp. 19–27 (1982).
49. D. E. Holmes, R. T. Chen, K. R. Elliott, and C. G. Kirkpatrick, Symmetrical contours of deep level EL2 in liquid encapsulated Czochralski GaAs, *Appl. Phys. Lett.*, **43**, 305–7 (1983).
50. M. R. Brozel, I. Grant, R. M. Ware, and D. J. Stirland, Direct observation of the principal deep level (EL2) in undoped semi-insulating GaAs, *Appl. Phys. Lett.*, **42**, 610–12 (1983).
51. M. S. Skolnick, M. R. Brozel, I. Grant, D. J. Stirland, and R. M. Ware, Inhomogeneity of EL2 in GaAs observed by direct infrared imaging, *Electronic Materials Conf.*, Burlington, VT, Abstracts, pp. 75–6 (1983).
52. M. Bonnet, N. Visentin, B. Gouteraux, and J. P. Duchemin, Homogeneity of LEC semi-insulating GaAs wafers for IC applications, *Proc. GaAs IC Symp.*, New Orleans, pp. 54–7 (1982).
53. M. R. Brozel, J. B. Clegg, and R. C. Newman, Carbon, oxygen, and silicon impurities in GaAs, *J. Phys. D: Appl. Phys.*, **11**, 1331–9 (1978).
54. C. P. Lee, R. Zucca, and B. Welch, Orientation effect on planar GaAs Schottky barrier field effect transistors, *Appl. Phys. Lett.*, **37**, 311–13 (1980).
55. P. M. Asbeck, J. Tandon, E. Babcock, B. Welch, C. Evans, and V. Deline, Effects of Cr redistribution on electrical characteristics of ion implanted semi-insulating GaAs, *IEEE Trans. Electron Devices Lett.*, **EL-1**, 35–7 (1980).
56. B. Tuck, G. A. Adegoboyega, P. R. Jay, and M. J. Cardwell, Out-diffusion of chromium from GaAs substrates, *Gallium Arsenide and Related Compounds 1978*, Inst. Phys. Conf. Ser. 45, Bristol and London, Institute of Physics, pp. 114–24 (1978).
57. P. M. Asbeck, D. L. Miller, R. A. Milano, J. S. Harris, Jr, G. R. Kaelin, and R. Zucca, (Ga,Al)As/GaAs bipolar transistors for digital integrated circuits, *IEEE IEDM Tech. Digest*, 629–32 (1981).
58. C. P. Lee, D. L. Miller, D. Hou, and R. J. Anderson, Ultra-high speed integrated circuits using GaAs/GaAlAs high electron mobility transistor, *41st Device Res. Conf.*, Burlington, VT, Abstracts, p. IIA-7 (1983).
59. D. L. Miller, P. M. Asbeck, R. J. Anderson, and F. H. Eisen, (GaAl)As/GaAs heterojunction bipolar transistors with graded composition in the base, *Electronics Lett.*, **19**, 367–8 (1983).

Gallium Arsenide
Edited by M. J. Howes and D. V. Morgan
© 1985 John Wiley & Sons Ltd

CHAPTER 3

Epitaxial Growth of GaAs

I. A. DORRITY, J. D. GRANGE, and D. K. WICKENDEN

3.1 INTRODUCTION

In this chapter we present a broad, general review of the three major epitaxial growth techniques for GaAs: liquid phase epitaxy (LPE), vapour phase epitaxy (VPE), and molecular beam epitaxy (MBE). The fundamentals of each technique are explained, the various growth models are presented, and typical GaAs material results (electrical and optical) are given. A brief intercomparison of the different techniques concludes this chapter.

3.2 LIQUID PHASE EPITAXY

Liquid phase epitaxy (LPE) of GaAs depends for its operation on the fact that the solubilities of As in Ga-rich solutions decreases with decreasing temperature, as is apparent from inspection of the schematic of the Ga–As phase diagram in Figure 3.1. As an example, the point labelled 'a' in the figure represents a solution saturated with As at temperature T_a. Cooling this solution to temperature T_b creates a driving force for the precipitation of GaAs until the new saturation condition at point 'c' is reached. Provided that the conditions are right, some of this precipitate may be deposited as an epitaxial layer on a GaAs substrate that is placed in contact with this solution. LPE is an extremely versatile growth technique, which, in addition to GaAs, has been used to prepare a wide variety of other III–V compound and alloy semiconductors. It remains the major source of material for double heterostructure AlGaAs–GaAs lasers.

LPE growth of GaAs was first reported using a tipping system,[1] of the type illustrated in Figure 3.2. In this method the substrate and solution are placed at opposite ends of a graphite boat, which in turn is placed inside a growth tube that contains a high-purity atmosphere, usually Pd-diffused hydrogen. The tube and contents are situated in a furnace that can be tipped to elevate either end of the boat. The solution consists of Ga and enough As (in the form of GaAs) for

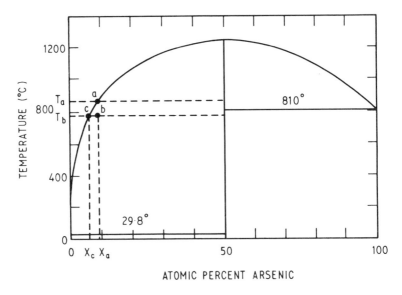

Figure 3.1 Schematic of Ga–As phase diagram

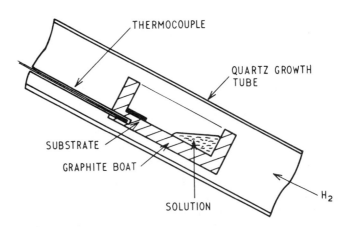

Figure 3.2 Schematic diagram of tipping LPE system

saturation at the growth temperature.[2] The temperature cycle, which is common to almost all LPE growth systems, is as shown in Figure 3.3. The furnace is initially heated to temperature T_0 with the substrate end high and out of the solution, and it is kept there for sufficient time to enable the melt to equilibrate and attain uniform saturation. After equilibration the temperature is ramped down at a linear rate, usually in the range 0.1–0.8°C min^{-1}. Growth is

Epitaxial growth of GaAs

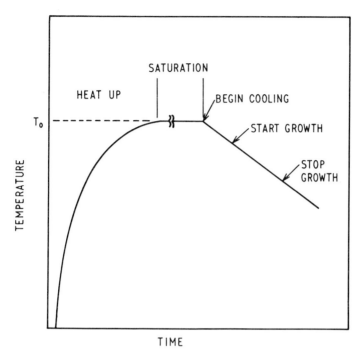

Figure 3.3 Temperature cycle for LPE growth

commenced at a supersaturation sufficient to ensure even nucleation of the epitaxial layer by tipping the furnace so that the melt covers the substrate and is terminated by tipping the furnace back to its original position.

The dipping technique[3] uses a vertical furnace and growth tube, with a crucible containing the solution at the lower end of the tube and the substrate fixed in a movable holder that is initially positioned just above the solution. Growth is initiated by lowering the holder to immerse the substrate in the solution and is terminated by raising the holder to its original position. In some systems the holder incorporates a sliding cover over the substrate that ensures equilibration in temperature prior to the initiation of growth.[4] Large dipping systems have been developed for commercial exploitation of GaP LEDs[5] and GaAs solar cells.[6]

Tipping and dipping systems are comparatively simple, cheap, and easy to operate. High-quality single layers have been produced by both methods. However, such systems become cumbersome with complex sample handling and furnace arrangements if used for the growth of multilayer structures. The principal LPE method, which overcomes these objections and is ideally suited to multilayer growth, is called the sliding technique,[7] and a horizontal sliding system is shown schematically in Figure 3.4. It consists of several wells in a graphite block

98   Gallium arsenide

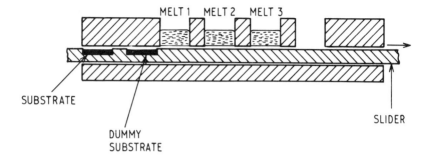

Figure 3.4 Schematic diagram of multiwell graphite sliding boat for GaAs LPE

with each well containing the melt constituents required for a single layer of structure. Each melt is brought into contact with the GaAs substrate in turn by sliding the substrate holder platform from well to well. A dummy substrate precedes the growth substrate such that it is brought into contact with each melt to ensure exact saturation of the solution just prior to growth. The clearance between the slider and the bottom of the wells is optimized to ensure a clean wipe-off of the melts and to prevent melt carryover from one well to another. Slider systems have also been developed that use a rotary graphite crucible in a vertical furnace.[8]

Alloy compositions, doping levels, and layer thicknesses are readily estimated from the temperature and melt composition using appropriate phase diagrams and solubility data, although it must be borne in mind that these equilibrium calculations do not take into account the kinetic effects that influence growth rate. It has been shown that in LPE growth by either supercooling, step-cooling or equilibrium cooling techniques the growth rate is determined by the rate of As diffusion to the crystal/melt interface.[9] The cooling rate and degree of initial supersaturation have significant effects on the surface morphology of the epitaxial layers.[10]

For the growth of III–V binary compounds, the composition of the epitaxial layer is not significantly altered by the change in the melt composition as the growth proceeds due to the near-stoichiometry of the deposit. This is not the case for the growth of ternary and quaternary alloys since the distribution coefficients relating the concentrations of the various elements in the solid to their concentrations in the solution may differ from each other. As a consequence the alloy composition may vary significantly as growth proceeds, with the rate of change of alloy composition depending on the initial composition of the melt and the growth temperature.[11] Other factors that influence the growth of heterostructures are the thermodynamic instability between the melt and the crystal surface when a new layer is commenced, and the lattice pulling effect which makes it difficult to grow graded layers between compounds or alloys with significantly different lattice parameters.

Large differences in distribution coefficient are also observed for dopant atoms added intentionally or otherwise to the melt. In the latter case it often results in the purity of the deposited GaAs being higher than that of the starting source materials, although this is generally assisted by systematic baking out of the melts prior to growth together with between-run loading of the boat in a controlled dry atmosphere.[12]

Tin and tellurium are the most common n-type dopants used in LPE GaAs and AlGaAs growth. Tin has a low distribution coefficient ($k \approx 10^{-4}$) and a low vapour pressure at usual growth temperatures and can be used to produce doped layers covering a range between 10^{15} and 8×10^{18} cm^{-3}.[13] Higher carrier concentrations of up to 7×10^{19} cm^{-3} can be obtained using Te dopant[14] but potential problems exist with a much higher vapour pressure and a distribution coefficient of $k \approx 1$ that makes it difficult to obtain homogeneously doped layers. Germanium is the most common p-type dopant because of its low vapour pressure and diffusion coefficient in GaAs.[15] Zinc has a similar distribution coefficient to Ge but its higher diffusion coefficient makes it impossible to produce sharp metallurgical junctions. Beryllium has an even larger diffusion coefficient and advantage is taken of this to produce a diffused p–n junction in GaAs while growing a p-AlGaAs window layer in solar cell structures.[16] Silicon is an amphoteric impurity in GaAs LPE layers and whether such a doped layer is n- or p-type depends on the growth conditions. For growth on a substrate of a particular orientation from a solution with a given Si concentration, the layer will be n-type above a certain transition temperature and p-type if grown below it.[17] Use is made of this for observation in the preparation of infrared emitting GaAs LEDs by a single-step growth process.

3.3 VAPOUR PHASE EPITAXY

Two methods are commonly used for the vapour phase epitaxial growth of GaAs for microwave device fabrication: (1) chloride transport and (2) metal-organic chemical vapour deposition (MOCVD). The former is most widely used since problems with the growth of high-purity material have limited the acceptance of the MOCVD technique, although its flexibility and versatility have meant that it has been widely exploited for the growth of indium- and aluminium-containing III–V alloys.

In this section we will briefly consider the principles and capabilities of each method.

3.3.1 Chloride transport

The trichloride and hydride processes are both variations of the chloride transport system in which Ga is transported in the form of GaCl and reacted with arsenic. Deposition occurs at temperatures typically in the range 700–800°C on a GaAs substrate according to the overall reaction

$$\text{GaCl}(g) + \tfrac{1}{4}\text{As}_4(g) + \tfrac{1}{2}\text{H}_2(g) \rightarrow \text{GaAs}(s) + \text{HCl}(g) \quad (3.1)$$

Growth is carried out in a hot-wall quartz reactor in a flowing H_2 ambient in both methods but the reactive species are generated in different ways. These processes are illustrated schematically in Figure 3.5.

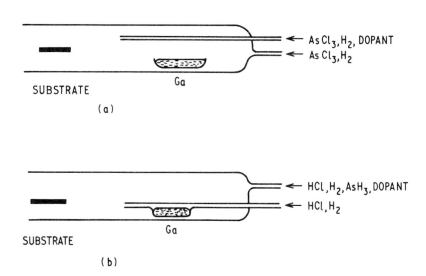

Figure 3.5 Schematic diagram of main features of GaAs VPE reactors

In the trichloride process, AsCl_3 is passed over a heated Ga source (850°C) where the following reactions take place:

$$\text{AsCl}_3(g) + \tfrac{3}{2}\text{H}_2(g) \rightarrow \tfrac{1}{4}\text{As}_4(g) + 3\text{HCl}(g) \quad (3.2)$$

$$\text{Ga}(l) + \tfrac{1}{4}\text{As}_4(g) \rightarrow \text{GaAs}(s) \quad (3.3)$$

$$\text{GaAs}(g) + \text{HCl}(g) \rightarrow \text{GaCl}(g) + \tfrac{1}{4}\text{As}_4(g) + \tfrac{1}{2}\text{H}_2(g) \quad (3.4)$$

The arsenic formed by the reduction of the AsCl_3 dissolves in the gallium until the solution becomes saturated, when a GaAs crust forms on the surface. Transport of the GaCl and As_4 to the growth zone is achieved by the reaction of HCl with the GaAs crust according to equation (3.4).

The source saturation step is critical[18,19] and it is vital that a flat temperature profile is maintained over the source as failure to do so results in partial dissolution of the GaAs crust and reaction of the exposed gallium with HCl. These lead to uncontrolled variations in the vapour composition which cause surface morphology problems and loss of growth rate control. This problem can be avoided by the use of a solid GaAs source[20] which gives better reproducibility but purity problems are encountered due to the lower purity of commercially available GaAs compared with Ga (6–7 N). However, precise control of the source temperature is still essential since the GaCl/HCl ratio and thus the growth rate are determined by the source equilibrium.

In the hydride process,[21] GaCl is generated directly by passing HCl over the gallium source and As_4 from the pyrolysis of arsine (AsH_3) according to

$$Ga(l) + HCl(g) \xrightarrow{>800°C} GaCl(g) + \tfrac{1}{2}H_2(g) \tag{3.5}$$

$$AsH_3(g) \rightarrow \tfrac{1}{4}As_4(g) + \tfrac{3}{2}H_2(g) \tag{3.6}$$

Free HCl is also injected directly into the deposition zone to reduce the gas phase supersaturation.

No source saturation step is required and, since the reaction of HCl with gallium goes almost to completion at temperatures in excess of 800°C,[22] critical temperature control of the source is unnecessary. Furthermore the independent generation of gas phase species allows variation of the Ga/As ratio, unlike in the trichloride method where it is fixed. Similarly the GaCl/HCl ratio can easily be adjusted.

There have been numerous theoretical and experimental studies[18,21,23-31] of the growth rate dependence on reactant concentrations and growth temperature. Growth is considered to take place via the following sequence of events: (i) transfer of reactants to substrate surface, (ii) absorption of reactants, (iii) surface reaction, (iv) desorption, and (v) transfer of products away from the surface. The slowest step dictates the growth rate, and either (i) or (iii) is generally the rate-limiting process.

Inspection of the growth rate dependence on temperature shows that at high temperatures the growth rate increases as the temperature decreases, as expected from thermodynamic considerations due to the exothermic nature of the deposition reaction (3.1). In this region the substrate–vapour interface is in equilibrium and growth is limited by the diffusion of reactants to the substrate.[23-25]

On decreasing the temperature still further the growth rate reaches a maximum and then decreases. Here the growth becomes kinetically limited and the rate-limiting step is considered to be the reduction of chemisorbed $(As-Ga-Cl)_s$ complexes to form GaAs.[27] Support for kinetic limitation of growth in this region comes from the strong growth rate dependence on substrate orientation.[32,33]

Studies of the growth rate[24] as a function of GaCl and As_4 partial pressures have shown that the growth rate increases with increasing GaCl pressure to a maximum and then decreases and is linearly dependent of As_4 pressure.[24,25] Theoretical modelling[27,29,30] has given results in good qualitative agreement with this behaviour and the decrease in growth rate at high GaCl concentrations is attributed to competitive absorption by GaCl for arsenic sites.[24,25]

Similar good agreement is observed between experimental[31] and theoretical[29,30] studies for the influence of GaCl/HCl ratio, where it is found that increasing HCl partial pressures decrease the growth rate until etching is achieved.

3.3.1.1 *Electrical properties*

Undoped layers Silicon is the principal residual impurity[34] in VPE GaAs due to the reaction of HCl with the quartz reactor tube forming volatile chlorosilanes:

$$SiO_2(g) + (4-n)H_2(g) + nHCl(g) \rightarrow SiH_{4-n}Cl_n(g) + 2H_2O(g) \quad (3.7)$$

$$\text{for } n = 1\text{-}4$$

In trichloride systems the incorporation of silicon can be suppressed by maintaining[35] high HCl partial pressures. This is achieved either by injecting free HCl into the growth zone via a second $AsCl_3$ bubbler[36] or by using high $AsCl_3$ mole fractions.[35] Using these techniques, very high-purity material can be obtained with carrier concentrations of 10^{13}–10^{14} cm^{-3} and liquid nitrogen mobilities up to 200 000 cm^2 V^{-1} s^{-1}.

These results are consistent with the model of DiLorenzo[34] who considered the silicon incorporation reaction to be

$$(n-2)H_2(g) + SiCl_nH_{4-n}(g) \rightarrow Si(\text{in GaAs}) + nHCl(g) \quad (3.8)$$

and showed the gas phase silicon activity to be inversely proportional to the HCl pressure.

However, contrasting behaviour is shown in hydride systems where the presence of HCl leads to increased background impurity levels.[37,38] Here the use of high reactant partial pressures is observed to decrease impurity levels.[37–39] It has been proposed that blocking of silicon incorporation[39] sites by As_4 molecules is the dominant mechanism but the reason for the non-applicability of the DiLorenzo model is not clear although it may indicate the presence of other impurities in the HCl. This factor was probably the reason for the inferior background levels obtained in hydride systems until recently. Carrier concentrations of 10^{15} cm^{-3} were typical although low 10^{14} cm^{-3} levels and liquid nitrogen mobilities of 100 000 cm^2 V^{-1} s^{-1} had been reported.[37] Attempts to improve the purity of HCl by cooling HCl cylinders to reduce impurity vapour pressures[40] or by *in situ* generation of HCl by $AsCl_3$ reduction had only limited success. However, very high-purity material has recently been grown[38,41] with carrier concentrations of about 10^{14} cm^{-3} and liquid nitrogen mobilities 160 000–200 000 cm^2 V^{-1} s^{-1}, demonstrating the improved purity of commercially available HCl. This is comparable with state-of-the-art trichloride material

Epitaxial growth of GaAs

and removes the principal reservation commonly held about the hydride technique.

Intentional doping Gaseous n-type dopant sources are most conveniently used since these can be injected directly into the growth zone, with H_2S[34] being the most favoured, although others such as H_2Se,[34] SiH_4,[42] and GeH_4[34] have been employed. Alternatively liquid sources like $SiCl_4$[43] and diethyltellurium,[44] held in temperature-controlled bubblers, are transported by bubbling hydrogen through them. Another method is to add the element, i.e. Sn[26] or Te,[45] directly to the source, but this makes doping control difficult for multilayer structures.

P-type doping is normally achieved by using elemental zinc in an ancillary furnace and varying the partial pressure by controlling the temperature.

High-resistivity layers have been produced by doping with compensating impurities such as Cr, using a CrO_2Cl_2 source.[46]

The influence of growth parameters on dopant incorporation has received some attention[45,47,48] particularly for sulphur, but the mechanism is not well understood. A linear dependence of carrier concentration is normally observed, in conflict with an equilibrium model of incorporation. However, the observation of high temperature and orientation dependence[47-49] is consistent with a kinetically limited process. These considerations have led to the proposal[50] that impurity incorporation is controlled by absorption on the surface and trapping by the growing layer. Clearly, further investigations of doping processes are warranted.

Uniformity There is considerable interest[51-54] in the growth of large-area GaAs layers with uniform thickness and doping owing to the requirements for GaAs IC fabrication. Non-uniformity arises from variations in surface concentrations during growth because of mass transfer of gas phase species across the non-uniform boundary layer. Improvements in uniformity can be achieved by optimization of reactor design or operation at low pressure to minimize the boundary layer thickness variation or by using growth conditions such that growth is under kinetic control. Excellent uniformity on 2 inch diameter substrates was achieved using this latter approach by Komeno *et al.*[53] who obtained FET I_{DSS} variations of less than 1%. The incorporation of substrate rotation in a horizontal reactor in order to minimize temperature and gas flow variations has produced variations below 1.5% in FET pinch-off voltages V_p over a 2 inch slice.[54]

3.3.2 Metal-organic chemical vapour deposition

MOCVD growth of GaAs involves the pyrolysis of a vapour phase mixture of arsine and, most commonly, trimethylgallium (TMG) or triethylgallium (TEG).

Free Ga atoms and As_4 molecules are formed and these species recombine on the hot substrate surface in an irreversible reaction to form GaAs:

$$R_3Ga(g) + AsH_3(g) \rightarrow GaAs(s) + 3RH(g) \qquad (3.9)$$

$$R = CH_3 \text{ or } C_2H_5$$

Growth is carried out in a cold-wall quartz reactor in flowing H_2 at atmospheric[55,56] or low pressure.[57,58] The substrate is heated to temperatures of 600–800°C, typically by RF heating of a graphite susceptor although IR radiation from quartz lamps has been employed. Transport of the metal-organics to the growth zone is achieved by bubbling H_2 through the liquid sources which are held in temperature-controlled bubblers. A schematic diagram of an MOCVD system is shown in Figure 3.6.

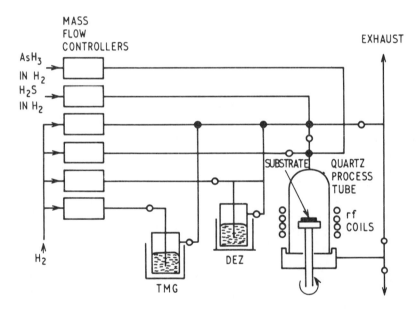

Figure 3.6 Schematic diagram of vertical MOCVD reactor for GaAs

Studies[55–59] of the growth rate dependence on the TMG and arsine fluxes have shown that the growth rate is independent of arsine partial pressure and linearly dependent on the TMG partial pressure under arsenic-rich conditions. Furthermore, the growth rate is observed to be temperature independent over the range 500–800°C. This behaviour indicates that growth is controlled by diffusion of the TMG species to the substrate rather than any surface kinetic reactions. Consequently, the TMG flux is the only parameter that must be precisely controlled for reproducible growth rates.

3.3.2.1 Electrical properties

Undoped layers Background doping levels in MOCVD have been shown[55,60,61] to be influenced by the growth temperature and As/Ga ratio, although the purity of the starting material is of paramount importance. Lower carrier concentrations are obtained with decreasing growth temperatures and the conduction type can be altered from p to n by increasing the As/Ga ratio.

Until recently, the purity of commercially available metal-organics and arsine limited the carrier concentrations obtainable to above 10^{15} cm^{-3}. However, improved purity of TMG sources has enabled layers with liquid nitrogen mobilities of 125 000–140 000 cm^2 V^{-1} s^{-1} and carrier concentrations of $(1-5) \times 10^{14}$ cm^{-3} to be achieved.[60,61] Far-infrared photoconductivity and photoluminescence measurements have identified the dominant residual impurities as Si, Ge, C, and Zn.[62]

Doped layers A wide variety of dopant sources have been used to obtain n- and p-type GaAs epilayers. These are either gases such as H_2S,[55] H_2Se,[63] SiH_4[58] or metal-organic sources such as dimethylzinc (DMZ),[55] diethylzinc (DEZ),[58] dicyclopentadienylmagnesium (Mg(Cp)$_2$),[64] tetramethyltin (Sn(CH$_3$)$_4$),[65] tetraethyltin (Sn(C$_2$H$_5$)$_4$),[66] and diethylberyllium (Be(C$_2$H$_5$)$_2$).[67] Gases are metered directly into the growth zone and metal-organics introduced by bubbling H_2 through the liquid sources. All group IV and VI elements act as donors and group II elements as acceptors.

High-resistivity layers have been prepaped using chromium hexacarbonyl Cr(CO)$_6$ as a source of chromium deep levels and compensation of carrier levels as high as 5×10^{16} cm^{-3} has been achieved.[68]

Studies[55,58,63] of dopant incorporation dependence on growth parameters show that incorporation is influenced by growth temperature and TMG and arsine fluxes. However, as is the case with chloride transport, the doping mechanism is not well established although the influence of growth temperature has been qualitatively described.[58] Furthermore, the growth rate has been observed to be affected by vapour phase dopant concentrations.[63,66]

3.4 MOLECULAR BEAM EPITAXY

Molecular beam epitaxy (MBE)[69-72] is the growth of elemental, compound, and alloy semiconductor films by the impingement of directed, thermal-energy atomic or molecular beams on a crystalline surface under ultra-high-vacuum conditions. This technique is basically a sophisticated extension of vacuum evaporation and is illustrated schematically in Figure 3.7. Modern practical MBE systems are generally multichamber apparatus comprising a fast entry load-lock, a preparation chamber, and a growth chamber. A typical system is illustrated in Fgure 3.8. Systems are of stainless steel construction pumped to UHV conditions using a

106   Gallium arsenide

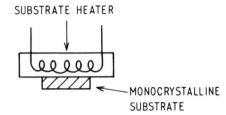

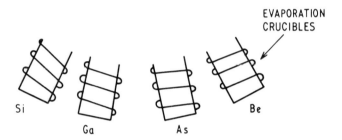

Figure 3.7 Schematic diagram of the basis of MBE

suitable combination of ion, cryo-, turbomolecular, diffusion, sorption, and sublimation pumps. Base pressures of 10^{-11}–10^{-10} Torr are obtained. A major attraction of MBE is that the use of UHV conditions enables the incorporation of high-vacuum based surface analytical and diagnostic techniques. Reflection high-energy electron diffraction (RHEED) is commonly employed to examine the substrate prior to and the actual epitaxial film during deposition. A (quadrupole) mass spectrometer is essential for monitoring the gas composition in the MBE growth chamber and also for leak detection. Other surface analysis and assessment techniques such as Auger electron spectroscopy (AES), x-ray photoelectron spectroscopy (XPS), and secondary ion mass spectrometry (SIMS) are not normally included in the growth chamber unless demanded by specific studies. They are placed in the preparation chamber to reduce contamination problems. Electron beam evaporators for dielectric or metal depositions can also be included as can gas handling facilities for introducing controlled amounts of known gases. The individually shuttered evaporation cells are usually made of boron nitride and heated using tantalum wire or foil resistive heaters. They are set in a cooled shroud to lessen impurity outgassing and thermal crosstalk between

Epitaxial growth of GaAs 107

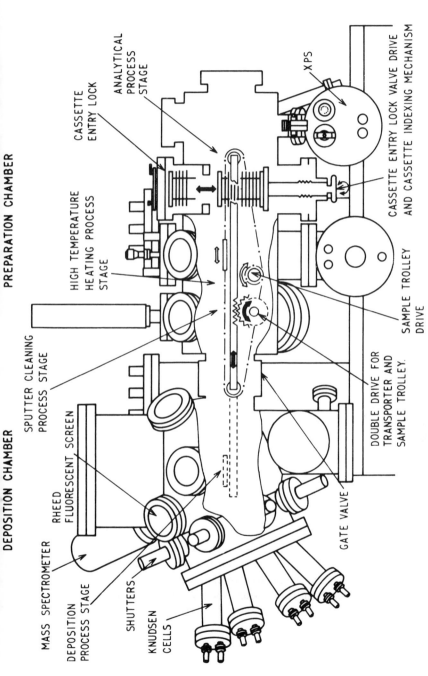

Figure 3.8 Typical multichamber MBE system (diagram courtesy of VG Semicon Ltd)

cells. The substrate heater is rotated during growth to improve the thickness and doping uniformity. Thickness variations of less than 1% over 5 cm have been achieved. A major advance in MBE, enabling the growth of high-purity material, was the introduction of load-locks, which meant that the growth chamber was not exposed to air and hence contamination for the loading of each substrate.

3.4.1 Model for the MBE growth of homoepitaxial GaAs

The model for the growth of MBE GaAs presented here is based on the work of Foxon and Joyce.[73-76] The Ga–As_4–(100)GaAs and Ga–As_2–(100)GaAs systems are considered. This is because MBE GaAs is prepared by the co-evaporation of elemental gallium in the atomic form (Ga) and arsenic in the form of either a dimeric molecule (As_2) or a tetramer (As_4).

In simplified terms it can be assumed that at low growth temperatures all the incident Ga atoms stick on the GaAs substrate and only enough As atoms adhere to produce stoichiometric GaAs. The excess As is desorbed. Thus the control of gross stoichiometry in the growth of MBE GaAs is not a difficult task. GaAs is, however, temperature unstable:

(a) above about 640°C (the congruent sublimation temperature[77]) As in the form of As_2 is preferentially desorbed; and

(b) at even higher temperatures ($\geq 700°C$) the evaporation of Ga becomes significant.

As a consequence of the above, in practice an excess As flux is provided during MBE growth to avoid non-stoichiometry due to a Ga excess, and at high growth temperatures the epitaxial layer thickness is less than expected from the Ga arrival rate because of re-evaporation.

More specifically, when GaAs is grown from Ga and As_2 (Figure 3.9a) the reaction is one of dissociative chemisorption of As_2 molecules on single Ga atoms.[74] The sticking coefficient of As_2 is proportional to the Ga flux (a first-order process). The excess As_2 is merely re-evaporated, leading to stoichiometric growth. For the growth of GaAs from Ga and As_4 (Figure 3.9b) the process is more complex.[73] Pairs of As_4 molecules react on adjacent Ga sites. Even with a Ga excess there is a desorbed As_4 flux. The maximum sticking coefficient for As_4 is 0.5. For low As/Ga flux ratios, when the As_4 surface concentration is small compared to the number of Ga sites, the growth rate-limiting step is the encounter and reaction probability between As_4 molecules (a second-order process). A more practical situation is $J_{As} > J_{Ga}$ and there is a high probability that the arriving As_4 molecules will find adjacent sites occupied by other As_4 molecules and the deposition rate becomes proportional to the number of molecules being supplied. Growth proceeds by adsorption and desorption of As_4 via a bimolecular interaction resulting in one As atom sticking for each Ga atom.

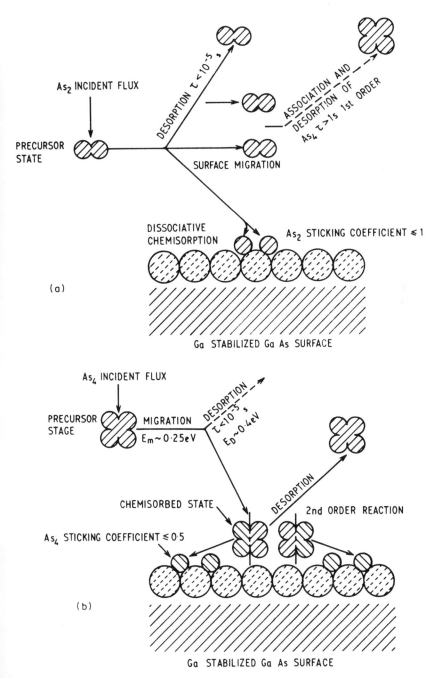

Figure 3.9 Diagram showing the growth models for MBE GaAs: (a) Ga–As$_2$–(100)GaAs system; (b) Ga–As$_4$–(100)GaAs system (diagram reproduced by kind permission of *Acta Electronica* and Dr C. T. Foxon)

110 *Gallium arsenide*

As the growth temperature is increased As_2 is lost by desorption (irrespective of the form of the incident As flux), resulting in an increase in the Ga surface population.

While wishing to emphasize clearly the fundamental role of kinetic studies to the understanding of MBE GaAs growth, it is important to make certain comments here:

(a) Most of the kinetic studies have been undertaken using very low incident flux rates and at temperatures well below those used for the growth of high-quality MBE GaAs. However, *a priori* one has no reason to suppose that there will be any radical departures from the model, except to note that at higher temperatures Ga re-evaporation has to be included.

(b) This model is a gross one, and it does not preclude deviations from stoichiometry that are slight but significant in terms of dopant levels, etc.

(c) The model is not predictive with regard to electrical and optical properties of the material.

(d) Thermodynamic aspects of MBE growth[78,79] are receiving much attention.

3.4.2 Unintentionally doped GaAs

By careful attention to substrate preparation, growth procedures, system design, and the quality of starting materials, residual carrier concentrations of 10^{14} cm^{-3} or less have been obtained.[80-82] This material is essentially semi-insulating but allows controlled doping in the 10^{14} cm^{-3} range. The residual impurities in most MBE systems are C and Si, giving rise to a residual p-type conduction of 10^{14}–10^{15} cm^{-3}. The origin of these dopants is not known at present. Carbon is in the residual UHV ambient as CO_2, CO, and CH_4. Silicon can be a persistent impurity even in refined gallium. Other system-related impurities often present are Mn (from hot stainless steel) and S (an impurity in elemental arsenic). For MBE GaAs the growth temperature is typically 580–640°C at a growth rate of about 1 μm h^{-1}.

3.4.3 Intentionally doped GaAs

Numerous n- and p-type impurities have been used for doping MBE GaAs, including: Sn,[83] Si,[84] S,[85] Se,[86] Ge,[87] Te,[88] Mn,[89] Be,[90] Mg,[91] Zn and Cd,[92] Cr,[93] Fe,[94] and C.[95] Of these Si is the most widely used n-type dopant and Be the favourite p-type dopant. Both give near bulk-like electrical properties, with little diffusion or segregation problems, for carrier concentrations of 10^{14} to 5×10^{18} cm^{-3}. The other shallow dopants either have complex incorporation mechanisms or are not readily incorporated in GaAs under normal MBE growth conditions. The specific nature of the difficulty can be found by reference to the

Epitaxial growth of GaAs 111

papers quoted or in general review articles.[70, 71] It should also be noted that many workers are interested in AlGaAs–GaAs structures and so a chosen dopant must be acceptable in both materials. Cr and Fe are of course deep-level dopants in GaAs used to produce insulating material for buffer layer applications, etc. As well as good electrical properties, Be and Si gave rise to high-quality optical GaAs (and AlGaAs) suitable for optoelectronic devices applications.

3.4.4 Deep levels in MBE GaAs

A large amount of data exist on traps or deep levels in GaAs in general (see, for example, ref. 96). Most of the electron traps observed by Lang et al.[97] in early MBE GaAs were identified as being due to impurity incorporation. However, three traps (0.19, 0.3, and 0.48 eV) do appear to be common to all MBE-grown GaAs. Typical total trap densities are 10^{12}–10^{16} cm^{-3}. These trap densities have been shown to be influenced by:

(a) the magnitude and form of the As species flux used[98, 99], and
(b) the substrate temperature used.[99]

Higher substrate temperatures and low As_2 (as opposed to As_4) fluxes produce lower trap densities. The origin and nature of these deep levels are unclear though arsenic antisites and gallium vacancy complexes are frequently considered. Hole traps have been less well studied and are considered at present to be primarily impurity related.

3.4.5 Optical properties of MBE GaAs

High-luminescence MBE GaAs[100, 101] is readily produced using a low As/Ga flux ratio and a growth temperature of 600–640°C. This is further supported by the ability of MBE layers to be fabricated into very low threshold current density AlGaAs–GaAs double heterostructure lasers.[102] Very sharp photoluminescence lines from GaAs deposited using As_2 (as opposed to As_4) have recently been reported.[103, 104] These results were comparable to the best LPE samples; however, it is possible that free electron screening caused the absence of the normally present defect-related lines.[105]

3.4.6 Selective area MBE

Because of the use of directed (line of sight) molecular and atomic beams it has been possible by suitable masking to produce regions of selective epitaxy by MBE. Two approaches have been adopted. In the first[106] polycrystalline and single-crystal GaAs were grown together giving a pattern of areas of single-crystal GaAs surrounded by electrically semi-insulating polycrystalline material. The

112 Gallium arsenide

original substrate was patterned with SiO_2. Single-crystal material grows where the SiO_2 is etched to reveal the underlying substrate. The GaAs deposited on the SiO_2 is polycrystalline and semi-insulating.

The second approach[107] is to use a mask made of Si or some refractory metal which is placed in front of the substrate producing isolated islands or mesas of MBE GaAs growth only where the holes exist in the mask. This technique can be extended[108] by moving the mask to give writing of patterned GaAs layers. Typical dimensions achieved by these techniques are 1–10 μm wide structures.

3.5 INTERCOMPARISON OF TECHNIQUES

The major disadvantage of LPE is that the surface morphology of the grown layers is inferior to that produced by MBE or VPE and so introduces difficulties in the subsequent processing of fine definition devices. Another is the restricted substrate size that can be used in sliding boat systems to maintain acceptable uniformity. However, LPE can routinely produce high electrical and optical quality GaAs (and AlGaAs). We consider that the surface quality of VPE and MBE GaAs are comparable, with the trichloride VPE routinely producing the best (least compensated lowest carrier concentration) material. MOCVD and MBE are highly attractive because, unlike VPE trichloride or hydride, they can be used for AlGaAs–GaAs structures, but unfortunately these are outside the scope of this chapter. For n-doped GaAs, and only GaAs, the well established trichloride and hydride techniques are more than satisfactory.

MOCVD is attractive for thinner layers (< 100 Å) and p- and n-type doping within one growth run. MBE can produce even thinner layer (approximating monolayers) with more abrupt changes in doping levels. It will be interesting to observe the coming competition between MBE and MOCVD as the proponents of each strive to push the frontiers of their respective techniques the furthest. The final choice of epitaxial technique will ultimately depend upon the application (materials, device—if so, which one—surface science), the quantities required, and the finance available. It is especially difficult to be definite in a chapter concerned only with GaAs and excluding the techniques of ion implantation and diffusion. The former is widely accepted as the technique for many GaAs integrated circuit applications.

REFERENCES

1. H. Nelson, Epitaxial growth from the liquid state and its applications to the fabrication of tunnel and laser diodes, *RCA Rev.*, **24**, 603 (1963).
2. R. N. Hall, Solubility of III–V compound semiconductors in column III liquids, *J. Electrochem. Soc.*, **110**, 385 (1963).
3. H. Rupprecht, New aspects of solution regrowth in the device technology of gallium arsenide, *Gallium Arsenide*, Inst. Phys. Conf. Ser. 3, London, Institute of Physics, p. 57 (1967).

4. G. S. Kamath, J. Ewan, and R. C. Knechtli, Large-area high efficiency (AlGa)As–GaAs solar cells, *IEEE Trans. Electron Devices*, **ED-24**, 473 (1967).
5. A. Mottram and A. R. Peaker, The growth of gallium phosphide layers of high surface quality by liquid phase epitaxy, *J. Cryst. Growth*, **27**, 193 (1974).
6. G. S. Kamath, GaAs solar cells for space applications, *Proc. 16th Intersociety Energy Conversion Engineering Conf.*, vol. 1, p. 416 (1981).
7. M. B. Panish, I. Hayashi, and S. Sumski, A technique for the preparation of low-threshold room-temperature GaAs laser diode structures, *IEEE J. Quantum Electronics*, **QE-5**, 210 (1969).
8. G. H. B. Thompson and P. A. Kirkby, Liquid phase epitaxial growth of six-layer GaAs/(GaAl)As structures for injection lasers with 0.04 μm thick centre layer, *J. Cryst. Growth*, **27**, 70 (1974).
9. J. J. Hsieh, Thickness and surface morphology of GaAs LPE layers grown by supercooling, step-cooling, equilibrium-cooling and two-phase solution techniques, *J. Cryst. Growth*, **27**, 49 (1974).
10. I. Crossley and M. B. Small, Some observations of the surface morphologies of GaAs layers grown by liquid phase epitaxy, *J. Cryst. Growth*, **19**, 160 (1973).
11. M. B. Panish and M. Ilegems, Phase equilibria in ternary III–V systems, *Prog. Solid State Chem.*, **7**, 39 (1972).
12. P. A. Houston, The growth of buffered GaAs MESFET structures by LPE, *J. Electron Mater.*, **9**, 79 (1980).
13. M. B. Panish, The system Ga–As–Sn: incorporation of Sn into GaAs, *J. Appl. Phys.*, **44**, 2659 (1973).
14. H. C. Casey, M. B. Panish, and K. B. Wolfshin, Influence of surface band bending on the incorporation of impurities in semiconductors: Te in GaAs, *J. Phys. Chem. Solids*, **32**, 571 (1971).
15. J. Vilms and J. P. Garrett, The growth and properties of LPE GaAs, *Solid-State Electronics*, **15**, 443 (1972).
16. K. Masu, M. Konagai, and K. Takahashi, Diffusion of beryllium into GaAs during liquid phase epitaxial growth of p-$Ga_{0.2}Al_{0.8}As$, *J. Appl. Phys.*, **54**, 1574 (1983).
17. B. H. Ahn, R. R. Shurtz, and C. W. Trussel, Dependence of growth properties of silicon-doped GaAs epitaxial layers upon orientation, *J. Appl. Phys.*, **42**, 4512 (1971).
18. D. W. Shaw, Kinetics of transport and epitaxial growth of GaAs with a Ga–$AsCl_3$ system, *J. Cryst. Growth*, **8**, 117 (1971).
19. T. H. Miers, The influence of source stability on the purity and morphology of GaAs grown in the Ga/$AsCl_3$/H_2 system, *Gallium Arsenide and Related Compounds 1982*, Inst. Phys. Conf. Ser. 65, London and Bristol, Institute of Physics, pp. 125–32 (1983).
20. J. C. C. Fan, *GaAs shallow-homojunction Solar Cells*, NASA Report CR-165167 (1980).
21. G. H. Olsen and T. J. Zamerowski, Crystal growth and properties of binary, ternary and quaternary (In,Ga)(As,P) alloys grown by the hydride vapour phase epitaxy technique, *Prog. Cryst. Growth Charact.*, **2**, 309 (1979).
22. D. J. Kirwan, Reaction equilibria in the growth of GaAs and GaP by the chloride transport process, *J. Electrochem. Soc.*, **117**, 1572 (1970).
23. D. W. Shaw, Influence of substrate temperature on GaAs epitaxial deposition rates, *J. Electrochem. Soc.*, **115**, 405 (1968).
24. D. W. Shaw, Epitaxial GaAs kinetic studies: [001] orientation, *J. Electrochem. Soc.*, **117**, 683 (1970).
25. D. W. Shaw, Kinetic aspects in the vapour phase epitaxy of III–V compounds, *J. Cryst. Growth*, **31**, 130 (1975).

26. L. Holland, J. M. Durand, and R. Cadoret, Influence of the growth parameters in GaAs vapour phase epitaxy, *J. Electrochem. Soc.*, **124**, 135 (1977).
27. R. Cadoret and M. Cadoret, A theoretical treatment of GaAs growth by vapour phase transport for [001] orientation, *J. Cryst. Growth*, **31**, 142 (1975).
28. J. L. Genter and R. Cadoret, Low pressure vapour phase epitaxy of GaAs—the growth rate limiting processes, *J. Physique*, **43**, C5, 111 (1982).
29. J. Korec and M. Heyen, Modelling of chemical vapour deposition, *J. Cryst. Growth*, **60**, 286 (1982).
30. M. Heyen and P. Balk, Epitaxial growth of GaAs in chloride transport systems, *Prog. Cryst. Growth Charact.*, **6**, 265 (1983).
31. M. Heyen and B. Balk, Vapour phase etching of GaAs in a chlorine system, *J. Cryst. Growth*, **53**, 558 (1981).
32. L. Hollan and C. Schiller, Study of the anisotropy in the vapour phase epitaxial growth of GaAs, *J. Cryst. Growth*, **13/14**, 319 (1972).
33. D. W. Shaw, Vapour phase epitaxial growth of GaAs. *Gallium Arsenide and Related Compounds 1968*, Inst. Phys. Conf. Ser. 7, London, Institute of Physics, p. 50 (1969).
34. J. V. DiLorenzo and G. E. Moore Jr, Effects of the $AsCl_3$ mole fraction on the incorporation of germanium, silicon, selenium and sulphur into vapour grown epitaxial layers of GaAs, *J. Electrochem. Soc.*, **118**, 1823 (1971).
35. J. V. DiLorenzo, Vapour growth of epitaxial GaAs: a summary of parameters which influence the purity and morphology of epitaxial layers, *J. Cryst. Growth*, **17**, 189 (1972).
36. T. Nozaki, M. Ogawa, and H. Watanabe, Multi-layer epitaxial technology for the Schottky barrier field-effect transistor, *Gallium Arsenide and Related Compounds 1974*, Inst. Phys. Conf. Ser. 24, London and Bristol, Institute of Physics, p. 46 (1975).
37. J. F. Kennedy, W. D. Potter, and D. E. Davies, The effect of the hydrogen carrier gas flow rate on the electrical properties of epitaxial GaAs prepared in a hydride system, *J. Cryst. Growth*, **24/25**, 233 (1974).
38. J. K. Abrokwah, T. N. Peck, R. A. Walterson, G. E. Stillman, T. S. Low, and B. Skromme, High purity GaAs grown by the hydride VPE process, *J. Electron. Mater.*, **12**, 681 (1983).
39. H. B. Pogge and B. M. Kemlage, Doping behaviour of silicon in vapour-grown III–V epitaxial films, *J. Cryst. Growth*, **31**, 183 (1975).
40. R. E. Enstrom and J. R. Appert, Improving the electrical properties of vapour grown III–V compounds by cooling the HCl liquid source, *J. Electrochem. Soc.*, **129**, 2566 (1982).
41. I. A. Dorrity, Unpublished Results.
42. L. Palm, H. Bruch, K.-H. Bachem, and P. Balk, Effect of oxygen injection during VPE growth of GaAs films, *J. Electron. Mater.*, **8**, 55 (1979).
43. M. Feng, V. Eu, T. Zielinski, H. B. Kim, and J. M. Whelan, Si-doped GaAs using a $SiCl_4$ technique in a $AsCl_3/Ga/H_2$ CVD system for MESFET, *Gallium Arsenide and Related Compounds 1980*, Inst. Phys. Conf. Ser. 56, London and Bristol, Institute of Physics, p. 1 (1981).
44. R. Sankaran, Te doping of vapour phase epitaxial GaAs, *J. Cryst. Growth*, **50**, 859 (1980).
45. E. Veuhoff, A. Sauerbrey, N. Pütz, M. Heyen, and P. Balk, Dopant incorporation during LP-VPE of GaAs, *J. Physique*, **C5**, 101 (1982).
46. D. Mizuno, S. Kikuchi, and Y. Seki, Epitaxial growth of semi-insulating gallium arsenide, *Jap. J. Appl. Phys.*, **10**, 208 (1971).
47. H. Heyen, H. Bruch, K.-H. Bachem, and P. Balk, Doping behaviour of sulphur during growth of GaAs from the vapour phase, *J. Cryst. Growth*. **42**, 127 (1977).

48. E. Veuhoff, M. Maier, K.-H. Bachem, and P. Balk, Sulphur incorporation in VPE GaAs, *J. Cryst. Growth*, **53**, 598 (1981).
49. J. L. Gentner, Anisotrophy in sulphur doping of GaAs grown by VPE, *J. Physique*, **C5**, 267 (1982).
50. J. B. Mullin, Element incorporation in vapour grown III–V compounds, *J. Cryst. Growth*, **42**, 77 (1977).
51. J. Komeno, S. Ohkawa, A. Miura, K. Dazai, and O. Ryuzan, Variation of GaAs epitaxial growth rate with distance along substrate within a constant temperature zone, *J. Electrochem. Soc.*, **124**, 1440 (1977).
52. J.-P. Chane, Analysis of the main factors influencing the thickness uniformity of VPE GaAs thin layers, *J. Electrochem. Soc.*, **127**, 913 (1980).
53. J. Komeno, M. Nogami, A. Shibatomi, and S. Ohkawa, Ultra-high uniform GaAs layers by vapour phase epitaxy, *Gallium Arsenide and Related Compounds 1980*, Inst. Phys. Conf. Ser. 56, London and Bristol, Institute of Physics, p. 9 (1981).
54. H. M. Cox, A. S. Prior, and V. G. Keramidas, High-throughput $AsCl_3/Ga/H_2$ vapour phase epitaxial system for growth of extremely uniform multilayer GaAs structures, *Gallium Arsenide and Related Compounds 1982*, Inst. Phys. Conf. Ser. 65, London and Bristol, Institute of Physics, p. 133 (1983).
55. S. J. Bass and P. E. Oliver, Controlled doping of gallium arsenide produced by vapour epitaxy using trimethylgallium and arsine, *Gallium Arsenide and Related Compounds 1976* (St Louis), Inst. Phys. Conf. Ser. 33b, London and Bristol, Institute of Physics, p. 1 (1977).
56. H. M. Manasevit and W. I. Simpson, The use of metal-organics in the preparation of semiconductor materials. 1: Epitaxial gallium-V compounds, *J. Electrochem. Soc.*, **116**, 1725 (1969).
57. J.-P. Duchemin, M. Bonnet, F. Koelsch, and D. Huyghe, A new method for the growth of GaAs epilayers at low H_2 pressure, *J. Cryst. Growth*, **45**, 181 (1978).
58. S. D. Hersee and J. P. Duchemin, Low pressure chemical vapour deposition, *Annu. Rev. Mater. Sci.*, **12**, 65 (1982).
59. M. R. Leys and H. Veenvliet, A study of the growth mechanism of epitaxial GaAs as grown by the technique of metalorganic vapour phase epitaxy, *J. Cryst. Growth*, **55**, 145 (1981).
60. T. Nakanisi, T. Udagawa, A. Tanaka, and K. Kamei, Growth of high-purity GaAs epilayers by MOCVD and their applications to microwave MESFETs, *J. Cryst. Growth*, **55**, 255 (1981).
61. P. D. Dapkus, H. M. Manasevit, K. L. Hess, T. S. Low, and G. E. Stillman, High purity GaAs prepared from trimethylgallium and arsine, *J. Cryst. Growth*, **55**, 10 (1981).
62. T. S. Low, B. J. Skromme, and G. E. Stillman, Incorporation of amphoteric impurities in high purity GaAs, *Gallium Arsenide and Related Compounds 1982*, Inst. Phys. Conf. Ser. 65, London and Bristol, Institute of Physics, p. 515 (1983).
63. R. W. Glew, H_2Se doping of MOCVD grown GaAs and GaAlAs, *J. Physique*, **C5**, 281 (1982).
64. C. R. Lewis, W. T. Dietze, and M. J. Ludowise, The growth of magnesium-doped GaAs by the OM-VPE process, *J. Electron. Mater.*, **12**, 507 (1983).
65. J. D. Parsons and F. G. Krajenbrink, Tin doping of gallium arsenide by metalorganic chemical vapour deposition (MOCVD), *J. Electrochem. Soc.*, **130**, 1780 (1983).
66. M. K. Lee, C. Y. Chang, and Y. K. Su, Investigation of Sn-doped GaAs epilayers grown by low pressure metalorganic chemical vapour deposition, *Appl. Phys. Lett.*, **42**, 88 (1983).

67. J. D. Parson and F. G. Krajenbrink, Use of diethylberyllium for metalorganic chemical vapour deposition of beryllium doped gallium arsenide, *J. Electrochem. Soc.*, **130**, 1782 (1983).
68. S. J. Bass, Growth of semi-insulating epitaxial gallium arsenide by chromium doping in the metal-alkyl + hydride system, *J. Cryst. Growth*, **44**, 29 (1978).
69. A. Y. Cho and J. R. Arthur, Molecular beam epitaxy, *Prog. Solid-State Chem.*, **10**, 157 (1975).
70. K. Ploog, Molecular beam epitaxy of III–V compounds, *Crystals: Growth, Properties and Applications*, ed. L. F. Boschke, Heidelberg, Springer-Verlag (1979).
71. C. E. C. Wood, Progress, problems and applications of MBE, *Phys. Thin Films*, **11**, 35 (1980).
72. A. Y. Cho, Recent developments in molecular beam epitaxy (MBE), *J. Vac. Sci. Technol.*, **16**, 275 (1979).
73. C. T. Foxon and B. A. Joyce, Interaction kinetics of As_4 and Ga on (100) GaAs surface using a modulated molecular beam technique, *Surf. Sci.*, **50**, 434 (1975).
74. C. T. Foxon and B. A. Joyce, Interaction kinetics of As_2 and Ga on (100) GaAs surfaces, *Surf. Sci.*, **64**, 293 (1977).
75. C. T. Foxon and B. A. Joyce, Surface processes controlling the growth of $Ga_xIn_{1-x}As$ and $Ga_xAs_{1-x}P$ alloy films by MBE, *J. Cryst. Growth*, **44**, 75 (1978).
76. C. T. Foxon, B. A. Joyce, and M. T. Norris, Compositional effects in the growth of $Ga(In)As_yP_{1-y}$ alloys by MBE, *J. Cryst. Growth*, **49**, 132 (1980).
77. C. T. Foxon, J. A. Harvey, and B. A. Joyce, The evaporation of GaAs under equilibrium and non-equilibrium conditions using a modulated beam technique, *J. Phys. Chem. Solids*, **34**, 1693 (1973).
78. R. Heckingbottom and G. J. Davies, Germanium doping of gallium arsenide grown by molecular beam epitaxy—some thermodynamic aspects, *J. Cryst. Growth*, **50**, 644 (1980).
79. R. Heckingbottom, C. J. Todd, and G. J. Davies, The interplay of thermodynamics and kinetics in molecular beam epitaxy (MBE) of doped gallium arsenide, *J. Electrochem. Soc.*, **127**, 444 (1980).
80. R. Dingle, C. W. Weishuck, H. L. Störmer, H. Morkoç, and A. Y. Cho, Characterisation of high purity GaAs grown by MBE, *Appl. Phys. Lett.*, **40**, 507 (1982).
81. J. C. Hwang, H. Tomkin, T. Brennan, and R. Frahn, Growth of high purity GaAs layers by MBE, *Appl. Phys. Lett.*, **42**, 66 (1983).
82. M. Heiblum, E. E. Mandez, and L. Osterling, High purity GaAs and AlGaAs grown by molecular beam epitaxy, Presented at *Fifth MBE Workshop*, Atlanta, GA (1983), *J. Vac. Sci. Tech.*, to be published.
83. C. E. C. Wood and B. A. Joyce, Tin doping effects in GaAs films grown by MBE, *J. Appl. Phys.*, **49**, 4854 (1978).
84. T. Shimanoe, T. Murotani, M. Nakatoni, M. Otsubo, and S. Mitsui, High quality Si-doped GaAs layers grown by molecular beam epitaxy, *Surf. Sci.*, **86**, 126 (1979).
85. D. A. Andrews, R. Heckingbottom, and G. J. Davies, The influence of growth conditions on sulphur incorporation in GaAs grown by molecular beam epitaxy, *J. Appl. Phys.*, **54**, 4421 (1983).
86. G. J. Davies, R. Heckingbottom, and D. A. Andrews, The influence of growth conditions of sulphur and selenium incorporation in GaAlAs grown by MBE, Presented at *Fifth MBE Workshop*, Atlanta, GA (1983), *J. Vac. Sci. Tech.*, to be published.
87. C. E. C. Wood, J. Woodcock, and J. J. Harris, Low compensation n-type and flat surface p-type Ge doped GaAs by MBE, *Gallium Arsenide and Related Compounds 1978*, Inst. Phys. Conf. Ser. 45, London and Bristol, Institute of Physics, p. 28 (1979).

Epitaxial growth of GaAs

88. G. Wicks, Unpublished Work.
89. M. Ilegems, R. Dingle, and L. Rupp, Optical and electrical properties of Mn-doped GaAs grown by molecular beam epitaxy, *J. Appl. Phys.*, **46**, 3059 (1975).
90. M. Ilegems, Be doping and diffusion in MBE of GaAs and AlGaAs, *J. Appl. Phys.*, **48**, 1278 (1979).
91. A. Y. Cho and M. B. Panish, Magnesium doped GaAs and AlGaAs by MBE, *J. Appl. Phys.*, **43**, 5118 (1972).
92. J. R. Arthur, Adsorption of zinc on GaAs, *Surf. Sci.*, **38**, 394 (1973).
93. H. Morkoc and A. Y. Cho, High-purity GaAs and Cr-doped GaAs epitaxial layers by MBE, *J. Appl. Phys.*, **50**, 6413 (1979).
94. D. Covington, J. Comas, and P. W. Yu, Iron doping in gallium arsenide by molecular beam epitaxy, *Appl. Phys. Lett.*, **37**, 1094 (1980).
95. M. Ilegems and R. Dingle, Acceptor incorporation in GaAs grown by MBE, *Gallium Arsenide and Related Compounds 1974*, Inst. Phys. Conf. Ser. 24, London and Bristol, Institute of Physics (1975).
96. A. Mircea and D. Bois, A review of deep level defects in III–V semiconductors, *Defects and Radiation Effects in Semiconductors 1978*, Inst. Phys. Conf. Ser. 46, London and Bristol, Institute of Physics, p. 82 (1979).
97. D. V. Lang, A. Y. Cho, A. C. Gossard, M. Ilegems, and W. Wiegmann, Study of electron traps in n-GaAs grown by molecular beam epitaxy, *J. Appl. Phys.*, **47**, 2558 (1976).
98. J. H. Neave, P. Blood, and B. A. Joyce, A correlation between electron traps and growth processes in n-GaAs prepared by molecular beam epitaxy, *Appl. Phys. Lett.*, **36**, 311 (1980).
99. R. A. Stall, C. E. C. Wood, P. D. Kirchner, and L. F. Eastman, Growth parameter dependance of deep levels in molecular beam epitaxial GaAs, *Electron. Lett.*, **16**, 171 (1980).
100. S. Gonda, Y. Matsushima, Y. Makita, and S. Mukai, Characterisation and substrate temperature dependence of crystalline state of GaAs grown by molecular beam epitaxy, *Jap. J. Appl. Phys.*, **14**, 935 (1975).
101. H. C. Casey, A. Y. Cho, and P. A. Barnes, Application of molecular-beam epitaxial layers to heterostructure lasers, *IEEE J. Quantum Electronics*, **QE-11**, 467 (1975).
102. W. T. Tsang, F. K. Reinhart, and J. A. Ditzenberger, The effect of substrate temperature on the current threshold of GaAs–AlGaAs double heterojunction lasers grown by molecular beam epitaxy, *Appl. Phys. Lett.*, **36**, 118 (1980).
103. H. Künzel, J. Knecht, H. Jung, K. Wünstel, and K. Ploog, The effect of arsenic vapour species on electrical and optical properties of GaAs grown by molecular beam epitaxy, *Appl. Phys. A*, **28**, 167 (1982).
104. H. Künzel and K. Ploog, The effect of As_2 and As_4 molecular beam species on photoluminescence of molecular beam epitaxially grown GaAs, *Appl. Phys. Lett.*, **37**, 416 (1980).
105. P. J. Dobson, G. B. Scott, J. H. Neave, and B. A. Joyce, The occurrence of sharp exciton-like features in low temperature photoluminescence spectra from MBE grown GaAs, *Solid-State Commun.*, **43**, 917 (1982).
106. A. Y. Cho and W. C. Ballamy, GaAs planar technology by molecular beam epitaxy, *J. Appl. Phys.*, **46**, 783 (1975).
107. W. T. Tsang and M. Ilegems, Selective area growth of GaAs/AlGaAs multilayer structures with molecular beam epitaxy using Si shadow masks, *Appl. Phys. Lett.*, **31**, 301 (1977).
108. W. T. Tsang and A. Y. Cho, Molecular beam epitaxial writing of patterned GaAs epilayer structures, *Appl. Phys. Lett.*, **32**, 491 (1978).

Gallium Arsenide
Edited by M. J. Howes and D. V. Morgan
© 1985 John Wiley & Sons Ltd

CHAPTER 4

Etching and Surface Preparation of GaAs for Device Fabrication

S. D. MUKHERJEE and D. W. WOODARD

4.1 INTRODUCTION

A complete GaAs device fabrication process may involve epitaxial growth, ion implantation, lithography, metallizations, alloying and annealing steps, as well as a number of etching and surface preparation steps. In fabricating lateral devices, etching is used for mesa isolating a device (although ion beam isolation is also available) and for recess etch prior to gate metallization in MESFET's. Etching is used also for the fabrication of vertical devices such as the permeable base transistor and the vertical FET. After etching, one needs to remove surface oxides in preference to the GaAs underneath prior to metallization. The presence of a thin interfacial dielectric layer causes poor ohmic contact formation and also causes non-reproducible, non-ideal Schottky barrier characteristics to appear.

As a result of its use in devices having diversely different structures, a phenomenological understanding of the different etching processes is needed.

In this chapter we shall describe the physical and chemical nature of the different etching-induced structuring processes available so that useful choices can be made for a given fabrication requirement.

4.1.1 General outline

Etching processes commonly used in microstructuring can be broadly divided into two groups: wet etching and dry etching. A wet or liquid-based etching process could be self-sustaining, could involve localized galvanic corrosion-induced material removal (Section 4.2), or might require an external supply of electrons, as in anodic etching (Section 4.3). Dry etching is a loosely formed generic term used for all processes where no liquid is involved. These include (Section 4.5) processes that are physical in nature with no chemical interaction taking place, such as diode sputter etching or ion beam etching, using heavy, chemically inert, gaseous atoms such as argon. Processes that are basically chemical in nature but are in need of some surface excitation processes, as in plasma-induced etching and reactive ion beam etching, also fall under the category of dry etching. However, etching of GaAs by gaseous HCl at a high temperature prior to vapour phase epitaxy (Chapter 3), called vapour phase etching, is excluded mainly because it is not used in fabricating devices. In plasma-assisted etching, both ionized and neutral chemical radicals are capable of reacting with the surface, while in reactive ion beam etching, only ionized species are allowed to take part in the reaction.

After etching, wet or dry, the surface has to be prepared for the next step in the device fabrication process, such as ohmic and Schottky metal contact deposition, dielectric deposition for surface passivation or for fabricating MIS structures (Chapter 7). Wet etching processes leave behind a thin, up to 200 Å, layer of reacted products, usually a mixture of lower oxides and other compounds, that are undesirable for ohmic contact and Schottky barrier formation. The residual oxide layer is removed prior to the metal deposition step in order to facilitate good ohmic contact formation through metal–GaAs interactions in the solid or liquid phase and to realize nearly ideal Schottky barrier characteristics (Chapter 6). These processes fall under the category of surface preparation (Section 4.4) and are of extreme importance for desired device operations. Ion beam and plasma-induced etching leave behind non-volatile residues and cause subsurface damage that creates traps for mobile carriers and changes the surface discontinuity to render Schottky barriers less ideal. Thus, a very small thickness of GaAs has to be etched away by a slow and gentle wet etching process to eliminate the degrading effects of the damaged layer. This leaves behind a thin oxide layer and subsequent surface preparation is therefore needed.

We shall conclude this chapter by comparing different etching/microstructuring techniques available today and indicate that a well optimized device fabrication process sequence may include a number of different process steps and may not depend on one kind alone (Section 4.6).

GaAs surfaces having different orientations etch at different rates in conditions where the surface reaction rate controls the etching phenomenon. These manifest themselves in etch profiles that could often be far from that required, and as a result one needs to alter such behaviour by controlling the etching conditions. This is due to the different chemical reactivities of the surfaces involved. In the next subsection, therefore, we shall discuss some aspects of the crystallography of GaAs that are pertinent to the understanding of the orientation dependence of etching processes.

4.1.2 The crystallography of GaAs

Following standard notations,[1] we shall use Miller indices enclosed between ⟨ ⟩ brackets to indicate axial directions of a form. Directions of the ⟨100⟩ type would include specific directions given by the [ ] bracket, such as [100], [010], [001], [$\bar{1}$00], [0$\bar{1}$0], and [00$\bar{1}$]. Miller indices enclosed within { } brackets would indicate planes of a form, with ( ) brackets indicating specific crystal planes. Hence, type {100} planes woud include (100), (010), (001), ($\bar{1}$00), (0$\bar{1}$0), and (00$\bar{1}$). In a cubic type crystal, such as a GaAs crystal, an axial direction [uvw] is perpendicular to the plane (hkl) if their Miller indices are identical, including sign.

GaAs has a zinc blende structure, i.e. an f.c.c. lattice of As with another f.c.c. lattice of Ga displaced by $\frac{1}{4}$, $\frac{1}{4}$, $\frac{1}{4}$ of the lattice constant (Figure 4.1a). Unlike an elemental semiconductor such as Si, GaAs is polar, since it is made up of two different kinds of atoms having different valences. As a result, except for the ⟨100⟩ and ⟨110⟩ type axial directions, all other axes are polar in nature, i.e. with unequal Ga–As separations. This is most clearly seen in the ⟨111⟩ case (Figure 4.1a). If we consider the sequence of (111) type atomic planes that repeat themselves, sheet by sheet, perpendicular to the [111] direction, we obtain an As plane (contained by As_{21-26})—Ga plane (Ga_{21-23})—As plane (As_{11-13})—Ga plane (Ga_1)—As plane (As_0)—and so on. Each (111)As plane is separated from its nearest (111)Ga plane by a distance 0.82 Å, since the three As atoms (As_{24-26}) are connected to the Ga_{23} atom by three ionic–covalent bonds making large angles to the plane itself. The next As atom (As_{13}) is attached to the Ga_{23} atom by a single bond with the characteristic tetrahedral separation distance of 2.45 Å. In short, if we designate a (111) atomic plane with a subscript p, then the asymmetry of the [111] polar axial direction can be indicated by As_p–Ga_p----As_p–Ga_p----As_p–Ga_p and so on. Likewise, the [$\bar{1}\bar{1}\bar{1}$] direction is given by Ga_p–As_p----Ga_p–As_p etc.

It is obvious that two closely spaced atomic planes that are held together by three bonds between group III and group V elements are separated by a weaker single bond perpendicular to the plane itself. It is also seen that for the [111] direction, a (111)Ga or group III atomic plane is exposed through these weak binding forces to the next atomic plane, while in the case of [$\bar{1}\bar{1}\bar{1}$] it is the ($\bar{1}\bar{1}\bar{1}$)As or group V atomic plane that is exposed to the open spaces in the crystal. Any discontinuity imposed upon the crystal in a ⟨111⟩ direction by breakage, cleavage or etching would cause the crystal to terminate with the strongly bound atoms exposed to the surface. In other words, the weak, single bond would be broken and the strongly bonded plane with atoms attached to the next plane by three bonds each would remain exposed. For the two specific directions, As atoms are exposed for a free ($\bar{1}\bar{1}\bar{1}$)As surface and Ga atoms are exposed for the (111)Ga surface. This exposure of different kinds of atoms for the two surfaces

122   *Gallium arsenide*

contributes to preferential etching of one plane with respect to another, which will be discussed later in Section 4.2.2.

Conventionally, group III is assigned the letter A and group V the letter B and these two directions are called {111}A (or Ga) and {111}B (or As) directions respectively. Following a three-fold rotational symmetry about a ⟨111⟩ type axis, Figure 4.1b indicates the different A and B directions in GaAs. Out of the eight possible directions, [111], [1̄1̄1], [1̄11̄], and [11̄1̄] are A or Ga types, and [1̄1̄1̄], [111̄], [11̄1], and [1̄11] are B or As types. Note, with reference to Figure 4.1a, that any analysis based upon the interatomic separation along the

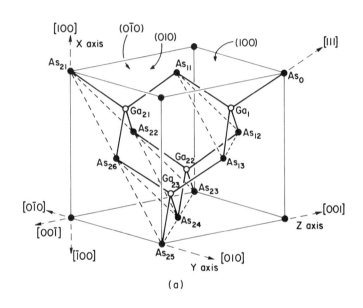

(a)

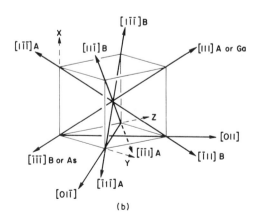

(b)

Etching and surface preparation of GaAs for device fabrication 123

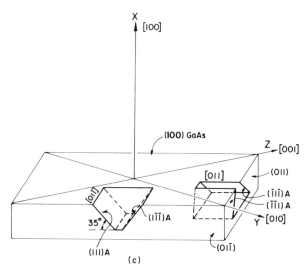

Figure 4.1 (a) GaAs crystal structure in which an f.c.c. Ga lattice is displaced from an f.c.c. As lattice by ($\frac{1}{4}$, $-\frac{1}{4}$, $\frac{1}{4}$) of the lattice parameter, $a_0 = 5.654$ Å. The heavy lines joining two neighbouring Ga and As atoms indicate chemical bonding, the bond length being equal to 2.45 Å. The (111) plane is given by either As_{21-26} or Ga_{21-23} and the ($\bar{1}\bar{1}\bar{1}$) plane by $As_{0,12,13,23-25}$ or $Ga_{1,22,23}$. These two planes intersect the ($01\bar{1}$) plane, given by $As_{21,11,0,24}$, perpendicularly. (b) The different $\langle 111 \rangle$ A (Ga) and $\langle 111 \rangle$ B (As) axial directions in GaAs. A plane perpendicular to such an axis will have Ga or As as the tightly bound element, respectively, towards the direction shown by the arrow. (c) A GaAs (100) wafer cleaved along ($01\bar{1}$) and (011) planes normal to the (100) surface, showing the differences between these two planes, so far as the orientations of the $\{111\}$A or Ga planes are concerned. Reaction-rate-controlled etchants that attack the $\{111\}$Ga type planes at the slowest rates would yield anisotropic etch profiles as shown. (See Section 4.2.2 for more detail)

[111] axis gives a totally erroneous inference regarding A and B directions; interatomic spacings along the [111] axis are irrelevant to the consideration of how strongly a sheet of Ga or As atoms is bound to the next atomic plane.

Almost all major electronic and electro-optical devices today are fabricated on (100) GaAs cut at about 2° away from the [100] axis towards the [110] axis. A GaAs (100) wafer cleaves along (011) and ($01\bar{1}$) planes (Figure 4.1c), the cleavage direction being given by the [$01\bar{1}$] and [011] axes, respectively, lying on the (100) plane. Etching with window edges parallel to these two axial directions can cause asymmetrical edge profiles, which will be discussed later.

4.2 WET ETCHING

A number of excellent review and research articles on the wet etching of semiconductors in general[2-4] and compound semiconductors in particular[5-7]

124   *Gallium arsenide*

are available in the literature. The topics of defect characterization in GaAs by etching and staining and of chemical polishing are subjects in their own rights and, in view of the excellent reviews already available,[5-7,16] will not be dealt with here. Instead, we shall focus on the understanding of etching processes that are useful for practical device fabrication only.

Preferential etching vs selective etching In discussing specific etching processes,[4] the word 'preferential' will be used to describe a process in which the etching of certain crystallographic planes occurs faster in preference to some others. The slowest etching crystal planes therefore dictate the final shape of the etched groove. Etching is then anisotropic with respect to crystal directions.

In contrast, the word 'selective' will be reserved for etches that remove one material significantly faster than another. The differences in etch rates could be caused by either different etch mechanisms or differences in contact potentials with respect to the etch solution. The layer of material in contact with GaAs is usually a metal, a dielectric, a photoresist or an electron beam resist. Since our focus is on GaAs devices, selective etching of GaAs with respect to $Al_xGa_{1-x}As$ will not be discussed.

In existing literature, however, some authors have used 'preferential' to mean 'selective' and vice versa.

4.2.1 General mechanisms

Most etchants for GaAs contain an oxidizing agent, a complexing agent, and a dilutant, such as water (Table 4.1). The oxidizers usually are Br_2,[8] H_2O_2,[9-16] $AgNO_3/CrO_3$,[6] HNO_3,[3] and $NaOCl$.[5] The oxidized layer is usually insoluble in water, and is made soluble by complex formation with the help of a solubilizing or complexing agent such as NH_4OH,[11,12,16,18] $NaOH$,[24] H_2SO_4,[12,14,15,24] HCl,[15] HF,[6,7] H_3PO_4,[10] and citric acid, $C_3H_4(OH)(COOH)_3H_2O$.[9] At least in

Table 4.1 Concentrations, specific gravities, molecular weights, molecular weight to chemical equivalent weight ratios (or valency), and normalities of stock solutions of reagents commonly used for wet (chemical) etching of GaAs

Stock solution	Concn in aq. sol. (wt %)	Specific gravity (gm cm^{-3})	Molecular weight	Valency	Normality (N)
H_2O_2	30	1.10	34	2	19.4
H_3PO_4	86	1.71	98	3	45.0
NH_3 (as NH_4OH)	30	0.90	15	1	18.0
H_2SO_4	96–98	1.84	98	2	36.4
HCl	37	1.19	36.5	1	12.1

Etching and surface preparation of GaAs for device fabrication 125

one case, Br_2–CH_3OH, the dilutant, methanol, dissolves the reaction products for GaAs–Br_2 reactions, mainly bromides.[8]

The physical and chemical mechanisms involved in a wet etching process can be understood in terms of the model presented in Figure 4.2. The model assumes that the oxidizing agent used is H_2O_2, which is the case for most practical etchants used for GaAs.

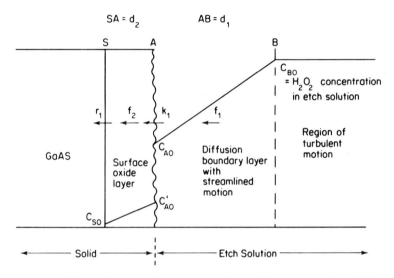

Figure 4.2 Schematic generalized representations of the concentration of the oxidizing component, H_2O_2, in the etch solution close to the surface and inside the thin surface oxide, during a wet chemical etching process

The various ongoing steady-state mechanisms are as follows:

(a) Diffusion of H_2O_2 towards the GaAs surface, with diffusing flux f_1. Let C_{AO} and C_{BO} be the concentrations of the oxidizing agent at A and B, then

$$f_1 = (C_{BO} - C_{AO})D_1/d_1 \qquad (4.1)$$

where D_1 is the appropriate diffusion coefficient.

(b) Adsorption of H_2O_2 molecules at active sites on the oxidized GaAs surface, with areal adsorption rate k_1. If a_1 is the appropriate adsorption rate constant, then

$$k_1 = C_{AO}a_1 \qquad (4.2)$$

(c) Diffusion of H_2O_2 or its decomposed product, such as an oxygen atom, through the wholly or partially oxidized GaAs surface layer SA to reach the oxide–GaAs interface, with diffusion flux f_2 given by

$$f_2 = (C'_{AO} - C_{SO})D_2/d_2 \tag{4.3}$$

where C'_{AO} is the volume concentration of the oxidizer just underneath the oxide surface and D_2 is the diffusion coefficient of the oxidizer molecules in the oxide.

(d) Oxidation of GaAs at the GaAs–oxide interface or of the unoxidized Ga or As atoms in the residual layer by the oxidizer, with a rate r_2 given by

$$r_2 = C_{SO}a_2 \tag{4.4}$$

where a_2 is the oxidation rate constant at the GaAs–oxide interface.

(e) Diffusion of the solubilizing/complexing agent towards the oxidized GaAs surface.

(f) Adsorption of the complexing agent molecules at active sites on oxidized GaAs surface.

(g) Complex formation and dissolution of the soluble complexed product into the dilutant (solvent) at the liquid–oxide interface.

(h) Diffusion of these complexed molecules away from the surface into the bulk of the etching solution.

For the sake of clarity we have not written out rate equations for (e) to (h). This in no way affects general conclusions to be drawn later.

Under steady state the four fluxes f_1, k_1, f_2, and r_2 must be equal. In reality, however, among (b), (c), and (d), it is the absorption rate that is the slowest, the diffusive flux f_2 inside the oxide and the oxidation rate r_2 keeping up with rising or falling C'_{AO}. Consequently, a steady-state C_{AO}, i.e. the oxidizer concentration in the solution just outside the oxide layer, is obtained by equating $f_1 = k_1$, i.e.

$$D_1(C_{BO} - C_{AO})/d_1 = a_1 C_{AO}$$

or

$$C_{AO} = D_1 C_{BO}/(a_1 d_1 + D_1) \tag{4.5}$$

Let us call D_1/d_1 the diffusion rate constant, having the same dimension as the adsorption and oxidation rate constants, namely that of velocity, cm/s. If $D_1/d_1 \gg a_1$, then $C_{AO} \simeq C_{BO}$, and so

$$f_1 = k_1 = a_1 C_{BO} \tag{4.6}$$

In such a case the thickness d_1 of the boundary layer becomes very small and the etching process becomes reaction-rate-limited. If, on the other hand, $D_1/d_1 \ll a_1$, then D_1 is neglected compared with $a_1 d_1$ giving

$$C_{AO} = D_1 C_{BO}/a_1 d_1 \tag{4.7}$$

and so

$$f_1 = k_1 = D_1 C_{BO}/d_1 \tag{4.8}$$

This is a case where the supply of H_2O_2 through diffusion does not keep up with

Etching and surface preparation of GaAs for device fabrication 127

the high surface adsorption (and hence oxidation) rate and etching becomes diffusion-limited.

Diffusion-limited reactions often occur in very viscous etch solutions containing high concentrations of the complexing agent and low concentrations of the oxidizing agent, such as H_2O_2. However, an etch solution need not be highly viscous in order to have diffusion-controlled properties. For example, the etching of GaAs and AlGaAs by the potassium ferrocyanide/ferricyanide based selective etches for GaAs[17] was found to be diffusion-limited for total solute concentrations between 0.2 and 0.45 mol l^{-1}. These solutions had rather low viscosities.

We shall now discuss the phenomenological differences between diffusion- and reaction-rate-limited etches with examples of etching solutions commonly used in device fabrication processes.

4.2.2 Diffusion- and reaction-rate-limited etching

4.2.2.1 *Agitation dependence*

In a reaction-rate-limited etching process, $C_{BO} \simeq C_{AO}$, or $d_1 \simeq 0$. Consequently, agitation does not change the surface adsorption rate significantly, causing virtually no change in etch rates. For a diffusion-limited etching condition, agitation-induced turbulence would decrease the boundary layer thickness d_1, causing the diffusive flux f_1 as well as C_{AO} to rise. This in turn would increase the surface adsorption rate k_1 and etch rate would increase[3,9,14] (Figure 4.3).

4.2.2.2 *Mask edge trenching*

A photoresist or a dielectric (SiO_2 or Si_3N_4) mask with window openings to expose GaAs to be etched has unused/fresh reactants over the masked areas. If the etching is diffusion-limited, then the availability of fresh, unreacted etchant from the masked area causes abnormally high etch rates to appear on GaAs adjacent to the mask edge. A flat etched profile can only be seen at distances as much as 50 to 300 μm away from the mask edge, depending upon the diffusion lengths involved. In other words, for diffusion-limited etchants, the etch rate for GaAs through a small opening in a mask is found to be considerably higher than that for a GaAs wafer with no masking. For reaction-rate-limited etching, the etch rate is independent of the mask opening dimensions[12] (Figure 4.4).

4.2.2.3 *Preferential (anisotropic) etching*

For tetrahedrally bonded elemental semiconductors such as Ge and Si, held together with covalent bonds, the major crystallographic planes exposed to a surface have different numbers of broken bonds, as shown in Table 4.2.[20] Surface oxidation rates are related to the number of reaction sites available on a given

Gallium arsenide

surface. For reaction-rate-controlled etching processes, the etch rate is roughly proportional to the areal density of dangling bonds on elemental semiconductor surfaces, as shown in Table 4.2.

GaAs, being polar and with an ionicity of 0.31,[21] behaves rather differently. As mentioned earlier, a free {111}Ga surface has Ga atoms attached firmly to three As atoms underneath and the valency of 3 of Ga is completely satisfied. The {111}As plane, on the other hand, contains As atoms that have two extra unbound electrons per atom owing to its valency of 5. Since oxidation involves

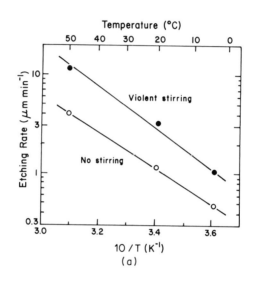

(a)

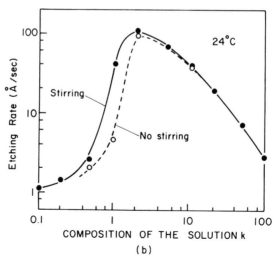

(b)

Etching and surface preparation of GaAs for device fabrication 129

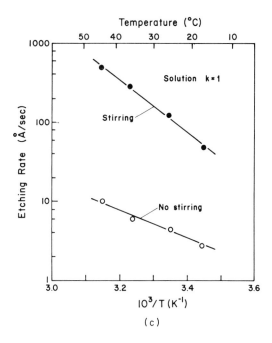

Figure 4.3 Agitation dependence of diffusion-controlled etching processes. (a) A substantial increase in (100) GaAs etch rate occurs using H_2SO_4–H_2O_2–H_2O (8:1:1), caused by violent stirring over an extended temperature range. (After Iida and Ito.[14]) (b) Stirring increases the etch rate of (100) GaAs for $k \lesssim 4$, where k is the volume ratio of concentrated (50 wt %) citric acid to undiluted (30 wt %) H_2O_2. This also shows that as k becomes lower, that is as the relative concentration of H_2O_2 increases, the etch rate increases in the reaction-rate-controlled regime, for $k \gtrsim 4$. As k decreases further, agitation dependence increases as the etching process becomes more and more diffusion-controlled. For small k, that is for smaller concentration of citric acid, it is suggested that the diffusion of either the complexing agent (citric acid) towards the surface or the complexed product away from the surface is the limiting process. (c) Temperature dependence of agitated and unagitated etching process for $k = 1$ as in (b). This shows that the activation energy ΔE for the unagitated case, $\Delta E = 8.4$ kcal mol^{-1} (0.36 eV/molecule), is lower than that for the agitated case, $\Delta E = 15.3$ kcal mol^{-1} (0.66 eV/molecule), indicating that diffusion-limited etching processes requires less activation and hence are less temperature dependent than reaction-rate-limited etching processes. (Both (b) and (c) are after Otsubo et al.[9]; see also Figure 4.6c)

loss of electrons, the As atoms present on a {111}As surface react much more readily with the oxidizer than do Ga atoms present on a {111}Ga surface. Once an As atom from the {111}As surface is removed by oxidation, the Ga atoms in the plane underneath, which are connected to the other underlying As atoms each by a single bond, are dislodged relatively easily by the oxidation process. Consequently, the {111}As etch rate is found to be by far the highest in GaAs for reaction-rate-limited etching processes. For diffusion-limited processes, the etch rate dependence on orientation almost disappears[10,14] (Figure 4.5).

Table 4.3 shows some examples of orientation effects of reaction-rate-limited etching processes. There are some general trends, but almost all of these tend to have exceptions.

For example, except etchant (k) none of these follow the etch rate trends for elemental semiconductors as indicated by Table 4.2. If an average of the two {111}Ga and {111}As etch rates is taken, then for etchant (k) the ratio of etch rates of {100}, {110}, and {111} type planes becomes 1.00:0.69:0.60, rather close to 1.00:0.71:0.58 as in Table 4.2.

All the etchants listed in Table 4.3 etch the {111}As surface faster than the {111}Ga surface, except (k).

Good reaction-rate-limited etchants, such as (b), (d), (f), (g), and (h), are strongly anisotropic. Others, such as (a), (c), and (j), are weakly anisotropic, but do not show any other characteristics of diffusion-limited etchants. Also note that the only diffusion-limited etchant, (e), etches {111}Ga at a rate slower than the other three planes. In spite of this, it exhibits almost all other characteristics of a

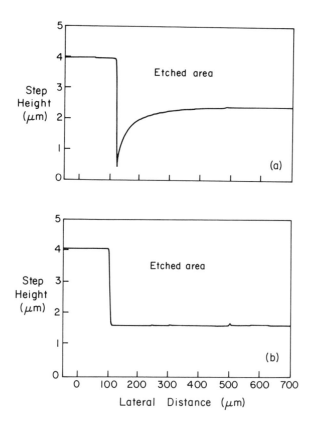

Etching and surface preparation of GaAs for device fabrication 131

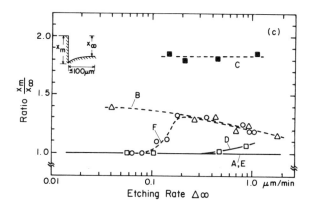

Figure 4.4 Trenching, or mask edge increase in etch rate, for diffusion-limited etching processes. (After Kohn.[12]) (a, b) Step height profilometer traces of two compositions of H_2SO_4–H_2O_2–H_2O etchants on GaAs, showing (a) enhanced etching at the mask edge for 10:1:1 and (b) a flat-bottomed profile for 1:1:16. The 10:1:1 etchant is diffusion-limited owing to the high concentration of H_2SO_4 and the 1:1:16 is, of course, reaction-rate-limited. Note that a flat profile is obtained only at a distance larger than about 400 μm from the mask edge for the diffusion-limited etchant. (After Kohn.[12]) (c) Plots of the ratio x_m/x_∞ where x_m is the etch depth at the mask edge and x_∞ is that at a point far away from the mask edge where the surface profile is flat. Here for diffusion-limited etching processes $x_m/x_\infty > 1$ and for reaction-rate-limited etching processes $x_m/x_\infty = 1$. The etchants are: A, NaOH–H_2O_2–H_2O (2:x:100, $1 < x_\leftarrow < 10$); B, NH_4OH–H_2O_2–H_2O (1:1:x, $16 < x_\leftarrow < 50$); C, H_2SO_4–H_2O_2–H_2O (x:1:1, $10 < x_\leftarrow < 20$); D, H_2SO_4–H_2O_2–H_2O (1:1:x, $10 < x_\leftarrow < 250$); E, citric acid–H_2O_2–H_2O (50:x:50, $1 < x_\leftarrow < 10$); and F, H_3PO_4–H_2O_2–H_2O (1:1:x, $18 < x_\leftarrow < 50$). The arrows following underneath x indicate directions of increasing etch rate. (After Kohn.[12])

Note: Although NH_4OH-based etchants were found to cause mask edge trenching in this study,[12] other studies[11] using NH_4OH–H_2O_2–H_2O (3:1:50) have found flat-bottomed etch craters with preferential etching-induced anisotropy for larger etch depths, indicating reaction-rate-limited behaviour. The anomaly remains unexplained.

Table 4.2 Densities of free bonds on germanium surfaces, and dissolution rates in oxygen-saturated water (from ref. 20)

Orientation	Density of Free bonds (cm^{-2})	Relative free bond density	Relative etch rate
{000}	1.25×10^{15}	1.00	1.00
{110}	8.83×10^{14}	0.71	0.89
{111}	7.22×10^{14}	0.58	0.62

132   *Gallium arsenide*

diffusion-limited etch[14] (Figure 4.5a). A number of unusual etched profiles have been obtained with the help of reaction-ratio-limited wet etching processes that are mostly useless for present-day devices. The reader is, nonetheless, encouraged to look into the relevant references out of interest.[9,10,14,15]

4.2.2.4 *Temperature dependence*

Etch rates always increase with etchant temperature because the rates of all the participating physical and chemical phenomena increase as $\exp(-\Delta E/kT)$ with temperature T, ΔE being the relevant activation energy. Most diffusion phenomena in common water-based etchants have activation energies ranging from 5 to

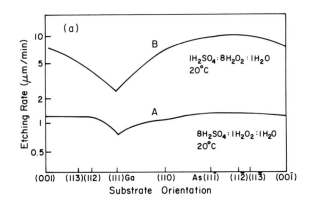

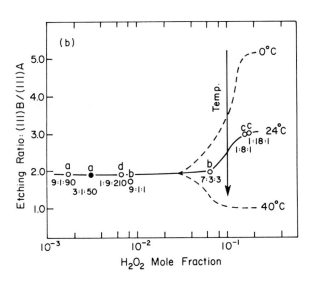

Etching and surface preparation of GaAs for device fabrication   133

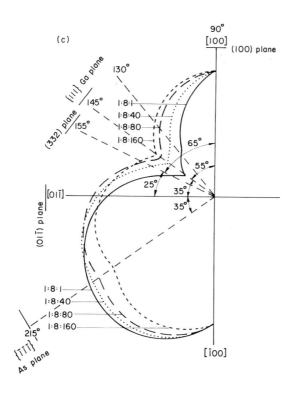

Figure 4.5 Etch rate as a function of GaAs crystal orientation. (a) For the H_2SO_4–H_2O_2–H_2O etchant, a reaction-rate-limited mixture (1:8:1) (curve B) is compared with a diffusion-limited one (8:1:1) (curve A). Note that the latter solution still etches {111}Ga planes at a slower rate than all others but does not etch {111}As any faster than any other direction. (After Iida and Ito.[14]) (b) The ratio of the etch rate for {111}As to that for {111}Ga plotted against H_2O_2 mole fraction for the H_3PO_4–H_2O_2–H_2O etchant at three different temperatures, 0, 24, and 40°C. The diffusion-limited region is below a H_2O_2 mole fraction of 3×10^{-2}, that is for very low H_2O_2 concentrations. In this regime the {111}Ga still etches at a rate lower than any other plane, including {111}As, this behviour being identical to that shown in (a) and applying to almost all other systems. Different activation energies, e.g. 0.43 eV/molecule and 0.27 eV/molecule, associated respectively with {111}Ga and {111}As etching in the reaction-rate-limited regime account for the lowering of the ratio with increased temperature. (After Mori and Watanabe.[10]) (c) Polar plots of normalized etch rates of GaAs for the H_2SO_4–H_2O_2–H_2O etchant having different dilutions, as indicated. Note that the etch rate ratio for {111}As/{111}Ga decreases with increased dilutions of the 'stock' solution 1:8:1, although even 1:8:1000 dilution (not shown here) does not yield an isotropic etchant. The system becomes reaction-rate-limited only when the H_2SO_4 concentration is increased, as in 8:1:1 (a). (After Shaw.[15])

Table 4.3 Etch rates at about 27°C of {111} As, {100}, {110}, and {111} Ga planes in GaAs for a few wet etchants

Etch solutions used	Etch rate (Å s^{-1}) {111}As	{100}	{110}	{111}Ga	Etch rate ratio and type[a]	Ref.
(a) Citric acid–H_2O_2 10:1	35	28.5		21	1/0.82/0.6 (R/D)	9
(b) H_3PO_4–H_2O_2–H_2O 1:9:1	715	525	785	250	1/0.73/1.1/0.35 (R)	10
(c) 3:1:50	13.5	14.5	15.5	7.2	1/1.07/1.15/0.53 (R/D)	10
(d) NH_4OH–H_2O_2–H_2O 20:7:150 ($\cong$ 3:1:150)	33	20		6.2	1/0.6/0.2 (R)	11
(e) H_2SO_4–H_2O_2–H_2O 8:1:1	195	190	167	115	1/0.97/0.86/0.63 (D/R)	14
(f) 1:8:1	1670	1325	1250	390	1/0.79/0.73/0.23 (R)	14
(g) 1:8:1	3015	2433		688	1/0.8/0.23 (R)	15
(h) 1:8:40	239	200		96	1/0.84/0.4 (R)	15
(i) 1:8:80	104	90		49	1/0.86/0.4 (R)	15
(j) 1:8:160	43	43		26	1/1/0.6 (R/D)	15
(k) H_2O–$AgNO_3$–CrO_3–HF 2 ml:8 mg:1 g:1 ml	1050	2100	1450	1490	1/2/1.38/1.42 (R)	6

[a] R, Reaction-rate-limited.
D, Diffusion-limited.
R/D, Mostly reaction-rate-limited, but shows mask edge trenching. No other characteristic of a diffusion-limited etchant is, however, visible.
D/R, Basically D, but still etches {111}Ga at a lower rate, indicating the R component dominates for {111}Ga.

8.5 kcal mol^{-1} (0.22 to 0.36 eV/molecule).[22] The surface adsorption process of the oxidizing molecules on partially oxidized GaAs surface has activation energies between 8 and 16 kcal mol^{-1} (0.35 to 0.70 eV/molecule). As a result, reaction-rate-limited etching processes generally show a stronger temperature dependence than diffusion-limited etching processes.

As mentioned earlier, vigorous agitation causes the thickness d_1 of the static, streamlined liquid layer in contact with the GaAs to become smaller, which increases C_{AO}. In extreme cases, d_1 can become so small that $C_{AO} = C_{BO}$, and the etching process becomes reaction-rate-limited. Figures 4.3a and c show examples of this where the shift from diffusion-limited to reaction-rate-limited etching process manifests itself with an increase in the activation energy.[9,14]

4.2.2.5 Time dependence

During a diffusion-limited etching process in a stagnant solution, the layers of solution close to the etched surface slowly become depleted of the molecules that are responsible for etching and so, with an increase in d_1, the etch rate drops following $E \propto t^{1/2}$. Depending upon the reaction and diffusion rates, a steady state is reached after prolonged etching. In contrast, reaction-rate-controlled etching processes show constant etch rates with time[10] (Figure 4.6a). If, for diffusion limited etching, vigorous agitation causes d_1 to shrink almost to zero, etching becomes reaction-rate-limited and the etch rate remains constant with time[23] (Figure 4.6b).

Some etchants such as citric acid–H_2O_2[9] show sublinear behaviour, but do not follow a $t^{1/2}$ relationship either. The process is therefore partly diffusion- and partly reaction-rate-limited (see Figure 4.6c). Sublinear behaviour has also been observed for reaction-rate-limited etchants containing ingredients that either dissociate or decompose (e.g. H_2O_2) with time, thus reducing the strength of the etchant, as in the case of NH_4OH–H_2O_2–H_2O based etchants[11] (Figure 4.6d).

4.2.2.6 Galvanic effect

A reaction-rate-limited etching mechanism depends upon the availability of electrons for oxidation to occur. A material having an electronegativity that is different from the electron affinity for GaAs, when immersed in the etch solution, would form a galvanic cell. The galvanic action would then supply electrons for the etching reactions to progress at a faster rate. Increases in etch rates are seen when the dissimilar objects are stainless steel tweezers or wafer carriers,[18] ohmic and Schottky metallizations on GaAs, and exposed $Al_xGa_{1-x}As$ in AlGaAs/GaAs heterostructure based devices.[25] Such an etch rate increase would be smaller for diffusion-limited etching processes provided that the physical separation of this dissimilar material from GaAs is much larger than the diffusion layer thickness, d_1. For metallizations on GaAs with unprotected edges, excessive etching at these boundaries is expected to occur, due to the close proximities involved.

4.2.2.7 Doping type and concentration dependence

For a similar reason, reaction-rate-limited etching is much more dependent upon the doping concentration and type owing to its dependence on the availability of electrons and holes (see Figure 4.6d).

4.2.2.8 Chemical polishing effects

If the surface has a protrusion, the tip would be exposed to a higher diffusion current of reacting agents and therefore a diffusion-limited etchant tends to make

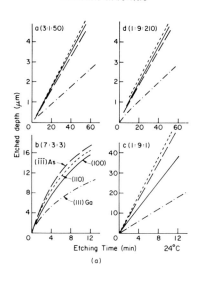
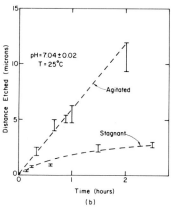
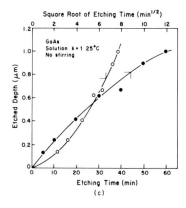

Etching and surface preparation of GaAs for device fabrication 137

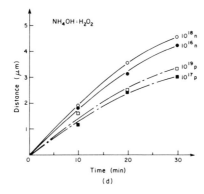

(d)

Figure 4.6 Time dependence of etch rates, and their causes. (a) Etch depth plotted against time for the four major directions in GaAs for the H_3PO_4–H_2O_2–H_2O etchant with different ratios. Undoubtedly the 7:3:3 solution is diffusion-limited, the rest are reaction-rate-limited. (After Mori and Watanabe.[10]) (b) In a diffusion-limited etchant, such as straight-from-the-bottle, 30 wt% H_2O_2 brought up to pH 7.04 ± 0.02 with the addition of a few drops of NH_4OH solution, the etch rate decreases with time owing to the depletion of NH_4OH from regions near the surface. When agitated, the supply of NH_4OH is increased and the etch rate remains constant with time at 6 μm h^{-1}. (After Logan and Reinhart.[23]) (c) Etch depth plotted against time and square root of time for (100) GaAs etching with citric acid–H_2O_2 (1:1). Although the time plot shows sublinear behaviour, the plot with square root of time yields a superlinear curve, indicating contributions from both diffusion and reaction rates. (After Otsubo *et al.*[9]) (d) Etching rate (slope) variation with time for the etching of (100) GaAs of p- and n-types with different carrier concentrations. The etchant is NH_4OH–H_2O_2–H_2O (3:1:150), which is basically a 0.35 N NH_4OH/0.12 N H_2O_2 solution. The sublinear behaviour, in this case, is believed not to be due to any diffusion-limited behaviour but to be caused by the gradual decomposition of H_2O_2. (After Gannon and Neuse.[11])

the surface smoother. A reaction-rate-limited etchant, however, would maintain the original surface topography, unless of course its anisotropy takes over. In such a case, it would etch the fast etching planes quickly and stop at the slowest etching planes, creating characteristic patterns.[9, 10, 14, 15]

4.2.2.9 Etch rate magnitudes

The reaction-rate-controlled etching condition occurs when the surface reaction rate, e.g. oxidation, is low due to the small concentration of the oxidizer in the solution. As the oxidizer concentration is increased, the etch rate increases and the complexing action keeps up with the oxidation rate, producing an almost linear relationship between the etch rate and the oxidizer concentration, in accordance with equation (4.2)[12] (see Figure 4.7). Eventually, the etch rate dependence on oxidizer concentration becomes sublinear, as the complexing action fails to keep

138 Gallium arsenide

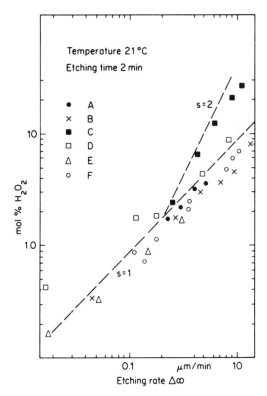

Figure 4.7 Correlation between H_2O_2 content and etching rates for various etch solutions. A to F refer to the same etching solutions as described in the caption to Figure 4.4c. Except for etchant C, H_2SO_4–H_2O_2–H_2O ($x:1:1$, $10 < x < 20$), which is diffusion-limited, all the others show etch rates proportional to the H_2O_2 concentration. This indicates surface-oxidation-limited etching processes, even though mask edge trenching was observed for a number of cases. (After Kohn.[12])

up with oxide growth. Consequently, one has to increase the concentration of the complexing agent as well in order to have an increased etch rate. However, the solution becomes viscous and the etching process becomes diffusion-limited. The etch rate flattens out and starts dropping owing to the inability of the oxidizer (H_2O_2 molecules) to reach the GaAs surface freely (Figure 4.3b and Table 4.4).

4.2.2.10 *Thickness of the residual oxide layers*

Generally, a high etch rate involves a fast surface oxidation rate and also a relatively fast complexing rate. In such a case the complexing agent works quickly to remove the partially oxidized surface layer formed. As a result, the residual (oxide) layer remains rather thin.

Etching and surface preparation of GaAs for device fabrication   139

Table 4.4 Properties of some commonly used wet chemical etchants for GaAs, using H_2O_2 as the oxidizing agent

Etchant	Volume ratio	Normality ratio	Etch type[a]	Etch rate for {100}GaAs (Å s)	Ref
H_3PO_4–H_2O_2–H_2O	1:9:1	4.1:16	R–D	525	10
	7:3:3	24.2:4.5	D	330	10
	1:1:25	1.7:0.7	R	55	19
	1:9:210	0.21:0.8	R	16	10
	3:1:50	2.5:0.36	R	13.5	10
NH_4OH–H_2O_2–H_2O	1:1:16	1:1.1	R	330	12
	3:1:50	1:0.3	R	100	11
	3:1:150	0.35:0.13	R	30	11
H_2SO_4–H_2O_2–H_2O	1:8:1	3.6:15.3	R	1300	14
	1:1:8	3.6:2.0	R	215	15
	8:1:1	29:2	D	200	15
	1:8:40	0.75:3.2	R	200	15
	1:8:80	0.4:1.75	R	90	15
	1:8:160	0.22:0.82	R	45	15
	20:1:1	33:0.9	D	30	12
	1:8:1000	0.036:0.15	R	7	15
HCl–H_2O_2–H_2O	1:1:16	1:1.1	R	330	12
	3:1:50	1:0.3	R	100	11
	3:1:150	0.35:0.13	R	30	11

[a] R, reaction-rate-limited; D, diffusion-limited.

In contrast, for slow etchants, thicker residual layers are observed. Slow etching can occur in both the reaction-rate- and diffusion-limited regimes. In the former case, the rate-limiting complexing action in a solution containing only a small concentration of the complexing agent does not dissolve away the oxide layer quickly. This results in a thicker residual oxide layer on terminating the etching process. Since the layer is predominantly formed by the oxidation process, it is composed mostly of a mixture of suboxides of Ga and As. Ellipsometric studies[12] show that the thicker the oxide layer, the lower the refractive index (Figure 4.8). Since the native oxides on GaAs have refractive indices of approximately 1.9, while for GaAs the real part of the complex refractive index is approximately 4.05, this indicates that a thicker residual oxide tends to look more like a native oxide than a metal-rich, partially oxidized GaAs surface, as if found after a fast etching process. It suggests that a longer exposure of the residual layer formed during a slow etching process to the oxidizer facilitates the formation of a nearly stoichiometric oxide layer which is slowly dissolved away in the solution containing low concentrations of the complexing agent. It also suggests that,

140 *Gallium arsenide*

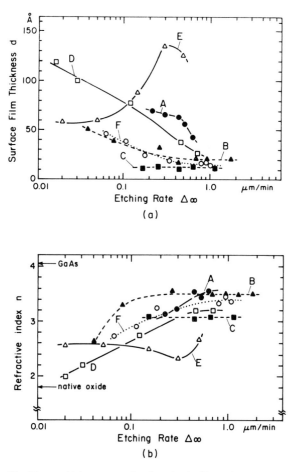

Figure 4.8 Residual layer thicknesses and refractive indices after etching as measured by ellipsometry on GaAs. (After Kohn.[12]) (a) The drop in residual layer thickness with increased etch rate. (b) The refractive indices of the thicker residual layers approach that of the native oxide on GaAs. The only exception is the citric acid based etchant, E. The descriptions for A to E can be found under Figure 4.4c

during fast etching processes, even a partially oxidized GaAs surface with a refractive index between about 3.0 and 4.0 succumbs to the complexing action of a concentrated solution, leaving behind a thinner residual layer of amorphous nature.

In the other extreme, in a diffusion-controlled etching solution containing very high concentrations of complexing agents, a thicker residual layer is known to form. A low concentration of H_2O_2, a high concentration of complexing agent, and a slow removal of the etched product suggest that the large residual layer thickness is due to the lack of removal of the complexed material. The residual layer, therefore, contains oxides as well as other compounds that are dependent upon the chemical nature of the complexing agent.

A few exceptions to this generalization can be found and these depend upon the specific nature of the chemical reactions involved. For citric acid–H_2O_2–H_2O (50:x:50; 1 < x < 50) etchants, the etch rate increased with increasing x, but so did the thickness of the residual layer. The layer itself was found to be porous and loosely bound to the GaAs surface, and could be wiped off with a cotton bud.[12] Diluted H_2O_2–H_2O solutions containing even lower concentrations of H_3PO_4[10] cause GaAs surface clouding, presumably by excessive surface oxidation rather than by removal of the oxide. However, such an effect is not observed with H_2O_2–H_2O solutions containing no complexing agents or with only a few drops of NH_4OH.[23] Presumably, the surface clouding observed with low concentrations of H_3PO_4 is associated with phenomena unique to the nature of reactions involved.

The major characteristic differences between diffusion- and reaction-rate-limited etching phenomena are summarized in Table 4.5 for the sake of quick reference and clarity.

4.2.3 Device applications

In mesa isolation of GaAs devices, one needs to have gradually sloping edges for the deposition of contact or interconnect metallizations without thinning out. For nearly vertical side-walls, evaporated metallic layers not only have reduced conformal thicknesses but also may suffer abrupt discontinuities. For etched side-walls with negative slopes,[14, 15] continuous interconnect metallization is an impossibility. Consequently, most reaction-rate-limited etching solutions that are known to produce etched profiles similar to those shown in Figure 4.9 are considered to be unsuitable.

This induces one to consider a diffusion-limited etchant for mesa isolation etching, but diffusion-limited etchants are known to cause severe trenching and are unsuitable as well.

In reality, however, there are a number of reaction-rate-limited etchants that tend to have planes other than {111}As as the fastest etching planes or planes other than {111}Ga as the slowest etching planes. For example, the H_2SO_4–H_2O_2–H_2O (1:8:1000) etchant reveals (111)Ga and ($\bar{1}\bar{1}\bar{1}$)Ga planes for etching with mask edges parallel to [011], as expected, yielding sloping edges. For mask edges parallel to [0$\bar{1}$1], it reveals (111)As and (11$\bar{1}$)As planes, again yielding sloping edges.[15] Also, HCl–H_2O_2–H_2O (80:4:1), a diffusion-controlled etchant, produces sloping edges but does not show pronounced trenching effects and is a suitable candidate for mesa etching. Another practical example for a mesa etch is H_3PO_4–H_2O_2–H_2O (1:1:25), with an etch rate of 55 Å s^{-1} at 25°C (see Figure 4.10).

Table 4.5 Major characteristic differences between diffusion-limited and surface reaction-rate-limited wet etching processes

	Parameter	Diffusion-limited	Reaction-rate-limited
(i)	Agitation dependence	agitation increases supply of etchants and increases etch rate	no increases observed
(ii)	Mask edge trenching	supply of fresh etchants from the masked side causes excessive etching at mask edges to form trenches	no such effect
(iii)	Preferential (anisotropic) etching	not pronounced; can be completely eliminated by increasing solution viscosity	very pronounced effect; $E\{111\}$ As $> E\{100\}$ $> E\{111\}$ Ga (E = etch rate)
(iv)	Temperature dependence	lower activation energies, hence relatively insensitive to temperature	higher activation energies, hence more sensitive to temperature
(v)	Time dependence	etch rate is roughly proportional to the square root of time and, if not, is sublinear	etch rate constant with time; it is sublinear if the etching reagents, such as H_2O_2, NH_4OH, dissociate during etching
(vi)	Galvanic effect	not very pronounced, except at metallization edges	galvanic cell formation enhances etch rates everywhere on wafer
(vii)	Doping type (n or p) and concentration dependence	very small	if each rate is surface oxidation-limited, then etch rate could increase with increased electron concentration
(viii)	Chemical polishing effects	surface protrusions experience higher diffusion current, hence higher etch rates; surface becomes smoother	surface topography is well maintained, unless affected by crystal anisotropic effects (iii)
(ix)	Etch rate magnitudes	highly viscous solutions have low etch rates, but very high rates in the transition region	usually rather low due to diluted solutions used in practice
(x)	Residual oxide layer thickness and composition	high etch rates leave thinner oxides; highly viscous, low etch rate solution would leave thicker, partially complexed/oxidized residues	low etch rates cause thicker oxide build-up that resembles native oxides, rarely less than 40 Å thick

Etching and surface preparation of GaAs for device fabrication 143

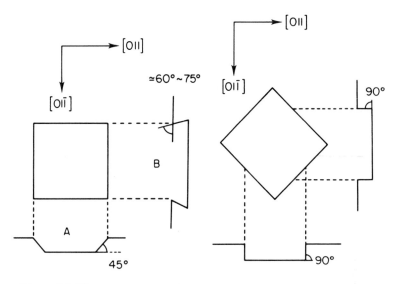

Figure 4.9 The appearance of etched wall for reaction-rate-limited etching on (100) GaAs, for window edges parallel to and at 45° to the cleavage planes, (011) and (01̄1). Compare with Figure 4.1c. (After Iida and Ito.[14])

Lastly, it should be pointed out that most of these unusual structures formed by the preferential (anisotropic) nature of reaction-rate-limited etchants become visible only after prolonged etching, when the etch depth exceeds 1 μm or so. Consequently, for modern devices with thin (< 3000 Å) active layers, one may choose a reaction-rate-limited etchant such as $(H_3PO_4$ or $NH_4OH)-H_2O_2-H_2O$ (3:1:50) having low etch rates and still obtain desirable mesa etch profiles.

One of the devices that has benefited from the negative slopes for mask edges parallel to the [01̄1] axis on (001) GaAs is the vertical transistor, where the overhang inhibits the gate evaporated metal from contacting the top ohmic contact metallization.[25]

We shall discuss the uses of wet etching for recess etching of both depletion and enhancement mode logic and power FETs in Section 4.4. A number of practical etch solution recipes are indicated in Tables 4.3 and 4.4, but the reader is advised to consult the original papers for more detailed information.

4.3 ANODIC ETCHING

Since a number of articles have been published on the subject of epilayer thinning by anodic oxidation, we shall limit the scope of this section to a general description of the process, and shall refer the reader to the literature[26-29] for both a more detailed study of the theory[26] and a more quantitative discussion of the achievable results.[27]

Figure 4.10 Planar Gunn logic device with mesa, 10 μm wide, 6 μm high, crossed by a 5 μm wide gate stripe. Mesa edges are outward sloping in both (011) and ($0\bar{1}1$) directions. The etch was H_3PO_4–H_2O_2–H_2O (1:1:25), 20 min at room temperature. (After D. W. Woodard, Unpublished Work, 1983)

A key parameter in the design of FET devices is the doping × thickness product of the active layer. Anodic oxidation is a process which, in cases where there are no other layers on top of the active layer, provides a reproducible means for adjusting the value of this parameter after growth. The reproducibility derives from the fact that the process is self-limited at a doping × thickness product that is fixed by the applied voltage and the ambient or applied light intensity. Thus not only is there reproducibility from layer to layer, but also the uniformity of doping × thickness product within a layer can be greater after the etch than it was before. This is a distinct advantage over all other etching techniques, which, at best, remove a fixed amount from each portion of the wafer thereby increasing, on a percentage basis, any existing non-uniformities.

Figure 4.11 is a schematic representation of the typical anodic oxidation apparatus. The essential features are a cathode material which is non-reactive in the electrolyte, a provision for contacting the epilayer with a conductor which is itself preferably insulated from the electrolyte, an electrolyte which is capable of oxidizing the wafer surface without dissolving the resultant oxide layer, a power

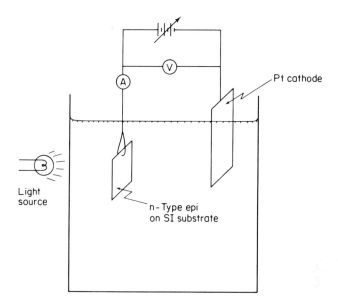

Figure 4.11 Apparatus for anodic etching

supply which is controlled either for voltage or current, and a variable-intensity light source.

For reasons discussed below, the self-limiting effect occurs when the material below the layer being etched has a higher resistivity. This is certainly true for FET wafers which are n-type on semi-insulating, as shown in the example of Figure 4.12.

The interface between the n-type semiconductor and the electrolyte forms a Schottky barrier which is reverse-biased when the sample is made the anode, that is, positive with respect to the electrolyte. If the applied voltage exceeds the breakdown voltage of the Schottky barrier, the depletion region avalanches producing a current that results in the growth of an oxide layer. Ionic conduction in the resulting oxide layer allows current to continue flowing as long as the applied voltage exceeds the sum of semiconductor breakdown voltage and the resistance drop in the oxide. If the applied voltage is held to some finite value, as it must be to avoid avalanching the oxide, then the oxide layer will become uniform in thickness while its growth rate asymptotically approaches zero. In growing the oxide to a uniform thickness, a uniform thickness of GaAs is consumed. Thus by repeatedly stripping the oxide and reoxidizing, one has the means to remove a uniform thickness of arbitrary magnitude. The amount removed per oxidation depends on the voltage and other conditions, but is highly reproducible and typically in the range 600 to 1200 Å, allowing a high precision for removal.

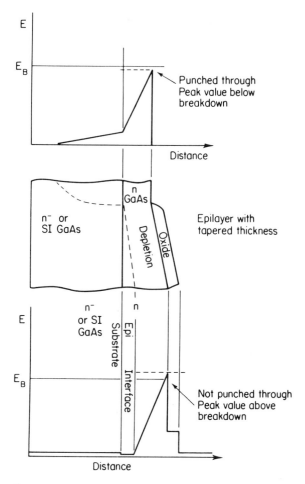

Figure 4.12 Field profiles through wafer with uneven epilayer thickness, as is often found on LPE-grown material, showing areas where depletion does (above) and does not (below) punch through

To understand the self-limiting effect which occurs when the layer being etched has higher n-type doping than the material underneath, we consider the situation shown in Figure 4.12, where the thickness or at least the doping × thickness product of the epilayer is assumed to be uneven. The depletion edge of the avalanching Schottky barrier first reaches the interface with the higher-resistivity material at the thin areas of the layer. In these areas, the width of the depletion region expands into the high-resistivity material, reducing the peak electric field to below the breakdown field and terminating further oxide growth. In the

Etching and surface preparation of GaAs for device fabrication 147

thicker areas, however, oxidation and thinning continue until the same condition of depletion punchthrough is reached. The final thickness of the layer is less than the depletion width at avalanche by a small amount that depends on the doping of the material underneath. The final thickness can be further reduced by application of light, which reduces the voltage and hence the depletion width required for fixed current.[26] With incident light, however, the process is not self-limited. To monitor the desired endpoint, one typically applies a constant avalanche current while measuring the voltage required. If the layer is uniform in thickness, depletion edge punchthrough is signalled by a rapid rise in the voltage, at which point the process can be terminated.

4.4 SURFACE PREPARATION

4.4.1 Electrical effects of surface dielectrics on GaAs

It is known that when a clean GaAs wafer is exposed to air an oxide layer grows to a thickness of about 30 Å after about four days[30, 31] and a then carbon overlayer covers the oxide layer when the steady-state thickness is reached.[32]

It has also been mentioned earlier that wet etching processes leave behind residual layers on GaAs surfaces. The thicknesses of these layers are dependent upon the etch rate: the higher the etch rate, the thinner the residual layer.[12] Although no studies of chemical composition for these layers were made, ellipsometric studies revealed refractive indices between 2 and 3.5, suggesting that the layer contained either a mixture of metal-rich native oxides or other compounds characteristic of a particular etch. An etching step often precedes an ohmic contact metallization step to reach a specific underlying layer or a Schottky gate metal deposition step in a MESFET fabrication process in order to control the channel conductance. We shall discuss here the effects of these dielectric layers on surface electrical characteristics.

Ohmic contacts to n-GaAs are usually formed by the liquid phase alloying of an eutectic mixture of Au-Ge (24 wt % or 12 at %) having a eutectic temperature of approximately 357°C. When an AuGe metallized n-GaAs sample is heated above the eutectic temperature, the AuGe liquid eutectic interacts with the GaAs in the liquid state. On solidification, complex intermetallic compounds coupled with highly n^+-type GaAs doped with Ge produce a low-resistance ohmic contact.

If, however, the surface oxide thickness is not reduced substantially, the liquid eutectic has to penetrate through the weak spots in the oxide layer to interact with GaAs. Non-uniform interactions occur, which yield poor specific contact resistances, degrade surface morphology, and reduce reproducibility substantially.

Schottky barriers are found to be more sensitive to surface dielectric thicknesses than ohmic contacts. The barrier height ϕ_B increases with surface oxide thickness d as[30]

$$\phi_B = \phi_i + Ad$$

where ϕ_i is the extrapolated barrier height at zero thickness, which varies between 0.60 and 0.75 V, and A is the proportionality constant, with values between 5×10^5 and 17×10^5 V cm^{-1}. As a result of the increased barrier heights, forward conduction does not follow a Schottky behaviour.[32] Also, reverse breakdown voltage decreases. Part of the applied reverse bias applies across the oxide and the depletion width in GaAs is reduced. When the Schottky gate in a GaAs MESFET behaves in this fashion, it reduces the transconductance of the device considerably. Lastly, avalanche breakdown as well as forward conduction show hysteresis caused by charge trapping in the interfacial layer,[33] which is unacceptable for devices such as Schottky barrier GaAs IMPATT diodes.

For good ohmic contact formation and for nearly ideal Schottky behaviour, the residual dielectric layer thickness has to be reduced to a minimum. In the next subsection we shall discuss the means for achieving this.

4.4.2 GaAs surface residual layer removal processes

A number of methods for the removal of surface residual layers on GaAs are available in the literature.[12, 30-32, 34, 35] We shall briefly describe the processes first and then we shall discuss their applications to device fabrication processes, such as the gate recess etch prior to gate metal deposition for GaAs MESFETs.

Among the oxides of Ga and As, all oxides of arsenic, such as As_2O_3 and As_2O_5, are highly soluble in water,[36] although the rate of dissolution is unknown. These oxides also dissolve in alcohols and acids, and As_2O_3 dissolves in alkaline solutions as well. In contrast, Ga_2O_3, $Ga_2O_3(H_2O)$, and Ga_2O (suboxide) are insoluble in water, slightly soluble in acids, and readily soluble in alkaline solutions. It is interesting that an attempt to remove a thick (120–140 Å) residual layer with HCl treatment after a slow H_2SO_4–H_2O_2–H_2O (1:1:250) etch, having an etch rate of about 3.5 Å s^{-1}, reduced the thickness only down to about 80 Å.[12] This layer was found to have a refractive index of approximately 2.0, resembling a native oxide. It is possible either that HCl did not dissolve Ga_2O_3 in the time involved or that the layer had other compounds that are insoluble in HCl. On the other hand, a 1:1:16 etch with an etch rate of about 75 Å s^{-1} had only a 50 Å thick residual layer that was reduced to about 10 Å with HCl treatment, presumably because of the presence of the suboxide Ga_2O, which dissolves more readily in HCl, and of As_2O_3, which is soluble in water.

Since water dissolves arsenic oxides, a short duration (< 5 min) soak removes surface As.[32] But a prolonged soak in water (≥ 1 h) increases oxide thicknesses to about 100 Å.[31] Long exposures to H_2SO_4 and H_3PO_4 also cause residual layer thicknesses to increase.

In a number of studies recipes have been given for reducing the oxide

Etching and surface preparation of GaAs for device fabrication 149

thickness[31, 34] down to 10–15 Å or for obtaining near-ideal Schottky behaviour with its correlation with strong light emission during avalanching.[37] Using these and other[32, 35] information we summarize the findings:

(a) Alkaline solutions such as strong NH_4OH solution,[8] NaOH solutions,[37] and NaOH-based fast etchants like $NaOH-H_2O_2-H_2O$ (1:3:30[8] and 1 M/0.76 M in water[34]) yield lower oxide thicknesses and excellent Schottky characteristics. Auger analysis of GaAs surfaces after a strong NH_4OH dip indicated the presence of As oxides[32] which correlated with non-ideal Schottky characteristics. A quick water rinse would have dissolved the arsenic oxides, probably with no increase in oxide thickness,[31] although results of such a study are unavailable. NH_4OH is known to remove residual carbon from surfaces, but only at higher concentrations.[32]

(b) In one study of GaAs surface preparation, a final oxide thickness of about 12 Å[31] was achieved with the GaAs cleaned in HCl, dipped in H_2SO_4, and then rinsed in methanol, presumably to dissolve away arsenic oxides. A sample dipped in HF and blow dried in N_2 also had a similar oxide thickness. However, diodes made on H_3PO_4 and HCl dipped GaAs, with no rinsing process prior to metallization, yielded non-ideal Schottky characteristics.[37] In view of the apparently slow etching the surface oxide by HCl, a prolonged ($>$ 5 min) dip is required.[35]

After at least 5 min HCl dip, a short, few seconds rinse in deionized water might be needed to remove arsenic oxides, which are believed to be solely responsible for non-ideal Schottky behaviour.[32, 33]

In view of the rather diversely different viewpoints about the usefulness of HCl in removing GaAs surface residual layers, NH_4OH solutions of various strengths, such as 1:15 to 1:2 are routinely used in many laboratories with good results.

4.4.3 Practical applications of surface treatment

For a GaAs wafer sitting on the shelf, the surface has to be cleaned to remove oily substances and carbon contamination before any processing step such as lithography can proceed. This first cleaning step varies from one laboratory to another and none of them seem to have any adverse effect upon subsequent etching processes. This is perhaps because prolonged (24 h) dips in various organic solvents fail to increase the oxide thickness.[31] A simple treatment of sequential vigorous rinsing in acetone, methanol, and deionized water seems to be as good as any other. The only precaution one needs to take is not to allow the wafer to become dry before commencing the subsequent rinsing, as the evaporation of one of the solvents would leave behind residues that may not be soluble in the solvent used in the next step.

A GaAs MESFET wafer after gate lithography may need a gate recess etch

150 Gallium arsenide

prior to metallization. The depth of the recess may vary from a few tens to a few hundred angstroms, depending upon the device layer structure. So, the recess etch has to have a low rate, should be non-preferential, should not seep under the resist and create an etch pit wider than intended, and should not leave residues behind. Among the etch ingredients in Table 4.1, H_2O_2, HCl, and NH_4OH are volatile, while citric acid, H_3PO_4, and H_2SO_4 are not. Hence care must be taken in using H_3PO_4 or H_2SO_4 based etches in removing the acids with a thorough water rinse.

A general precaution prior to any etching process is to degrease the surface and remove surface carbon.[32] Of course, after a lithographic step one cannot use a solvent for surface cleaning purposes, but a strong NH_4OH-H_2O (1:2) solution is capable of degreasing as well as surface carbon removal prior to a critical etching process, such as gate recess etch in fabricating short gate GaAs MESFETs. Almost any slow etchant (Table 4.4) can be used, and the residual surface layer can be removed using either an $HCl-H_2O$ (1:1) dip or an $NH_4OH-H_2O_2$ (1:2) or (1:7) dip with the necessary precautions mentioned earlier. Figure 4.13 shows an example of a study of recess etch on GaAs.

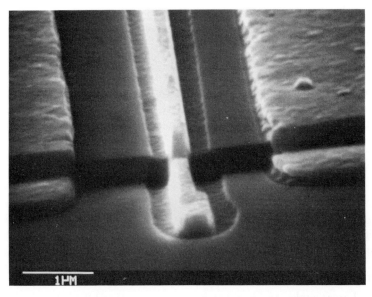

Figure 4.13 GaAs low-noise microwave MESFET with 0.5 μm gate: mesa edge shows recess etch with lifted-off Ti/Pt/Au gate at recess centre. Mesa etch used was $H_3PO_4-H_2O_2-H_2O$ (1:1:25), etch rate approximately 55 Å s^{-1}. The recess etch was slow etch, $H_3PO_4-H_2O_2-H_2O$ (3:1:50), etch rate approximately 15 Å s^{-1}. (After Camnitz.[19])

Etching and surface preparation of GaAs for device fabrication   151

4.5 DRY ETCHING

4.5.1 Sputtering and ion beam-induced etching

A number of excellent articles are available on the fundamentals and uses of sputter-induced etching.[38-41] Consequently, we shall briefly mention the physical processes involved and discuss their applications to GaAs etching.

In a diode sputter etch system, the wafer to be patterned is usually placed on the lower electrode, which is powered with an RF source of 13.56 MHz through a capacitor. The upper electrode is grounded. In an argon plasma, low mass electrons leave the plasma to reach the powered lower electrode, causing it to become negatively biased. This attracts heavy, positively charged argon ions towards it to sputter the material away from the wafer on the atomic scale at a controlled rate.[40,41]

In an ion beam etching (IBE) equipment, a separate ion source, such as the Kaufman type,[42] is used, with or without an ion neutralization facility.

Sputter etching using a parallel-plate RF sputtering unit usually takes place in the argon gas pressure range of 10–100 m Torr. The separation between the two plates remains within 5 to 10 cm and 30% of redeposition is common.[38] In an ion beam etching equipment, the gas is injected into the ion gun while the rest of the chamber is maintained at a much lower pressure, such as 1×10^{-5} to 1×10^{-4} Torr. Also, a much larger distance, such as 50–100 cm, is maintained between the ion source and the substrate. Consequently, the possibility of localized redeposition of material sputtered away from the wafer is reduced in the ion beam etching vacuum chamber. Because of operations at a lower pressure, cleanliness is better maintained. As a result of this and also because of the reduced redeposition of unwanted material, most of the reports of device fabrications using sputter etching of GaAs available in the literature involve ion beam etching.[43]

The etch mask that is commonly used is a thin (1000–2000 Å) layer of titanium. Titanium has a much lower etch rate in pure argon, approximately 200 Å min^{-1} for 500 eV average argon ion energy at about 1 mA cm^{-2} of current density, while GaAs etches at about 2000 Å min^{-1} or less.[45] Much improvement in etching selectivity is achieved from the fact that both stoichiometric and partially oxidized Ti etch at a slower rate than pure Ti. For example, if during the sputter etching process oxygen gas is leaked into the system at roughly 5 to 10 vol %, the Ti mask etch rate becomes less than 50 Å min^{-1}.[25]

For normal incidence of argon ions with moderate energies, such as 500 eV, the etched side-walls usually yield angles around 70 to 75° with respect to the crystal surface. Larger angles do not occur because, although the sputter rate peaks at this angle, the ion current density (or flux) upon the side-wall surface drops drastically with increasing angle. This aspect is discussed with clarity elsewhere[40,41] and the reader should consult the references for details. A higher

152 *Gallium arsenide*

energy would produce higher-angle side-walls at the expense of trenching caused by forward-scattered energetic ions off the side-walls.

A Kaufman-type ion source can also be used to generate reactive ions that would etch the GaAs both physically by the sputtering process and chemically by combining with Ga and As to form volatile components. While Ga and As fluorides have very low vapour pressures in the 0–200°C temperature range, $GaCl_3$ and $AsCl_3$ are highly volatile.[44] Therefore etching of GaAs has been achieved with the help of ion beam etching systems using CCl_4[45] or Ar ions with Cl_2 bled into the vacuum chamber.[46] This process of using a collimated ion beam containing reactive gases for the etching of GaAs and other materials is called reactive ion beam etching (RIBE).[45-49]

Unlike in a parallel-plate diode sputtering unit, the sample can be held at any angle with respect to the incident beam in an ion beam or reactive ion beam etching apparatus. This facilitates the control of etch profiles by minimizing trenching (which occurs at 0° incidence) and localized redeposition from the sidewall upon the etched surface adjacent to it (which occurs at larger angles of incidence, such as 25° and above) during Ar ion etching. In chemically assisted RIBE, the ability to control the angle of incidence allows one to change the slope of the etched wall independently of the ion energy used.

All dry etching processes are capable of contaminating the GaAs wafer with sputtered metallic ions from the inside walls or other vacuum components that are exposed to the plasma. The problem increases in an IBE or RIBE system as the ions from the ion source carry with them atoms sputtered off the different components of the source. Some of these problems can be eliminated by proper design of the ion source and the chamber, and some with the help of organic or dielectric coatings on the inside of the chamber itself.[49]

Table 4.6 summarizes some of the studies of ion beam and reactive ion beam etching of consequence to GaAs device fabrication.

4.5.2 Reactive plasma-assisted etching

Plasma-assisted etching using a reactive gas that forms volatile compounds as the etched product has become fairly well established in silicon technology.[50] Presently plasma-assisted etching of GaAs is being explored in a number of laboratories.[51-58] All the reported studies involve parallel-plate diode etching with chlorinated etch gases.

When the RF power is applied to the top electrode and the wafer is placed on the grounded bottom electrode, the wafer remains at a moderately low negative bias with respect to the plasma. At higher operating pressures, about 100 m Torr or more, the mean free paths of the colliding gas molecules and ions become smaller than the interelectrode spacing. The wafer then sits in a sea of randomly moving ions, excited radicals, and molecules, and is etched mostly by surface chemical action of the ions and excited species. Unless the process is carefully

Etching and surface preparation of GaAs for device fabrication   153

Table 4.6 A summary of some of the ion beam etching (IBE) and reactive ion beam etching (RIBE) studies of GaAs

Ion used	Energy (eV)	Current density (mA cm^{-2})	Etch rates and electrical effects	Ref.
Ar–O$_2$	500	1.0	1400 Å min^{-1} etch rate; after 2000 Å of ion beam etching another 2000–3000 Å of n-GaAs was liquid wet etched to remove ion damage for vertical FET fabrication; see Figure 4.14.	25
CCl$_4$	500	1.0	4600 Å min^{-1} etch rate for CCl$_4$; 2000 Å min^{-1} for Ar etching. Note: This is about four times larger than that reported in ref. 35	33
Ar	500	1.0		
Ar Ar–Cl$_2$	500	1.0	2000 Å min^{-1} etch rate; uses same Varian RIBE equipment as in ref. 33; increased Cl$_2$ partial pressure in vacuum increases GaAs etch rates	34
Ar–CCl$_4$	500	1.0	drastic etch rate increase observed	34
Ar	500	1.2	18°tilt, etch rate $\cong$ 520 Å min^{-1}; Schottky barrier degraded. Note: In this case Cr–Au Schottky barriers were ion beam (sputter) deposited, which is known to yield low ϕ_B and high n caused by subsurface damage; Schottky barrier diodes on control samples that were chemically etched had $\phi_B \cong 0.67$ eV and $n \cong 1.2$, which indicate rather poor performance anyway; 400°C anneal yields $\phi_B = 0.53$ eV, $n = 1$.	35
CF$_4$–H$_2$ (1:1)	600	?	etch rate $\sim$ 50 Å min^{-1}; $\phi_B = 0.59$ eV, $n = 1.32$ after etching; after 300°C 30 min anneal, $\phi_B = 0.83$ eV, $n = 1.09$, which indicate complete recovery	36
Ar	250	0.25	ϕ_B goes down from 0.76 to 0.60 eV, n remains constant at $\cong 1.05$;	37
	500	0.50	V_{BR} decreases from 38 to 16 V;	
	1000	0.50	totally leaky Schottky barrier diodes with $V_{BR} < 50$ mV	

controlled, isotropic etching results. Traditionally, plasma-assisted etching performed at relatively higher pressures, at lower powers, and with less associated ion bombardments on the wafer surface is referred to simply as plasma etching or PE.

154 *Gallium arsenide*

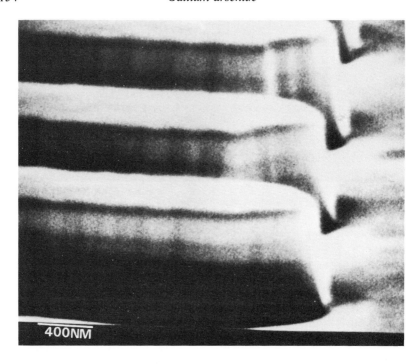

Etching and surface preparation of GaAs for device fabrication 155

When the lower electrode is powered, as in a parallel-plate sputter etching process, it becomes negatively biased. This draws positively charged chemically active ions away from the plasma, which at low enough pressures ($\lesssim 50$ m Torr) impinge in a perpendicular direction upon the wafer placed on the bottom electrode. Neutral radicals and excited molecules float freely around and impinge upon the surface in random directions as well. Because of the energy deposited on the surface by the impinging positive ions, surface reaction rate is enhanced considerably more at the bottom of the etched crater than on the side-walls, and anisotropic etching results: the bottom etches much faster than the side-walls, and nearly vertical etch profiles can be obtained. This situation is commonly referred to as reactive ion etching (RIE). Note that the etch mechanism for RIBE is very similar to that for RIE except that in RIBE neutral radicals and excited radicals and molecules do not take part in the etching process.

In PE and RIE, whenever halocarbons are used as an etching gas, two competing processes occur: polymerization of the halocarbon upon the surface, and surface disruption-induced removal of the polymer together with etching of GaAs. In many instances the nearly vertical side-wall is obtained because the polymerized deposits inhibit side-wall etching, where the ion flux-induced deposition of energy is very small. Complete polymerization and no etching were observed for halocarbons such as C_2Cl_4, CBr_2Cl_2, and $CHCl_3$. Even PH_3 was found to polymerize at 300–400 m Torr and high (1–5 W cm^{-2}) power densities.[51]

A number of pioneering research papers on the PE/RIE of GaAs are available in the literature that are useful for device fabrication. Some of these are summarized in Table 4.7.

4.5.3 Practical uses of dry etching processes

Unlike liquid etching, PE/RIE of GaAs is new to the world of device fabrication, and almost no data on the electrical properties of chlorine-based plasma and reactive ion etched GaAs are currently available. In comparison, the phenomenon of sputter etching has been known for a longer period of time and some useful applications can be found.[25,47]

Figure 4.14 GaAs vertical electron transistor (VET) at two stages during fabrication. The upper SEM picture shows the ohmic contact fingers over roughly 3000 Å GaAs after Ar ion beam etching (IBE) of regions unmasked by the Ti layer on top. Note enhanced GaAs etching at regions close to the side-walls caused by forward-scattered Ar ions off the wall. The lower micrograph shows the same fingers after wet etching of GaAs to remove damaged layer and also to create overhangs using NH_4OH–H_2O_2–H_2O (2:1:100) etching solution with etch rate of approximately 50 Å s^{-1}. This etchant is, of course, reaction-rate-limited and different GaAs undercut profiles are obtained with the fingers parallel to (01$\bar{1}$): trapezoidal; or to (011): nearly vertical. (After Mishra et al.[25])

Table 4.7 Plasma etching (PE) and reactive ion etching (RIE) studies of GaAs—a selective survey

Gas used	Type of system used	Etch conditions	Observations, etch rates, surface and electrical effects	Ref.
CCl_2F_2, CCl_4, PCl_3, HCl	parallel plate, PE	power supplied to both electrodes; 200–500 mτ	etched both GaAs and oxides; GaAs etch rate much higher than oxide etch rate	39
Cl_2, $COCl_2$	same	same	GaAs was etched, oxide was not; etch rate $\sim 5\,\mu$m min^{-1} at 5 W cm^{-2} for Cl_2, very rough surface	39
CCl_2F_2/Ar/O_2 (:1:1)	RF–RIE	0.5 W cm^{-2}; 5 mτ, 500 V self-bias; 10 mτ, 650 self-bias	at 5 mτ, 3800 Å min^{-1} etch rate; at 10 mτ, (0 Å min^{-1} etch rate; very sharp angle etch walls	40
Ar	DC–RIE	-3 kV on bottom; 40 mτ	400 Å min^{-1} etch rate	41
Ar/CCl_2F_2 (9:1)	same	25–60 mτ	500–2000 Å min^{-1} etch rate	
CCl_2F_2	same	10–40 mτ	2000–8000 Å min^{-1} etch rate; surface carbon build-up observed and ion bombardment helps remove it; rate-limiting factor is GaF_3 sputter removal	
CCl_2F_2/He (1:1)	RF–RIE	40 mτ, 0.18 W cm^{-2}	GaAs to AlGaAs etch rate = 200:1, with ~ 20 Å min^{-1} for AlGaAs; more surface degradation observed for lower He	42
CCl_2F_2/O_2 (1:1)	RF–RIE	0.1 mτ, 50 W (area unknown)	7 μm min^{-1} GaAs etch rate	43
CCl_3F/O_2	55 kHz –PE	300°C, 160 mτ	addition of O_2 binds carbon and releases Cl and F ions for GaAs etching; high O_2 concn lowers etch rate due to GaAs surface oxidation	45
Br_2	0.1–14 MHz	300 mτ, 30 SCCM Br_2, 100°C, $\lesssim 0.5$ W cm^{-2}	for < 0.3 W cm^{-2}, etch rates $\{111\}$ Ga $< \{110\} < \{100\} < \{111\}$ As; first demonstration of preferential crystallographic plasma etching of GaAs	46
CF_4 CHF_3	RF-RIE-	500 V self-bias	ϕ_B and V_{BR} increase, but new deep levels appear	37

mτ = mTorr.

Etching and surface preparation of GaAs for device fabrication 157

In spite of being a relatively new technique, PE/RIE have definite advantages over sputter etching. Since PE/RIE have large chemical etching components, it is possible to reduce the self-bias voltage and the power density so that the sputter component of etching can be reduced to a minimum. In this circumstance, subsurface damage would be very small. Consequently, only a simple surface treatment to remove the residual etch products would be adequate prior to subsequent processing steps. Also, as a result of a reduced sputter component of etching, the redeposition of foreign material upon the GaAs device wafer would be very small.

PE/RIE is being used for a number of high-speed electron devices, such as the GaAs power FET,[59] AlGaAs/GaAs high electron mobility transistor (HEMT),[54] the permeable base transistor (PBT),[60,61] where RIE is used to etch the tungsten grid as well as a small quantity of GaAs for better epitaxial overgrowth,[62] and for metal/AlGaAs/GaAs FET(MAGFET).[63] There is no doubt that its high resolution and good uniformity would make its use more universal in the near future.

4.6 CONCLUSIONS

We have attempted to describe the different etching and surface preparation processes that are available and pertinent to practical GaAs device fabrication processes. The subject is large and, owing to the number of diversely different physical or chemical process descriptions required to discuss even a single etching process, our descriptions had to be somewhat superficial. Also, the number of different wet etching processes that have been used, at one time or another, is much larger than those described here. This is because GaAs process technologies are beginning to mature in the way silicon technology did perhaps about 15 years ago. At present, there is a tendency towards using simpler, well characterized wet etchants that have H_2O_2 as the oxidizer. In contrast, plasma-assisted etching of GaAs is at its early stage of development. Because of this and also the fact that any two plasma etching machines behave considerably differently from one another, we have not suggested process recipes. More work is needed to understand the ways in which dry etching processes can be used successfully for device fabrication. Special emphasis has to be made in developing processes that include both dry and wet etching processes so that the vast knowledge available in the literature for good surface preparation steps using wet chemicals can be used.

ACKNOWLEDGEMENTS

The authors wish to acknowledge the help of Dr A. Chandra for his critical comments and many suggestions pertaining to the wet etching section. Thanks are also due to L. A. Camnitz, J. D. Berry, and Drs U. K. Mishra and E. Kohn for

stimulating discussions, to Mrs E. Weaver for typing the manuscript, and to Professor L. F. Eastman for his continued encouragement.
The work was supported by a JSEP grant from US–AFOSR under Contract number F49620-81-C-0082.

REFERENCES

1. B. D. Cullity, *Elements of X-ray Diffraction*, Reading, Mass., Addison Wesley, pp. 37–49 (1956).
2. W. Kern, Chemical etching of silicon, germanium, gallium arsenide and gallium phosphide, *RCA Rev.*, **39**, 278–308 (1978).
3. B. Tuck, The chemical polishing of semiconductors, *J. Mater. Sci.*, **10**, 321–39 (1975), and references therein.
4. R. Tijburg, Advances in etching of semiconductor devices, *Phys. Technol.*, Sept. 202–7, (1976).
5. D. J. Stirland and B. W. Straughan, A review of etching and defect characterization of gallium arsenide substrate material, *Thin Solid Films*, **31**, 139–70 (1976).
6. M. S. Abrahams and C. J. Buiocchi, Etching of dislocation on low-index faces of GaAs, *J. Appl. Phys.*, **36**, 2855–63 (1965).
7. J. Nishizawa, Y. Oyama, H. Tadano, K. Inokuchi, and Y. Okuno, Observations of defects in LPE GaAs revealed by new chemical etchant, *J. Cryst. Growth*, **47**, 434–6 (1979).
8. Y. Tarui, Y. Komiya, and T. Yamaguchi, Self-aligned GaAs Schottky barrier gate FET using preferential etching, *J. Jap. Soc. Appl. Phys.*, Suppl., **42**, 78–87 (1973).
9. M. Otsubo, T. Oda, H. Kumabe, and H. Miiki, Preferential etching of GaAs through photoresist masks, *J. Electrochem. Soc.*, **123**, 676–80 (1976).
10. Y. Mori and N. Watanabe, A new etching solution system, H_3PO_4–H_2O_2–H_2O, for GaAs and its kinetics, *J. Electrochem. Soc.*, **125**, 1510–14 (1978).
11. J. J. Gannon and C. J. Nuese, A chemical etchant for the selective removal of GaAs through SiO_2 masks, *J. Electrochem. Soc.*, **121**, 1215–19 (1974).
12. E. Kohn, A correlation between etch characteristics of GaAs etch solutions containing H_2O_2 and surface film structure, *J. Electrochem. Soc.*, **127**, 505–8 (1980).
13. S. P. Yenigalla and C. L. Ghosh, Fabrication of via holes in 200 μm thick GaAs wafer, *J. Electrochem. Soc.*, **130**, 1377 (1983).
14. S. Iida and K. Ito, Selective etching of GaAs crystals in H_2SO_4–H_2O_2–H_2O system, *J. Electrochem. Soc.*, **118**, 768–71 (1971).
15. D. W. Shaw, Localized GaAs etching with acidic peroxide solutions, *J. Electrochem. Soc.*, **128**, 874–80 (1981).
16. L. I. Greene, A new defect-revealing etch for GaAs, *J. Appl. Phys.*, **48**, 3739–41 (1977).
17. R. P. Tijburg and T. Van Dongen, Selective etching of III–V compounds with redox system, *J. Electrochem. Soc.*, **123**, 687–91 (1976).
18. K. Kenefick, Selective etching characteristics of peroxide/ammonium-hydroxide solutions for GaAs/$Al_{0.16}Ga_{0.89}$As, *J. Electrochem. Soc.*, **129**, 2380–2 (1982).
19. (a) L. Camnitz, Unpublished Work, Cornell University (1984). (b) L. Camnitz, Performance of electrically short 0.5 micron gate MESFET, *WOCSEMMAD*, 20–21 February, 1983, San Antonio.
20. W. W. Harvey and H. C. Gatos, The reaction of germanium with aqueous solutions, *J. Electrochem. Soc.*, **105**, 654–9 (1958).

Etching and surface preparation of GaAs for device fabrication   159

21. J. C. Phillips, *Bonds and Bands in Semiconductors*, New York, Academic Press, Chapter 2 (1973).
22. To convert kcal mol^{-1}, multiply by 0.0434 to obtain eV/molecule.
23. R. A. Logan and F. K. Reinhart, Optical waveguides in GaAs–AlGaAs epitaxial layers, *J. Appl. Phys.*, **44**, 4172–6 (1973).
24. I. Shiota, K. Motoya, T. Ohmi, N. Miyamoto, and J. Nishizawa, Auger characterization of chemically etched GaAs surfaces, *J. Electrochem. Soc.*, **124**, 155–7 (1977).
25. U. Mishra, E. Kohn, and L. F. Eastman, Submicron GaAs vertical electron transistor, *Proc. IEDM*, 594–7 (1982) and other reports pertaining to this work.
26. A. Shimano, H. Takagi, and G. Kano, Light controlled anodic oxidation of n-GaAs and its application to preparation of specified active layers for MESFETs, *IEEE Trans. Electron Devices*, **11**, 1690 (1979).
27. F. H. Doerbeck, Materials technology for X-band power GaAs FETs with uniform current characteristics, *GaAs and Related Compounds 1978*, Inst. Phys. Conf. Ser. 45, London and Bristol, Institute of Physics, p. 335 (1979).
28. W. C. Niehaus and B. Schwartz, A self limiting anodic etch-to-voltage (AETV) technique for fabrication of modified Read-Impatts, *Solid State Electronics*, **19**, 175 (1976).
29. D. L. Rode, B. Schwartz, and J. V. DiLorenzo, Electrolytic etching and electron mobility of GaAs for FETs, *Solid State Electronics*, **17**, 1119 (1974).
30. B. R. Pruniaux and A. C. Adams, Dependence of barrier height of metal semiconductor (Au–GaAs) on thickness of semiconductor surface layers, *J. Appl. Phys.*, **43**, 1980–3 (1972).
31. A. C. Adams and B. R. Pruniaux, Gallium arsenide surface film evaluation by ellipsometry and its effect on Schottky barriers, *J. Electrochem. Soc.*, **120**, 408–14 (1973).
32. C. M. Garner, C. Y. Su, W. A. Saperstein, K. G. Jew, C. S. Lee, G. L. Pearson, and W. E. Spicer, Effect of GaAs or $Ga_xAl_{1-x}As$ oxide composition on Schottky-barrier behavior, *J. Appl. Phys.*, **50**, 3376–82 (1979).
33. E. H. Nicollian, B. Schwartz, D. J. Coleman, R. M. Ryder, and J. R. Brews, Influence of a thin oxide layer between metal and semiconductor on Schottky diode behavior, *J. Vac. Sci. Technol.*, **13**, 1047–55 (1976).
34. D. W. Woodard, H. Morkoç, and A. Chandra, Unpublished Results, Cornell University (1976). (In this experiment, GaAs wafers held securely in a Teflon holder were immersed in a mixture of conc. HCl–methanol (1:1) for 5 min then dipped for 1 s each in two beakers containing deionized water and immediately blown dry in dry nitrogen. SIMS results indicated that atmospheric air containing moisture decomposed the surface chloride within 25 s and oxides formed. In vacuum or in dry N_2 flow the chloride remained for a longer period of time, about 10 min.)
35. A. Chandra, The growth, characterization and applications of high purity $Ga_{1-x}Al_xAs$, Ph.D. Thesis, Cornell University (1979).
36. CRC, *CRC Handbook of Chemistry and Physics*, 62nd edn, Cleveland, Ohio, Chemical Rubber Co. Press, pp. B79–102 (1981–2).
37. H. Adachi and H. L. Hartnagel, GaAs Schottky light emitters for the study of surface avalanching and electroluminescence, *J. Vac. Sci. Technol.*, **19**, 427–30 (1981).
38. C. M. Melliar-Smith and C. J. Mogab, Plasma assisted etching techniques for pattern delineation, *Thin Film Processes*, eds J. L. Vossen and W. Kern, New York, Academic Press, Chapter V-2, pp. 497–556 (1978).
39. B. Chapman, *Glow Discharge Processes—Sputtering and Plasma Etching*, New York, Wiley-Interscience (1980).
40. C. M. Melliar-Smith, Ion etching for pattern delineation, *J. Vac. Sci. Technol.*, **13**, 1008–22 (1976).

41. T. C. Tisone and P. D. Cruzan, Low-voltage triode sputtering with a confined plasma: Part V—Application to back sputter definition, *J. Vac. Sci. Technol.*, **12**, 677–88 (1975).
42. (a) P. D. Reader and H. R. Kaufman, Optimization of an electron-bombardment ion source for ion machining applications, *J. Vac. Sci. Technol.*, **12**, 1344–7 (1975), and references therein. (b) P. D. Reader, D. P. White, and G. C. Isaacson, Argon plasma bridge neutralizer operation with a 10-cm-beam-diameter ion etching source, *J. Vac. Sci. Technol.*, **15**, 1093–5 (1978).
43. R. E. Lee, Microfabrication by ion-beam etching, *J. Vac. Sci. Technol.*, **16**, 164–70 (1979).
44. CRC, *CRC Handbook of Chemistry and Physics*, 62nd edn, Cleveland, Ohio, Chemical Rubber Co. Press, pp. D170–1 (1981–2).
45. R. A. Powell, Reactive ion beam etching of GaAs in CCl_4, *Jap. J. Appl. Phys.*, **21**, L170–2 (1982).
46. J. D. Chinn, A. Fernandez, I. Adesida, and E. D. Wolf, Chemically assisted ion beam etching of GaAs, Ti and Mo, National Submicron Facility Report, Cornell University (1983).
47. C.-L. Chen and K. D. Wise, Gate formation in GaAs MESFETs using ion beam etching technology, *IEEE Trans. Electron Devices*, **ED-29**, 1522–9 (1982).
48. Y. Yamane, K. Yamasaki, and T. Mizutani, Annealing behavior of damage introduced in GaAs by reactive ion beam etching, *Jap. J. Appl. Phys.*, **21**, L537–8 (1982).
49. S. W. Pang, Dry etching induced damage in Si and GaAs, *Solid State Technol.*, April, 249–56 (1984), and references therein.
50. J. W. Coburn, *Plasma Etching and Reactive Ion Etching*, Am. Vac. Soc. Monograph Ser., New York, American Institute of Physics (1982).
51. G. Smolinsky, R. P. Chang, and T. M. Mayer, Plasma etching of III–V compound semiconductors and their materials, *J. Vac. Sci. Technol.*, **18**, 12–16 (1981).
52. E. L. Hu and R. E. Howard, Reactive-ion etching of GaAs and InP using $CCl_2F_2/Ar/O_2$, *Appl. Phys. Lett.*, **37**, 1022–4 (1980).
53. R. E. Klinger and J. E. Greene, Reactive ion etching of GaAs in CCl_2F_2, *Appl. Phys. Lett.*, **38**, 620–2 (1981).
54. K. Hikosaka, T. Mimura, and K. Joshin, Selective dry etching of AlGaAs–GaAs heterojunction, *Jap. J. Appl. Phys.*, **20**, L847–50 (1981).
55. C. B. Burstell, R. Y. Hung, and P. G. McMullin, Preferential etch scheme for GaAs–AlGaAs, *IBM Tech. Disclosure Bull.*, **20**, 2451 (1977).
56. R. H. Burton and G. Smolinsky, CCl_4 and Cl_2 plasma etching of III–V semiconductors and the role of added O_2, *J. Electrochem. Soc.*, **129**, 1599–604 (1982).
57. R. H. Burton, C. L. Hollien, L. Marchut, S. M. Abys, G. Smolinsky, and R. A. Gottscho, Etching of gallium arsenide and indium phosphide in rf discharges through mixtures of trichlorofluoromethane and oxygen, *J. Appl. Phys.*, **54**, 6663–71 (1983).
58. D. E. Ibbotson, D. L. Flamm, and V. M. Donnelly, Crystallographic etching of GaAs with bromine and chlorine plasmas, *J. Appl. Phys.*, **54**, 5974–81 (1983).
59. L. A. D'Asaro, A. D. Butherus, J. V. DiLorenzo, D. E. Iglesias, and S. H. Wemple, Plasma etched via connectors to GaAs FETs, Bell Laboratories Report (1981).
60. C. O. Bozler and G. D. Alley, The permeable base transistor and its application to logic circuits, *Proc. IEEE*, **70**, 46–52 (1982).
61. B. A. Vojak and J. P. Salerno, Transmission electron microscopy of GaAs grown over submicrometer-period tungsten gratings, *Appl. Phys. Lett.*, **41**, 1151–3 (1982).
62. R. Calawa, Private Communications, MIT Lincoln Laboratories. (1984).
63. P. M. Solomon, Private Communications, IBM Research Laboratories. (1984).

Gallium Arsenide
Edited by M. J. Howes and D. V. Morgan
© 1985 John Wiley & Sons Ltd

CHAPTER 5

Ion Implantation and Damage in GaAs

D. V. MORGAN and F. H. EISEN

5.1 INTRODUCTION

The technology for the fabrication of semiconductor devices and integrated circuits requires the ability to produce p- and n-type doped layers with precise and controlled three-dimensional geometry within the semiconductor. This can be achieved by thermal diffusion or ion implantation. This chapter is concerned with the latter of these techniques, namely, ion implantation doping, with particular reference to the III–V semiconductor gallium arsenide. With ion implantation doping, the desired impurities can be placed into the near-surface region of the solid by accelerating them to a high velocity and directing them as a beam onto the semiconductor surface. The ions are then able to penetrate the solid and are gradually brought to rest. The mass, velocity, and particle flux of the implanted ions are used to control the doping density and its depth distribution. Ion implantation doping offers a number of advantages in the fabrication of semiconductor devices. Among these are independent control of the doping level and the thickness of the doped layer, good uniformity and reproducibility of doping, the ability to achieve doping profiles not easily obtained by other techniques, and the possibility of selectively doping designated areas of semiconductor material with the aid of suitable masking techniques. Furthermore, ion implantation is more directional than diffusion, with a sideways scatter outside the masking shadow of the order of the straggling in the ion range. Clearly this offers great potential in the development of VLSI technology.

Ion implantation is now an established production technology,[1–5] and in the case of silicon the advantages described above are sufficient to justify the large capital investment necessary for commercial exploitation. These advantages are of even greater importance in gallium arsenide, for which a diffusion doping technology has not been developed.

5.2 ION–CRYSTAL INTERACTIONS

5.2.1 Ion ranges

When an energetic ion enters a solid, it loses energy by collisions with the nuclei and electrons of the target atom, until it is finally brought to rest. If the target material is amorphous, then the stopping process for any particular ion of an incident monoenergetic beam will be random and the distribution of implanted ions can be shown to have an approximately Gaussian or normal shape, characterized by an average projected range $\bar{R}_p$ and its standard deviation $\overline{\Delta R_p}$ (Figure 5.1). The most widely used theory for computing the ranges of ions in solids is that of Lindhard et al.[6, 7] Extensive tables of average projected range and its standard deviation for ions implanted into GaAs are readily available for device applications.[8]

In crystalline targets the distribution of implanted ions is dependent on the orientation of the substrate during the implantation. If the incident beam is

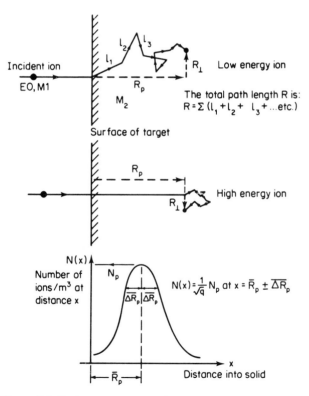

Figure 5.1 Basic range concepts for low- and high-energy ions[1]

Ion implantation and damage in GaAs 163

aligned with one of the major crystal axes or planes, then the phenomenon of channelling may take place[1-5] whereby some of the implanted ions penetrate to depths far greater than predicted for amorphous targets. Hence the final distribution of ions is characterized by a deeply extending *tail* with a possible secondary build-up of ions. This problem is generally minimized by misaligning the target crystal so that the ion beam is incident in a non-channelling direction, so that the distribution of the implanted species is similar to that in an amorphous target. However, channelling cannot be avoided altogether owing to some ions being scattered trajectories between into the crystal planes, resulting in tails to otherwise Gaussian distributions.

5.2.2 Ion-induced damage and annealing

During the implantation of crystalline materials, target atoms may be displaced from their lattice sites as a result of collisions with incident ions and, if the energy and mass of the ion beam species is great enough, cascades of displacements and hence zones of amorphization may be produced.[9] For large doses, these zones may overlap, and the implanted layer of the crystalline material is rendered into a state of total disorder. If the doping of semiconductors by implantation is to be a successful method of device fabrication, the lattice damage has to be removed and this is achieved by thermal annealing. For GaAs it is found that annealing at temperatures up to and above 900°C is necessary to minimize sufficiently the number of defects, which otherwise would mask the effect of doping atoms, although it has been found that an annealing stage exists at around 150–200°C.[10-12] It has been observed that, by implanting at temperatures above 150°C, no amorphous layers are formed even for high doses. The radiation damage produced during the room-temperature implantation is also found to depend on the dose and dose rate. Initially, the damage increases with dose; then, for sufficiently large beam currents and doses, the energy dissipated in the GaAs substrate may be large enough for the first thermal annealing stage (150–200°C) to be reached and thus avoid the formation of amorphous layers.[13] However, despite the high-temperature annealing of 'hot' implanted GaAs, certain carrier compensating defects remain, which may be in the form of Ga vacancy–donor atom complexes which have been detected by photoluminescence[14] and cathodoluminescence measurements.[15]

5.2.3 Surface passivation and annealing of GaAs

The high-temperature annealing of GaAs to remove radiation damage has been one of the main obstacles in advancing the use of ion implantation technology. However, the annealing process has not impeded the application of ion implantation in silicon. The difficulty with GaAs lies in the fact that above about 600°C the surface of GaAs crystals dissociates rapidly with As evaporating at a far

164   *Gallium arsenide*

greater rate than Ga.[16] Three techniques have been developed to preserve the GaAs surface during the annealing process, each of which has some disadvantages (Figure 5.2):

(a) passivation using a dielectric encapsulating film (i.e. a cap);[17-19]
(b) heating in an arsenic overpressure;
(c) transient annealing using a laser or scanned electron beam (this technique is sometimes carried out with a cap).

Initial work on GaAs used pyrolytically deposited SiO_2[17] and, although this encapsulant withstood high-temperature annealing, it was found that gallium

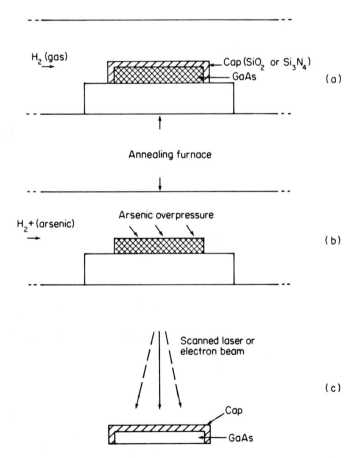

Figure 5.2 A schematic diagram illustrating the range of surface passivation schemes for annealing ion-implanted GaAs

Ion implantation and damage in GaAs

tended to outdiffuse leaving a non-stoichiometric GaAs surface.[18] The most widely used encapsulant has been Si_3N_4 which, when plasma deposited, was found to be superior to SiO_2 for the passivation of Te-implanted GaAs.[19] Other techniques of depositing Si_3N_4 that have been employed are chemical vapour deposition (CVD), RF sputtering, and reactive sputtering. The latter of these, depending on the deposition process, has produced dielectrics of varying quality and composition and has occasionally led to blistering and loss of adhesion of Si_3N_4 films.[20] Nevertheless, Si_3N_4 encapsulation has permitted the successful annealing of implanted GaAs to temperatures as high as 950°C[21] in a conventional furnace and 1000°C for transient annealing. Another encapsulant that has been found to be suitable for the passivation of GaAs surfaces is RF sputter deposited oxygen-rich AlN.[22] Better activation of high doses of n-type dopants have been observed when this encapsulant was used than when Si_3N_4 deposited by the same technique was employed. The difference in behaviour of the two caps may be associated with the magnitudes of the expansion coefficients; there is a large thermal mismatch between Si_3N_4 and GaAs but very little difference between the thermal expansion coefficients of AlN and GaAs.[23] Nishi et al.[24] have recently reported on the use of AlN with very little oxygen contamination as an encapsulant. They suggest that such AlN may be very promising for integrated circuit applications.

Aluminium encapsulation of implanted GaAs has also been shown to prevent thermal dissociation,[25] although apart from pulse annealing[26] it is restricted to temperatures up to 700–750°C above which considerable Al indiffusion may occur. Since aluminium melts at 660°C, the metal acts as a liquid encapsulant, thus avoiding the problem of adhesion, blistering or cracking occasionally encountered with dielectrics.

The annealing procedures described above all involve holding the sample at the required temperature for around 15 min. An alternative is to use a transient process.[27] Here the sample is raised to a higher temperature (up to 1000°C) for a very short time. Incoherent light beams,[28, 29] multiply scanned electron beams,[30] and graphite strip heaters[31] have been used to carry out transient annealing. There is as yet limited data on the application of these techniques to GaAs, but there is some indication that it may be possible to obtain better results than by conventional thermal annealing. A possible problem is that of ensuring that the implanted layer temperature is able to rise to the substrate temperature in the short annealing time interval employed.

Another passivation technique that has recently been found to be suitable for GaAs at temperatures as high as 900°C is that of annealing in an arsenic vapour overpressure sufficiently great to prevent As evaporation.[32-34] The attraction of this method is that since there is no encapsulation involved, no indiffusion of unwanted impurities takes place, the loss of dopant atoms by outdiffusion is less likely, and the whole annealing process is considerably simplified. In practice, this technique yields variable results owing to the lack of control of the actual overpressure near the sample surface. This situation can be improved by the use of a system with a controlled AsH_3 overpressure,[35] but at present only a limited effort has occurred in this direction.

Alternative annealing techniques which are at present under study are those of electron beam annealing[36] and laser annealing.[37, 38] These two techniques are very successful in silicon and are now beginning to yield good results for GaAs. Laser annealing, in particular, is producing very good results in the high-dose regime where capped techniques are inadequate. These points will be taken up later in this chapter.

Diffusion of implanted impurity atoms during high-temperature annealing has been found to be the main process for modifying their theoretical Gaussian distribution, although some enhanced diffusion may actually take place during the implantation. This latter diffusion may be a result of a substitutional process via radiation damage-induced vacancies and would be more evident during the subsequent high-temperature anneal. For simple diffusion, where the diffusion coefficient is a constant and the encapsulant is considered to be a continuation of the GaAs substrate, annealing will result in a broader and lower Gaussian distribution. The peak of the new profile will still remain at the same depth but the new standard deviations will be broader. The substrate–encapsulant boundary condition is an important parameter which, depending on the film used, has a considerable influence on the final distribution of implanted ions. Most encapsulants behave as barriers to outdiffusion, leading to a build-up of impurity atoms at the substrate surface where it is possible for some to act as sinks. Lattice defects may also play an important role in the diffusion mechanisms of impurity atoms. It may be possible to minimize diffusion effects by employing transient annealing techniques.

5.3 RADIATION-INDUCED DAMAGE AND SEMI-INSULATING LAYERS

In the preceding discussion, radiation-induced defects were thought of as one of the undesirable problems associated with the technique of ion implantation, which had to be removed to activate the desired doping effect. Consequently, much of the early research was directed to the problem of determining the optimum implantation conditions and annealing sequence needed to remove the damage and thus to allow the implanted atoms to dope the semiconductor. In GaAs, radiation-induced defects are observed to produce effective compensating centres, stable at room temperature, which convert doped GaAs into a semi-insulating form.[39-44] Some idea of the significance of this problem may be gained when one considers that for light ions such as B^+ implanted into GaAs at 1 MeV, 200 carriers are removed per implanted ion.[45] Thus when implanting dopant ions, even if all the implanted ions become electrically active, one residual defect per implanted ion (i.e. 0.5 %) is sufficient to compensate all of the doping effect. It was

Ion implantation and damage in GaAs

appreciated early on that this damage technique could be valuable for device isolation applications and could thus be used in device technology.

During implantation, many different types of defects occur in damage clusters around the path of the implanted particle, for example, vacancies, interstitial atoms, substitutional atoms, and combinations of these with impurities. The exact class of defect or defect complex to be found depends on many factors, including the mass and energy of the incident particles and the mass of the target atom. Theoretical studies of the dependence of the damage profiles upon the incident ion energy and atomic number show good agreement with measured profiles for a wide range of projectiles and targets.[46] An important feature to note is that the damage profile does not in general coincide with the implanted ion profile, the degree of disparity between these two being a function of the mass difference between the bombarding on and the lattice atoms. For light ions such as protons in GaAs, the two profiles show close correlation. Figure 5.3 shows the theoretical damage and ion density distributions for 60 keV boron ions incident on silicon in a non-channelled direction.[46] For comparison, Figure 5.4 shows a carrier concentration profile created by proton bombardment in GaAs. The proton range is plotted as a function of energy in Figure 5.5. In order to obtain uniform semi-insulating layers using proton bombardment, it is necessary to

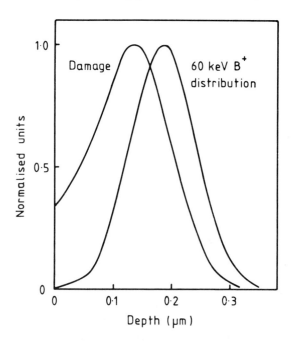

Figure 5.3 Theoretical damage and ion density distributions for 60 keV B$^+$ ions incident on 'amorphous' silicon[46]

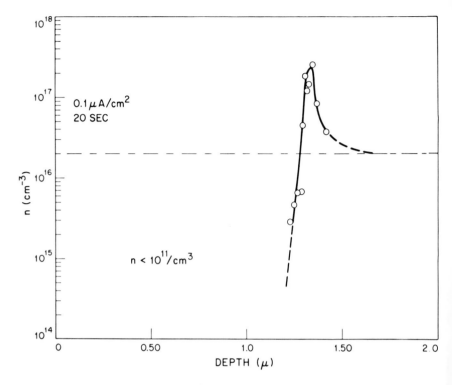

Figure 5.4 Carrier concentration profile created by room-temperature proton bombardment of n-GaAs ($n \sim 2 \times 10^{16}$ cm^{-3}) with a 100 keV beam of dose 10^{13} cm^{-2} [39]

conduct a sequence of implantations with various energies in much the same way as that used for implantation doping.[47]

One very interesting effect of the ion damage is to reduce the near-surface sheet resistivity of semi-insulating material (Figure 5.6). This curve, measured with a four-point probe, suggests that the near-surface region begins to conduct and the value of sheet resistance falls from its initial 10^7 Ω/□ to a final saturated value of 2×10^4 Ω/□.

The thermal stability of the damage may be considered initially by reference to electron and neutron bombardment. Whereas the stability of damage produced by proton bombardment of GaAs is not fully understood, the annealing of material damaged by electrons[49–55] and neutrons[54,55] has been investigated more fully by many authors. The electron damage in GaAs has been found to anneal out in two stages,[50,53,55] one at 150–200°C and the other at 200–300°C, although the lower stage is often not resolved. Electron damage generally shows complete recovery above 300°C for a moderate flux of particles ($< 10^{16}$ cm^{-2}) and above 600°C for the higher fluences. This is consistent with the view that the damage consists initially of simple isolated defect structures. Conversely, the

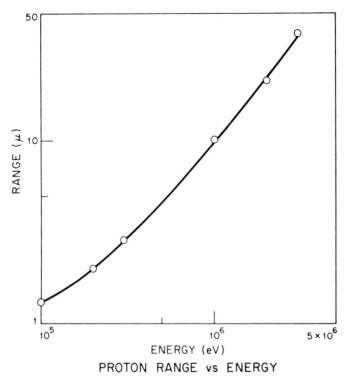

Figure 5.5 A plot of proton range versus energy[39]

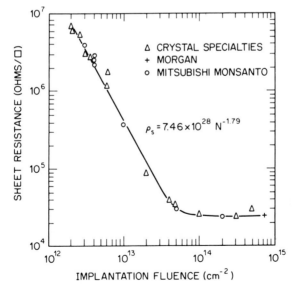

Figure 5.6 The sheet resistance of a semi-insulating GaAs crystal plotted as a function of the silicon implant fluence[48]

Gallium arsenide

more complex damage induced by neutron bombardment shows only 10% recovery below 300°C, but additional annealing stages are present at 450°C and 600–700°C. Studies of the damage produced by ion implantation have revealed annealing stages similar to those observed for the neutron-induced damage,[56] the 200–300°C and 600–700°C stages being observed most consistently. Thus, if stable semi-insulating layers are required for device applications, the heavier ions are preferable to electrons. Indeed, electron damage could only be considered at the high fluence range of the spectrum. Electrons, however, with their deep penetration range, offer features that would complement the shallower layers more typical of heavy ions, although it may be difficult to develop suitable masking techniques. Further studies on the stability of electron damage would be particularly valuable in assessing its full potential for device work.

5.3.1 Device applications

In the case of GaAs, damage-induced semi-insulating layers have found a number of device applications, including both active and passive components. Proton damage, for example, has been used to improve the efficiency of GaAs IMPATT diodes as compared with conventional mesa structures.[57, 58] In this case the fully isolated IMPATT diodes (both p^+–n and Schottky diode) have been fabricated from relatively thick (15 μm) GaAs layers. The proton-isolated diodes were superior to mesa structures fabricated from the same material. The diode structure used is illustrated schematically in Figure 5.7. The feasibility of using proton bombardment to fabricate a semi-insulated gate FET has been demonstrated, but no high-frequency results have been reported.[59] Another good example of the use of proton bombardment is in the formation of stripe geometry heterostructure junction lasers.[43] A schematic drawing of a double heterostructure laser formed by proton bombardment is shown in Figure 5.8. The devices produced in this manner yield more reproducible mode patterns and lower threshold currents than structures using oxide insulation. Proton damage has also yielded significant results in the production of optical waveguide components.[60, 61] These include simple waveguide structures, directional couplers, and waveguide detectors. The waveguiding properties arise from the fact that damaged regions have a smaller free carrier concentration than the substrate. Thus the smaller plasma contribution to the dielectric constant of the damaged region ensures that its dielectric constant exceeds that of the substrate. As a consequence of this effect, the implanted stripe regions can be used to guide optical signals and thus to produce a range of waveguide components. With regard to the waveguide detectors, the defect levels give rise to optical absorption at wavelengths greater than the nominal 0.9 μm edge. Thus an implanted region placed inside the depletion region of a Schottky diode can be used to produce an effective photodetector at these long wavelengths.[62]

At the present time there is considerable interest in the development of

Ion implantation and damage in GaAs

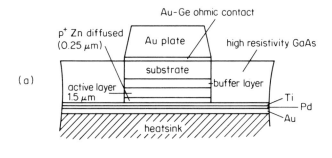

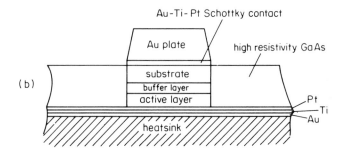

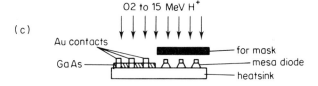

Figure 5.7 Cross sectional view of (a) planar passivated diffused junction GaAs IMPATT, (b) planar passivated Schottky barrier IMPATT, and (c) the planar mesa sample[58]

monolithic GaAs circuits. Proton isolation is one procedure that may be used for isolating component devices in such structures and the technique is anticipated to be of considerable use for future development in this area. Recent application of proton isolation to GaAs integrated circuits has resulted in significant improvement in isolation between devices and reduction in back-gating effects.[63]

5.4 ION IMPLANTATION DOPING OF GaAs

The most important application of ion implantation to date is in the p- or n-type doping of the material in order to fabricate semiconductor devices and integrated

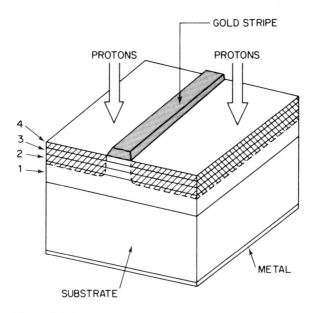

Figure 5.8 Schematic diagram of a stripe geometry double heterostructure laser with resistive layers formed by proton bombardment. The epitaxial layers are (1) n-AlGaAs, (2) p-GaAs, (3) p-AlGaAs, and (4) p-GaAs. The substrate in GaAs and the highly resistive regions produced by proton bombardment are indicated by the cross hatching[43]

circuits. In this context annealing must be carried out in order to minimize the radiation damage and to promote the dopant atoms to occupy a substitutional lattice site as was discussed in Section 5.2.

5.4.1 Implanted semi-insulating layers

Semi-insulating layers in GaAs similar to those produced by ion damage can also be produced by the introduction of deep-lying levels arising from the implantation of ions such as oxygen.[64-66] Oxygen-implanted samples are usually annealed at 650–800°C, whereby most of the radiation damage is removed leaving deep-lying donor levels situated 0.63 eV below the conduction band edge[67] and associated with the implanted oxygen atoms. Favennec et al.[64, 65] found that, after annealing n-type GaAs at 800°C, two electrons were removed per implanted oxygen atom, indicating the existence of a double trap. In contradiction Gecim et al.[66] found that the number of carriers removed per O^+ ion implanted into semi-insulating n-GaAs was very much dose-dependent with no significant electrical compensation taking place for doses less than 10^{13} O^+ ions/cm^2 at an energy of 400 keV.

5.4.2 N-type doping

N-type doping of GaAs by ion implantation can be achieved in two ways.[67-85] Dopant ions from either column VI elements (Se, Te, S) or column IV elements (Si, Sn) may be used. For the implanted ions to become electrically active they must occupy substitutional lattice sites, but since GaAs is a binary compound this introduces the extra complication of placing the implanted ions on the correct sublattice. It has been found that for high dose ($\gtrsim 10^{14}$ ions/cm^2) the implantation of the column VI ions mentioned above must be carried out into heated substrates ($\geqslant 150°$C) in order for the dopants to be substitutionally located following post-implantation annealing. In comparison with implanting GaAs at room temperature, 'hot' high-dose implantations result in considerably higher doping efficiencies, as illustrated in Figure 5.9 for 400 keV Se$^+$ implantation into GaAs.[44] In contrast to the situation for such high-dose implants, the activation of low doses ($\lesssim 5 \times 10^{12}$ ions/cm^2) is not dependent on the temperature of the substrate during implantation. Column IV impurity atoms, however, behave in a

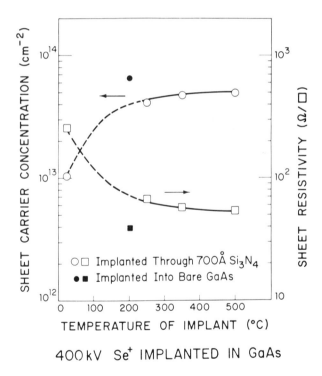

Figure 5.9 Sheet carrier concentration and sheet resistivity against temperature of implant for 1×10^{14} cm^{-2} 400 keV Se$^+$ ion-implanted GaAs annealed at 900°C for 15 min with pyrolytic Si$_3$N$_4$ encapsulation[68]

different manner since in order to act as donor atoms they must occupy gallium sites. Implantation of Si^+ at elevated temperatures has been found to result in a smaller increase in doping efficiency than is observed for column VI dopants[15, 21] However, Sn was found to behave like group VI ions, showing substantial increases in electrical activity after 'hot' implantation.[15]

The doping efficiency of n-type impurities is found to be a function of not only the substrate temperature during implantation and the subsequent anneal temperature (Figure 5.10) but also the energy and dose. Low doses can lead to a high proportion of the implanted impurities becoming electrically active, whereas increasing the dose results in a general decrease in doping efficiency (Figure 5.11). This decreases in doping efficiency as the dose and hence the impurity concentration is increased is attributed to carrier compensating residual damage and defects remaining after high-temperature annealing rather than solubility limits and the effects of degeneracy.[68]

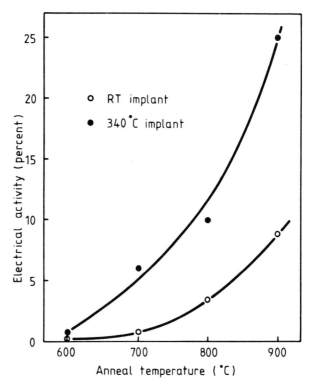

Figure 5.10 Percentage electrical activity against anneal temperature for a 1×10^{15} cm^{-2} dose of 50 keV Si^+ ions implanted into Cr-doped SI GaAs. The wafers were annealed with Si_3N_4 encapsulant[71]

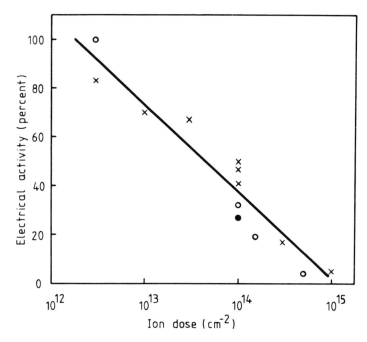

Figure 5.11 Percentage electrical activity against dose for 400 keV Se^+ ions implanted into Cr-doped SI GaAs at 350°C. The wafers were annealed at 900°C for 10–15 min with Si_3N_4 encapsulation: ×, ref. 21; •, ref. 73; o, ref. 81

The resulting carrier concentration profile after thermal annealing is also strongly dependent on the diffusion behaviour of implanted ions. Tellurium-implanted layers have been found to be stable, thus indicating a low value of diffusion coefficient, whereas implanted sulphur profiles have shown considerable broadening during the annealing process.[69] While the $C-V$ profiles for Si and Se are close to the calculated values, those for S show a maximum electron concentration approximately an order of magnitude below that of theory. The difference in behaviour is illustrated by the sequence of curves shown in Figure 5.12. From these results we see that the implanted sulphur atoms diffuse deeply into the GaAs during annealing. It appears that the comparatively high level of diffusion exhibited by sulphur atoms could be responsible for the wide variation in reported electrical activities. Radioactive S^+ tracer techniques have also shown considerable outdiffusion into a surface layer between the GaAs substrate and the Si_3N_4 encapsulant, the actual amount being dependent on the depth of the implant. This implies that the method of surface treatment and the subsequent encapsulant deposition process are critical factors in the outdiffusion of the implanted sulphur.

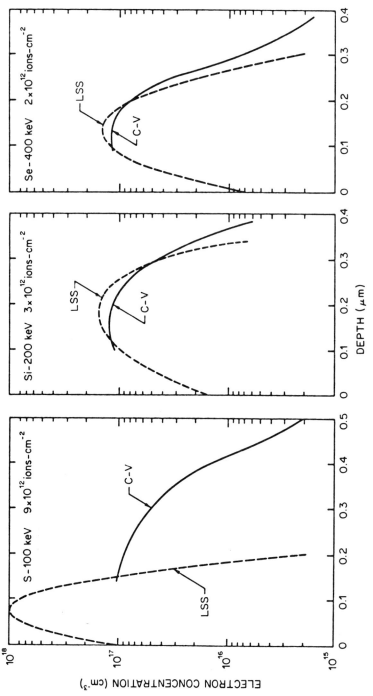

Figure 5.12 Comparison of electron concentration profiles derived from C–V measurements for S,[110] Si, and Se[110] implanted in GaAs with the indicated dose and energy with estimates of the profile of the implanted ions derived from LSS range parameters.[6,8] (Silicon C–V data from A. A. Immorlica Jr, private communication)

Ion implantation and damage in GaAs

In order to fabricate relatively low-cost digital integrated circuits in GaAs, ion implantation into bulk-grown chromium-doped semi-insulating substrates has been investigated. This technology has shown some considerable success but it is affected by the redistribution of the chromium that occurs during thermal annealing. The consequence of this redistribution, which results in a deficiency of chromium near the surface, is to form a surface n-type layer when the background donor density exceeds the chromium density. One way of minimizing the effects of this distribution may be to anneal in an arsenic overpressure with a dielectric cap. Under these conditions the decrease in chromium concentration at the GaAs surface is lower than with no As overpressure.[70] In addition, it has been possible to grow undoped semi-insulating GaAs that is not chromium-doped and exhibits less tendency to surface conversion.

In the results published thus far, the capped and arsenic overpressure annealing techniques produce poor results in terms of doping efficiencies for ion doses greater than 10^{14} cm^{-2}. Some care must be exercised in comparing the actual dose levels, since the implanted-ion concentration and damage density depend on the implant energy as well as the dose. For implant doses above this level, the doping level begins to saturate at a value of $(5-8) \times 10^{18}$ cm^{-3}. Figure 5.13 summarizes some recent results[48] that illustrate this saturation process.

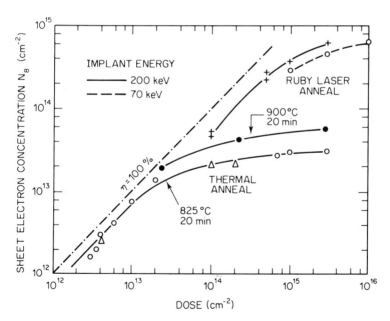

Figure 5.13 The sheet electron concentration of silicon-implanted samples plotted as a function of implantation dose. The lower curves are for the thermally annealed samples and the upper curves are for those that were laser annealed[48]

In view of the difficulties involved in removing radiation damage from high-dose implanted layers, two alternative methods that have recently been investigated are pulsed laser beam annealing[37, 86-90] and pulsed electron beam annealing.[75, 91] Pulsed laser beam annealing consists of applying a beam of light, usually supplied by a Q-switched ruby laser, to the implanted layer for a very short period of time ($\sim$ 20 ns). It has been found that there exists a threshold energy which must be applied before the as-implanted amorphous region is converted into a single-crystal state. Below this threshold, the implanted GaAs layer remains in a polycrystalline state, the grain size increasing as the threshold energy is approached. This pulse energy required before total recrystallization is realized is dependent on the depth of the amorphous layer. As an example, for a 1×10^{15} cm^{-2} dose of 400 keV tellurium ions implanted into semi-insulating GaAs at room temperature, which produced an amorphous region approximately 2300 Å thick, the threshold energy density was 1 J cm^{-2}. After total recrystallization, Rutherford back-scattering and channelling measurements showed that the implanted species had almost entirely occupied substitutional lattice sites.

The reported work on laser annealing has so far gained its main success in the region of high-dose implants (i.e. $> 10^{14}$ cm^{-2}) where conventional annealing does not result in high doping efficiency. This is shown by the results plotted in Figure 5.13. As with the thermal annealing, the curve again saturates at high doses but still yields acceptable doping efficiencies for doses 20 times greater than thermal annealing. These sheet carrier concentrations correspond to doping levels of around 5×10^{19} cm^{-3}, which takes the doping level into those required for ohmic contact formation without alloying. In these high doped levels the electron mobility and doping efficiencies were found to be lower than expected theoretically. One other important problem with laser annealing is that the doping efficiencies again fall off at low doses, suggesting that the process leaves undesirable levels of residual defects in the material. The technique therefore has not yet yielded results for GaAs comparable to those for silicon.

Isothermal electron beam annealing has, as yet, not been studied in such detail. There are, however, very encouraging results beginning to appear.[75, 91, 92] The technique has produced doping efficiencies of 50 to 60% for low-dose silicon implants ($\sim 6 \times 10^{12}$ cm^{-2}), with comparable results for selenium. These early results show some future promise for this technique. In the high-dose region, the results are comparable to the best reported laser work.[48]

5.4.3 P-type doping

Initial investigations of p-type doping of GaAs by ion implantation were found to be successful, with high values of doping efficiency ($\sim 100\%$) being reported for doses up to 10^{14}ions/cm^2.[78, 93-95] For larger doses, a saturation effect was observed with a decrease in doping efficiency, which was generally attributed to solubility limits. High values of electrical activation were obtained for anneals at about 800°C mainly using pyrolytically or sputter deposited SiO_2 as the encapsulant. P-type doping of GaAs usually requires the implantation of group III ions, and the following species have been successfully activated: Zn, Be, Cd, Mg.[93-109] In addition, the activation of carbon (group IV) has been reported. Figure 5.14 shows the percentage of 60 keV Zn^+ ions implanted into n-type GaAs and electrically activated after a 800°C 10 min anneal as a function of dose. The encapsulation used was 2000 Å thick SiO_2. From these results, it can be seen that increasing the substrate temperature to 400°C during the implantation has no effect at low doses and only marginally increases the doping efficiency at high doses. For p-type impurities the optimum anneal temperatures is found to be about 800°C, although annealing at 600°C has produced p-type layers.[78, 94, 96]

In general the early investigations of p-type doping did not include measurement of hole concentration and mobility profiles, and so little was known about doping levels and diffusion. More recent work on the implantation of Be^+ ions involving SIMS measurements showed that for a beam energy of 250 keV and doses above 6×10^{14} ions/cm^2, significant redistribution of implanted impurities took place during annealing with a build-up of Be atoms at the surface.[97, 98]

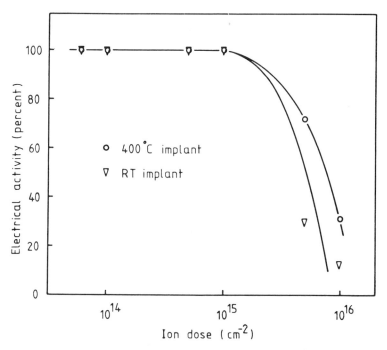

Figure 5.14 Percentage electrical activity against dose for 60 keV Zn^+ ion-implanted GaAs annealed at 800°C for 10 min with SiO_2 encapsulation[95]

Implantation of Zn has also been shown to produce high hole concentrations with peak values in the range of $(1-2) \times 10^{19}$ cm^{-3} and corresponding mobilities of 80–90 cm^2 V^{-1}s^{-1}.[99] Cd implantation has produced similar peak hole concentrations,[100] although in general there is still only limited information on profiles for these two acceptor impurities.

Pulsed laser beam annealing has also been applied to high-dose p-type implants in GaAs.[38, 109] Peak carrier concentrations of $(3-7) \times 10^{19}$ cm^{-3} and corresponding mobilities of 40–80 cm^2 V^{-1}s^{-1} were reported by Kular et al.[38] for zinc-implanted GaAs using an Si$_3$N$_4$ encapsulant. Similar results were obtained by Davies et al.[28] for zinc implantation annealed for ~1 s using light from incandescent lamps.

5.5 ION-IMPLANTED SEMICONDUCTOR DEVICES AND INTEGRATED CIRCUITS

Ion implantation has been used in fabricating a wide variety of devices in GaAs. Some examples are field-effect transistors (FETs)[110–114] and integrated circuits employing such transistors,[115–117] solar cells,[118] varactor diodes,[119, 120] transferred electron devices, IMPATT diodes,[121] and bipolar transistors.[122] The widest application of implantation in GaAs has occurred in the development of GaAs integrated circuits and this topic will be discussed later in this section. Consider first the application to the fabrication of discrete devices.

5.5.1 Discrete devices

Ion implantation doping offers considerable advantages when used in the fabrication of transferred electron devices (TEDs). For high-frequency (> 10 GHz) devices, shallow hyperabrupt n$^+$-doped contact layers are required to reduce thermal resistance and microwave losses. Using 116 keV S$^+$ implantation, Lee et al.[123] produced n$^+$ contact layers less than 1 μm thick in K$_a$-band GaAs Gunn effect diodes. In this case n-type epitaxial material was implanted with a 2.3×10^{13} cm^{-2} dose of S$^+$ ions and annealed at 800°C for 20 min with an SiO$_2$ encapsulation. The resultant near-surface carrier concentration profile (Figure 5.15) has a peak of 8×10^{17} cm^{-3} and was characterized by a tail extending further into the active layer than predicted by ion range parameters.[8] In CW operation, oscillators incorporating these mesa structured devices exhibited a maximum DC to RF conversion efficiency of 3.1% at 35.1 GHz. This result is comparable with results of state-of-the-art transferred electron oscillators.

The ability to implant discrete n-type layers in semi-insulating material is of particular value in the fabrication of planar TEDs. Since no mesa etch isolation is required, device structures are attainable that allow Schottky barrier gate lengths of less than 1 μm to be used for operation above 20 GHz with improved Gunn domain trigger sensitivity. The major problem encountered with this fabrication

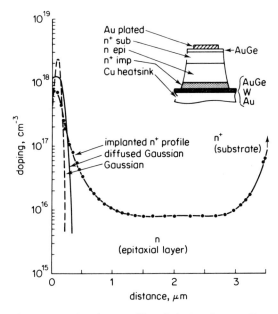

Figure 5.15 Doping profile of GaAs Gunn effect diode with shallow (< 1 μm) sulphur-implanted n⁺ contacts[123]

technology is the ability to achieve low net donor concentrations reproducibly in semi-insulating GaAs. Anderson et al.[124] found that only bulk (liquid encapsulation Czochralski technique) or epitaxial undoped semi-insulating material was suitable for activation of multiple energy Si⁺ implants. Little or no doping was obtained for implanted Cr- or O-doped semi-insulating material. This result is directly attributed to excessive concentrations of compensating impurities. Transferred electron devices were produced by implanting Si⁺ ions at energies of 40, 120, 300, and 600 keV with respective doses of 0.23×10^{12}, 0.62×10^{12}, 1.3×10^{12}, and 2.3×10^{12} cm^{-2} to form a 6500 Å thick active layer. The implanted GaAs was annealed at 800°C for 30 min using Si_3N_4 passivation. The net theoretical distribution of implanted impurities[8] and a typical measured carrier concentration profile are shown in Figure 5.16. Three terminal TEDs fabricated on the activated Si-implanted layers had gate-to-anode lengths of 20–100 μm with Schottky barrier gate lengths between 2 and 5 μm. Such planar device structures exhibited negative resistance in the range 2–7 GHz with Gunn domain triggering by the gate producing DC negative resistance current dropbacks between 20 and 40%.

S⁺ implantation has also been used to fabricate GaAs planar Gunn effect digital devices.[125, 126] Here, the relatively high mobility of this species in GaAs at

Gallium arsenide

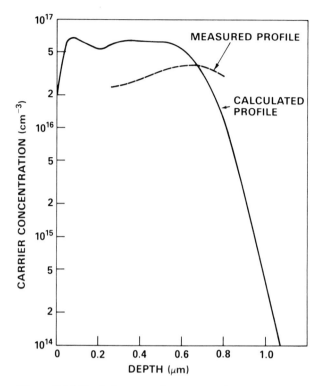

Figure 5.16 Typical measured average carrier concentration profile and theoretical distribution of Si ions implanted into undoped bulk semi-insulating GaAs[124] (see text for details)

high anneal temperatures was used to form diffused active layers 0.6–1.1 μm in depth with average carrier concentrations of about 4×10^{16} cm^{-3}.

5.5.1.1 *IMPATT diodes*

Further applications of ion implantation doping have been demonstrated in the fabrication of IMPATT diodes.[121, 127] Bozler et al.[121] used Si$^+$ implantation to produce the hi-region of lo-hi-lo GaAs Schottky barrier IMPATT diodes, resulting in devices with very uniform electrical characteristics. Doses of (3.2–3.6) × 10^{12} 250 keV Si$^+$ ions/cm^2 were implanted at room temperature into n-type epitaxial GaAs grown on an n$^+$ substrate. Annealing the implanted wafers at 850°C for 15 min with Si$_3$N$_4$ passivation produced hi-regions which had peak carrier concentrations of about 10^{17} cm^{-3} but were deeper and broader than predicted by LSS theory. In comparison with epitaxial growth techniques, ion implantation doping was found to give significantly higher yields of devices having oscillator DC to RF conversion efficiencies greater than 30%. GaAs Read-type IMPATT diodes with implanted hi-regions showed similar improvement over epitaxially grown devices.[127] Oscillator CW output powers of 1.1 W with 25% conversion efficiencies at 11 GHz were reported.

5.5.1.2 P–N junctions

The properties of p–n junctions have been reported by several authors.[122, 128] Junctions were formed by the implantation of Be into an n-type epitaxial layer[128] or by implantation of both Be and Se.[122] The diodes were found to exhibit an ideality factor of 2 and to have low leakage currents. The use of implantation to form p–n junctions is potentially of interest in the fabrication of such GaAs devices as IMPATT diodes and solar cells. Work has been done on the fabrication of bipolar transistors in GaAs using both Se and Be implantation.[122] Low current gains were observed in these devices. This effect is thought to be due to an electron lifetime in the base region of approximately 10^{-10} s, which is not inconsistent with the lifetime measured in bulk GaAs. More recent work using Si and Be implantation has resulted in current gains of about 100 for buried-emitter heterojunction devices.[129]

Planar hyperabrupt varactor diodes have been fabricated in GaAs using a Be implant to form the p–n junction and S implantation into a low-doped n-type epitaxial layer to obtain an n-type doping profile which would give hyperabrupt diode characteristics.[119] Other ions have also been used to form p–n junctions. Barnoski et al.[130] produced p–n junction laser diodes by implanting a 1×10^{16} cm^{-2} dose of 20 keV Zn$^+$ ions into n-type GaAs ($n \sim 1 \times 10^{18}$ cm^{-3}) and annealing the material at 900°C for 3 h using an SiO$_2$ encapsulant. Diffused p-type layers were formed with junction depths of 1–2 μm. Contacted diode chips, which were processed from the implanted wafer, were reported to lase with threshold current densities as low as 2×10^3 Å cm^{-2} when pulsed at 77 K. These results are comparable to those for Zn-diffused devices, which have similar junction depth.[131]

Although ion implantation doping offers considerable advantages over conventional growth techniques, its use in the production of p–n junction solar cells has been limited. Initial work employed laser annealing to activate the implanted n$^+$ layer in shallow homojunction n$^+$–p–p$^+$ structures.[118] Although the conversion efficiencies of up to 12% at AM1 for these cells are somewhat lower than obtained for all-CVD grown devices, it is anticipated that optimization of implantation and annealing parameters should lead to significant improvements.

5.5.1.3 Metal–semiconductor field-effect transistors (MESFETs)

A high density of surface states is usually observed when an insulating layer is grown or deposited on GaAs.[132] For this reason it has not been possible thus far to fabricate MIS-type field-effect transistors which are usable for microwave or IC applications. The FETs made in GaAs have been Schottky barrier devices (MESFETs) and therefore require a high-quality n-type layer for their active region. A schematic diagram showing the evolution of GaAs depletion mode MESFET technology is presented in Figure 5.17.[133] Early MESFETs were fabricated using n-type epitaxial layers grown on semi-insulating GaAs with mesa etching employed to isolate the devices. Later developments involved the use of implantation of n-type dopants into epitaxial buffer layers or directly into semi-insulating GaAs to achieve continuous n-type regions with mesa etching again employed for device isolation.[110] With the development of depletion mode FET (DMESFET) based integrated circuits, localized implantation of n-type dopants has been employed to form the active regions of FETs and diodes[134] as will be discussed in more detail below. The advent of localized implantation techniques has also made it possible to achieve a high carrier concentration in the contact regions of the FETs by employing a second implantation. Such a process replaces the one illustrated in Figure 5.17a where the growth of a second epitaxial layer for the purpose of achieving lower source and drain resistance is depicted.

The n-type layer required for FET channels ranges from about a thousand to several thousand angstroms in thickness with an electron concentration of about 10^{17} cm^{-3}. N-type layers with such a thickness and doping level can readily be

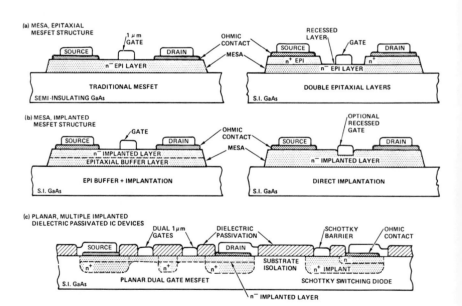

Figure 5.17 Schematic diagrams showing the evolution of GaAs FET and IC technology from the mesa epitaxial depletion mode MESFET to the planar ion-implanted GaAs IC[133]

fabricated in GaAs by the implantation of Si, S, Se, or Te. The first three of these dopants have all been employed in the fabrication of FETs using ion implantation.[110, 114]

It is possible to fabricate reproducibly uniform high-quality n-type layers suitable for FETs or integrated circuits by implanting suitable dopants into semi-insulating GaAs. However, the quality of the semi-insulating materials is critical in achieving the desired layer properties. As a result of the chromium outdiffusion discussed earlier, chromium-doped semi-insulating GaAs sometimes develops an n-type layer at the surface when that material is annealed in the same manner as an implanted sample would be annealed[110] (Figure 5.18). Even if a measurable n-type layer is not observed after annealing, it is possible that undesirable tails may be seen on the electron concentration profiles in implanted material.[110, 135] This problem has been circumvented by some workers by implanting into a semi-insulating buffer layer grown on the surface of chromium-doped substrate material.[117] However, in high-volume production of integrated circuits, it is desirable to avoid the use of a buffer layer if possible. For this reason, considerable attention has been given to the development of processes for FET and integrated circuit fabrication that employ ion implantation directly into semi-insulating GaAs. It has been found that it is possible to identify satisfactory material for ion implantation processes by the use of a suitable series of qualification tests on wafers taken from each end of a boule of semi-insulating GaAs. These qualification tests, which are likely to vary somewhat between

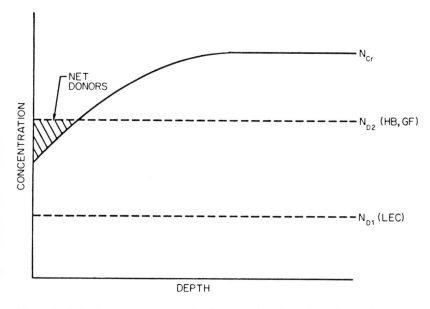

Figure 5.18 Surface conversion mechanism resulting from chromium redistribution

different laboratories, may involve measurements of the resistivity of unimplanted semi-insulating samples which have been annealed under the same conditions as implanted wafers are annealed. If their resistivity after annealing is below some minimum value, the material is rejected for IC applications. The properties of implanted layers may also be studied, using implantation parameters suitable for FET channels. Criteria for acceptance or rejection of semi-insulating material may involve: (a) doping efficiency in the implanted layer within a selected range, (b) electron mobility in the implanted layer above a selected minimum value, and (c) absence of undesirable deep doping tails on the electron concentration profiles for the implanted layer.

The use of semi-insulating material which is not chromium-doped may give a better chance of obtaining high-mobility n-type layers than if chromium-doped materials were used. Experimental results to date seem to confirm this expectation[136, 137] and it may be that the undoped semi-insulating GaAs grown by the LEC method will provide a good source of material for devices fabricated using ion implantation (see Chapter 2).

S,[111, 112] Si,[113, 114] and Se[110] have all been employed as implanted dopants in the fabrication of low-noise FETs. However, since better doping profile control can be obtained with Si and Se, most of the applications of implantation to FET active layers have involved either Se or Si as the implanted dopant. It has been possible to achieve high-frequency performance characteristics in implanted FETs that are competitive with those found for devices fabricated using other techniques.[138] Good uniformity of FET characteristics over a given processed wafer and reproducibility from wafer to wafer have also been observed.[110] Not only has implantation been employed in realizing the FET active layer but it has also been used to dope the source and drain regions of low-noise FETs, resulting in decreased source series resistance and in improved minimum noise figure and associated gain.[139] Ion implantation has also been used in the fabrication of GaAs power FETs both to form the FET channel and to enhance the doping in the source and drain contact regions.[140, 141] Such source and drain doping is important for power FETs since it results in higher drain source breakdown voltage. One interesting example of this n^+ doping involves co-implantation of Si and P at room temperature, resulting in typical peak carrier concentrations of 2.5×10^{18} cm^{-3}.[140] The doping profile flexibility afforded by implantation has been utilized to fabricate power FETs of superior gain linearity. This work involving the use of Si implantation resulted in significantly lower third-order intermodulation distortion than for FETs with uniformly doped active layers.[141]

5.5.2 Integrated circuits

The fabrication of GaAs integrated circuits is potentially the largest scale application of ion implantation in GaAs to device fabrication. The uniformity of doping over relatively large areas which can be achieved with ion implantation is essential for complex integrated circuits where uniformity of device characteristics must be attained in order to achieve operation of such ICs. The development of GaAs digital integrated circuits is under way in a number of laboratories using a variety of different approaches. This is discussed in detail in Chapters 12 and 13. Implantation is also being employed in the development of monolithic microwave integrated circuits.

5.6 CONCLUSIONS

The utilization of ion damage and implantation is now established in two main areas of GaAs device fabrication, electrical isolation and impurity doping. The first of these processes, damage-induced isolation, is particularly suited to planar technology and there are a number of examples where this technique has been found superior to conventional chemical etching. These include IMPATT diodes, lasers, optical waveguide structures, FET devices, and integrated circuits. In the future, it is expected that this technique, employing both proton and heavier-ion bombardment, will find increasing use in submicron device fabrication.

Ion implantation doping of GaAs has necessitated the use of surface passivation techniques for annealing the implanted samples to remove the radiation-induced damage and to activate the implanted ions electrically. There are many annealing techniques described in the literature, and these include dielectric capping, arsenic overpressure (capless) passivation and, more recently, pulsed laser and electron beam annealing, and transient thermal annealing. To a limited extent all these techniques claim a certain degree of success, while at the same time the need for annealing temperatures in excess of 800°C results in a distinct lack of consistency in the results obtained. It is not certain, for example, how much the variability is due to lack of control of the passivation and how much is due to the crystalline and electrical quality of the starting material. Considerable research effort is being devoted to these problems at the present time.

For the fabrication of inexpensive GaAs integrated circuits, implantation into bulk-grown semi-insulating substrates is desirable. A particular problem experienced with chromium-doped semi-insulating material is the redistribution of the near-surface Cr impurity, which may result in an n-type surface conversion. The recent development of chromium-free high-purity semi-insulating layers is one possibility being investigated to overcome this problem. However, it must be recognized that thermal conversion problems may also be possible in such material.

In terms of specific GaAs device applications, ion implantation has established a clear technological role. The most important achievements to date appear to be in the area of monolithic ICs. Here the technique has opened the way to the fabrication of digital and monolithic microwave ICs from bulk-grown semi-insulating wafers, an important milestone in the realization of inexpensive GaAs

ICs. Other GaAs applications include p–n junctions for IMPATT and solar cells, doping profiles for hyperabrupt varactors, and doping notches for transferred electron devices.

ACKNOWLEDGEMENTS

We are very grateful to all the authors who have allowed us to reproduce their published figures, and to our colleagues at the University of Leeds and Rockwell International for their constructive criticism of this review.

REFERENCES

1. G. Carter and W. A. Grant, *Ion Implantation of Semiconductors*, London, Edward Arnold (1976).
2. J. F. Gibbons, Ion implantation in semiconductors—Part I: Range distribution theory and experiments, *Proc. IEEE*, **56**, 295–320 (1968).
3. J. F. Gibbons, Ion implantation in semiconductors—Part II: Damage production and annealing, *Proc. IEEE*, **60**, 1062–97 (1972).
4. G. Dearnaley, J. H. Freeman, R. S. Nelson, and J. Stephen, *Ion Implantation*, Amsterdam, North-Holland (1973).
5. D. V. Morgan, *Channeling*, New York Wiley (1973).
6. J. Lindhard, M. Scharff, and M. E. Schiott, Range concepts and heavy ion ranges, *K. Danske Vidensk. Selsk., Mat.-Fys. Medd.*, **33**, 14 (1963).
7. J. Lindhard, V. Nielsen, and M. Scharff, Approximation method in classical scattering by screened Coulomb fields, *K. Danske Vidensk. Selsk., Mat.-Fys. Medd.*, **36**, 10 (1968).
8. J. F. Gibbons, W. S. Johnson, and S. W. Mylroie, *Projected Range Statistics: Semiconductors and Related Materials*, Stroudsburg, Penn., Dowden, Hutchinson and Ross (1975).
9. J. A. Brinkman, On the nature of radiation damage in metals, *J. Appl. Phys.*, **25**, 961 (1953).
10. B. J. Sealy, An examination of tellurium ion-implanted GaAs by transmission electron microscopy, *J. Mater. Sci.*, **10**, 683–91 (1975).
11. J. R. Brawn and W. A. Grant, Lattice location studies of GaAs implanted with Te, *Applications of Ion Beams to Materials 1975*, Inst. Phys. Conf. Ser. 28, London, Institute of Physics, pp. 59–63 (1975).
12. M. Takai, K. Gamo, K. Masuda, and S. Namba, Lattice site location of cadmium and tellurium implanted in gallium arsenide, *Jap. J. Appl. Phys.*, **14**, 1935–41 (1975).
13. W. J. Anderson and Y. S. Park, Flux and fluence dependence of implantation disorder in GaAs substrates, *J. Appl. Phys.*, **8**, 4568–70 (1978).
14. Y. Kushiro and T. Kobayashi, The effects of ion dose and implantation temperature of enhanced diffusion in selenium ion-implanted gallium arsenide, *Ion Implantation in Semiconductors*, ed. S. Namba, New York, Plenum, pp. 47–53 (1975).
15. J. M. Woodcock, J. M. Shannon, and D. J. Clark, Electrical and cathodoluminescence measurements on ion implanted donor layers in GaAs, *Solid State Electronics*, **18**, 267–75 (1975).
16. C. T. Foxon, J. A. Harvey, and B. A. Joyce, The evaporation of GaAs under equilibrium and non-equilibrium conditions using a modulated beam technique, *J. Phys. Chem. Solids*, **34**, 1693–701 (1973).

Ion implantation and damage in GaAs 189

17. A. G. Foyt, J. P. Donnelly, and W. T. Lindley, Efficiency doping of GaAs by Se$^+$ ion implantation, *Appl. Phys. Lett.*, **14**, 372–4 (1969).
18. J. Gyulai, J. W. Mayer, I. V. Mitchell, and V. Rodriguez, Outdiffusion through silicon oxide and silicon nitride layers on gallium arsenide, *Appl. Phys. Lett.*, **17**, 332–4 (1970).
19. J. S. Harris, F. H. Eisen, B. Welch, J. D. Haskel, R. D. Pashley, and J. W. Mayer, Influence of implantation temperature and surface protection on tellurium implantation in GaAs, *Appl. Phys. Lett.*, **21**, 601–3 (1972).
20. F. H. Eisen, J. S. Harris, B. Welch, R. D. Pashley, D. Sigurd, and J. W. Mayer, Properties of tellurium implanted gallium arsenide, *Ion Implantation in Semiconductors and Other Materials*, ed. B. L. Crowder, New York, Plenum, pp. 631–40 (1973).
21. J. P. Donnelly, W. T. Lindley, and C. E. Hurwitz, Silicon- and selenium-ion-implanted GaAs reproducibly annealed at temperatures up to 950°C, *Appl. Phys. Lett.*, **27**, 41–3 (1975).
22. F. H. Eisen, B. M. Welch, H. Muller, K. Gamo, T. Inada, and J. W. Mayer, Tellurium implantation in GaAs, *Solid-State Electronics*, **20**, 219–23 (1977).
23. R. D. Pashley and B. M. Welch, Tellurium-implanted N$^+$ Layers in GaAs, *Solid-State Electronics*, **18**, 977–81 (1975).
24. H. Nishi, S. Okamura, T. Inada, and H. Hashimoto, AlN encapsulant for fabrication of implanted GaAs ICs, *Gallium Arsenide and Related Compounds 1981*, Inst. Phys. Conf. Ser. 63, Bristol and London, Institute of Physics, pp. 365–70 (1982).
25. B. J. Sealy and R. K. Surridge, A new thin film encapsulant for ion-implanted GaAs, *Thin Solid Films*, **26**, L12–22 (1975).
26. R. K. Surridge, B. J. Sealy, A. D. E. D'Cruz, and K. G. Stephens, Annealing kinetics of donor ions implanted into GaAs, *Gallium Arsenide and Related Compounds 1976* (Edinburgh), Inst. Phys. Conf. Ser. 33a, Bristol and London, Institute of Physics, pp. 161–7 (1977).
27. B. J. Sealy, R. K. Surridge, and K. G. Stephens, Pulse annealing of ion implanted GaAs, *Defects and Radiation Effects in Semiconductors 1978*, Inst. Phys. Conf. Ser. 46, Bristol and London, Institute of Physics, Chapter 8, pp. 476–81 (1979).
28. D. E. Davies, P. J. McNally, J. P. Lorenzo, and M. Julian, Incoherent annealing of implanted layers in GaAs, *IEEE Electron Device Lett.*, **EDL-3**, 25–6 (1982).
29. M. Arai, K. Nishiyama, and N. Watanabe, Radiation annealing of GaAs implanted with Si, *Jap. J. Appl. Phys.*, **20**, L124–6 (1981).
30. M. Bujatti, A. Cetronia, R. Nipoti, and E. Olzi, Multiply scanned electron beam annealing of Si implanted GaAs, *Appl. Phys. Lett.*, **40**, 334–6 (1982).
31. J. C. C. Fan, B.-Y. Tsaur, and M. W. Geis, Transient heating with graphite heaters for semiconductor processing, *Laser and Electron-beam Interaction with Solids*, eds B. R. Appleton and F. K. Cellen, New York, Elsevier, pp. 741–8 (1982).
32. B. J. Sealy, Transient annealing of ion implanted GaAs, *Microelectronics J.*, **13**, 1, 21–8 (1982).
33. A. A. Immorlica and F. H. Eisen, Capless annealing of ion-implanted GaAs, *Appl. Phys. Lett.*, **29**, 94–5 (1976).
34. R. M. Malbon, D. H. Lee, and J. M. Whelan, Annealing of ion-implanted GaAs in a controlled atmosphere, *J. Electrochem. Soc.*, **123**, 1413–15 (1976).
35. J. Kasahara, M. Ari, and N. Watanabe, Capless anneal of ion-implanted GaAs in controlled arsenic vapor, *J. Appl. Phys.*, **50**, 541–3 (1979).
36. J. L. Tandon and F. H., Eisen, Pulsed annealing of implanted semi-insulating GaAs, *Laser–Solid Interaction and Laser Processing 1978*, AIP Conf. Proc. 50, American Institute of Physics, pp. 616–22 (1979).

37. B. J. Sealy, S. S. Kular, K. G. Stephens, R. Croft, and A. Palmer, Electrical properties of laser-annealed donor-implanted GaAs, *Electronics Lett.*, **14**, 720–1 (1978).
38. S. S. Kular, B. J. Sealy, K. G. Stephens, D. R. Chick, Q. V. Davis, and J. Edwards, Pulsed laser annealing of zinc implanted GaAs, *Electronics Lett.*, **14**, 85–7 (1978).
39. A. G. Foyt, W. T. Lindley, C. M. Wolfe, and J. P. Donnelly, Isolation of junction devices in GaAs using proton bombardment, *Solid State Electronics*, **12**, 209–14 (1969).
40. J. C. Dyment, J. C. North, and L. A. D'Asaro, Optical and electronic properties of proton bombarded p-type GaAs, *J. Appl. Phys.*, **44**, 207 (1973).
41. B. R. Pruniaux, J. C. North, and G. L. Miller, Compensation of n-type GaAs by proton bombardment, *Proc. 2nd Int. Conf. on Ion Implantation in Semiconductors*, eds I. Ruge and J. Gravi, Berlin, Springer-Verlag, pp. 212–21 (1971).
42. H. Haradu, and M. Fujimoto, Electrical properties of proton bombarded n-type GaAs, *Proc. 4th Int. Conf. on Ion Implantation in Semiconductors*, New York, Plenum, pp. 73–81 (1975).
43. J. C. Dyment, L. A. D'Asaro, J. C. North, B. I. Miller, and J. E. Ripper, Proton bombardment formation of strip geometry heterostructure lasers for 300 K cw operation, *Proc. IEEE*, **60**, 726 (1972).
44. J. P. Donnelly, Ion implantation in GaAs, *Gallium Arsenide and Related Compounds 1976* (St Louis), Inst. Phys. Conf. 33b, Bristol and London, Institute of Physics (1977).
45. D. E. Davies, J. F. Kennedy, and A. C. Yang, Compensation from implantation in GaAs, *Appl. Phys. Lett.*, **23**, 615–16 (1973).
46. D. K. Brice, Spatial distribution of energy deposited in atomic collision processes, *Ion Implantation Range and Energy Distribution*, Vol. I, New York, Plenum (1975).
47. J. P. Donnelly and F. J. Leonberger, Multiple energy proton bombardment in n$^+$ GaAs, *Solid State Electronics*, **20**, 183–9 (1977).
48. S. G. Liu, E. C. Douglas, C. P. Wu, C. W. Magee, S. Y. Narayan, S. T. Jolly, F. Kolondra, and S. Jain, Ion implantation of sulfur and silicon in GaAs, *RCA Rev.*, **41**, 227–62 (1980).
49. J. K. O'Brien and J. C. Correlli, Photoconductivity of Cr-compensated GaAs after irradiation by 0.8 MeV electrons, *J. Appl. Phys.*, **44**, 1921 (1973).
50. L. M. Aukerman and R. D. Graft, Anealing of electron irradiated GaAs, *Phys. Rev.*, **127**, 1576 (1976).
51. A. K. Kalma and R. A. Bergen, Electrical properties of electron irradiated GaAs, *IEEE Trans. Nucl. Sci.*, **NS-19**, 209–14 (1972).
52. H. Schade, Studies of electron bombardment damage in GaAs by TSC measurements, *J. Appl. Phys.*, **40**, 2613 (1969).
53. F. L. Vook, Change in thermal conductivity upon low temperature irradiation of GaAs, *Phys. Rev.*, **135**, A1742 (1964).
54. H. J. Stein, Electrical studies of low temperature neutron and electron irradiated epitaxial n-type GaAs, *J. Appl. Phys.*, **40**, 5300 (1969).
55. L. W. Aukerman, P. W. Davies, R. D. Graft, and J. S. Shilliday, Radiation effects in GaAs, *J. Appl. Phys.*, **34**, 3590 (1963).
56. D. Lang, Review of radiation induced defects in III–V compounds, *Radiation Effects in Semiconductors 1976*, Inst. Phys. Conf. Ser. 31, Bristol and London, Institute of Physics (1977).
57. R. A. Murphy, W. T. Lindley, D. F. Peterson, A. G. Foyt, C. H. Wolfe, C. F. Horwitz, and J. P. Donnelly, Proton guarded GaAs IMPATT diodes, *Gallium Arsenide and Related Compounds 1972*, Inst. Phys. Conf. Ser. 17, Bristol and London, Institute of Physics, pp. 224–30 (1973).

58. J. D. Speight, P. Leigh, N. McIntyre, I. G. Grove, and S. O'Hara, High efficiency proton isolated GaAs IMPATT diodes, *Electronics Lett.* **10**, 98–9 (1974).
59. B. R. Pruniaux, J. C. North, and A. V. Payer, A semi-insulated gate GaAs field effect transistor, *IEEE Trans. Electron Devices*, **ED-19**, 672–4 (1972).
60. E. Gamire, H. Stoll, A. Yariv, and R. G. Hunsperger, Optical waveguiding in proton implanted GaAs, *Appl. Phys. Lett.*, **21**, 87–8 (1972).
61. S. Somekh, E. Garmire, A. Yariv, H. L. Garvin, and R. G. Hunsperger, Channel optical waveguide directional couplers, *Appl. Phys. Lett.*, **22**, 46–7 (1973).
62. H. Stoll, A. Yariv, R. G. Hunsperger, and G. L. Tangonan, Proton implanted optical waveguide detectors in GaAs, *Appl. Phys. Lett.*, **23**, 664–5 (1973).
63. D. C. D'Avanzo, Proton isolation for GaAs integrated circuits, *IEEE Trans. Electron Devices*, **ED-29**, 1051–9 (1982).
64. P. N. Favennec, G. P. Pelous, M. Binet, and P. Baudet, Compensation of GaAs by oxygen implantation, *Ion Implantation in Semiconductors and Other Materials*, ed. B. Crowder, New York, Plenum, pp. 621–30 (1973).
65. P. N. Favennec, Semi-insulating layers of GaAs by oxygen implantation, *J. Appl. Phys.*, **47**, 2532–6 (1976).
66. S. Gecim, B. J. Sealy, and K. G. Stephens, Carrier removal profiles from oxygen implanted GaAs, *Electronics Lett.* **14**, 306–8 (1978).
67. S. M. Sze and J. C. Irvin, Resistivity, mobility and impurity levels in GaAs, Ge and Si at 300 K, *Solid State Electronics*, **11**, 599–602 (1968).
68. J. F. Gibbons and R. E. Tremain, The effects of degeneracy on doping efficiency for n-type implants in GaAs, *Appl. Phys. Lett.*, **26**, 199–201 (1975).
69. F. H. Eisen and B. M. Welch, Radiotracer profiles in sulphur implanted GaAs, *Ion Implantation in Semiconductors and Other Materials*, eds F. Chernow, J. A. Borders, and D. K. Brice, New York, Plenum, pp. 97–106 (1977).
70. M. Feng, V. Eu, H. Kanber, and W. B. Henderson, Study of Cr and Mn distribution in SI GaAs after annealing with and without SiO_2 in an H_2–As_4 atmosphere, *J. Electronic Mater.* **10**, 6, 973–86 (1981).
71. T. Miyazaki and M. Tamura, Implantation of silicon into gallium arsenide, *Ion Implantation in Semiconductors and Other Materials*, ed. S. Namba, New York, Plenum, pp. 41–6 (1975).
72. J. L. Tandon, M. A. Nicolet, and F. H. Eisen, Silicon implantation in GaAs, *Appl. Phys. Lett.*, **34**, 165–7 (1979).
73. K. Gamo, T. Inada, S. Krekeler, J. W. Mayer, F. H. Eisen, and B. M. Welch, Selenium implantation in GaAs, *Solid-State Electronics*, **10**, 213–17 (1977).
74. T. Inada, H. Miwa, S. Kato, E. Kobayashi, T. Hara, and M. Mihara, Annealing of Se implanted GaAs with an oxygen-free CVD Si_3N_4 encapsulant, *J. Appl. Phys.*, **49**, 4571–3 (1978).
75. T. Inada, K. Tokunaga, and S. Taka, Pulsed electron-beam annealing of selenium implanted gallium arsenide, *Appl. Phys. Lett.*, **35**, 546–8 (1979).
76. B. J. Sealy, R. K. Surridge, E. C. Bell, and A. D. E. D'Cruz, Donor activity in ion-implanted GaAs, *Applications of Ion Beams to Materials 1975*, eds G. Carter, J. S. Colligon, and W. A. Grant, Inst. Phys. Conf. Ser. 28, Bristol and London, Institute of Physics, pp. 69–74 (1976).
77. J. D. Sansbury and J. F. Gibbons, Conductivity and Hall mobility of ion implanted silicon in semi-insulating gallium arsenide, *Appl. Phys. Lett.*, **14**, 311–13 (1969).
78. R. G. Hunsperger and O. J. Marsh, Electrical properties of Cd, Zn and S ion-implanted layers in GaAs, *Radiation Effects*, **6**, 263–8 (1970).
79. D. E. Davies, J. K. Kennedy, and C. E. Ludington, Comparison of group IV and VI doping by implantation in GaAs, *J. Electrochem. Soc.*, **122**, 1374–7 (1975).
80. T. J. Harris, B. J. Sealy, and R. K. Surridge, Effects of channeling on the electrical

properties of donor implanted GaAs, *Electronics Lett.*, **12**, 664–5 (1976).
81. F. H. Eisen, B. M. Welch, K. Gamo, T. Inada, H. Mueller, M. A. Nicolet, and J. W. Mayer, Sulphur, selenium and tellurium implantation in GaAs, *Applications of Ion Beams to Materials 1975*, eds G. Carter, J. S. Colligon, and W. A. Grant, Inst. Phys. Conf. Ser. 28, Bristol and London, Institute of Physics, pp. 64–8 (1976).
82. B. K. Shin, Electrical properties of Te implanted GaAs, *J. Appl. Phys.*, **47**, 3612–17 (1976).
83. J. P. Donnelly, C. O. Bozler, and W. T. Lindley, Low-dose n-type ion implantation into Cr-doped GaAs substrates, *Solid-State Electronics*, **20**, 273–6 (1977).
84. R. K. Surridge and B. J. Sealy, Active layers for device applications by using high-energy selenium implantation into GaAs, *Electronics Lett.*, **13**, 233–4 (1977).
85. M. Fujimoto, H. Yamazaki, and T. Honda, Sulphur ion implantation in gallium arsenide, *Ion Implantation in Semiconductors*, eds F. Chernow, J. A. Borders, and D. K. Brice, New York, Plenum, pp. 89–96 (1977).
86. J. A. Golovchenko and T. N. C. Venkatesan, Annealing of Te-implanted GaAs by ruby laser irradiation, *Appl. Phys. Lett.*, **32**, 147–9 (1978).
87. S. U. Campisano, I. Catalano, G. Foti, E. Rimini, F. Eisen, and M. A. Nicolet, Laser reordering of implanted amorphous layers in GaAs, *Solid-State Electronics*, **21**, 485–8 (1978).
88. E. Rimini, P. Baeri, and G. Foti, Laser pulse energy dependence of annealing in ion implanted Si and GaAs semiconductors, *Phys. Lett.*, **65A**, 153–5 (1978).
89. J. L. Tandon, M. A. Nicolet, W. F. Tseng, F. H. Eisen, S. U. Campisano, G. Foti, and E. Rimini, Pulsed-laser annealing of implanted layers in GaAs, *Appl. Phys. Lett.*, **34**, 597–9 (1979).
90. S. U. Campisano, G. Foti, E. Rimini, F. H. Eisen, W. F. Tseng, M. A. Nicolet, and J. L. Tandon, Laser pulse annealing of ion-implanted GaAs, *J. Appl. Phys.* **51**, 295–8 (1980).
91. R. L. Mozzi, W. Fabian, and F. J. Piekarski, Nonalloyed ohmic contacts to n-GaAs by pulse-electron-beam-annealed selenium implants, *Appl. Phys. Lett.*, **35**, 337–9 (1979).
92. N. J. Shah, R. A. McMahon, and H. Ahmed, Isothermal electron beam processing of GaAs, *IEEE Colloq. on GaAs Material for IC Application*, London, March, Digest No. 1982/28 (1982).
93. R. G. Hunsperger, R. G. Wilson, and D. M. Jamba, Mg and Be ion implanted GaAs, *J. Appl. Phys.*, **43**, 1318–20 (1972).
94. M. A. Littlejohn, J. R. Hauser, and L. K. Montieth, The electrical properties of 60 keV zinc ions implanted into semi-insulating gallium arsenide, *Radiation Effects*, **10**, 185–90 (1971).
95. Y. Yuba, K. Gamo, K. Masuda, and S. Namba, Hall effect measurements of Zn implanted GaAs, *Jap. J. Appl. Phys.*, **13**, 641–4 (1974).
96. C. L. Anderson and H. L. Dunlap, Low temperature annealing behaviour of GaAs implanted with Be, *Appl. Phys. Lett.*, **35**, 178–80 (1979).
97. J. Comas, L. Plew, P. K. Chatterlee, M. V. McLevige, K. V. Vaidyanathan, and B. G. Streetman, Impurity distribution of ion-implanted Be in GaAs by SIMs, photoluminescence and electrical profiling, *Ion Implantation in Semiconductors and Other Materials*, eds F. Chernow, J. A. Borders, and D. K. Brice, New York, Plenum, pp. 141–8 (1977).
98. S. Nojoma and Y. Kawasaki, Annealing characteristics of Be ion implanted GaAs, *Jap. J. Appl. Phys.*, **17**, 1845–50 (1978).
99. S. S. Kular, B. J. Sealy, and K. G. Stephens, Electrical profiles from zinc implanted GaAs, *Electronics Lett.*, **14**, 2–4 (1978).
100. B. K. Shin, D. C. Llok, Y. S. Park, and J. E. Ehret, Hall effect measurements in Cd-implanted GaAs, *J. Appl. Phys.*, **47**, 1574–9 (1976).

101. J. W. Mayer, O. J. Marsh, R. Mankarious, and R. Bower, Zn and Te implantations in GaAs, *J. Appl. Phys.*, **38**, 1975–6 (1967).
102. R. G. Hunsperger and O. J. Marsh, Electrical properties of zinc and cadmium ion implanted layers in gallium arsenide, *J. Electrochem. Soc.*, **116**, 488–92 (1969).
103. J. D. Sansbury and J. F. Gibbons, Properties of ion implanted silicon, sulphur and carbon in gallium arsenide, *Radiation Effects*, **6**, 269–76 (1970).
104. R. B. Benson, M. A. Littlejohn, K. Lee, and R. E. Ricker, Some structural and electrical characteristics of GaAs annealed after implantation with Be, Mg, Zn and Cd, *Ion Implantation in Semiconductors and Other Materials*, eds F. Chernow, J. A. Borders, and D. K. Brice, New York, Plenum, pp. 131–9 (1976).
105. J. P. Donnelly, F. J. Leonberger, and C. O. Bozler, Uniform-carrier-concentration p-type layers in GaAs produced by beryllium ion implantation, *Appl. Phys. Lett.*, **28**, 706–8 (1976).
106. S. S. Kular, B. J. Sealy, and K. G. Stephens, Comparison of electrical profiles from hot and cold implantation of zinc ions into GaAs, *Electronics Lett.*, **14**, 22–3 (1978).
107. Y. K. Yeo, Y. S. Park, and P. W. Yu, Electrical measurements and optical activation studies in Mg-implanted GaAs, *J. Appl. Phys.*, **50**, 3274–81 (1979).
108. S. Kwun, W. G. Spitzer, C. L. Anderson, H. L. Dunlap, and K. V. Vaidyanathan, Optical studies of Be-implanted GaAs, *J. Appl. Phys.* **50**, 6873–80 (1979).
109. T. Inada, S. Kato, Y. Maeda, and K. Tokunaga, Doping profiles in Zn-implanted GaAs after laser annealing, *J. Appl. Phys.*, **50**, 6000–2 (1979).
110. J. A. Higgins, R. L. Kuvas, F. H. Eisen, and D. R. Ch'en, Low-noise GaAs FETs prepared by ion implantation, *IEEE Trans. Electron. Devices*, **ED-25**, 587–96 (1978).
111. J. A. Higgins, B. M. Welch, F. H. Eisen, and G. D. Robinson, Performance of sulphur ion-implanted GaAs FETs, *Electronics Lett.*, **12**, 17–19 (1976).
112. R. G. Hunsperger and N. Hirsh, GaAs field-effect transistors with ion-implanted channels, *Electronics Lett.*, **9**, 577–8 (1973).
113. T. Nozaki and K. Ohata, Sub-micron gate GaAs MESFETs with ion-implanted channels, *Jap. J. Appl. Phys.*, **16**, 111–14 (1977).
114. J. K. Kung, R. M. Malbon, and D. H. Lee, GaAs FETs with silicon implanted channels, *Electronics Lett.*, **13**, 187–8 (1977).
115. R. C. Eden, B. M. Welch, and R. Zucca, Planar GaAs IC technology: applications for digital LSI, *IEEE J. Solid-State Circuits*, **SC-13**, 419–26 (1978).
116. F. L. Troeger, A. F. Behle, P. E. Friebertshauser, K. L. Hu, and S. H. Watanabe, Fully ion implanted planar GaAs E-JFET process, *Int. Electron Devices Meeting, Tech. Digest*, pp. 497–500 (1979).
117. R. L. Vantuyl, C. A. Liechti, R. E. Lee, and E. Gowen, GaAs MESFET logic with 4-GHz clock rate, *IEEE J. Solid-State Circuits*, **SC-12**, 484–96 (1977).
118. J. C. C. Fan, R. L. Chapman, J. P. Donnelly, G. W. Turner, and C. O. Bozler, Ion-implanted laser-annealed GaAs solar cells, *Appl. Phys. Lett.*, **34**, 780–2 (1979).
119. A. A. Immorlica Jr and F. H. Eisen, Planar passivated GaAs hyperabrupt varactor diodes, *Proc. Sixth Biennial Cornell Electrical Eng. Conf.*, pp. 151–9 (1977).
120. I. Niikura, T. Nobuyuki, Y. Shimura, K. S. Yokoyama, M. Mihara, T. Hayashi, and T. Hara, GaAs varactor diodes for UHF TV tuners fabricated by ion implantation, *Electronics Lett.*, **14**, 9–10 (1978).
121. C. O. Bozler, J. P. Donnelly, R. A. Murphy, R. W. Laton, and W. T. Lindley, High-efficiency ion-implanted lo-hi-lo GaAs IMPATT diodes, *Appl. Phys. Lett.*, **35**, 74–6 (1979).
122. H. T. Yuan, F. H. Doerbeck, and M. V. McLevige, Ion implanted GaAs bipolar transistors, *Electronics Lett.*, **16**, 637–8 (1980).

123. D. H. Lee, J. J. Berenz, and R. L. Bernick, Ion implanted n^+ contacts for K_a band GaAs Gunn-effect diodes, *Electronic Lett.*, **11**, 189–91 (1975).
124. W. T. Anderson, H. B. Dietrich, E. W. Swiggard, S. H. Lee, and M. L. Bark, Gallium arsenide transferred-electron devices by low-level ion implantation, *J. Appl. Phys.* **51**, 3175–7 (1980).
125. T. Mizutani and K. Kurumada, GaAs planar Gunn digital devices by sulphur ion implantation, *Electronics Lett.*, **11**, 638–9 (1975).
126. T. Mizutani, S. Ishida, K. Kurumada, and M. Ohmori, Selectively ion implanted Gunn-FET integrated circuits for pulse regeneration, *Electronics Lett.*, **14**, 294–5 (1978).
127. J. J. Berenz, R. S. Ying, and D. H. Lee, CW operation of ion-implanted GaAs Read-type IMPATT diodes, *Electronics Lett.*, **10**, 157–8 (1974).
128. M. J. Helix, K. V. Vaidyanathan, and B. G. Streetman, Properties of Be-implanted planar GaAs p–n junctions, *IEEE J. Solid-State Circuits*, **SC-134**, 426–9 (1978).
129. W. V. McLevige, H. T. Yuan, H. T. Duncan, W. R. Frensley, F. H. Doerbeck, H. Morkoc, and T. J. Drummond, GaAs/GaAlAs heterojunction bipolar transistors for integrated circuit applications, *IEEE Electron Devices Lett.*, **EDL-3**, 43–5 (1982).
130. M. K. Barnoski, R. G. Hunsperger, and A. Lee, Ion implanted GaAs injection laser, *Appl. Phys. Lett.*, **24**, 627–8 (1974).
131. C. J. Hwang and J. C. Dyment, Correlation of GaAs junction laser thresholds with photoluminescence measurements, *Gallium Arsenide, Inst. Phys. Conf. Ser.* **7**, London, Institute of Physics, pp. 83–90 (1969).
132. E. Kamieniecki, T. E. Kazior, T. E. Lagowski, and H. C. Gatos, Study of GaAs–oxide interface by transient capacitance spectroscopy: discrete energy interface, *J. Vac. Sci. Technol.*, **17**, 1041–4 (1980).
133. R. C. Eden and B. M. Welch, GaAs digital integrated circuits for ultra high speed LSI/VLSI, *Very Large Scale Integration (VLSI) Fundamentals and Applications*, ed. D. F. Barbe, New York, Berlin, Heidelberg, Springer-Verlag, pp. 128–77 (1980).
134. B. M. Welch and R. C. Eden, Planar GaAs integrated circuits fabricated by ion implantation, *Int. Solid State Circuits Conf., Tech. Digest*, pp. 205–8 (1977).
135. P. M. Asbeck, J. Tandon, B. M. Welch, C. A. Evans Jr, and V. R. Deline, Effects of Cr redistribution on electrical characteristics of ion-implanted semi-insulating GaAs, *IEEE Electron Device Lett.*, **EDL-1**, 35–7 (1980).
136. R. N. Thomas, H. M. Hobgood, D. L. Barrett, and G. W. Eldridge, Large diameter, undoped semi-insulating GaAs for high mobility direct ion implanted FET technology, *Semi-insulating III–V Materials*, ed. G. J. Rees, Nottingham, Shiva, pp. 76–82 (1980).
137. H. M. Hobgood, G. W. Eldridge, D. L. Barrett, and R. N. Thomas, High-purity semi-insulating GaAs material for monolithic microwave integrated circuits, *IEEE Trans. Electron Devices*, **ED-28**, 140–9 (1981).
138. M. Feng, V. K. Eu, H. Kanber, E. Watkins, J. M. Schellenberg, and H. Yamasaki, Low noise GaAs metal–semiconductor field-effect transistor made by ion implantation, *Appl. Phys. Lett.*, **40**, 802–4 (1982).
139. K. Ohata, K. Nozaki, and N. Kawamura, Improved noise performance of GaAs MESFETs with selectively ion-implanted n^+ source regions, *IEEE Trans. Electron Devices*, **ED-24**, 1129–30 (1977).
140. E. Stoneham, T. S. Tan, and J. Gladstone, Fully ion implanted GaAs power FETs, *IEDM Tech. Digest*, pp. 330–3 (1977).
141. A. A. Immorlica Jr, J. A. Higgins, W. A. Hill, G. D. Robinson, and R. L. Kuvas, Ion implanted GaAs power FETs, *IEDM Tech. Digest*, pp. 368–72 (1978).

Gallium Arsenide
Edited by M. J. Howes and D. V. Morgan
© 1985 John Wiley & Sons Ltd

CHAPTER 6
Metallizations for GaAs Devices and Circuits

C. J. PALMSTRØM and D. V. MORGAN

6.1 INTRODUCTION

Metallization systems are a fundamental component to all semiconductor devices and integrated circuits. They provide the correct electrical link between the active region of the semiconductor and the external circuit. In the case of an integrated circuit, they also interconnect, with great precision, the circuit elements to provide the desired conducting paths. The primary consideration in the choice of a metallization system is to ensure that the metal chosen has the desired electrical properties. In terms of metal/semiconductor contacts, there are two distinct requirements made by the semiconductor designer. These are the ohmic (low-resistance) and the Schottky (rectifying) type contacts.[1] In the former it is necessary to ensure that the series resistance presented at the metal/ semiconductor interface is very small and can be ignored during device operation. The Schottky barrier contact is quite the reverse. Here a charge dipole, with its associated depletion region, is a necessary prerequisite and the resulting contact exhibits rectification properties which in many respects are similar to those of a p–n junction diode. This chapter will concentrate on contacts to n-type GaAs, which is the primary interest of current device and integrated circuit structures. The convention used in this chapter is that Mo/Ti/GaAs represents a layer of molybdenum on top of a titanium layer on GaAs, and Mo–Ti/GaAs represents an alloy of molybdenum and titanium on GaAs.

6.2 ELECTRICAL PROPERTIES OF METAL/GaAs SYSTEMS

When a metal makes intimate contact with a semiconductor, a barrier to electron and hole flow is formed. Ideally, the magnitude of this barrier is dependent on the metal work function, ϕ_m, and the semiconductor electron affinity, χ_{sc}.[1] The electron barrier height, ϕ_{bn}, is given by

$$\phi_{bn} = \phi_m - \chi_{sc}$$

Gallium arsenide

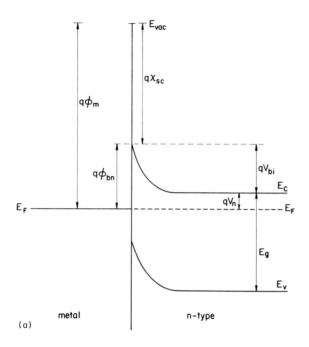

(a)

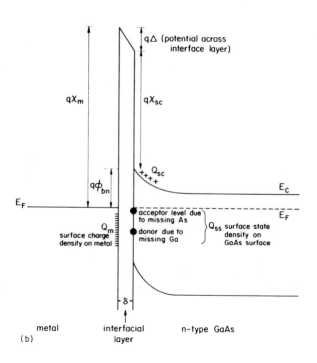

(b)

Metallizations for GaAs devices and circuits

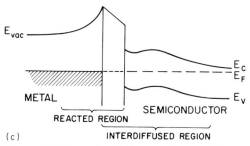

(c)

Figure 6.1 Energy band diagrams for metal/semiconductor contacts: (a) ideal Schottky contact; (b) defect model type contact; (c) extended metal/semiconductor interface model (From Brillson[26]).

and the hole barrier height, ϕ_{bp}, by

$$\phi_{bp} = \frac{E_g}{q} - (\phi_m - \chi_{sc})$$

where E_g is the band gap of the semiconductor (see Figure 6.1a) and q the electronic charge. Thus for an n-type semiconductor, an ideal ohmic contact is obtained when $\phi_m \leq \chi_{sc}$ and $\phi_{bn} \leq 0$. Conversely, an ideal Schottky barrier (rectifying contact) results when $\phi_m > \chi_{sc}$ and $\phi_{bn} > 0$. In practice, however, this ideal behaviour is not observed, in that the barrier height for a metal/GaAs system is found to be virtually independent of the metal work function. This is thought to be due to 'Fermi level pinning' at the semiconductor surface, which occurs when the surface state density is 10^{13} cm^{-2} or more.[2] Tamm[3] and Shockley[4] have shown that these surface states could be intrinsic to the semiconductor surface. However, recent results on 'well cleaved' (110) GaAs surfaces showed that there are no detectable intrinsic surface states in the band gap of GaAs.[5,6] Pinning of the Fermi level results from submonolayer coverages of different metals or oxygen, this pinning level being virtually independent of the nature of the contamination and roughly at about 0.8 eV below the conduction band minimum.[7] This result has led to the 'unified defect model' for Schottky barrier formation to compound semiconductors,[8,9] in which surface states are associated with native defects (acceptor or donor), for example, arsenic and gallium vacancies, in the GaAs produced by the interaction of the contaminant and the GaAs (see Figure 6.1b). Grant et al.[10] have proposed that the surface states (acceptor and donor) are associated with a single defect (single-defect model). Support for the defect models comes from results of silver contacts on (100) InP.[11] When the polar face is saturated with phosphorus prior to silver deposition, the Schottky barrier height is very small (~ 0.1 eV). However, when no special surface treatment is done to the InP prior to metal deposition, the normal barrier height of 0.4 eV is obtained. Hence, it can be argued that the excess phosphorus inhibits the formation of the 0.4 eV level associated with phosphorus vacancies.[8,9]

Brillson[12] argues that, in general, metal/semiconductor interfaces are not atomically abrupt, but contain reacted interfacial regions of finite width with different dielectric properties and/or interdiffused regions with electrically active sites extending into the semiconductor. The Schottky barrier behaviour is thus governed by the electrical properties of the reacted and interdiffused regions. A schematic energy band diagram proposed for the contact is shown in Figure 6.1c. A possible extension of Brillson's model to three dimensions is the 'effective work function model' proposed by Woodall and Freeouf.[13] They argue that the Fermi energy position at the interface is not governed by surface state density but rather is related to the work functions of microclusters of one or more interface phases resulting from either oxygen contamination or metal–semiconductor reaction occurring during metallization. This means that the effective work function is a weighted average of the work functions of different interface phases. Figure 6.2 shows a schematic of the model, which predicts that the measured barrier height ϕ_{bn} is partly dependent on the measurement technique (e.g. different weighted averages for I–V and C–V measurements). In the case of metal barriers deposited in conventional deposition systems (not UHV), the effective work function ϕ_{eff} is primarily due to the work function of the group V element (in III–V compounds)

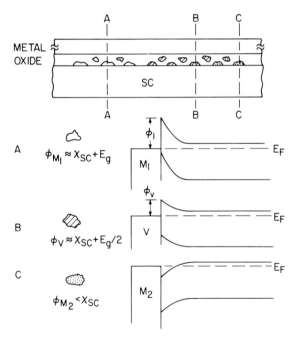

Figure 6.2 Schematic of the effective work function model of metal/semiconductor contact formation. Cross sections A, B, and C show the barrier height associated with phases M_1, V, and M_2 of arbitrarily different work functions, respectively. (From Woodall and Freeouf[13])

and hence the Schottky barrier height depends weakly on the metal work function. The dominance of the group V element work function, ϕ_V, arises from the fact that excess group V element is generally observed at the interface of native oxides on III–V compounds[14,15] (one exception is GaP[14]). Woodall and Freeouf associate the high 'interface state densities' in GaAs MOSFET structures to arsenic clusters at the interface, which act as Schottky barriers with $\phi_{bn} \sim 0.8$ eV embedded in an oxide matrix.[13] This view is supported by the observation of excess arsenic at the interface.[16]

Further evidence of the importance of the metal/semiconductor interface comes from the following:

(a) Ga/GaAs contacts are ohmic when prepared with no interfacial oxides but become rectifying after exposure to air.[17]

(b) The Al/GaAs Schottky barrier height is dependent on whether the surface is Ga- or As-stabilized.[18]

(c) Molecular beam epitaxial grown Al/AlAs Schottky barriers have a barrier height[19] that is very different from the reported barrier heights of Au/AlAs.[20] Similarly, Al/Ga$_{0.5}$In$_{0.5}$As contacts deposited *in situ* after growth by MBE are ohmic, whereas if Al is evaporated after air exposure the barrier height is of the order of 0.2–0.3 eV.[21]

(d) The barrier height of thin gold layers (few monolayers) deposited on UHV cleaved GaAs show a barrier height of about 1.1 eV or more.[22]

(e) The barrier height of Al/GaAs can be reduced from about 0.8 eV to 0.4 eV by adsorption of H$_2$S on the MBE grown GaAs surface prior to *in situ* deposition of aluminium.[23]

The above discussion indicates clearly the importance of the detailed nature of the metal/semiconductor interface in determining the barrier properties. However, there still remains disagreement in the literature about which model is most applicable for the range of experimental results reported.[24,25] Further discussions on recent results and several of the current models can be found in a review article by Brillson[26] and refs 8, 12, 13, and 27.

For device applications the preferred orientation is {100} GaAs. The implanted or epitaxially grown device layers are prepared for contact deposition by chemical cleaning (i.e. seldom is a contact surface prepared by cleaving in vacuum). Grant et al.[10] have reported that the interface Fermi level position on both {100} and {110} GaAs samples is a function of surface treatment. Massies et al.[28] observed a high density of surface states ($> 10^{12}$ cm^{-2}) in MBE grown (001) GaAs. For Schottky barriers fabricated in conventional vacuum there is at present little control over the interface, films, and impurities; the only certainty is that the GaAs will have a thin native oxide on the surface prior to metal deposition. From these comments it is evident that a contaminated interface will strongly influence the contact behaviour. Since this interface layer will not, in general, be uniform, reproducible Schottky barrier heights (within ±0.01 eV) across a wafer are difficult to obtain. In silicon technology the problem of the interface oxide in contact metallization is overcome by reacting the metal layer with the underlying silicon. The Schottky barrier is thus formed between the metal silicide and the clean silicon while the oxide layer remains near the surface of the reacted contact layer.[29] What is also evident from metal–Si reactions is that a thick surface oxide can retard the silicide formation[30] and affect the uniformity of the reacted layer.[31] The silicide reaction will start at pinholes or weak areas in the oxide and, hence, produce a non-uniform interface.[31] These observations suggest that the interfacial oxide layer always present in conventional metal/GaAs contacts may govern not only the electrical but also the metallurgical behaviour of the interface (i.e. the metal–GaAs reactions and interdiffusion). A further problem arises from the purity of the deposited metal films. Deposition is normally carried out in conventional evaporators working at a relatively modest vacuum of 10^{-6}–10^{-7} Torr. This results in the introduction of oxygen into the metal films, which may alter the silicide reaction rates and also the phase growth sequence.[32, 33]

The previous discussion shows that at present it is not possible to make metal contacts to GaAs without a dipole layer forming. Hence, ohmic contacts with $\phi_{bn} = 0$ cannot be made to low doped GaAs by simply choosing a metal with the appropriate work function. To fabricate ohmic contacts the general procedure therefore is to dope a thin layer of GaAs beneath the contact to as high a level as possible. This results in a narrow depletion region and a thin electron barrier (Figure 6.3). At sufficiently high doping levels the barrier becomes thin enough for most of the conduction to occur by tunnelling. For good ohmic behaviour this

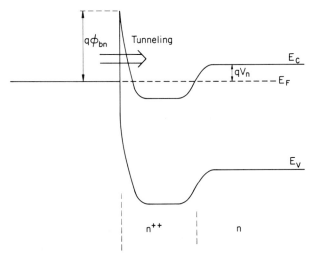

Figure 6.3 Energy band diagram for a tunnelling metal-/semiconductor contact

n^{++} layer must be doped to a level of 5×10^{19} donors/cm^{-3} or more. Achieving these high doping levels in GaAs is not an easy task, as discussed in Section 6.4.

Instead of using an n^{++} layer beneath the contact, Sebestyen[34] proposed the use of an amorphous or disordered layer with a large number of states in the band gap between the metal and semiconductor. The band diagram for this contact is shown in Figure 6.4. In this contact the conduction occurs by electrons hopping

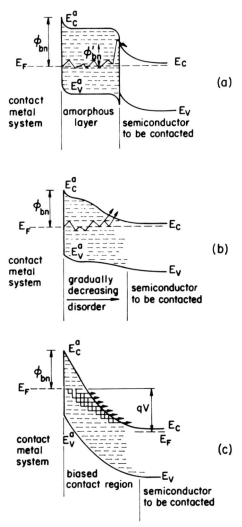

Figure 6.4 Energy band diagrams for metallized (a) abrupt amorphous/crystalline junction and (b, c) graded amorphous/crystalline junction at thermal equilibrium (b) and in biased state (c). Typical electron paths are shown by arrows and dashes illustrate the localized states in the mobility gap. (From Sebestyen[42])

between the mobility gap states near the Fermi level[35-37] or, alternatively, it can be thought of as multistep tunnelling and trap-assisted recombination through trap states.[38] An abrupt amorphous/crystalline semiconductor junction has, generally, some associated barrier to electron flow[39] (Figure 6.4a), which, however, may disappear if the degree of disorder gradually decreases from the metal towards the crystalline surface (Figure 6.4b).

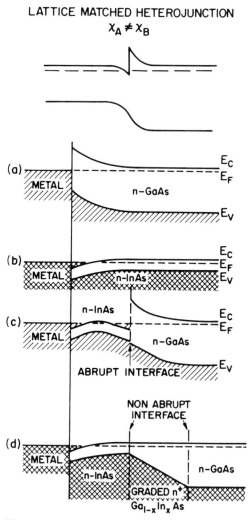

Figure 6.5 Energy band diagrams for various semiconductor interfaces: (a) metal/n-GaAs; (b) metal/n-InAs; (c) metal/n-InAs/n-GaAs; (d) metal/InAs/graded $Ga_{1-x}In_xAs$/GaAs (From Woodall et al.[76])

Metallizations for GaAs devices and circuits 203

One way of reducing the effective Schottky barrier height to GaAs involves the use of heterojunctions[40, 41] (Figure 6.5). The concept here is to grow another semiconductor layer on top of the GaAs and deposit the metal onto it. The semiconductor chosen must have an associated metal/semiconductor barrier height $\phi_{bn} \leq 0$ and be lattice-matched to GaAs. As in the case of abrupt amorphous/crystalline interfaces, an abrupt heterojunction will, in general, have an associated barrier (Figure 6.5c). In order to overcome this and to improve the lattice matching, a graded heterojunction must be used[42–46] (Figure 6.5d).

For most applications involving Schottky barriers, the primary object is to obtain good rectification, which implies a high barrier height. As is shown in Figure 6.6 the majority of metals can produce Schottky barriers with only a small spread in barrier height. Consequently, the choice of metal hinges on other considerations, such as its stability and resistance to interdiffusion and degradation during device operation. This can mean combinations of elevated temperatures ($\leq 300°C$), high electric fields, and high current densities.

The handling of high current densities results in its own special problem, namely electromigration. In circumstances where a metal film transports a current density in excess of about 10^5 A cm^{-2}, the electrons can impart sufficient momentum to the atoms to enable them to move in the direction of the electron flow. This problem is particularly acute at points of high current density and particularly so in the interconnects of integrated circuits. Figures 6.7 and 6.8 show the result of material transport in an ohmic contact on a transferred electron device and the Schottky gate of a MESFET.

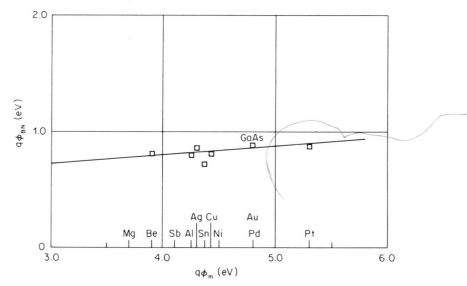

Figure 6.6 A plot of metal/GaAs barrier heights for various metal systems. (After Sze[1])

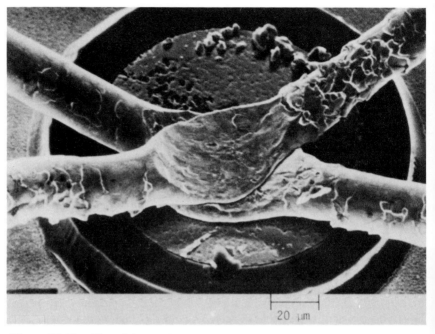

Figure 6.7 The effect of electromigration on an ohmic contact to a transferred electron device

The metallization design criteria are in reality more complex than those described above since other demands are made upon the choice. These include the following summary:

(a) contacts must produce the correct electrical behaviour;
(b) good stability during operation;
(c) good resistance to metallurgical reactions, oxidation, and corrosion;
(d) potential of selective etchability between metal and GaAs;
(e) reasonable ductility to allow wire bonding;
(f) good adherence to the GaAs surface.

It is very difficult to find metallizations that satisfy all the criteria described above and this has led to the development of multilayer metallizations which seek to obtain the optimum metal system satisfying the above demands. A typical example of such a structure is a three-layer metallization scheme; the first metal layer will be used to produce the correct metal/semiconductor characteristics with good adhesion; ductile metal such as gold is used to optimize the performance needs of the outer surface; and a barrier layer is placed between these two to keep the fast-diffusing gold from the interface metal.

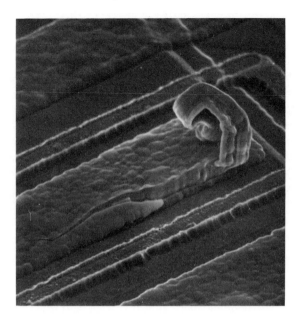

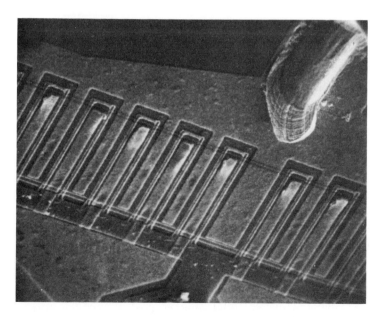

Figure 6.8 The effect of electromigration on a GaAs power MESFET gate electrode. (After Davey[57])

6.3 SCHOTTKY BARRIER METALS

The majority of metals when deposited on GaAs result in the formation of a dipole layer. When the doping density of GaAs is less than 10^{19} cm^{-3}, a rectification characteristic will always result. For the special case of high doping ($N_D > 10^{19}$ cm^{-3}) a barrier is still formed at the interface, but it is sufficiently thin to allow tunnelling to occur, and consequently a low-resistance non-rectifying contact is formed. Such ohmic contacts are discussed in Section 6.4. The fact that all the common metals, deposited on low and medium doped GaAs, result in Schottky contacts suggests considerable freedom in the choice of metallization. Figure 6.6 shows, for example, that gold (with a work function of approximately 4.8 eV) produces a barrier height of 0.9 eV while aluminium (with a work function of 4.25 eV) results in the slightly smaller barrier of 0.85 eV. To identify the reasons for choice we must look further.

6.3.1 Single-layer contacts

Au/GaAs As an illustration consider the case of Au/GaAs. From the point of view of rectification, bonding, thermal heat sinking, and resistance to chemical

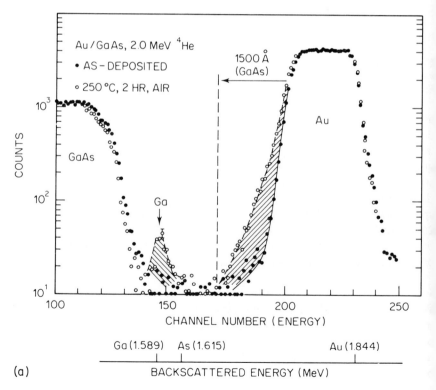

(a)

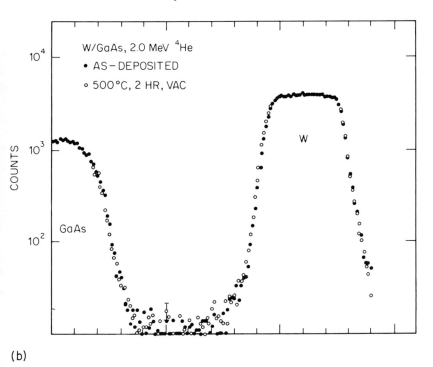

(b)

Figure 6.9 (a) RBS spectra for Au/GaAs, as deposited and after a 250°C anneal for 2 h in air. (b) RBS spectra for W/GaAs, as deposited and after annealing at 500°C for 2 h. (After Sinha and Poate[47])

attack, gold would be an ideal choice. However, problems begin to emerge when devices have to operate at elevated temperatures. The Au/GaAs interfaces interdiffuse and degrade at the relatively low temperature of 250°C. Furthermore, gold in GaAs produces deep levels which degrade device performance. Indeed, such levels located in the active channel of a MESFET would drastically degrade the device performance. Figure 6.9a shows the Rutherford back-scattering spectra for an Au/GaAs structure in the as-deposited form and after annealing at 250°C for 2 h in air. This shows that gold diffuses into the GaAs while Ga diffuses out through the gold and accumulates on the surface. This poor thermal stability of gold on GaAs restricts its use to a component in a multilayer structure.

W/GaAs Figure 6.9b shows the corresponding back-scattering spectra for W/GaAs as deposited and after annealing at 500°C for 2 h in air. In this case the interface does not degrade after the anneal. However, one problem often observed is that at temperatures in excess of 400°C the tungsten metallization tends to peel off owing to thermal expansion mismatch.[49] Similar results have been observed for molybdenum (see Table 6.1).

208   *Gallium arsenide*

Pt/GaAs Another material that exhibits good resistance to chemical attack is platinum. In this case, however, interdiffusion and compound formation with GaAs begin to occur in the relatively low temperature range of 250–300°C. This is illustrated in the sequence of back-scattering data shown in Figure 6.10 for temperatures up to 500°C. The compounds in Figure 6.10 were determined by a combination of back-scattering yields of the constituent elements and a knowledge of GaAs compound formation obtained from x-ray diffraction studies. (Details of these results are discussed in ref. 50, pp. 151–89.) Other metals that exhibit similar compound formation with GaAs are palladium and nickel, the thermal stability of such layers being summarized in Table 6.1.

Al/GaAs Three metals with good stable thermal properties on GaAs are aluminium, titanium, and tantalum, although unfortunately all suffer the problem of surface oxidation. In spite of the surface oxidation problem, however, aluminium is suitable for use as a Schottky metal for temperatures up to 400°C and has been used for MESFET fabrication. This system has been studied extensively and the main conclusion is that interdiffusion occurs even at temperatures as low as 125°C. The outdiffusion of arsenic and gallium causes stoichiometric imbalance and interfacial layers of AlAs, Al_2O_3, and $Al_xGa_{1-x}As$ may form. The reaction, however, is not large-scale as in the case of gold and the contact is considered relatively inert. The interfacial compounds are thought to increase the barrier height. However, the gallium outdiffusion constitutes a major failure mechanism in small-signal FETs characterized by gate short or high leakage. Furthermore, for the FETs, there is the possibility of electromigration or lateral diffusion of the aluminium Schottky gate metallization resulting in serious degradation.

Mo/GaAs Metal–GaAs reaction, although usually considered an undesirable problem, can, if controlled, be put to good use. For example, annealing molybdenum contacts in an arsenic overpressure at temperatures of 800 and 920°C results in Mo_5As_4 formation. A recent study of molybdenum contacts annealed in an N_2 ambient has shown some interesting results.[60] No interdiffusion of Mo and GaAs was observed for anneals below 700°C, whereas annealing at 700–800°C resulted in the formation of Mo_5As_4 and the contact becoming ohmic. Annealing at 800°C also resulted in the formation of Mo_2As_3, Mo_3Ga, and traces of Mo_3GaAs_2. After annealing at 1000°C a ternary phase believed to be Mo_3GaAs_2 with only a trace of Mo_5As_4 formed and the contact was rectifying. The ohmic behaviour after annealing at 700–800°C was attributed to dissociation of GaAs, the formation of As–Mo compounds, and the presence of 'free' gallium from GaAs substrate on the surface of the composite. The Schottky behaviour after annealing at 1000°C was attributed to the formation of the new ternary compound.

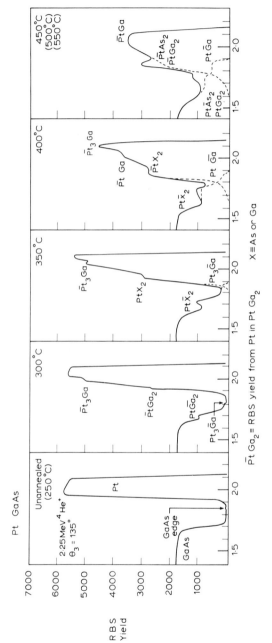

Figure 6.10 A sequence of RBS spectra for a **Pt/GaAs contact**[177]

Table 6.1 Metal/GaAs Schottky barrier systems

Metallization	Temperature (°C)	ϕ_{bn} (eV)	Analysis technique	Comments on degradation	Refs
Al	RT	~0.7	μAES		51, 52
	200	~0.8	SIMS	Slight Ga and As outdiffusion	53, 54
			XRD		55, 125
	250	~0.9	SEM	Ga outdiffusion, Al indiffusion	173, 174
	350	~0.87	TEM	More Ga outdiffusion, slight As outdiffusion	
	450—500	~0.8	RBS	More Ga and As outdiffusion, Al indiffusion 20–40 mm	
				GaAs + Al precipitates near GaAs surface, Al–As crystallites at interface	
	550			More Ga and As outdiffusion	
Ag	RT		SEM	Poor ohmic for p$^+$	175, 190
	300			Faceted holes below contact in GaAs	199,
	400–750				
Au	RT	~0.9	RBS channelling	Ga outdiffuses to the surface to form Ga_2O_3; low concentration of As in Au film; Au diffuses slightly into GaAs (40–150 nm)	47, 48,
	200–300	0.7	AES		51, 119,
			EPM		126, −137,
	350–450	0.5	XRD	More Ga outdiffusion; As outdiffusion and evaporation a Au–Ga (21 at% Ga) phase with Au segregation	176, 184,
	>450		SIMS		191, 192,
			TEM	Non-uniform Au diffusion into GaAs; Au precipitates extend 300 nm into GaAs producing a high dislocation density in GaAs;	195, 199
Cr	RT–300		RBS, SEM	No reaction	199,
	425		SIMS	2–3 at% Ga outdiffusion into Cr	177
	550			Intermetallic formation Cr_2X (X = Ga and/or As) Cr indiffusion	

Metallizations for GaAs devices and circuits

Metal	Temperature		Techniques	Observations	Refs
Mo	RT–450		RBS, SIM SEM AES XRD	Stable—no reaction; Bubbles on film surface owing to thermal expansion	60, 177 199,
	600/30 min	rectifying	AES, SIMS	No interdiffusion	
	700/30 min	ohmic	SEM XRD	Mo_5As_4 forms	
	800/30 min	ohmic		Mo_5As_4 major phase; Mo_2As_3, Mo_3Ga, and Mo_3GaAs_2 also detected	
	1000/30 min N_2	rectifying		Mo_3GaAs_2 major phase;	
Ni	250		RBS	Formation of Ni_2GaAs	175, 177
	400		SEM XRD AES	Ni_2GaAs decomposes to form NiAs and β'-NiGa	180 239–241
Pd	RT		UPS, XPS, RBS	Pd_2GaAs formation, Po	119, 236
	250		XRD, AES, TEM	Pd_2GaAs decomposes, PdGa forms	242
	>350				
Pt	RT	~0.9	RBS, SIM, AES XRD, TEM	$PtAs_2$ and PdGa form, Layered structure $Pt/Pt_3Ga/Pt$–$PtGa_2/PtAs_2/GaAs$ also reported	47, 51 59, 66 119, 177, –182 184, 237
	300–350	~0.9 0.86			
	500	1.0		$PtGa$–$PtGa_2/PtAs_2/GaAs$ forms; Pt indiffusion	
	600			Brittle alloy forms	
Ta	RT–300		RBS, SIMS μAES EPM	Stable—no reaction	177
	425–450			Ta indiffusion ~ 200–300 nm, Ga outdiffusion	
	500			More Ga and As outdiffusion; Ta indiffusion and possible TaX formation at interface (X–Ga and/or As)	

Table 6.1 Continued

Metallization	Temperature (°C)	ϕ_{bn} (eV)	Analysis technique	Comments on degradation	Refs
Ti	RT 450	~0.82 ~0.84	RBS, SIMS XRD EPM	TiAs forms;	66, 119 177, 199 293
	500 550	0.80	AES	Intermetallics have been reported TiAs, Ti$_5$As$_3$, Ti$_2$Ga$_3$, Ti$_5$Ga$_4$	
W	RT–500	~0.65 ~0.67	RBS, SIMS SEM–EDS	Stable—no reaction; slight increase in ϕ_b with temperature; layers tend to peel off ($T > 400°C$) owing to thermal expansion mismatch	47, 70 119, 177 184
	800/10 min	0.73	XRD, RBS	W–GaAs interaction occurs	
	950/10 min 900/3–30 min	0.74		Lateral diffusion of implanted dopants in a self-aligned MESFET structures Less spread if gates parallel to [01$\bar{1}$] than [0$\bar{1}$1] direction	
Au/Pt	350		AES	Ga and As outdiffuse; Ga edge leads As edge; Au diffuses through Pt	178, 179
	500			More Ga than Pt in Au; Au/Pt–Ga/Pt–Ga–As/GaAs	
Au/W	RT–500		XRD	Stable—no reaction	244
Ni/Au	RT–350		RBS, EPM	No Ni accumulation at Au/GaAs interface	142
Pt/Ni	RT–200 380–480 > 500	0.88 ~0.95	XRD	No reaction GaPt, PtAs$_2$, Ga$_2$Pt formed Ga$_2$Pt, PtAs$_2$, GaPt Schottky barrier degraded	180
Pt/Ti	RT–350 500	~0.81 ~0.86	RBS XRD	No reaction Pt$_3$Ga/TiAs/PtAs$_2$/GaAs formed	66
Pt/W	RT–500	~0.7	XRD	No reaction	244
Au/Pt/Ti	RT 250			Ga outdiffusion into Ti, slight As outdiffusion, Pt diffusion through Ti and into GaAs	57

Metallizations for GaAs devices and circuits

Au/Ti–W	as deposited		SEM–EDS	Leaky owing to damage induced by sputter deposition	64,
	700/10 min	0.8		No degradation	
	800/10 min	0.9		Surface starts to look silver in colour; poor breakdown characteristics	
	860/10 min		AES	Surface morphology unchanged	58
	as deposited	0.65		Leaky owing to damage induced by sputter deposition	
	500/24 h	0.7			
	600/15 h	0.75		Diode becomes leaky; Au detected throughout Ti–W	
Au/αW–Si	400/16 h		RED, AES, SEM Electron channelling	Amorphous W–Si layer; no interdiffusion; $D < 3 \times 10^{-18}$ cm^2 s^{-1} for Au, Ga, and As	62
Au/αTi–W–Si	300/944 h	0.79	RED, AES, SEM Electron channelling	Amorphous Ti–W–Si layer; good MESFET reliability No degradation	62
Au/Mo/Ti	as deposited	0.8	SEM–EDS	TiAs forms	64
	500/10 min	0.8		Ga detected at surface; contact appears silver in colour	
	550/10 min	0.9			
Ti–W	as deposited	0.6	RBS, SIMS	Ti–W–GaAs interaction starts	65
	750/15 min	0.68		Ga and As outdiffusion and Ti indiffusion	
	850/15 min	0.5			
	as deposited	0.8	SEM–EDS		64
	860/10 min	0.9			
Ti–W/Ti	as deposited	0.8	SEM–EDS	High leakage current owing to damage induced by sputter deposition, lower leakage currents for evaporated metallization	64
	500/10 min	0.74			
	680/10 min	0.8			
W/Si/W	as deposited		RED, AES, SEM Electron channelling	Weak channelling pattern; polycrystalline W near surface; an amorphous region forms at W/Si interface	62
	500/4 h			Metallization becomes amorphous	
Ti–W/Si/Ti–W	500/4 h		RED, AES, SEM Electron channelling	Metallization becomes amorphous	62

Table 6.1 *Continued*

Metallization	Temperature (°C)	ϕ_{bn} (eV)	Analysis technique	Comments on degradation	Refs
Ti–W–Si	as deposited 850/1 hr 800/10 min	0.75 0.85	RBS, SIMS	Composition close to Ti–W silicide No interaction observed Used for activating implants for self-aligned MESFETs	65
TaSi$_2$	as deposited 450/2 min 500/30 min 800/20 min	0.79 0.78	XRD, AES TEM, RBS	Amorphous with $R_s \sim 32\,\Omega\,\text{cm}^{-2}$; high leakage current owing to damage induced by sputter deposition Amorphous; withstands alloy cycle for ohmic contacts in MESFETs Crystallization starts and R_s decreases Crystalline $R_s \sim 8\,\Omega\,\text{cm}^{-2}$; no observable interdiffusion	56, 71 73
Au/αNi–Nb	as deposited 400/2 h air	0.51	AES, XRD	Amorphous Ni$_{45}$Nb$_{55}$ alloy film Au diffusion; Ni–As reactions at Ni–Nb/GaAs interface; soft reverse characteristics	63
Au/αTa–Ir	as deposited 500/24 h 600/24 h 700/24 h	0.79		Amorphous Ta$_{56}$Ir$_{44}$ alloy No observable interdiffusion Some Ga outdiffusion Complete chemical intermixing	63
Al/Ti/GaAs	400	0.75	XRD, AES	Al$_3$Ti GaAlAs formation at interface	238

Metallizations for GaAs devices and circuits 215

Amorphous metallizations: W–Si/GaAs, Ti–W–Si/GaAs, Ni–Nb/GaAs, and Ta–Ir/GaAs Another approach to forming stable contacts has been the use of amorphous metallization.[61–63] Christou et al.[62] have studied amorphous W–Si and Ti–W–Si contacts obtained by depositing alternate layers of W or Ti–W and Si and annealing near the glass transition temperature, T_g (~ 500°C), when these polycrystalline metal films turned amorphous. These also form good diffusion barriers between gold and GaAs, with no detectable interdiffusion observed by Auger electron spectroscopy after annealing at 400° C. Kelly et al.[63] investigated the stability of contacts formed using Ni–Nb and Ta–Ir amorphous alloys deposited by sputtering. Ta–Ir was found to be the most stable metallization with no interdiffusion being observed for the Au/Ta–Ir/GaAs structure after annealing at 500°C.

6.3.2 Multilayer contacts

Au/Pt/Ti/GaAs The Au/Pt/Ti and Au/Pd/Ti systems have proved very successful in silicon beam-lead technology but as yet have had limited success with GaAs. Although stable Au/Pt/Ti contacts can be realized for temperatures less than 50°C, above 200°C there is increased interdiffusion between GaAs and Ti, resulting in the formation of TiO_2 at the surface and a non-stoichiometric interface. This seriously degrades the device performance. Despite these limitations, the Au/Pt/Ti system is widely adopted in the industry for the gates of MESFETs.

Au/Ti–W/GaAs There is, however, a growing shift towards the Au/Ti–W system, which maintains its stability up to 600°C for times up to 15 h. An advantage of the gold/refractory metal contacts is that they can withstand adverse sodium and chloride ion contamination and exposure to a high relative humidity for prolonged periods. The main limitations of the Au/Ti–W structure, as with other refractory metallizations, is the thermal expansion difference between the Ti–W and GaAs, which causes adhesion problems, and the introduction of defects in the semiconductor beneath the contact. In MESFET devices this causes a reduction in the mobility and degradation of the transconductance, g_m.

Au/Pt/GaAs A different metallization system has been used for GaAs IMPATT diodes. In these devices an Au/Pt system bonded at a diamond heat sink is preferred, platinum being used in preference to tungsten or titanium because its larger barrier height is better at the device's higher operating temperature. A disadvantage of platinum, however, is that it allows the gold to migrate through it to the GaAs to form metallic spikes which then short-circuit the junction. Although this situation can be improved by increasing the thickness of platinum, the new reaction products consume the GaAs epitaxial layer and upset the tailored doping profile which is the key to the IMPATT efficiency.

Au/Pt/W/Pt/GaAs and Au/Ti–W/Pt/GaAs A dramatic improvement has been achieved by reacting a thin film a platinum on the GaAs to form a layered structure of PtGa/PtAs$_2$/GaAs; this is then followed by an Au/Pt/W metallization system. However, this structure has not completely solved the reliability problem as tests indicate some instability in the PtGa/PtAs$_2$ layers which drastically degrades the efficiency to less than 15% when annealed at 175°C for 1000 h. Other systems, particularly the Au/Ti–W/Pt, show some measure of success but the presence of Pt still presents an unsolved reliability problem.

6.3.3 High-temperature stable Schottky contacts

The move towards smaller device geometry has resulted in a need for self-aligned implantation in GaAs. Figure 6.11 shows how a Schottky barrier metallization may be used as an implantation mask to enable n$^+$ regions to be formed close to the gate electrodes of a FET transistor. This technology requires a Schottky barrier metallization which is stable during the annealing stage used for activating the implanted atoms. From Table 6.1 it is clear that few metallizations are stable

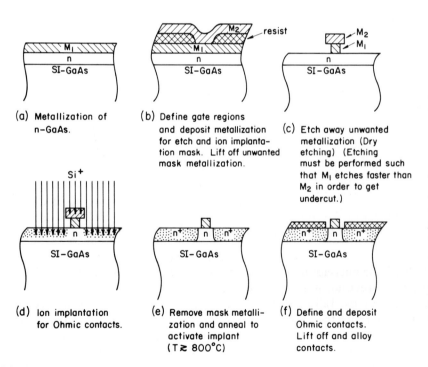

Figure 6.11 Schematic of processing steps for fabrication of a self-aligned MESFET structure

on GaAs during anneals at temperatures up to 800–950°C. This annealing is usually carried out with a dielectric cap of SiO_2 or Si_3N_4 (or both) or alternatively in an ambient with an arsenic overpressure.

Ti–W/GaAs Recently, Ti–W alloy (10 wt % Ti) has been found to be stable in its electrical characteristics up to about 700°C.[64] Annealing above about 750°C results in degraded Schottky barriers and outdiffusion of gallium and arsenic together with indiffusion of titanium.[65] This is perhaps not too surprising since TiAs is known to form in Ti/GaAs structures above 500°C.[66] However, Sadler et al.[67] have used Ti–W metallization for self-aligned FET structures, the annealing being done with an arsenic overpressure at 800°C for 15 min.

Ti–W silicide/GaAs, W silicide/GaAs, and W/GaAs Yokoyama et al.[65] found that co-depositing Ti–W alloy with silicon to form a mixed Ti–W silicide after annealing resulted in a metallization stable for annealing temperatures up to 850°C. Pure tungsten Schottky barriers also exhibit stable electrical characteristics, but with metallurgical reactions, when annealed in an arsenic overpressure to temperatures up to 950°C.[68] Ti–W silicide,[65] W silicide,[69] and tungsten[70] have been used successfully for self-aligned FET device structures.

Ta silicide/GaAs Another metallization that has been shown to be stable is tantalum silicide.[56,71-73] Figure 6.12a shows the back-scattering spectra of a layered Si/Ta/Si/GaAs structure before and after annealing at 800°C and 920°C for 20 min in an arsenic overpressure ambient. Note that there is no change in the GaAs front edge in the 800°C annealed spectrum, which indicates that little metallization–GaAs interdiffusion has taken place. The drop in the Ta peak height and change in the Si peaks (formation of one broad peak rather than two individual peaks) arise from the formation of $TaSi_2$ on top of the GaAs, as is shown in the insert of Figure 6.12a. Complete $TaSi_2$ formation is observed after the 920°C 20 min anneal. The GaAs front edge shows some broadening, which can arise from either a non-uniform $TaSi_2$ layer or slight $TaSi_2$–GaAs interdiffusion. The cross sectional TEM micrograph in Figure 6.12b shows that the $TaSi_2$/GaAs interface is non-uniform. The peak appearing around 2.4 MeV in the RBS spectrum is probably due to arsenic diffusing from the annealing ambient into the $TaSi_2$ and to In at the $TaSi_2$/GaAs interface, arising from In diffusing through the $TaSi_2$ from the ambient (InAs was used as the source of As_2 and As_4 for the arsenic overpressure). These preliminary results are clear evidence that self-aligned implant technology will be feasible for future GaAs technology. Furthermore, the development of these high-temperature stable Schottky barriers will presumably also result in high-reliability contacts.

218 Gallium arsenide

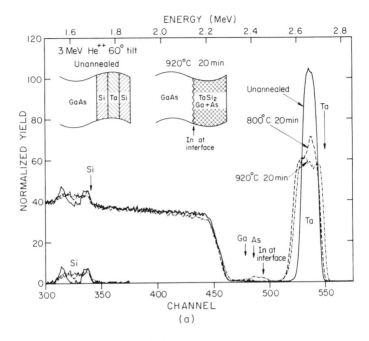

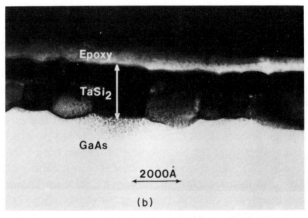

Figure 6.12 (a) Back-scattering spectra for a Si/Ta/Si/GaAs structure unannealed and after annealing at 800 and 920°C for 20 min in an arsenic overpressure furnace. (b) Cross sectional TEM micrograph of a Si/Ta/Si/GaAs structure after annealing at 920°C for 20 min in an arsenic overpressure furnace

6.3.4 Schottky barrier height tailoring

Earlier in this section it was stated that the barrier height of a metal on GaAs is nearly independent of the metal work function. Consequently the barrier height cannot be used as a design parameter. However, the effective barrier height can be modified by: (i) incorporation of a thin highly doped layer beneath the contact and (ii) the use of a heterojunction. The use of a thin highly doped zone within the depletion region has been used to tailor the effective barrier height of Schottky diodes,[74,75] in that, for an n-type semiconductor, a p-type zone results in an increase in effective barrier height while an n^+ zone results in a decrease.

The alternative technique for barrier height control is to use a heterostructure. For example, Woodall et al.[76] have suggested the use of a graded $Ga_{1-x}In_xAs$ heterojunction for controlled barrier height formation. The technique works by growing a graded $Ga_{1-x}In_xAs$ layer on top of GaAs. The growth is started with $x = 0$ at the GaAs surface, thereafter increasing x until the desired band gap, and hence barrier height, is obtained.[77]

The behaviour of reacted Mo/GaAs contacts[60] (ohmic when annealed at 700–800°C and Schottky for 1000°C anneals) suggests that selective compound formation on GaAs could possibly be used for controlled barrier height formation.

From the recent result of Massies et al.[23] with H_2S adsorption prior to Al deposition on UHV cleaved GaAs, which results in $\phi_{bn} \sim 0.4\,eV$ rather than $0.8\,eV$ associated with Al/GaAs, one can speculate that interface chemistry may well be able to be used for tailoring of the barrier heights in the future.

6.4 OHMIC CONTACTS

During the past few years a number of review articles dealing with ohmic contacts to III–V compound semiconductors have appeared in the literature.[50,78–81] In general these have been concerned with the technology involved in producing conventional 'alloyed' ohmic contacts. The present survey will emphasize the principles of each technique rather than the particular recipe for producing an 'ideal' ohmic contact. It will also concentrate on ohmic contacts to n-type GaAs.

6.4.1 Tunnelling contacts: metal/n^+-GaAs

In order to achieve good ohmic behaviour by depositing a metal layer directly onto n-type GaAs without any subsequent annealing, the GaAs immediately beneath the metallization must be heavily doped to a level of 5×10^{19} donors/cm^3 or more.[82] This can be seen from Figure 6.13, which shows the specific contact resistivity versus $(N_d)^{-1/2}$ for both GaAs and Ge, as calculated from tunnelling theory. The maximum obtainable donor concentrations for conventional epitaxial growth techniques (liquid phase epitaxy (LPE), vapor phase epitaxy (VPE), metal-organic chemical vapor deposition (MOCVD), and molecular beam epitaxy (MBE)) is generally limited to below about $(1–2) \times 10^{19}$ donors/cm^3. However, donor levels up to 5×10^{19} cm^{-3} have been reported for MBE-grown layers.[83] In most cases the problem is not simply to obtain high concentrations of

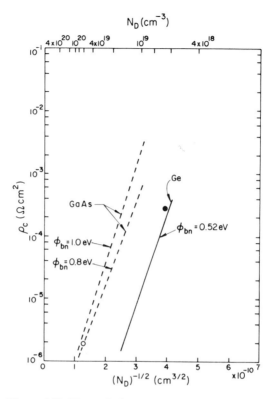

Figure 6.13 Theoretical contact resistivity, ρ_c, plotted as a function of doping level for metal contacts on Ge and GaAs. The conduction is assumed to be by field emission. The points represent experimental data for Ge (●) and GaAs (o). (Adapted from Devlin et al.[117])

impurity atoms in the GaAs, but also to make them electrically active. Woodall[27] has discussed the effect of arsenic overpressure on the doping behaviour of Si. He found that for high donor activation the annealing or growth of the GaAs must be done at high arsenic overpressure ($P_{As}/P_0 \sim 10$–1000, where P_0 is the As vapour pressure from GaAs).

Electrical activation is also the limiting factor for doping by ion implantation. Conventional annealing for activating an implant is generally limited to doping levels up to about $(2$–$5) \times 10^{18}$ cm^{-3},[84] although concentrations of $(1$–$2) \times 10^{19}$ cm^{-3} have been reported.[85,86] Pulsed annealing by either laser or electron beams can result in higher doping levels, $(2$–$4) \times 10^{19}$ cm^{-3}. Pulsed annealing techniques are discussed in refs 87–93. Recent review articles by Anderson,[94] Williams and Harrison,[95] and Williams[96,97] provide excellent overviews of thermal and transient annealing of GaAs.

Metallizations for GaAs devices and circuits

High doping concentrations of 1×10^{19} cm^{-3} or more have also been achieved by depositing a thin film containing a dopant on the GaAs surface prior to pulsed annealing.[98,99] Overlayers have included Ge, Si, Sn, In, Si_3N_4, SnO_2–SiO_2, and As_2S_3.[98-104] The exact structure of these doped layers is somewhat unclear. Greene et al.[105] found that for scanned CW laser annealing of Ge/GaAs, a $Ge_x(GaAs)_{1-x}$ alloy could be formed with high doping levels of 10^{20}–10^{21} cm^{-3}. Similarly, $In_xGa_{1-x}As$ ternary compound formation has been suggested to form after pulsed laser annealing of In/GaAs.[103] Sn_3As_2 has been observed after CW laser annealing SnO_2–SiO_2/GaAs.[104]

Unfortunately, although pulsed annealing can give rise to high doping levels, the layers produced have poor mobilities. Extended line defects and planar defects are thought to be primarily responsible for this low mobility.[94] A recent review article by Fan et al.[106] provides an overview of the correlation between electrical properties and extended crystallographic defects. Thus, only a narrow range of energies can activate implants and give defect-free layers[107] and good mobilities.[108] Furthermore, the high doping levels are in general achieved by having dopant concentrations above the solid solubility in the GaAs. This metastable situation may lead to decreased carrier concentrations during further annealing or during device operation at elevated temperature, resulting in increased contact resistivity. Pianetta et al.[109,110] found that the donor level of about 4×10^{19} cm^{-3} obtained by pulsed electron beam annealing of Se$^+$- or Te$^+$-implanted GaAs decreased dramatically as a result of thermal annealing. They suggest that the decrease during low-temperature anneals (~ 200–$315°C$) arises in the case of Te from Te–vacancy complex formation and during high-temperature annealing ($> 450°C$) from Ga_2Te_3 precipitate formation.[110] Table 6.2 contains a summary of ohmic contacts formed by direct deposition of contact metal on n^{++}-GaAs. Note that by annealing after contact deposition a reduction in the specific contact resistivity is generally observed.[111]

A variation of the metal/n^{++}-GaAs contact (also included in Table 6.2) is to deposit the metal layer prior to implantation and implant the dopant ions through the contact layer to form an n^{++} region beneath it.[112] A subsequent anneal is required to activate the implant. Christou and Davey[112] suggest the use of a refractory metallization (such as Ti–W, which does not react with the GaAs during the activating anneal. It is of course a question of whether the relatively low temperature of roughly 700°C (see Table 6.2) used for activating the implant is sufficient to anneal out all the radiation damage in the underlying GaAs. However, the promising results of metal silicides discussed in Section 6.3.3 suggest that these metallizations could be used for anneals up to about 900°C.

6.4.2 Low barrier height Schottky contacts: heteroepitaxy

Ge/GaAs Low barrier height Schottky contacts to GaAs are difficult to obtain owing to the associated Fermi level pinning discussed earlier. However, low

Table 6.2 Homoepitaxial and heteroepitaxial ohmic contacts (unreacted contacts)

Dopant/ion/layer Ion implantation (keV, atoms/cm^2) Annealing	Metallization	ρ_c (Ω cm^2)	N_d of contact layer (cm^{-3})	N_d beneath contact (cm^{-3})	Analysis technique	Comments	Refs
HOMOEPITAXIAL LAYERS							
MBE-grown layrs							
Sn-doped layer	Pt/Ti	1.86×10^{-6}	6×10^{19}	2×10^{18}			83
		$10^{-3} - 10^{-4}$	1×10^{19}	2×10^{18}			
	Au/Sn	7.8×10^{-5}	8×10^{18}	4×10^{16}		Sn *in situ* deposited	185
		2.9×10^{-5}	8×10^{18}	1×10^{18}			
	Au/Ag/Au–Ge	$\sim 6 \times 10^{-6}$	5×10^{19}			ρ_c reduced upon < 360°C anneal	111
	Au/Ag/Au–Ge/Sn	$\sim 1 \times 10^{-6}$	5×10^{19}				
	Au/Ni	4.6×10^{-6}	1×10^{19}	8×10^{16}		ρ_c reduced upon 250°C anneal	98
Ion implantation							
Si$^+$							
70–200, (1–3) $\times 10^{15}$ Q-switched ruby 0.8–1.5 J cm^{-2}	Au/Au–Ge	$(3–90) \times 10^{-7}$			AES	Little metallization–GaAs reaction	108
	Au/Ni/Au–Ge Au/Pt/Ti	$(3–90) \times 10^{-7}$ poor ohmic				No difference with Ni	
100, 1 $\times 10^{15}$ Q-switched ruby ~ 0.65 J cm^{-2}	Au/Ni	$(1.4–4) \times 10^{-6}$	$> 2 \times 10^{19}$			Si$_3$N$_4$ encapsulant for anneal; 500 Å GaAs removed prior to metallization	98
Se$^+$							
250, 5 $\times 10^{15}$ PEBA ~ 0.75 J cm^{-2}	Al	$(2.3–5.8) \times 10^{-6}$	$> 2 \times 10^{19}$		SIMS	Hot implant (350°C); doping concentration decreased during 250°C anneal	109, 186

Metallizations for GaAs devices and circuits

Ion; energy (keV), dose (cm^{-2}), anneal	Metallization	ρ_c (Ω cm^2)	N_D (cm^{-3})	N_s (cm^{-3})	Analysis	Comments	Ref.
50, 5×10^{15} PEBA ~ 0.9 J cm^{-2}	Al	$\sim 5.5 \times 10^{-6}$	$> 1 \times 10^{19}$				
120, 5×10^{15} PEBA ~ 0.71 J cm^{-2}	Au/Pt/Ti	3×10^{-7}	$> 1 \times 10^{19}$	$\sim 9 \times 10^{16}$		~ 750 Å GaAs removed prior to metallization	187
Te$^+$							
50, 1×10^{16} Q-switched Nd:YAG ~ 2.5 J cm^{-2}	Pt/Ti	$\sim 2 \times 10^{-5}$	$\sim 2.5 \times 10^{19}$	3×10^{18}	RBS–channelling SEM	Excess Ga on GaAs surface after pulsed anneal; 50 Å etched off prior to metallization	188, 189
In$^+$							
30–60, $(1$–$10) \times 10^{15}$ pulsed ruby	Probed	$\sim 1 \times 10^{-4}$	2×10^{18}		RBS–channelling	In$_x$Ga$_{1-x}$As surface alloy forms; 5 at % In on lattice sites beneath surface	97, 103
Sn$^+$							
150, 1×10^{15} Q-switched ruby ~ 0.7 J cm^{-2}	Au/Ni	$\sim 4 \times 10^{-6}$	$> 2 \times 10^{19}$			Si$_3$N$_4$ cap for anneal; 500 Å GaAs removed prior to metallization	98
Ion implantation through contact metallization							
Si$^+$							
60, $> 1 \times 10^{14}$, 700 °C/30 min	Au/Ti–W	$\sim 1 \times 10^{-6}$	$\sim 1 \times 10^{18}$			Implant through Ti–W layer and deposit Au after annealing implant	112
Se$^+$ or Ge$^+$ $> 10^{14}$							
Pulsed annealing of overlayers							
Si							
Q-switched ruby ~ 0.8 J cm^{-2}	Au/Ni	1.4×10^{-4}	$\sim 2 \times 10^{19}$			Si$_3$N$_4$ encapsulation for anneal; ~ 500 Å GaAs removed prior to metallization	98

Table 6.2 Continued

Dopant/ion/layer Ion implantation (keV, atoms/cm^2) Annealing	Metallization	ρ_c (Ω cm^2)	N_d of contact layer (cm^{-3})	N_d beneath contact (cm^{-3})	Analysis technique	Comments	Refs
Q-switched Nd:YAG ~ 0.74 J cm^{-2}		5×10^{-7}			RBS–channelling SIMS	SiO$_2$ encapsulation for anneal; possible Si-GaAs compound formation; contacts degrade upon long-term annealing at 300°C	100
Ge							
Q-switched ruby ~ 0.7 J cm^{-2}	Au/Ni	6.4×10^{-6}	$> 2 \times 10^{19}$			Si$_3$N$_4$ encapsulation for anneal; ~ 500 Å GaAs removed prior to contact metallization	98
Q-switched Nd:YAG	Au	$> 1 \times 10^{-6}$		2×10^{18}	SEM–EDS	Ge droplets ~ 5000 Å in diameter form on the surface after annealing	102
Sn							
Q-switched ruby ~ 0.7 J cm^{-2}	Au/Ni	$(1.8–2.6) \times 10^{-5}$	$\sim 3 \times 10^{19}$			Si$_3$N$_4$ encapsulation used for anneal; ~ 500 Å GaAs removed prior to metallization	98
In							
pulsed ruby	Probed	$< 1 \times 10^{-4}$				Possible In$_x$Ga$_{1-x}$As compound formation	103
CW, CO$_2$ from back-side	Probed	$< 1 \times 10^{-4}$				Back-side annealing enables thicker In layers to be used	
Si$_3$N$_4$ Q-switched ruby ~ 0.7 J cm^{-2}	Au/Ni	$(3.5–5) \times 10^{-5}$	$\sim 1 \times 10^{19}$			~ 500 Å GaAs removed prior to metallization	98

Material	Metallization	ρ_c (Ω cm²)			Technique	Comments	Ref.
SnO_2-SiO_2 900°C and CW argon	Au/Pt/Ti	$(1-2) \times 10^{-6}$				SiO_2 encapsulation for anneal; substrate held at 350°C during laser anneal; thermal anneal resulted in Sn_3As_2 and β-Ga_2O_3 formation; ρ_c increased ~ 50% after annealing at 320°C for 1000 h	104
HETEROEPITAXIAL LAYERS							
Ge MBE grown	Au	$< 1 \times 10^{-7}$	$> 1 \times 10^{20}$	1×10^{19}	RED, SIMS	Ge epitaxial layer doped $> 1 \times 10^{20}$ As/cm³	113, 114, 116, 117
		$< 1 \times 10^{-7}$	$> 1 \times 10^{20}$	1×10^{18}	AES		
		$\sim 1.5 \times 10^{-7}$	$> 1 \times 10^{20}$	1×10^{17}			
	Au/Mo	$\sim 9 \times 10^{-6}$	$> 1 \times 10^{20}$	1×10^{17}			
$Ga_{1-x}In_xAs$ MBE grown	Ag	$(5-50) \times 10^{-7}$		2×10^{17}		Graded heteroepitaxial layer $x = 0.01$ at interface and $x = 1$ at surface (layer 2500 Å thick); Ag *in situ* deposited	76

barrier heights can be obtained with other semiconductors, for example, germanium, where barrier heights of about 0.5 eV can be obtained. Furthermore, n-type Ge can be doped to higher levels (i.e. $\sim 10^{20}$ donors/cm^3) than GaAs. From Figure 6.13 it is clear that lower specific contact resistivities are obtainable to germanium. Stall et al.[113,114] have utilized this to form ohmic contacts to GaAs by using an intermediate heteroepitaxial Ge layer. In this technique an epitaxial Ge layer was grown directly onto the GaAs epitaxial layer by MBE. The proposed band structure of this contact is shown in Figure 6.14. Note that there is a small electron barrier at the Ge/GaAs interface. The low metal/Ge barrier height (~ 0.5 eV) and high doping level in the germanium ($\sim 10^{20}$ cm^{-3}) result in a very low specific contact resistivity ($\lesssim 1 \times 10^{-7}$ Ω cm^2) of this ohmic contact to GaAs. The same principle has been used by Anderson et al.[115] but in their case the germanium was deposited on heat-cleaned GaAs by vacuum deposition. The use of Ge/GaAs heterojunctions for ohmic contacts in FET device structures has been demonstrated by Devlin et al.[116,117]

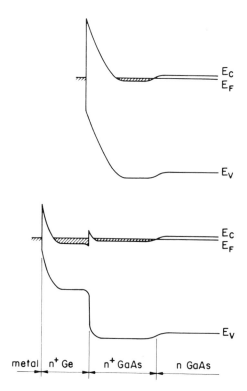

Figure 6.14 Energy band diagram for a metal/n$^+$-Ge/n$^+$-GaAs/n-GaAs contact. (From Devlin et al.[117])

Metallizations for GaAs devices and circuits

$Ga_{1-x}In_xAs/GaAs$ Woodall et al.[76] have used an MBE-grown epitaxial layer of $Ga_{1-x}In_xAs$ instead of germanium. In order to reduce the interfacial barrier and the interfacial strain produced by the large lattice mismatch between InAs and GaAs ($\sim 7\%$), the heteroepitaxial layer was graded in composition. It was grown by MBE on top of the GaAs, the composition being varied during growth such that $x = 0$ at the GaAs interface and $x \simeq 0.8$ to 1.0 at the surface. The proposed band structure of this contact is shown in Figure 6.5d. The resultant specific contact resistance of these systems was in the range of approximately 5×10^{-7} to $5 \times 10^{-6} \Omega \, cm^2$ (for $Ag/n-Ga_{1-x}In_xAs/GaAs$). Recently, a further development of heteroepitaxial contact layers has been shown by Woodall et al.[118] They suggest that the pinning of the GaAs Fermi level at the heteroepitaxial layer/GaAs interface arises from the formation of misfit dislocations. By growing thin layers of $Ga_{1-x}In_xAs$ (< 1000 Å) on GaAs, they were able to produce pseudomorphic layers. Contacts made to these pseudomorphic layers showed ohmic behaviour. Conversely, when thicker layers were grown, misfit dislocations formed and the contacts were rectifying. These results show great promise for the use of heteroepitaxial contact layers for ohmic contacts. However, the lattice mismatch between the heteroepitaxial contact layer (e.g. InAs) and GaAs may result in a reliability problem even for graded layers, but as yet there are no reliability data available.

6.4.3 Conventional contacts

In conventional contact technology the basic idea is to deposit a metallization, which contains a dopant, onto the GaAs. Then the complete structure is heated and interdiffusion between the metallization and GaAs occurs. The dopant is believed to dope heavily a thin layer of GaAs immediately beneath the metallization. Hence, a $metal/n^{++}$-GaAs contact is formed. These conventional contacts fall into two categories: sintered and alloyed. In the former contacting scheme the metallization–GaAs interaction occurs in the solid phase, whereas in the latter it takes place in the liquid phase.

6.4.3.1 *Sintered contacts: solid phase reactions*

Pd/Ge/GaAs Table 6.3 contains a summary of sintered ohmic contacts. In the case of Pd/Ge/GaAs the contacts become ohmic after sintering at temperatures greater than roughly 400°C,[119,120] during which Pd_2Ge, PdGe, $PdAs_2$, and PdGa are formed. The ohmic behaviour is attributed to a combination of the doping action of Ge and fast diffusion of Pd.

Au–Ge/GaAs Recently, the commonly used Au–Ge alloyed contact metallization has also been used for a sintered contact (i.e. heating below the Au–Ge

Table 6.3 Sintered ohmic contacts

Metallization	Annealing conditions °C	time	ambient	ρ_c ($\Omega\,cm^2$)	N_d (cm^{-3})	Analysis	Comments	Refs
Pd/Ge	200	20 min	N_2	not ohmic	1.1×10^{16}	AES, XRD, SIMS, ESCA	Pd/Pd_2Ge–$PdGe/GaAs$ contact structure formation Reactions with GaAs starts; formation of Pd–Ge/Pd–As/Pd–Ga/GaAs layered structure; all Pd is reacted	120
	300	>10 min	N_2	not ohmic	1.1×10^{16}			
	>400	110 min	N_2		1.1×10^{16}		Ge redistributes throughout contact; surface becomes slightly rough	
	600	20 min	N_2	4×10^{-4}	1.1×10^{16}		Ge distributes uniformly in contact and some may be incorporated in GaAs immediately beneath contact	
	500	2 h	vacuum	3×10^{-4}	1.2×10^{16}	AES, XRD	Contact has a textured surface; all Pd reacts to form PdGe, PdGa, and $PdAs_2$	119
Pt/Ge	500	2 h	vacuum	not ohmic				
W/Ge	500	2 h	vacuum	not ohmic				
Ni/Ge	550	5 min	vacuum, H_2–N_2	3×10^{-5}	1.1×10^{17}	AES, RED, SEM	Ge deposited epitaxially on GaAs; extensive inter-diffusion of contact metallization and GaAs; Ge penetration into GaAs is deeper than Ni; contact surface is smooth	115
Ta/Ge	650	5 min	vacuum, H_2–N_2	5×10^{-4}	1.1×10^{17}		Ge deposited epitaxially on GaAs	
Mo/Ge	750	5 min	vacuum, H_2–N_2	5×10^{-4}	1.1×10^{17}		Ge deposited epitaxially on GaAs	
Au–Ge	300		vacuum	4.2×10^{-5}	2×10^{18}		Substrate held at 300°C during deposition of Au–Ge	121
	275	2 min	H_2–N_2	2.5×10^{-5}	2×10^{18}		Contacts are relatively stable for anneals at 225°C in air (alloyed contacts give $\rho_c \sim 4 \times 10^{-6}\,\Omega\,cm^2$)	122
	315	>3 h	N_2	2×10^{-5}	1×10^{18}	SIMS, SEM	Laterally non-uniform contact surface; Ge, Au, Ga and As interdiffuse	123
	330	1 h	N_2	3×10^{-6}	1×10^{18}			

eutectic temperature of 356°C).[129-131] As can be seen from Table 6.3 sintered ohmic contacts do not form contacts with particularly low specific contact resistances to low doped GaAs. In general, they are only used for GaAs with doping levels of 10^{18} cm^{-3} or higher.

Au/Ag/Au–Ge/GaAs Sintered contacts can form low specific resistivity contacts when they are used in conjunction with tunnelling contacts (Section 6.4.1). DiLorenzo et al.,[111] using this technique, have formed contacts to n-GaAs with doping of 8×10^{16} cm^{-3}. Using MBE they grew an n^{++}-GaAs layer doped with 5×10^{19} cm^{-3} Sn atoms on top of an n-GaAs that had a doping of 8×10^{16} atoms/cm^3. Some layers had *in situ* deposited Sn prior to Au/Ag/Au–Ge contact deposition, which was performed after a break in vacuum. These contacts exhibited ohmic behaviour for both the structures that contained the *in situ* Sn layer and those without. Furthermore, in both cases reduction in specific contact resistivity was observed after short anneals at temperatures in the range 250–300°C.

6.4.3.2 Alloyed contacts

Alloyed ohmic contacts have been extensively studied and are generally the favoured contact in present device technology. A number of excellent review articles are found in refs 78–81. Table 6.4 contains a summary of the metallizations used, together with their references in the literature. Most contacts presently in use are based upon the Au–Ge eutectic metallization scheme. In principle, this contact is formed in the following way: The Au–Ge/GaAs structure is heated above the Au–Ge eutectic temperature ($\sim 356°$C). The Au–Ge metallization melts and GaAs is dissolved in the melt. Upon cooling the GaAs regrows epitaxially from the melt at the melt/GaAs interface. This regrown GaAs layer incorporates a high concentration of Ge, making it heavily doped n-type, hence forming a 'tunnelling' type contact. In reality the contact formation is not so simple, the contact tending to 'ball up' and become non-uniform both laterally and in depth.[124] Improvements have been made by adding Ni,[124] Pt,[79] or In[125] to the Au–Ge metallization, either in the eutectic mixture or as a separate layer. Since these contacts are of such technological importance, they will be discussed in detail.

The Au/GaAs system In an attempt to understand alloyed contacts it is instructive to consider the Au/GaAs system, which has been studied extensively and reported in the literature.[47,48,51,126-137] Annealing at low temperatures (~ 250–350°C) results in both gallium and arsenic diffusing into the gold film and gold diffusing into the GaAs.[47,48,126,127] The low solubility for arsenic in gold results in arsenic evaporation from the contact surface.[128,137] A region of damage and strain forms in the GaAs beneath the metallization[127,129] and the

Table 6.4 Alloyed ohmic contacts

Metallization	°C	time	ambient	ρ_c ($\Omega\,cm^2$)	N_d (cm^{-3})	Analysis	Comments	Refs
In	350–360		H_2	$\sim 2 \times 10^{-4}$	1×10^{17}	SEM	Holes form in GaAs beneath contact; poor reliability; $\rho_c \propto N_d^{-1}$	157, 175, 190–192
Sn	400		N_2, H_2	4.8×10^{-4}	1.5×10^{16}	SEM-EDS	Large amounts of GaAs consumed during alloying; faceted holes form beneath contact; $\rho_c \propto N_d^{-1}$	124, 157, 159, 190, 191, 193–195
	400		N_2	4.4×10^{-4}	8.7×10^{17}			
Ag	~ 550		vacuum	2×10^{-5}	$0.0033\,\Omega\,cm$		Ni deposited after *in situ* annealing	196
Ag/In	200	15 min		5×10^{-2}	5×10^{17}	AES	In and Ga diffuse to surface; growth of Ga–In–Al–As crystals on GaAs surface	197, 198
Ag/In	500	10 min	H_2	1.1×10^{-2}	5×10^{15}			199
				9.6×10^{-5}	1×10^{17}			
Ag–Sn				$\sim 3 \times 10^{-3}$	$\sim 5 \times 10^{14}$			159
Ag/Sn	>420		vacuum			QMA	As_2 evaporates during alloying	140, 200
	450	1 min					Balling; thicker layers result in lower ρ_c	201
	550–650	30 s	H_2–N_2	$<3 \times 10^{-4}$	$(1-3) \times 10^{15}$		Erratic results	191
					6.8×10^{18}			
	500	10 min	H_2	3.6×10^{-3}	5×10^{15}			199
				8×10^{-6}	1×10^{17}			
	600	10 min	H_2			SEM	(111) facets from in GaAs $\sim 0.5\,\mu m$ deep	190
Sn/Ag	320–450							183

Contact	Temp (°C)	Time	Ambient	ρ_c	Carrier conc	Techniques	Comments	Ref
Ag–In–Ge		as deposited					In accumulation at contact/GaAs interface and Ge at contact surface	202
	640		He		5×10^{17}	EPM, AES, XPS	Ga outdiffusion and surface accumulation; Ge accumulates at contact/GaAs interface; In uniformly distributed in metallization; As/Ga ratio at interface > 1; etch pits form in GaAs beneath metallization	
	600	1 min	H_2–N_2	1×10^{-3}	2.6 Ω cm		Etch pits form in GaAs	203
				6×10^{-4}	0.3 Ω cm			
				1×10^{-4}	0.15 Ω cm			
	~630	30 s	H_2–N_2	$<1 \times 10^{-4}$	0.1 Ω cm	SEM, x-ray topography	Large degree of strain in GaAs around contacts, which is believed to cause high-resistivity region below contact	194, 204
Ag/In/Ge	620	1 min	N_2, H_2–N_2	1×10^{-5}	2×10^{16}	AES, electron channelling	Contact contains 'solid phase epitaxy' of (100) GaInGeAs particles aligned with {100} directions on GaAs surface and agglomerations containing Ag–Ga	143, 204
							Ge accumulates at contact/GaAs interface; ρ_c depends upon the size of the 'solid phase epitaxial' particles	
Ag/Ge/In	500	10 min	H_2	9×10^{-4}	1×10^{15}			199
				9.5×10^{-5}	1×10^{17}			
	620	3 min	H_2		2×10^{18}	SEM, QMA	Most Ge at GaAs surface; As_2 evaporation	200, 201
Ag/Ge/Ga/In	600	40 min	As–H_2	$<1 \times 10^{-4}$	1.5×10^{15}	QMA	Ga and As outdiffusion increases with In content Deep penetration of Ag and In (~1.0 μm); contact structure becomes Ag–In–Ge/$In_{0.4}Ga_{0.6}$ As(Ge + Ag)/Ge–As–Ga–Ag/$In_{0.4}Ga_{0.6}$As/ GaAs; large amount of interfacial strain beneath contact	163, 164, 201, 205
Ag/Ga/Ge/In	600		As_2				Balling (possibly due to lack of optimization)	206

Table 6.4 *Continued*

Metallization	Annealing conditions			ρ_c (Ω cm^2)	N_d (cm^{-3})	Analysis	Comments	Refs
	°C	time	ambient					
Ag/Ge/Sn	~600		vacuum			QMA	High As$_2$ yield from surface during initial alloying	140
	500	10 min	H$_2$	1×10^{-3}	1×10^{15}			199
				2.3×10^{-5}	1×10^{17}			
Au/In	~550	30 s	H$_2$–N$_2$	$<3 \times 10^{-4}$	6.8×10^{18}	XRD, RBS, SEM-EDS	Mounding of contact; In$_2$Au formation; poor device reliability	79, 191 192, 207
	~500	2 min	H$_2$–N$_2$					
Au–Si	425		H$_2$				Balling occurs for alloying above 450°C	79, 208
Au/Sn	<300		H$_2$–N$_2$		3.9×10^{15}	RBS,SEM ED, XRD	Several Au–Sn phases detected in as-deposited contact AuSn, AuSn$_2$, and AuSn$_4$; contacts are rectifying	209
	350–700						Holes form in GaAs; higher ρ_c than Au/In contacts	79, 190, 191
Au/Te	500			poor				79
Au/Ni	500			~4.4×10^{-2}	2.3×10^{17}	SIMS, XRD	Clusters form on surface and Ni and Au indiffusion	210
						SEM	Balling	51
Au–Ge	350–450					EPM	Au–Ga contact layer formation with Ge segregating in grains	79, 124, 139, 158, 194
								211

Contact	Temp (°C)	Time	Ambient	ρ_c (Ω·cm²)	Carrier conc. (cm⁻³)	Analysis	Comments	Ref.
Au/Ge	>360	15 min	H$_2$–N$_2$			RBS, SIMS SEM-EDS	Au–Ge–As precipitates form on GaAs surface; Au–Ge metallization penetrates >3000 Å into GaAs; contact matrix consists of Au, Ga, and Ge	138
	400–500		N$_2$			SEM, RBS-channelling	Non-uniform alloying; large amounts of disorder in GaAs surface region which increases with time and temperature of anneal; maximum Au concentration occurs 300–400 Å beneath contact surface (500°C/15 min)	129
				1.5×10^{-3}	6.8×10^{18}		Higher ρ_c than Au/In contacts	191
Au/Au–Ge	340	2.5 min	H$_2$–N$_2$			AES, EBIC, SEM-EDS	Ge accumulation at contact/GaAs interface with possible Ge penetration into GaAs	155
	366	2.5 min	H$_2$–N$_2$				Au and Ge indiffusion and Ga outdiffusion; precipitates form in contact	
	450	2.5 min	H$_2$–N$_2$	9×10^{-7}	3×10^{17}		Rectangular precipitates containing Au–Ga at the surface and Au–Ge–As at the GaAs interface form; Au–Ga (<5 at%) agglomerations form; ρ_c has a strong dependence upon surface coverage of rectangular precipitates	
	450	2.5 min rapid	N$_2$	1×10^{-6}	3×10^{16}	EPM	Islands containing Ge form on GaAs surface; rapid heating results in smaller and higher density of islands and lower ρ_c	144
	450	2.5 min slow	N$_2$	1×10^{-5}	3×10^{16}			
Ni–Au–Ge	480	3 min	H$_2$	$\sim 1 \times 10^{-3}$	$\sim 5 \times 10^{15}$		Non-uniform contact	159
Ni/Au–Ge	300	5 min	H$_2$			AES, XRD	Contact consists of Ni–Ge/Au–Ga/Ni–As/Au–Ge/GaAs structure with an irregular interface; NiAs, β-AuGa (21at% Ga), α-AuGa (13at% Ga) and Au phases detected	148, 211
	400		H$_2$				Contact becomes ohmic; Au–Ga/Ni–Ga–Ge/Au–Ga/Ni–As–Ge/ Au–Ga/ GaAs structure forms	
	500	5 min	H$_2$				Ni–As–Ge rich and Au–Ga rich grains form at contact/GaAs interface; good reliability	

Table 6.4 *Continued*

Metallization	Annealing conditions °C	time	ambient	ρ_c ($\Omega\,cm^2$)	N_d (cm^{-3})	Analysis	Comments	Refs
Ni/Au–Ge *Continued*	450	15 s	H_2		3×10^{15}		High dislocation density beneath contact; large doping tail extends ~2–3 μm into GaAs beneath contact	212
	450–480	45 s	vacuum				More uniform than Au–Ge	124
	600	2 min				SIMS	Ge and Ni penetration into GaAs is faster than Au; $D_{Ga} \sim 10^{-13} – 10^{-12}\,cm^2\,s^{-1}$ (out diffusion) and $D_{Au} \sim 1.5 \times 10^{-12}\,cm^2\,s^{-1}$ (indiffusion)	213
	450	20 s	He				If thickness of Ni > 50% of Au–Ge then contact is not ohmic	214
	450	2 min	He	$\sim 1 \times 10^{-6}$	$\sim 1 \times 10^{17}$		Contact degradation $E_a \sim 1.7$ eV, higher initial rate for thin contacts	
	450	1 min	N_2, H_2–N_2	1×10^{-6}	2×10^{16}	AES, electron channelling	'Solid phase epitaxial' particles form on GaAs surface up to ~3 μm in size; the contact matrix consists of Au–Ga; Ni accumulates at particle/GaAs interface; Ge migrates into GaAs; ρ_c found to be a minimum for maximum epitaxial particle size	125, 143
				5×10^{-7}	$> 1 \times 10^{18}$		Dual Si^+ implant (40 keV, $7 \times 10^{13}\,cm^{-2}$ and 100 keV, $1 \times 10^{14}\,cm^{-2}$) annealed at 800°C prior to metallization	215
	352 >360	2 min	N_2			AES, SEM	Ni-Ge/Au–Ga/NiAs + Ge/GaAs structure forms Ni diffuses and accumulates at GaAs surface; Ga outdiffusion $D_0 \sim 2 \times 10^{-12}\,cm^2\,s^{-1}$, $E_a \sim 0.1$ eV	139, 216
	503	2 min	N_2	$\sim 2 \times 10^{-4}$	2.5×10^{15}			
	600	2 min	N_2	1×10^{-6}	3×10^{17}		Balling occurs	

Contact	Temp (°C)	Time	Ambient			Method	Comments	Ref.
	>350 600 475	30 s	vacuum As_2 H_2	$\sim 8 \times 10^{-4}$ $\sim 4 \times 10^{-6}$	$\sim 1 \times 10^{15}$ $\sim 3 \times 10^{15}$ $(\sim 3 \times 10^{15})$	QMA	As_2 evaporation occurs Lower ρ_c than without As_2 flux	140, 200 206 217
	420		H_2	2.9×10^{-7}	1.8×10^{19}	AES	For S diffusion into GaAs prior to metallization, which results in $\sim 1 \times 10^{18}$ donors/cm^3 ~ 1–2 μm beneath contact	86
	480	30 s	H_2	5–7×10^{-5}	5×10^{17}	EPM	Hot implant (400°C) of Se$^+$ (4.4×10^{14} cm^{-2}) and Ga$^+$ (5×10^{14} cm^{-2}) followed by 950°C anneal prior to contact metallization; alloy depth ~ 1100 Å	218
	480	3 min	H_2	$\sim 1 \times 10^{-6}$			Various overlayers result in degradation of contacts after annealing at 327°C; Ag diffuses to the contact/GaAs interface, but ρ_c remains stable; Pt, Ti and Cu diffuse through the metallization and react to form a rectifying contact; Cu diffuses very fast	219
Ni/Au–Ge/Ga/Au–Ge							Relatively stable with Au/Ti, Au/Pt/Ti and Au/Ti–W overlayers after annealing at 250°C in air	219
	630		As_2		$\sim 1 \times 10^{15}$		Smooth contact surface	206
Au–Ge/Ni							Similar behaviour to Ni/Au–Ge	139
Ni/Au–Ge/Ni	460	4 min	H_2–N_2	3×10^{-5} 4×10^{-4} 1×10^{-6}	2×10^{16} 2×10^{15} 1×10^{18}		$\sim 2.5 \times$ reduction in ρ_c if Ar$^+$ sputter-clean GaAs prior to metallization; $\rho_c \propto 1/N_d$	158
Ni/Au/Ge	450	5 min	vacuum			RBS, XRD SEM-EDS	Au–Ga matrix with Ni–Ge–As crystallites forms	142

Table 6.4 Continued

Metallization	Annealing conditions			ρ_c (Ω cm^2)	N_d (cm^{-3})	Analysis	Comments	Refs
	°C	time	ambient					
Au/Ni/Au–Ge	430			1.5×10^{-6}	5×10^{16}			220
	~390	90 s	H$_2$–Ar	$>1 \times 10^{-4}$	2×10^{17}	STEM–EDS, XTEM	Initially formation of NiAs (Ga + Ge) grains at Au–Ge/GaAs interface, Au (Ga + As) grains at interface and throughout contact, and NiGe (Ga + As) at Ni/Au–Ge interface	147
	~440	115 s	H$_2$–Ar	~1×10^{-6}	2×10^{17}		Ge diffuses from NiGe to contact/GaAs interface where Ni$_2$GeAs forms; NiGe grains contain less Ga than for 390°C alloyed contact	
	~450	200 s	H$_2$–Ar	~5×10^{-6}	2×10^{17}		Grain growth, less Ni$_2$GeAs at GaAs surface and more Au (Ga + As) grains than for 440°C alloyed contact; penetration into GaAs <1000 Å	
Ni/Ge/Au/Ni	450	60 s	H$_2$	7×10^{-5} 6×10^{-6}	2×10^{16} 3×10^{17}	SEM		162
SiO$_2$/Ge/Au/Ni	450	60 s	H$_2$	1×10^{-5} 5×10^{-7}	2×10^{16} 3×10^{17}		~10× lower ρ_c for SiO$_2$ covered contacts during alloying	
	450	30 s	H$_2$–N$_2$	1.5×10^{-5}	2×10^{16}	SIMS, EPM	Ge–Ni (+ Au) clusters form in contact, the smaller and higher density of which result in lower ρ_c; Ni and Ge diffuse >3000 Å into GaAs	221
	450	30 s	H$_2$–N$_2$	2.9×10^{-7}	3×10^{17}	SEM	High cooling rates of substrate produces contacts with higher density of clusters and lower ρ_c than rapid cooling of contact directly	222

Contact	Temp (°C)	Time	Atmosphere	ρ_c		Method	Comments	Ref
Pt/Au-Ge	480	3 min	H_2	$\sim 1 \times 10^{-6}$			Contact relatively stable for anneals at 250°C in air Overlayers of Au/Ti; Au/Pt/Ti or Au/Ti-W result in a rapid increase in ρ_c ($>$)10^{-4} Ω cm^2) upon 250°C annealing	219
	500	30 s	H_2			EPM	For thick Pt contact structure Pt-Ga-Ge/ Au-Ga/Pt-As-Ge/Au-Ga/GaAs and for thin Pt films Pt-As-Ge/Au-Ga/GaAs	211
Ag/Au/Ge	500	10 min	H_2	7.6×10^{-3} 3.8×10^{-5}	1×10^{15} 1×10^{17}			199
Au/Ag/Au-Ge	360		H_2			SEM-EDS	Melting of Au-Ge occurs and holes appear in contact	133
	380		H_2				Contact turns silver colour	
	480		H_2				Complete melting of contact and Ag goes into solution	
	490		H_2	$1-2 \times 10^{-6}$			Rectangular etched areas form in GaAs	
Au/Ni/Sn-Ni	420	10 s	H_2-N_2	5×10^{-5}	2.3×10^{17}	AES	Au, Sn, Ni and Ga at contact surface; Sn concentration in contact is low; Ni penetrates deeper in GaAs than Sn	210
Au/Sn-Ni/Au	300	3 min	H_2-N_2	7×10^{-5}	3.5×10^{18}		Metallic droplets (~ 5 μm diameter) form on contact surface	223
Au-Ge-In							Contacts with Pt, Ti or Cu overlayers degrade as a result of annealing at 327°C; Ag overlayers result in relatively stable contacts	218
Au-Ge/In	520	1 min	H_2-N_2			AES, electron channelling	Formation of faceted regions at surface in a Au-Ga-In matrix; epitaxial particles consisting of Ge-In ($\sim 1:1$) with $\sim 40\%$ Ga form at the GaAs surface; around these particles a thin Ge-In layer forms on the GaAs surface; some Ge migration into GaAs occurs	143

Table 6.4 *Continued*

Metallization	Annealing conditions			ρ_c ($\Omega\,cm^2$)	N_d (cm^{-3})	Analysis	Comments	Refs
	°C	time	ambient					
Au/In–Ge	450					TEM–EDS	Single crystal Au–Ge–As precipitates, with composition approximately 1:1:1 ~2000 Å thick and ~2 μm diameter form on GaAs surface; around these the contact metallization (Au–Ga) 'chews' into the GaAs to a depth >1000 Å; the precipitates cover ~25% of contact area; in other regions of interface a thin skin of oxides or contamination is observed; high In containing metallization results in the skin being punctured and rectangular patches elongated along ⟨110⟩ directions form in the surface of the GaAs; long alloying times result in the Au–Ge–As phase disappearing and deterioration in contact performance; high Ge content contacts result in the formation of a cell structure of Ge dendrites with thick and more angular Au–Ge–As particles embedded in the cell walls; long alloying times produce rectangular epitaxial Ge particles elongated along the ⟨110⟩ direction and Au–Ga particles on the GaAs; the presence of the epitaxial Ge particles results in good ohmic contacts; a large amount of strain is present in the GaAs beneath the contact	145, 146
						SEM–EDS	Best contacts for pulsed TEDs (poor ohmic) were ones with low Ge (~1%) content and Au–Ge–As precipitate and Au–Ga phase formation	224

400–430	3 min		Au–Ge–As metastable and Au–Ge metastable phases form	149, 150
450	3 min	TEM–EDS, SEM	Formation of Au grains, Au$_7$In (ξ-phase) grains, Au–Ge–As phase and As$_2$Ge lath precipitates; As$_2$Ge and Au–Ge–As can cover up to 50% of GaAs surface; ~1 at% Au and traces of Ge observed in GaAs surface; prolonged annealing results in the Au–Ge–As phase disappearing	
500	3 min		Good ohmic properties; balling occurs; formation of Au$_7$Ga$_2$ and Au$_2$Ga phases occurs	

electrical properties of the Au/GaAs Schottky barrier degrade.[48,129] Annealing to temperatures up to 450°C results in increased gallium and arsenic outdiffusion together with gold indiffusion. The gold reacts with gallium to form an Au–Ga phase. Vandenberg and Kingsborn[132] reported the formation of the β phase (which could be indexed as hexagonal Au_7Ga_2). Zeng and Chung,[131] however, did not observe this phase below about 460°C, but did observe an orthorhombic AuGa phase. The interaction is somewhat non-uniform, resulting in a laterally non-uniform contact.[128] Complete melting of the metallization occurs in the range of approximately 460–500°C.[129–133,135] The contact becomes pitted and laterally very non-uniform with regions of deep Au penetration into the GaAs ($\geq$ 3000 Å). A disordered region with high dislocation density forms beneath the contact[129,134] and the intermetallic Au–Ga hexagonal β phase ($\sim$ 21 at% Ga) forms.[131,134,135] With cooling rates of about 2.5–40°C min^{-1} the metastable Au_2Ga compound has also been observed.[131] At higher cooling rates ($\sim$ 600°C min^{-1}) the formation of the α phase (Au–Ga solid solution) results.[131] Zeng and Chung[131] found that the Au_2Ga compound appears as irregularly shaped protrusions with jagged boundaries and topography, the α phase as aligned rectangular protrusions with rounded corners, and the β phase as aligned rectangular protrusions with sharp corners. Although the actual detailed results may disagree, it is clear that the Au–GaAs metallurgical interaction is complex, particularly after melting of the metallization has occurred. Furthermore, it is exceedingly difficult to understand the electrical behaviour of such a non-uniform contact. The contact becomes more ohmic with annealing, but does not become truly ohmic.[48,129] The amount of interaction is found to decrease if an Au–Ga alloy is used instead of pure Au for the metallization,[136] which indicates that the driving force in the interaction is probably an Au–Ga reaction.

Au–Ge/GaAs, Au–Ge–Ni/GaAs, and Au–Ge–In/GaAs contacts The normal 'alloying' of these contacts is achieved by heating the structure to temperatures of 360°C or more (usually $\sim$ 450°C), for approximately 30 s to 3 min. After this heat treatment, ohmic behaviour is observed. The specific contact resistivity is found to depend upon alloying time and temperature and also upon the heating and cooling rates. In general, the optimum alloy time and temperature must be determined experimentally for each alloying system used. Too long an alloy time or too high a temperature result in a degraded ohmic contact. Table 6.4 contains a summary of studies on alloyed contacts.

Since the alloying involves heating above the Au–Ge eutectic temperature, the reactions observed must clearly involve some liquid phase although the whole contact may not melt. As in the case of pure gold on GaAs, interreaction between the contact metallization and GaAs occurs[129,138,139] with arsenic evaporation from the contact surface.[140] The gold penetration into the GaAs is deep ($\gtrsim$ 1000–3000 Å), and its extent increases with alloying time, temperature, and amount of gold in the contact film. This penetration is very non-uniform.

Metallizations for GaAs devices and circuits 241

generally in the form of spikes.[129,138] The contact surface is non-uniform and it tends to 'ball up'. The balling is reduced by the addition of nickel, indium, or platinum. Early studies attributed this to the nickel acting as a wetting agent for the Au–Ge.[139] However, later studies on inert substrates have shown that nickel reacts with germanium to form Ni_2Ge and $NiGe$, and that the ratio of nickel to germanium is important for uniformity of the layers.[141,142] Island or precipitate formation occurs at the contact/GaAs interface.[138,143–145] Palmstrøm et al.[138] found that the precipitates formed after alloying the Au/Ge/GaAs structure contained gold, germanium, and arsenic. These precipitates could be the same as those observed by transmission electron microscopy (TEM) studies of alloyed Au/Ge–In/GaAs contacts,[145,146] which were identified to have a monoclinic structure composed of gold, germanium, and arsenic in approximately equal proportions.[145] Around these precipitates the contact metallizations 'eat' into the GaAs. Longer alloying times result in increased dissolution of GaAs, the AuGeAs phase disappears and contact performance deteriorates.[145,146] In contacts with higher germanium content, a cell structure composed of germanium dendrites with thicker and more angular AuGeAs precipitates embedded in the cell walls forms. Longer alloying times result in the formation of rectangular patches of germanium, elongated along $\langle 110 \rangle$ directions of the GaAs.[146] Christou[143] using scanning Auger electron spectroscopy (AES) has observed epitaxial particles at the contact/GaAs interface for both Au–Ge/In/GaAs and Ni/Au–Ge/GaAs metallizations. Around these epitaxial particles the matrix consists of a mixture of Au–Ga phases containing some germanium and nickel or indium. Kuan et al.[147] identified the epitaxial particles by cross sectional TEM in Au/Ni/Au–Ge/GaAs alloyed contacts to be a ternary phase Ni_2GeAs. During the initial stages of the reaction, NiAs forms at the contact/GaAs interface, but the contacts do not become ohmic until the NiAs converts to Ni_2GeAs.[147] This ternary Ni_2GeAs phase has a hexagonal structure with lattice parameters very close to that of NiAs.[147] Hence, the phase identified by x-ray diffraction as NiAs observed in alloyed Ni/Au–Ge/GaAs contacts[148] could in fact be Ni_2GeAs. A large number of other phases have been observed in the Au/In–Ge/GaAs alloyed contacts, which include Au_7In (Au–In ξ phase), As_2Ge, an Au–Ge metastable phase, Au_7Ga_2, Au–Ga (10 at% Ga), and Au_2Ga.[149,150]

The early concept of germanium doping producing a uniform n^{++}-GaAs layer immediately beneath the contact, discussed earlier, seems too simplistic. Clearly, the ohmic contact formation is more complicated. Furtheremore, growth of GaAs from GaAs-saturated Au–Ge[151,152] and Au–Ge–Ni[152] melts is unable to produce layers with carrier concentrations higher than about 5×10^{18} cm^{-3}, which is insufficient for a low-resistivity tunnelling contact. However, since the alloying is rapid enough for incomplete saturation of the metallization with GaAs, the situation may be different during alloying. The large amount of gallium remaining in the metallization after alloying may result in an excess of gallium vacancies in the GaAs immediately beneath the contact, which would cause preferential donor behaviour of the amphoteric germanium.[151] Jaros and Hartnagel[153] have proposed that a large amount of disorder or strain in the GaAs crystal may suppress the germanium acceptor levels, resulting in higher effective donor concentrations. Large amounts of disorder are present in the GaAs beneath the contact.[129] However, although Christou observed germanium[143] and Grovenor gold with traces of germanium[149,150] in the GaAs beneath the metallization, there is no clear evidence that this GaAs is responsible for the ohmic behaviour. The deep Au-rich spikes that form will have high electric fields associated with them and, hence, conduction by thermionic field emission could take place. The disordered GaAs beneath the contact may result in a conduction mechanism similar to that proposed for a graded amorphous/crystalline contact by Sebestyen[34,42] and discussed in Section 6.2. A possible contact system using damaged GaAs has been proposed by Sullivan.[154] This technique involves damaging the GaAs surface by ion implantation or plasma etching prior to contact deposition and annealing. A number of studies have observed a correlation between the number and size of the germanium-containing precipitates or germanium epitaxial particles at the contact/GaAs interface and the ohmic behaviour.[143,144,146,147,155] Thus, yet another model for the ohmic behaviour could be that these epitaxial particles form highly conducting regions due either to the formation of a low barrier height heterojunction (such as Ge/GaAs described in Section 6.4.2) or to preferential Ge doping of the GaAs beneath the particles.[155] Braslau[156] has used this concept to propose a model for the alloyed ohmic contact. It does not involve the specifics of the conduction mechanism (i.e. tunnelling, low barrier, etc.), but it assumes that conduction occurs through small localized regions (e.g. these epitaxial precipitates). In the model shown in Figure 6.15, the main contribution to the contact resistivity arises from the spreading resistance in the GaAs and not in the epitaxial particle/GaAs interface. By taking 'typical' values for the size of the precipitates Braslau predicts the specific contact resistivity, as a function of doping, to be proportional to N_D^{-1}. Figure 6.15 shows his calculated contact resistivity versus doping concentrations together with a number of measured contact resistivities obtained from the literature. This N_D^{-1} dependence has been observed previously[157–159] and a further model for explaining this behaviour was proposed by Popovic.[160] This model assumes that the contact consists of a thin (compared to the carrier mean free path) n^+ layer beneath the metallization. He derives a contact resistance proportional to N_D^{-1}, the doping in the underlying region, since only those ballistic electrons with energies exceeding the n^+/n barrier height can contribute to the current. A similar model has also been proposed by Dingfen and Heime.[161] Pure tunnelling theory would predict an $N_D^{-0.5}$ behaviour.[82]

Overlayers: SiO_2, Ag, and Au/Ag Attempts at producing more uniform contacts have involved the use of overlayers, such as SiO_2[162] and Ag,[133] and pulsed annealing. Pulsed annealing will be discussed later in Section 6.4.4. The use of

Metallizations for GaAs devices and circuits

overlayers does improve the uniformity and electrical performance of the contact, although in essence the contact remains the same.[133,156] Devices now are often fabricated with multilayer metallizations such as Au/Ag/Au–Ge/Ni/GaAs[133] and Au/Ti/Au/Ag/Au–Ge–Ni/GaAs. The Ag is believed to act as a diffusion barrier for outdiffusing Ga. Sebestyen et al.[163,164] have proposed the idea of thin-phase epitaxy in order to produce more uniform and controlled contacts. The concept is that by depositing Ga with the metallization and alloying in an As overpressure, controlled growth of a thin GaAs epitaxial layer doped with the metallization dopant occurs during alloying. As yet this technique has not been widely used.

Interface oxide The question that still remains to be answered is, why is the contact so non-uniform? A possible cause is the existence of the GaAs native oxide. This is typically 10–50 Å thick, depending upon surface preparation, and is probably sufficient to inhibit a uniform reaction between the contact metallization and GaAs. The reaction starts at 'weak' areas in the oxide layer and, since the interaction involves the liquid phase, it will be rapid in these areas, resulting in a non-uniform interface. It is generally believed that the precipitates that form at the interface form in areas where the interaction started.[155] Steeds et al.[146] present evidence of a thin oxide–carbon layer at other areas of the interface. Further evidence comes from experiments of the relationship between the contact resistivity and the degree of exposure of the GaAs to air after etching but prior to contact deposition. A doubling in the specific contact resistivity was observed for a 2 min exposure to air as compared with immediate loading into the evaporator.[165] Heime et al.[158] found an improvement in the contact performance and reproducibility when the GaAs surface was sputter cleaned prior to contact deposition. The dominant roles of the interface oxide, contact composition and thickness, alloy times, and heating and cooling rates explain why there are so many differences in contact performance and structure reported in the literature (see Table 6.4).

Interface mixing of Au–Ge and Au–Ge–Ni/GaAs contacts One interesting way of producing a more uniform reaction is by ion beam mixing. A schematic of the technique is shown in Figure 6.16a. After metallization, the contact is ion-implanted, the ion species and energy of implantation being chosen such that the projected range is close to the metallization/GaAs interface. If ohmic contacts or barrier height tailoring are desired, electrically active implants may be used (e.g. dopant ions such as Se, Te, Zn, Si, Ge, or Sn). If only mixing is wanted, implantation of As or Ga ions may be used. After implantation, the contact is heated so that a metallization–GaAs reaction occurs. Either the extent of the reaction must be such that the damaged GaAs is consumed by the metallization–GaAs reaction, or else the annealing must be such that the damaged GaAs regrows in order for the GaAs beneath the contact to have the

244   *Gallium arsenide*

desired electrical characteristics. Figures 6.16b and c show SEM micrographs of Au–Ge/GaAs structure for an unimplanted and a 340 keV argon-implanted sample after alloying at 455°C for 5 min. Comparing the SEM micrographs for the two cases, it is clear that the implanted sample produces a more uniform contact after alloying. Another example of this technique is shown in Figures

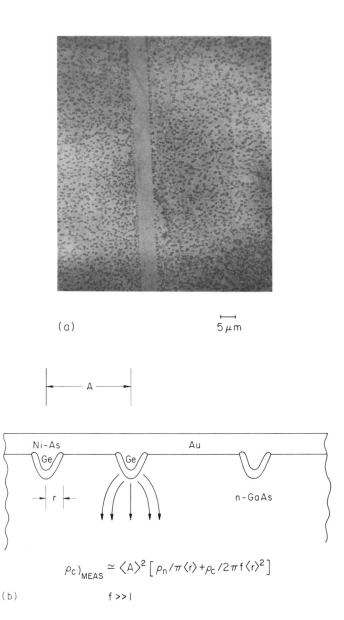

(a)   5 μm

(b) $\rho_c)_{MEAS} \simeq \langle A \rangle^2 \left[\rho_n/\pi \langle r \rangle + \rho_c/2\pi f \langle r \rangle^2 \right]$

f >> 1

Metallizations for GaAs devices and circuits 245

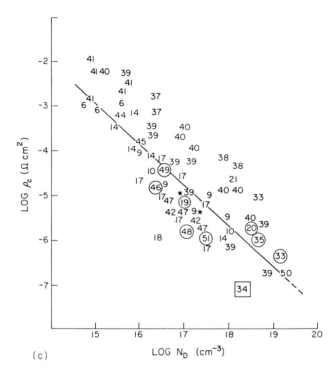

Figure 6.15 Localized conduction (spreading resistance) model for ohmic contact formation. (a) Optical micrograph of an Au–Ge–Ni alloyed contact after the metallization has been removed, showing irregular pitting of GaAs surface. (b) Schematic of the model. Conduction is through a parallel array of Ge-rich protrusions of negligible contact resistance compared to the spreading resistance in series with them (f is a field enhancement factor for field emission for protruding electrodes and ρ_n is the resistivity of the n-GaAs). (c) Observed contact resistance plotted as a function of n-GaAs doping. Numbers refer to identity of reference numbers (see ref. 156) for the cited points. Circled points are laser or electron beam annealed. The point in the square is an MBE-grown heterojunction. The straight line is the contact resistance predicted from the localized conduction model with the mean values of 'A' and 'r' taken from the micrograph in (a). (From Braslau[156])

6.17a and b, which shows SEM micrographs of an Au–Ge–Ni/GaAs contact for both an unimplanted and a 150 keV arsenic-implanted sample, after alloying at 420°C for 30 s. The structure of the films is similar; however, the implanted contact was not ohmic, presumably due to residual damage in the GaAs left by the implantation. Figures 6.17c and d show SEM micrographs of these contacts after they had been alloyed through a number of sequential alloying steps, the last one being 560°C for 15 min. The unimplanted contact shows severe balling and pitting whereas the implanted contact remains more uniform. After these alloying

Gallium arsenide

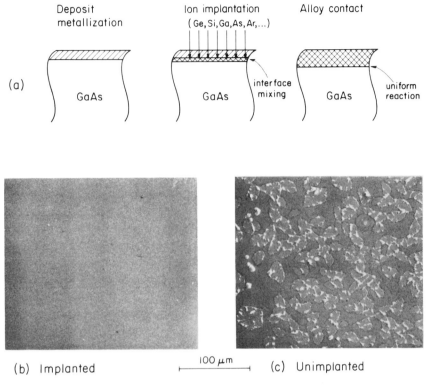

Figure 6.16 Ion implantation for producing more uniformly reacted contact. (a) Schematic of technique. (b, c) SEM micrographs of Au–Ge/GaAs structures after alloying at about 450 °C for 30 s: implanted with 340 keV 2×10^{15} Ar ions/cm^2 (b) and unimplanted (c)

cycles the implanted contact was ohmic. The reason for the improved reaction uniformity for the implanted structures may arise either from interface mixing (breaking-up of interface oxide and mixing contact metallization and GaAs) or from the possible amorphization of the GaAs beneath the contact. In metal–Si reactions it has been observed that amorphous silicon reacts more readily than crystalline silicon. Furthermore, it is established that the Al–Si reaction can be made more uniform by ion implantation.[166] Thus, although the detailed mechanism is not clearly understood, ion implantation can be used for inducing uniform reactions of metal films on Si and GaAs. By choosing the ion species and the dose, doped layers beneath the metallization may be formed. Thus, a method for making ohmic contracts to GaAs may involve an implantation of an n-type dopant such as Si, Ge, Se or Te. When platinum or palladium reacts with arsenic-implanted silicon, the arsenic atoms are 'snowploughed' by the advancing metal silicide/Si interface, resulting in an increase in the dopant concentration and

Metallizations for GaAs devices and circuits

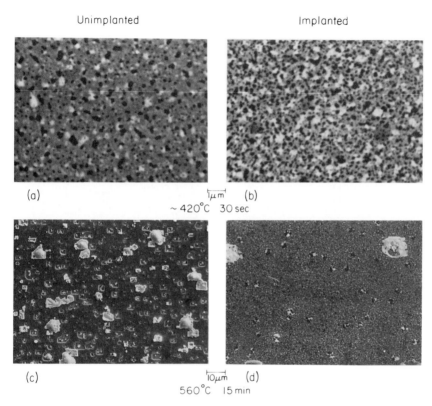

Figure 6.17 SEM micrographs of Au–Ge–Ni/GaAs structures after alloying unimplanted (a, c) and implanted (b, d) 150 keV 1×10^{15} As ions/cm^2: (a, b) after alloying at 420°C for 30 s; (c, d) after sequential alloying cycles—420°C/30 s, 460°C/15 min, and 560°C/15 min

hence carrier concentration, immediately beneath the contact.[167] As yet this has not been studied in GaAs. However, if the same effect occurs it may result in the formation of the high doping levels immediately beneath the contact required for ohmic behaviour.

6.4.4 Laser and electron beam alloyed contacts

Methods of applying laser and electron beam annealing for production of ohmic contacts are illustrated schematically in Figure 6.18. In this section, pulsed 'alloying' (transient annealing) will be considered in more detail. Excellent review articles on pulsed annealed contacts can be found in refs 97 and 168. In general, transient annealed contacts have comparable or lower specific contact resistance, better surface uniformity (no balling, etc.), more limited metallization–GaAs

248   *Gallium arsenide*

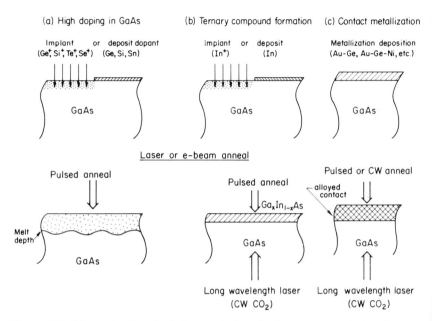

Figure 6.18 Schematic of methods for producing ohmic contacts to GaAs using laser or electron beam annealing

reaction, and poorer adhesion than conventionally alloyed contacts. Table 6.5 contains a summary of both laser and electron beam alloyed contacts. From this table it is clear that to date most studies have involved transient annealing of conventional metallization schemes used for thermal alloyed contacts. The rapid alloying time for pulsed annealing (≤ 1 μs) suggests that the interactions occur in the liquid phase. This rapid alloying leads to very high quench rates and hence to the possible formation of metastable phases. The stability of such phases is not known and they may be the cause of the poor thermal stability observed for pulse electron beam alloyed Pt/Au–Ge contacts.[169] Typical laser annealing times range from a few picoseconds to around 100 ns for pulsed systems and approximately 1 ms for scanned CW lasers. Pulsed electron beams have a typical pulse duration of approximately 100 ns. The main difference between laser and electron beam annealing arises from the differences in energy deposition. Metallic surfaces are good reflectors of light and therefore a significant part of the incident laser energy is reflected from the contact surface. Furthermore, the reflectivity depends upon the physical state of the surface (solid or molten, rough or smooth) and also upon its temperature. Hence, it is difficult to determine how much energy is coupled into the contact. The deposited energy as a function of depth depends upon the absorption coefficient of the contact material. During laser annealing most of the energy is deposited at the contact surface. Thus melting starts at the surface and

Table 6.5 Transient annealed ohmic contacts

Metallization	Annealing conditions				N_d (cm^{-3})	Analysis	Comments	Refs
	λ (nm)	time (ns)	energy (J cm^{-2})	ρ_c (Ω cm^2)				
Ge/Au	free-running pulsed ruby 0.69	~1000	15	2×10^{-6}	1×10^{17}	AES	Complete Au–Ge mixing and little Ga and As outdiffusion Excellent surface morphology; poor reproducibility	225
	CW argon						Comparable results to pulsed ruby, but problems of balling at contact edge; improves with Ni added to metallization	
Au/Ge	Q-switched Nd: YAG ~30			5×10^{-5}	2×10^{18}		Microcracks form	102
Au–Ge	Q-switched ruby 0.69	~15	~1.02	$<7 \times 10^5$	3×10^{16}		More uniform and lower ρ_c than thermally alloyed contacts (420°C/3 min, $\rho_c \sim 5.4 \times 10^{-4}$ Ω cm^2)	226
	pulsed ruby 0.69 0.69	25 ~20	0.3–0.7 0.7–0.9	poor poor	2×10^{17}	SEM–EDS SEM	Poor surface morphology; addition of Ni makes a large improvement Good surface morphology	227 228
	Q-switched Nd: YAG 1.06	~23	~1.5	$\sim 3 \times 10^{-6}$	7×10^{17}	SIMS, SEM	Deep penetration of Au and Ge into GaAs; ohmic contacts can still be formed after ~1 μm of GaAs has been etched off without further annealing	170, 229
	1.06	~23	2.65	$\sim 1 \times 10^{-6}$	7×10^{17}		After laser alloying and etching off contact and ~1200 Å GaAs and then redepositing contacts	

250   Gallium arsenide

Table 6.5 Continued

Metallization	Annealing conditions			ρ_c ($\Omega\,cm^2$)	N_d (cm^{-3})	Analysis	Comments	Refs
	λ (nm)	time (ns)	energy (J cm^{-2})					
	1.06	30	~1.6	~4×10^{-6}	2×10^{18}	TEM, AES	Little Ga and As outmigration to contact surface; Ge segregates at grain boundaries; ρ_c decrease with increase in contact grain size	230
	pulsed Nd: YAG 1.06 ~18		0.39	poor	2×10^{7}	SEM–EDS	Poor surface morphology; improvement is made by addition of Ni	227
	'multiple' Nd: YAG from back-side 1.06 120		~0.55	poor	2×10^{17}		Poor surface morphology; improvement is made by addition of Ni	227
	Q-switched Nd: glass from back-side 1.06 20		0.3–0.5	$<2 \times 10^{-5}$		RBS	Complete Au–GaAs mixing after 1.5 J cm^{-2}	231
Au/Au–Ge	pulsed ruby 0.69 ~20		0.5–1.5	poor				228
	pulsed Nd: YAG 1.06 ~40		~0.6	poor			Poor surface morphology	
	pulsed CO$_2$ 10.0 1000		~1.0	poor				
Ge/Au/Ni	free-running pulsed ruby 0.69 ~1000		~15			AES	Ge at GaAs surface after anneal; similar results to Ge/GaAs	225

Metallizations for GaAs devices and circuits

	0.69	~25	~0.6			change in GaAs sheet resistivity beneath contact
	PEBA	~100	~0.55	~1×10⁻⁶		Excellent morphology; Au–Ge 'outflow' around contact and damage to GaAs beneath contact may occur; small energy window exists where alloying of contact occurs without destroying Al/GaAs Schottky contacts
Au/Ni/Au–Ge	pulsed ruby 0.69 20		~0.8	~1×10⁻⁵	2×10¹⁷	SEM	Thermally alloyed 450°C/30 s, $\rho_c \sim 2 \times 10^{-5}\,\Omega\,\text{cm}^2$	228
	Q-switched Nd:YAG 1.06 40		~0.3	poor	2×10¹⁷		Damage in GaAs is often observed
	pulsed CO₂ 10.0 1000		~2.1	poor	2×10¹⁷		
	CW argon 0.51 4.3 mm/s	~640		4.8×10⁻⁶	~1×10¹⁷	SEM-EDS	Only slight redistribution of elements; FETs produced had better performance than ones with thermally alloyed ohmic contacts	168, 228
	PEBA ~10	~100	~0.28	3.5×10⁻⁵	~1×10¹⁷			232
Pt/Au–Ge	PEBA ~20 kV	~100	~0.4	4×10¹⁷	7×10¹⁷	AES	Little contact–GaAs intermixing; small amounts of Ga at surface and possibly some Pt–Ge reaction; sheet resistivity of GaAs beneath contact increases for energy densities $> 0.5\,\text{J}\,\text{cm}^{-2}$; thermally alloyed contacts 470°C/3 min give $\rho_c \sim 4 \times 10^{-6}\,\Omega\,\text{cm}^2$; post-annealing of PEBA contacts at 250°C for 500 h results in interdiffusion and an increase in ρ_c ($\sim 3 \times 10^{-4}\,\Omega\,\text{cm}^2$)	169, 232, 233

Table 6.5 Continued

Metallization	Annealing conditions			ρ_c ($\Omega\,cm^2$)	N_d (cm^{-3})	Analysis	Comments	Refs
	λ (nm)	time (ns)	energy (J cm^{-2})					
Au/Pt/Au–Ge	CW argon ~0.51	4.3 mm/s	~610	1.5×10^{-5}	1×10^{17}			168
	PEBA ~10 keV	~100	~0.28	8.6×10^{-5}	1×10^{17}		ρ_c depends upon metal X in Au/X/Au–Ge contacts alloyed using CW argon laser annealing, but is not very dependent on X in PEBA contacts	
Au/Ag/Au–Ge	CW argon ~0.51	4.3 mm/s	~660	2×10^{-4}	1×10^{17}			
	PEBA ~10 keV	~100	~0.32	2.3×10^{-5}	1×10^{17}			
Au/Ti/Au–Ge	CW argon ~0.51	2.3 mm/s	~1350	1.8×10^{-5}	1×10^{17}			
	PEBA ~10 keV	~100		not ohmic	1×10^{17}		Sample destroyed	
Au–Ge/In	CW argon ~0.51	4.3 mm/s	~565	$\sim 1 \times 10^{-6}$	1×10^{17}	SEM–EDS	Ge segregates at surface after deposition; relatively uniform depth distribution after anneal; thermally alloyed contacts give $\rho_c \sim 1 \times 10^{-6}\,\Omega\,cm^2$	168, 234
	PEBA ~10 keV	~100		not ohmic	1×10^{17}		Contact destroyed	
Ag/Sn	Pulsed ruby 0.69	1000		$\sim 3.2 \times 10^{-4}$	4.6×10^{17}			235
	Q-switched CO$_2$ 10.6	~0.5 μs		$\sim 3.1 \times 10^{-4}$	4.6×10^{17}			

propagates towards the GaAs. The high absorption coefficient of the contact can result in a large degree of superheating of the liquid and contact evaporation.[170] In the case of electron beams there are fewer problems of reflected energy and it is therefore easier to determine the deposited energy. The deposited energy with depth depends upon the electron stopping power in the material. Mayer et al.[171] have compared pulsed laser and pulsed ion beam annealing of silicon. For pulsed ion beams they calculate that the temperature is approximately constant in the silicon to a depth about equal to the projected range of the ions (~ 2 μm for 300 keV H^+) during the anneal. A similar behaviour can be expected during pulsed electron beam annealing. However, in the case of pulsed laser annealing the temperature of the silicon decreased rapidly with depth. Since most of the contacts studied contain Au–Ge, which has a low eutectic temperature compared with GaAs, or Au–GaAs, the contact will probably melt first during electron beam annealing. For both laser and electron beam annealing, once the metallization has melted, GaAs will be dissolved into the melt. Upon cooling rapid resolidification from the melt will occur, the regrowth velocity being dependent upon the temperature gradient at the liquid/solid interface. The more uniform energy deposition with depth of an electron beam could, with the choice of other metallization schemes, be used for interface melting (i.e. melting at the interface without melting the contact surface) which has been observed for pulsed ion beam annealing of metallizations on Si.[172] Long-wavelength laser light (such as a CO_2 laser), which is not absorbed by GaAs, incident from the back of the GaAs wafer could also be used for interface melting.[97] This interface melting may be sufficient for ohmic contact behaviour. A possible use may be to form an ohmic contact using a metal/Ge/GaAs system by melting at the Ge/GaAs interface to produce a metal/Ge/Ge$_x$(GaAs)$_{1-x}$/GaAs structure. Furthermore, interface melting may also be used to initiate a uniform solid phase reaction (similar to ion implantation discussed in Section 6.4.3).

Transient annealing of conventional alloy metallization schemes has been successful in producing low-resistivity contacts with limited contact–GaAs reaction and good surface morphology. However, the stability of these contacts is still a concern. Perhaps the metallization schemes used to date are not the optimal ones for transient processing. It would be preferable not to melt the complete metallization. The potential of interface melting by pulsed electron or ion beams and long-wavelength lasers from the GaAs back-side may enable better contacts to be made using new metallization schemes.

6.5 CONCLUSIONS

The present-day contact technology has enabled production of GaAs discrete devices and integrated circuits. However, the future trend towards smaller device sizes and large-scale integration is limited by current contact technology. Reproducibility and stability of the contacts are key issues. The biggest problem in present technology is probably the alloyed ohmic contact. It is a relatively crude method for contact fabrication as the degree of alloy penetration and the contact uniformity are fairly uncontrollable. The non-uniform interface makes it difficult for contacting thin layers, and the lateral non-uniformities do not aid small contact geometries. The reliability of these contacts are also of major concern. In Schottky barrier contact technology problems that may be identified are the lack of barrier height tailoring and reproducibility. These problems probably arise from an interfacial layer of oxide, excess As or 'dirt'. Variations in this interfacial layer result in variations of effective barrier height. A further uncertainty is the stability of these contacts during device operation.

Recent studies are showing improving trends in Schottky barrier technology. In the case of ohmic contacts, however, there is still a need for further definitive work. The possibility of producing Schottky barriers stable during annealing up to 800–900°C is a major breakthrough and has led to an improved self-aligned implantation technology. The importance of interfacial chemistry has been demonstrated clearly by the work on H_2S adsorption prior to contact deposition, which resulted in a factor of 2 lowering in barrier height. Interfacial reactions between contact metallization and GaAs, with possible use of ion implantation for improved uniformity, may lead to improved contact reproducibility and variations in barrier height. Homoepitaxial layers with or without thin doping layers, heteroepitaxial layers, and ion implantation show promise for both tailoring of the contact effective barrier height and for fabricating ohmic contacts of the tunnelling and low barrier height variety. Controlled interfacial melt reactions using pulsed electron beams or back-side long-wavelength laser annealing on new metallization schemes may be successful in producing reliable and reproducible ohmic contacts.

ACKNOWLEDGEMENTS

We acknowledge, with pleasure, our interactions and discussions with colleagues at Cornell, IBM Research Laboratories, and the University of Leeds, and we are particularly thankful to J. W. Mayer, L. F. Eastman (Cornell), J. Woodall, and N. Braslau (IBM). Permission from K. L. Kavanagh, S. Chen, and S. D. Mukherjee to reproduce some of our collaborative work is gratefully appreciated. Supported in part by Defence Advanced Research Project Agency under contract # N6600-1-83-6-0304 monitored by Naval Ocean System Center.

REFERENCES

1. S. N. Sze, *Physics of Semiconductor Devices*, New York, Wiley Interscience, p. 24. (1981).
2. J. Bardeen, *Phys. Rev.*, **71**, 717 (1947).
3. J. Tamm, *Phys. Z. Sowjetunion*, **1**, 733 (1932).

4. W. Shockley, *Phys. Rev.*, **56**, 317 (1939).
5. J. Van Laar, A. Huijser, and T. L. Van Rooy, *J. Vac. Sci. Technol.*, **14**, 894 (1977).
6. W. E. Spicer, P. W. Chye, P. R. Skeath, C. Y. Su, and I. Lindau, *Insulating Films on Semiconductors 1979*, Inst. Phys. Conf. Ser. 50, London and Bristol, Institute of Physics, p. 216 (1980) and references therein.
7. W. E. Spicer, I. Lindau, P. Skeath, and C. Y. Su, *J. Vac. Sci. Technol.*, **17**, 1019 (1980).
8. W. E. Spicer, S. Eglash, I. Lindau, C. Y. Su, and P. R. Skeath, *Thin Solid Films*, **89**, 447 (1982).
9. W. E. Spicer, P. W. Chye, P. R. Skeath, C. Y. Su, and I. Lindau, *J. Vac. Sci. Technol.*, **16**, 1422 (1979).
10. R. W. Grant, J. R. Waldrop, S. P. Kowalczyk, and E. A. Kraut, *J. Vac. Sci. Technol.*, **19**, 477 (1981).
11. R. F. C. Farrow, A. G. Cullis, A. J. Grant, and J. F. Patterson, *J. Cryst. Growth*, **45**, 292 (1978).
12. L. J. Brillson, *J. Vac. Sci. Technol.*, **20**, 652 (1982).
13. J. M. Woodall and J. L. Freeouf, *J. Vac. Sci. Technol.*, **19**, 794 (1981).
14. G. P. Schwartz, G. J. Gualtieri, J. E. Griffiths, C. D. Thurmond, and B. Schwartz, *J. Electrochem. Soc.*, **127**, 2488 (1980).
15. R. L. Farrow, R. K. Chang, S. Mroczkowski, and F. H. Pollak, *Appl. Phys. Lett.*, **31**, 768 (1977).
16. H. H. Wieder, *J. Vac. Sci. Technol.*, **15**, 1498 (1978).
17. J. M. Woodall, C. Lanza, and J. L. Freeouf, *J. Vac. Sci. Technol.*, **15**, 1436 (1978).
18. A. Y. Cho and P. D. Dernier, *J. Appl. Phys.*, **49**, 3328 (1978).
19. K. Okamoto, C. E. C. Wood, and L. F. Eastman, *Appl. Phys. Lett.*, **15**, 636 (1981).
20. C. A. Mead and W. G. Spitzer, *Phys. Rev.*, **134**, A713 (1964).
21. K. H. Hsieh, M. Hollis, G. Wicks, C. E. C. Wood, and L. F. Eastman, *GaAs and Related Compounds*, Int. Phys. Conf. Ser. 65, London and Bristol, Institute of Physics, p. 165 (1983).
22. P. Skeath, C. Y. Su, I. Hino, I. Lindau, and W. E. Spicer, *Appl. Phys. Lett.*, **39**, 349 (1981).
23. J. Massies, J. Chaplart, M. Laviron, and N. T. Linh, *Appl. Phys. Lett.*, **38**, 693 (1981).
24. L. J. Brillson, *Thin Solid Films*, **89**, L27 (1982).
25. S. Eglash, W. E. Spicer, and I. Lindau, *Thin Solid Films*, **89**, L35 (1982).
26. L. J. Brillson, *Surf. Sci. Rep.*, **2**, 123 (1982).
27. J. M. Woodall, Ph.D. Thesis, Cornell University (1982).
28. J. Massies, P. Devoldere, and N. T. Linh, *J. Vac. Sci. Technol.*, **16**, 1244 (1979).
29. K. N. Tu and J. W. Mayer, *Thin Films—Interdiffusion and Reactions*, eds J. M. Poate, K. N. Tu, and J. W. Mayer, Princeton NJ, Electrochemical Society, p. 359 (1978).
30. J. F. Ziegler, J. W. Mayer, C. J. Kircher, and K. N. Tu, *J. Appl. Phys.*, **44**, 3851 (1973).
31. H. Föll and P. S. Ho, *J. Appl. Phys.*, **52**, 5510 (1981).
32. C. A. Crider, J. M. Poate, J. E. Rowe, and T. T. Sheng, *J. Appl. Phys.*, **52**, 2860 (1981).
33. P. J. Grunthaner, F. J. Grunthaner, D. M. Scott, M.-A. Nicolet, and J. W. Mayer, *J. Vac. Sci. Technol.*, **19**, 641 (1981).
34. T. Sebestyen, *Amorphous Semiconductors '76*, ed. I. Kosa Somozyi, Budapest, Akadémiai Kiado, p. 321 (1977).
35. A. K. Johnscher and R. M. Hill, *Phys. Thin Films*, **8**, 196 (1975).
36. H. Y. Way, *Phys. Rev.*, **B13**, 3495 (1976).

37. K. E. Peterson and D. Adler, *IEEE Trans. Electron Devices*, **ED-23**, 471 (1976).
38. A. R. Riben and D. L. Feucht, *Int. J. Electronics*, **20**, 1359 (1966).
39. M. H. Brodsky and G. H. Dohler, *Crit. Rev. Solid State Sci.*, **5**, 591 (1975).
40. A. G. Milnes and D. L. Feucht, *Heterojunctions and Metal-Semiconductor Junctions*, New York, Academic Press (1972).
41. B. L. Sharma and R. K. Purohit, *Semiconductor Heterojunctions*, Oxford, Pergamon. (1974).
42. T. Sebestyen, *Solid State Electronics*, **25**, 543 (1982).
43. R. M. Raymond and R. E. Hayes, *J. Appl. Phys.*, **48**, 1359 (1977).
44. J. F. Womac and R. H. Rediker, *J. Appl. Phys.*, **43**, 4129 (1972).
45. D. T. Cheung, S. Y. Chiang, and G. L. Pearson, *Solid-State Electronics*, **18**, 263 (1975).
46. W. G. Oldham and A. G. Milnes, *Solid-State Electronics*, **6**, 121 (1963).
47. A. K. Sinha and J. M. Poate, *Appl. Phys. Lett.*, **23**, 666 (1973).
48. C. J. Madams, D. V. Morgan, and M. J. Howes, *Electron. Lett.*, **11**, 574 (1975).
49. S. D. Mukherjee, D. V. Morgan, M. J. Howes, J. G. Smith, and P. Brook, *J. Vac. Sci. Technol.*, **16**, 138 (1979).
50. D. V. Morgan, *Reliability and Degradation*, eds M. J. Howes and D. V. Morgan, Chichester, Wiley, p. 151 (1981).
51. H. B. Kim, G. G. Sweeney, and T. M. S. Heng, *Gallium Arsenide and Related Compounds 1974*, Inst. Phys. Conf. Ser. 24, London and Bristol, Institute of Physics, p. 307 (1975).
52. A. Christou and H. M. Day, *J. Appl. Phys.*, **47**, 4217 (1976).
53. O. Wada, S. Yanagisawa, and H. Takanashi, *Jap. J. Appl. Phys.*, **12**, 1814 (1973).
54. K. Sleger and A. Christou, *Solid-State Electronics*, **21**, 677 (1978).
55. N. M. Johnson, T. J. Magee, and J. Peng, *J. Vac. Sci. Technol.*, **13**, 838 (1976).
56. K. L. Kavanagh, S. H. Chen, C. J. Palmstrom, G. B. Carter and S. D. Mukherjee 'Thin Films and Interfaces II' Ed J. E. E. Baglin, O. R. Campbell and W. K. Chu. New York and Amsterdam, North Holland, p. 143 (1984).
57. J. E. Davey and A. Christou, *Reliability and Degradation*, eds M. J. Howes and D. V. Morgan, Chichester, Wiley, p. 237 (1981).
58. H. M. Day, A. Christou, and A. C. Macpherson, *J. Vac. Sci. Technol.*, **14**, 939 (1977).
59. G. E. Mahoney, *Appl. Phys. Lett.*, **27**, 613 (1975).
60. K. Suh, H. K. Park, and K. L. Moazed, *J. Vac. Sci. Technol.*, **B1**, 365 (1983).
61. J. D. Wiley, J. H. Perepezko, J. E. Nordmann, and Kang-Jin Guo, *IEEE Trans. Industrial Electronics*, **ED-29**, 154 (1982).
62. A. Christou, W. T. Anderson Jr, M. L. Bark, and J. E. Davey, *IEEE 20th Annu. Proc. Rel. Phys. Symp.*, p. 188 (1982).
63. M. J. Kelly, A. G. Todd, M. J. Sisson, and D. K. Wickenden, *Electron. Lett.*, **19**, 474 (1983).
64. E. Kohn, *IEDM Tech. Digest*, p. 469 (1979).
65. N. Yokoyama, T. Ohnishi, K. Odani, H. Onodera, and M. Abe, *IEDM Tech. Digest*, p. 80 (1981).
66. A. K. Sinha, T. E. Smith, M. H. Read, and J. M. Poate, *Solid-State Electronics*, **19**, 489 (1976).
67. R. A. Sadler and L. F. Eastman, *IEEE Electron. Devices Lett.*, **EDL-4**, 215 (1983).
68. K. Matsumoto, N. Hashizume, H. Tanoue, and T. Kanayama, *Jap. J. Appl. Phys.*, **21**, L393 (1982).
69. N. Yokoyama, T. Ohnishi, H. Onedera, T. Shinoki, A. Shibatomi, and H. Ishikawa, *Proc. IEEE Int. Solid State Circuits Conf.*, p. 44 (1983).
70. K. Matsumoto, N. Hashizume, N. Atoda, K. Tomizawa, T. Kurosu, and M. Iida,

Gaas and Related Compounds, Inst. Phys. Conf. Ser. 65, London and Bristol, Institute of Physics, p. 317 (1983).
71. W. F. Tseng and A. Christou, *IEDM Tech. Digest*, p. 174 (1982).
72. W. F. Tseng and A. Christou, *Electron, Lett.*, **19**, 330 (1983).
73. W. F. Tseng, B. Zhang, D. Scott, S. S. Lau, A. Christou, and B. R. Wilkins, *IEEE Electron. Devices Lett.*, **EDL-4**, 207 (1983).
74. R. J. Malik, T. R. Aucoin, R. L. Ross, C. E. C. Wood, and L. F. Eastman, *Electron. Lett.*, **16**, 836 (1980).
75. C. E. C. Wood, *J. Vac. Sci. Technol.*, **19**, 808 (1981).
76. J. M. Woodall, J. L. Freeouf, G. D. Pettit, T. Jackson, and P. Kirchner, *J. Vac. Sci. Technol.*, **19**, 626 (1981).
77. K. Kajiyama, Y. Mizushima, and S. Sakata, *Appl. Phys. Lett.*, **22**, 458 (1973).
78. B. L. Sharma, *Semiconductors and Semimetals*, vol. 15, New York, Academic Press, p. 1 (1981).
79. V. L. Rideout, *Solid-State Electronics*, **18**, 541 (1975).
80. M. N. Yoder, *Solid-State Electronics*, **23**, 117 (1980).
81. A. Piotrowska, A. Gruivarc'h, and G. Pelous, *Solid State Electronics*, **26**, 179 (1983).
82. C. Y. Chang, Y. K. Fang, and S. M. Sze, *Solid-State Electronics*, **14**, 541 (1971).
83. P. A. Barnes and A. Y. Cho, *Appl. Phys. Lett.*, **33**, 651 (1978).
84. D. V. Morgan, F. H. Eisen, and A. Ezis, *IEEE Proc. Pt 1*, **128**, 109 (1981).
85. T. Inada, H. Miwa, S. Kato, E. Kobayashi, T. Hara, and M. Mihara, *J. Appl. Phys.*, **49**, 4571 (1978).
86. T. Inada, S. Kato, T. Hara, and N. Toyoda, *J. Appl. Phys.*, **50**, 4466 (1979).
87. S. D. Ferris, H. J. Leamy, and J. M. Poate (eds), *Laser–Solid Interactions and Laser Processing—1978*, AIP Conf. Proc. 50, New York, American Institute of Physics (1979).
88. C. L. Anderson, G. K. Celler, and G. A. Rozgonyi (eds), *Laser and Electron Beam Processing of Electronic Materials*, Proc. Vol. 80–1, Princeton NJ, Electrochemical Society (1980).
89. C. W. White and P. S. Peercy (eds), *Laser and Electron Beam Processing of Materials*, New York, Academic Press (1980).
90. J. F. Gibbons, L. D. Hess, and T. W. Sigmon (eds), *Laser and Electron-Beam Solid Interactions and Materials Processing*, New York and Amsterdam, North-Holland (1981).
91. B. R. Appleton and G. K. Celler (eds), *Laser and Electron-Beam Interactions with Solids*, New York and Amsterdam, North-Holland (1982).
92. J. Narayan, W. L. Brown, and R. A. Lemons (eds), *Laser–Solid Interactions and Transient Thermal Processing of Materials*, New York and Amsterdam, North-Holland (1983).
93. J. M. Poate and J. W. Mayer (eds), *Laser Annealing of Semiconductors*, New York, Academic Press (1982).
94. C. L. Anderson, in ref. 91, p. 653.
95. J. S. Williams and H. B. Harrison, in ref. 90, p. 209.
96. J. S. Williams, in ref. 92, p. 621.
97. J. S. Williams, in ref. 93, p. 383.
98. J. M. Woodcock, in ref. 91, p. 665.
99. D. E. Davies, T. G. Ryan, J. P. Lorenzo, and E. F. Kennedy, in ref. 88, p. 247.
100. Y. I. Nissim, M. Greiner, R. J. Falster, J. F. Gibbons, P. Chye, and C. Huang, in ref. 91, p. 677.
101. P. J. Topham, M. A. Shahid, and B. J. Sealy *Microscopy of Semiconducting Materials* 1981, Inst. Phys. Conf. Ser. 60, London and Bristol, Institute of Physics, p. 133 (1982).
102. G. Badertscher, R. P. Salathé, and W. Luthy, *Electron. Lett.*, **16**, 113 (1980).

103. H. B. Harrison and J. S. Williams, in ref. 89, p. 481.
104. Y. I. Nissim, J. F. Gibbons, and R. B. Gold, *IEEE Trans. Electron. Devices*, **ED-28**, 607 (1981).
105. J. E. Greene, K. C. Cadien, D. Lubben, G. A. Hawkins, G. R. Erikson, and J. R. Clarke, in ref. 91, p. 701.
106. J. C. C. Fan, R. L. Chapman, J. P. Donnelly, G. W. Turner, and C. O. Bozler, in ref. 90, p. 261.
107. D. H. Loundes and B. J. Feldman, in ref. 91, p. 689.
108. S. G. Liu, C. P. Wu and C. W. Magee, in ref. 89, p. 341.
109. P. A. Pianetta, C. A. Stolte, and J. L. Hansen, in ref. 89, p. 328.
110. P. Pianetta, J. Amano, G. Woolhouse, and C. A. Stolte, in ref. 90, p. 239.
111. J. V. DiLorenzo, W. C. Niehaus, and A. Y. Cho, *J. Appl. Phys.*, **50**, 951 (1979).
112. A. Christou and J. E. Davey, Ion-implanted, improved ohmic contacts for GaAs semiconductor devices, US Patent 4263605 (1981).
113. R. Stall, C. E. C. Wood, K. Board, and L. F. Eastman, *Electron. Lett.*, **15**, 800 (1979).
114. R. A. Stall, C. E. C. Wood, K. Board, N. Dandekar, L. F. Eastman, and J. Devlin, *J. Appl. Phys.*, **52**, 4062 (1981).
115. W. T. Anderson, Jr, A. Christou, and J. E. Davey, *IEEE J. Solid-State Circuits*, **SC-13**, 430 (1978).
116. W. J. Devlin, R. Stall, C. E. C. Wood, and L. F. Eastman, *Proc. 7th Biennial Cornell Conf.*, p. 189 (1979).
117. W. J. Devlin, C. E. C. Wood, R. Stall, and L. F. Eastman, *Solid-State Electronics*, **23**, 823 (1980).
118. J. M. Woodall, G. D. Pettit, T. N. Jackson, C. Lanza, K. L. Kavanagh, and J. W. Mayer, *Phys. Res. Lett.*, **51**, 1783 (1983).
119. A. K. Sinha, T. E. Smith, and H. J. Levinstein, *IEEE Trans. Electron Devices*, **ED-22**, 218 (1975).
120. H. R. Grinolds and G. Y. Robinson, *Solid-State Electronics*, **23**, 973 (1980).
121. B. R. Pruniaux, *J. Appl. Phys.*, **42**, 3575 (1971).
122. J. G. Werthen and D. R. Scifres, *J. Appl. Phys.*, **52**, 1127 (1981).
123. O. Aina, W. Katz, B. J. Baliga, and K. Rose, *J. Appl. Phys.*, **53**, 777 (1982).
124. N. Braslau, J. B. Gunn, and J. L. Staples, *Solid-State Electronics*, **10**, 381 (1967).
125. A. Christou and K. Sleger, *Proc. 6th Biennial Cornell Conf.*, p. 169 (1977).
126. C. J. Todd, G. W. B. Ashwell, J. D. Speight, and R. Heckingbottom, *Metal–Semiconductor Contacts*, Inst. Phys. Conf. Ser. 22, London and Bristol, Institute of Physics, p. 171 (1974).
127. T. J. Magee, J. Peng, J. D. Hong, V. R. Deline, and C. A. Evans, Jr, *Appl. Phys. Lett.*, **35**, 615 (1979).
128. E. Kinsborn, P. K. Gallagher, and A. T. English, *Solid-State Electronics*, **22**, 517 (1979).
129. J. Gyulai, J. W. Mayer, V. Rodriguez, A. Y. C. Yu, and H. J. Gopen, *J. Appl. Phys.*, **42**, 3578 (1971).
130. S. Leung, A. G. Milnes, and D. D. L. Chung, *Thin Solid Films*, **104**, 109 (1983).
131. X.-F. Zeng and D. D. L. Chung, *Thin Solid Films*, **93**, 207 (1982).
132. J. M. Vandenberg and E. Kingsborn, *Thin Solid Films*, **65**, 259 (1980).
133. D. C. Miller, *J. Electrochem. Soc.*, **127**, 467 (1980).
134. T. J. Magee and J. Peng, *Phys. Status Solidi*, **32a**, 695 (1975).
135. K. Kumar, *Jap. J. Appl. Phys.*, **18**, 713 (1979).
136. S. Guha, B. M. Arora, and V. P. Salvi, *Solid-State Electronics*, **20**, 431 (1977).
137. S. Leung, L. K. Wong, D. D. L. Chung, and A. G. Milnes, *J. Electrochem. Soc.*, **130**, 462 (1983).

Metallizations for GaAs devices and circuits

138. C. J. Palmstrøm, D. V. Morgan, and M. J. Howes, *Nucl. Instrum. Meth.*, **150**, 305 (1978).
139. G. Y. Robinson, *Solid-State Electronics*, **18**, 331 (1975).
140. T. Sebestyen, M. Menyhard, and D. Szigethy, *Electron. Lett.*, **12**, 96 (1976).
141. M. Wittmer, R. Pretorius, J. W. Mayer, and M.-A. Nicolet, *Solid-State Electronics*, **20**, 433 (1977).
142. T. G. Finstad, *Thin Solid Films*, **47**, 279 (1977).
143. A. Christou, *Solid-State Electronics*, **22**, 141 (1979).
144. N. Yokoyama, S. Ohkawa, and H. Ishikawa, *Jap. J. Appl. Phys.*, **14**, 1071 (1971).
145. J. E. Loveluck, G. M. Rackham and J. W. Steeds, *Developments in Electron Microscopy and Analysis 1977*, Inst. Phys. Conf. Ser. 36, London and Bristol, Institute of Physics, p. 297 (1977).
146. J. W. Steeds, G. M. Rackham, and D. Merton-Lyn, *Microscopy of Semiconducting Materials 1981*, Inst. Phys. Conf. Ser. 60, London and Bristol, Institute of Physics, p. 387 (1981).
147. T. S. Kuan, P. E. Batson, T. N. Jackson, H. Rupprecht, and E. L. Wilkie, *J. Appl. Phys.*, **54**, 6952 (1983).
148. M. Ogawa, *J. Appl. Phys.*, **51**, 406 (1980).
149. C. R. M. Grovenor, *Solid-State Electronics*, **24**, 792 (1981).
150. C. R. M. Grovenor, *Thin Solid Films*, **104**, 409 (1983).
151. A. M. Andrews and N. Holonyak, Jr, *Solid-State Electronics*, **15**, 601 (1972).
152. M. Otsubo, H. Kumabe, and H. Miki, *Solid-State Electronics*, **20**, 617 (1977).
153. M. Jaros and H. L. Hartnagel, *Solid-State Electronics*, **18**, 1029 (1975).
154. A. B. J. Sullivan, *Electron. Lett.*, **12**, 133 (1976).
155. A. Iliadis and K. E. Singer, *Solid-State Electronics*, **26**, 7 (1983).
156. N. Braslau, *J. Vac. Sci. Technol.*, **19**, 803 (1981).
157. Yu. A. Gold'berg and B. V. Tsarenkov, *Sov. Phys. Semicond.*, **3**, 1447 (1970).
158. K. Heime, U. König, E. Kohn, and A. Wortmann, *Solid-State Electronics*, **17**, 835 (1974).
159. W. D. Edwards, W. A. Hartmann, and A. B. Torrens, *Solid-State Electronics*, **15**, 387 (1972).
160. R. S. Popovic, *Solid-State Electronics*, **21**, 1133 (1978).
161. W. U. Dingfen and K. Heime, *Electron. Lett.*, **18**, 940 (1982).
162. F. Vidimari, *Electron. Lett.*, **15**, 674 (1979).
163. T. Sebestyen, H. Hartnagel, and L. H. Herron, *Electron. Lett.*, **10**, 372 (1974).
164. T. Sebestyen, H. Hartnagel, and L. H. Herron, *IEEE Trans. Electron. Devices*, **ED-22**, 1073 (1975).
165. L. F. Eastman, Private Communication.
166. L. S. Hung, J. W. Mayer, M. Zhang and E. D. Wolf, *App. Phys. Lett.*, **43**, 1123 (1983).
167. M. Wittmer and T. E. Seidel, *J. Appl. Phys.*, **49**, 5827 (1978).
168. G. Eckhardt, in ref. 89, p. 467.
169. C. P. Lee, B. M. Welch, and T. L. Tandon, *Appl. Phys. Lett.*, **39**, 556 (1981).
170. O. Aina, J. Norton, W. Katz, and G. Smith, in ref. 91, p. 671.
171. J. W. Mayer, R. Fastow, G. Galvin, L. S. Hung, M. Nastasi, M. O. Thompson, and L. R. Zheng, *Metastable Materials Formation by Ion Implantation*, eds S. T. Picraux and W. J. Choyke, New York and Amsterdam, North-Holland, p. 125 (1982).
172. C. J. Palmstrøm and R. Fastow, in ref. 92, p. 715.
173. S. D. Mukherjee, D. V. Morgan, and M. J. Howes, *J. Electrochem. Soc.*, **126**, 1047 (1979).
174. B. J. Sealy and R. K. Surridge, *Thin Solid Films*, **26**, L19 (1975).
175. K. L. Klohn and L. Wandinger, *J. Electrochem. Soc.*, **116**, 507 (1969).

176. A. Hiraki, A. Shuto, S. Kim, W. Kammura, and M. Iwami, *Appl. Phys. Lett.*, **31,** 611 (1977).
177. D. V. Morgan and S. D. Mukherjee, Internal University of Leeds Report for SERC (1977).
178. C. C. Chang, S. P. Murarka, V. Kumar, and G. Quintana, *J. Appl. Phys.*, **46,** 4237 (1975).
179. S. P. Murarka, *Solid-State Electronics*, **17,** 869 (1974).
180. M. Ogawa, D. Shinoda, N. Kowamura, T. Nozaki, and S. Asanabe, *NEC Res. Dev. Rep.*, **22,** 1 (1971).
181. V. Kuman, *J. Phys Chem Solids*, **36,** 535 (1975).
182. D. J. Coleman, W. R. Wisseman, and D. W. Shaw, *Appl. Phys. Lett.*, **24,** 355 (1974).
183. T. B. Ramachandran and R. P. Santosuosso, *Solid-State Electronics*, **9,** 733 (1966).
184. A. K. Sinha and J. M. Poate, *Jap. J. Appl. Phys.*, Suppl. 2, 841 (1974).
185. W. T. Tsang, *Appl. Phys. Lett.*, **33,** 1022 (1978).
186. P. A. Pianetta, C. A. Stolte, and J. L. Hansen, *Appl. Phys. Lett.*, **36,** 597 (1980).
187. R. L. Mozzi, W. Fabian, and F. J. Piekarski, *Appl. Phys. Lett.*, **35,** 337 (1979).
188. P. A. Barnes, H. J. Leamy, J. M. Poate, S. D. Ferris, J. S. Williams, and G. K. Celler, in ref. 87, p. 647.
189. P. A. Barnes, H. J. Leamy, J. M. Poate, S. D. Ferris, J. S. Williams, and G. K. Celler, *Appl. Phys. Lett.*, **33,** 965 (1978).
190. J. Basterfield, M. J. Josh, and M. R. Burgess, *Acta Electronica*, **15,** 83 (1972).
191. S. Knight and C. R. Paola, *Ohmic Contacts to Semiconductors*, ed. B. Schwartz, Princeton NJ, Electrochemical Society, p. 102 (1969).
192. C. R. Paola, *Solid-State Electronics*, **13,** 1189 (1970).
193. B. Schwartz and J. C. Sarace, *Solid-State Electronics*, **9,** 859 (1966).
194. R. H. Cox and T. E. Hasty, *Ohmic Contacts to Semiconductors*, ed. B. Schwartz, Princeton NJ, Electrochemical Society, p. 88 (1969).
195. M. McColl, M. F. Millea, and C. A. Mead, *Solid-State Electronics*, **14,** 677 (1971).
196. W. A. Schmidt, *J. Electrochem. Soc.*, **113,** 860 (1966).
197. A. K. Kulkarni and T. J. Blankinship, *Thin Solid Films*, **96,** 285 (1982).
198. M. F. Healy and R. J. Mattauch, *IEEE Trans. Electron Devices*, **ED-23,** 374 (1976).
199. H. Matino and M. Tokunaga, *J. Electrochem. Soc.*, **116,** 709 (1969).
200. T. Sebestyen, I. Mojzes, and D. Szigethy, *Electron. Lett.*, **16,** 504 (1980).
201. H. Hartnagel, K. Tomizawa, L. H. Herron, and B. L. Weiss, *Thin Solid Films*, **36,** 393 (1976).
202. G. E. McGuire, W. R. Wisseman, R. D. Ragle, and J. H. Tregilgas, *J. Vac. Sci. Technol.*, **16,** 141 (1979).
203. R. H. Cox and H. Strack, *Solid State Electronics*, **10,** 1213 (1967).
204. T. Hasty, R. Stratton, and E. L. Jones, *J. Appl. Phys.*, **39,** 4623 (1968).
205. B. L. Weiss and H. L. Hartnagel, *Electron. Lett.*, **11,** 263 (1975).
206. I. Mojzes, T. Sebestyen, P. B. Barna, G. Gergely, and D. Szegethy, *Thin Solid Films*, **61,** 27 (1979).
207. T. G. Finstad, T. Andreassen, and T. Olsen, *Thin Solid Films*, **29,** 145 (1975).
208. L. K. J. Vandamme and R. P. Tijburg, *J. Appl. Phys.*, **47,** 2056 (1976).
209. L. Buene, T. Finstad, K. Rimstad, O. Lönsjö, and T. Olsen, *Thin Solid Films*, **34,** 149 (1976).
210. A. Aydinli and R. J. Mattauch, *J. Electrochem. Soc.*, **128,** 2635 (1981).
211. K. Ohata and M. Ogawa *IEE 12th Ann. Proc. Rel. Phys. Symp.*, p. 278 (1974).
212. J. S. Harris, Y. Nannichi, G. L. Pearson, and G. F. Day, *J. Appl. Phys.*, **40,** 4575 (1969).
213. P. D. Vyas and B. L. Sharma, *Thin Solid Films*, **51,** L21 (1978).

Metallizations for GaAs devices and circuits   261

214. H. M. Macksey, *Gallium Arsenide and Related Compounds 1976* (St Louis), Inst. Phys. Conf. Ser. 33b, London and Bristol, Institute of Physics, p. 254 (1977).
215. K. Ohata, T. Nozaki, and N. Kawamura, *IEEE Trans. Electron. Devices*, **ED-24**, 1129 (1977).
216. G. Y. Robinson and N. L. Jarvis, *Appl. Phys. Lett.*, **21**, 507 (1972).
217. B. P. Johnson and C. I. Huang, *J. Electrochem. Soc.*, **125**, 473 (1978).
218. K. Chino and Y. Wada, *Jap. J. Appl. Phys.*, **16**, 1823 (1977).
219. C. P. Lee, B. M. Welch, and W. P. Fleming, *Electron. Lett.*, **17**, 407 (1981).
220. M. Otsubo, H. Kumabe, and H. Miki, *Solid-State Electronics*, **20**, 617 (1977).
221. M. Heiblum, M. I. Nathan, and C. A. Chang, *Solid-State Electronics*, **25**, 185 (1982).
222. M. I. Nathan and M. Heiblum, *Solid-State Electronics*, **25**, 1063 (1982).
223. W. M. Kelly and G. T. Wrixon, *Electron. Lett.*, **14**, 80 (1978).
224. G. M. Rackham and J. W. Steeds, *Microscopy of Semiconducting Materials 1981*, Inst. Phys. Conf. Ser. 60, London and Bristol, Institute of Physics, p. 397 (1981).
225. R. B. Gold, R. A. Powell, and J. F. Gibbons, in ref. 87, p. 635.
226. S. Margalit, D. Fekete, D. M. Pepper, C.-P. Lee, and A. Yariv, *Appl. Phys. Lett.*, **33**, 346 (1978).
227. G. M. Martin, A. Mitonneau, M. Cathelin, S. Makram-Ebeid, C. Venger, D. Barbier, and A. Laugier, in ref. 90, p. 299.
228. G. Eckhardt, C. L. Anderson, L. D. Hess, and C. F. Krumm, in ref. 87, p. 641.
229. O. Aina, W. Katz, and K. Rose, *J. Appl. Phys.*, **52**, 6997 (1981).
230. O. Aina, S. W. Chiang, Y. S. Liu, F. Bacon, and K. Rose, *J. Electrochem. Soc.*, **128**, 2183 (1981).
231. A. H. Oraby, K. Murakami, Y. Yuba, K. Gamo, S. Namba, and Y. Musuda, *Appl. Phys. Lett.*, **38**, 562 (1981).
232. J. L. Tandon, C. G. Kirkpatrick, B. M. Welch, and P. Fleming, in ref. 89, p. 487.
233. C. P. Lee, J. L. Tandon, and P. J. Stocker, *Electron. Lett.*, **16**, 849 (1980).
234. G. Eckhardt, C. L. Anderson, M. N. Coborn, L. D. Hess, and R. A. Jullens, in ref. 88, p. 445.
235. R. S. Pounds, M. A. Saifi, and W. C. Hahn, Jr, *Solid-State Electronics*, **17**, 245 (1974).
236. J. O. Olowolafe, P. S. Ho, H. J. Hovel, J. E. Lewis, and J. M. Woodall, *J. Appl. Phys.*, **50**, 955 (1979).
237. C. Fontaine, T. Okumura, and K. N. Tu, *J. Appl. Phys.*, **54**, 1404 (1983).
238. Y. Wada and K. Chino, *Solid State Electronics*, **26**, 559 (1983).
239. M. Ogawa, *Thin Solid Films*, **70**, 181 (1980).
240. T. G. Finstad and J. S. Johannessen *Proc. 10th Nordic Semiconductor Meeting, 9–11 June 1982*, Elsinore, Denmark.
241. L. J. Chen and Y. F. Hsieh, *Proc. 41st Annu. Met. Electron Microscopy Soc. of America*, G. W. Bailey (Ed), San Francisco, p. 156 (1983).
242. P. Oelhaten, J. L. Freeouf, T. S. Kuan, T. N. Jackson and P. E. Batson, *J. Vac. Sci. Technol.*, **B1**, 588 (1983).
243. O. Wada, S. Yanagisawa and H. Takanashi, *Appl. Phys. Lett.*, **29**, 263 (1976).
244. A. K. Sinha, *Appl. Phys. Lett.*, **26**, 171 (1975).

Gallium Arsenide
Edited by M. J. Howes and D. V. Morgan
© 1985 John Wiley & Sons Ltd

CHAPTER 7

Metal–Insulator–GaAs Structures

D. L. LILE

7.1 INTRODUCTION

Dielectrics serve many functions in modern solid state devices and circuits including their use for metal crossover isolation as well as for active gate insulation in FETs and for passivation of circuit chips from the adverse effects of the environment. The importance of dielectric based technologies is perhaps best dramatized by noting that the insulated gate NMOS technology on Si now exceeds in value the competitive Si bipolar approach and is expected to widen its lead in the years ahead.[1] Moreover, all current VLSI designs are based on insulated gate devices in Si.

In view of the importance of this class of materials to the Si technology, it is perhaps not surprising that, over the last half-dozen years or so, a considerable effort has been devoted to an attempt to identify and develop insulators for use on semiconductors other than Si. It is well known that the state of perfection of the interface of this latter semiconductor when in contact with its thermal oxide is high, with surface state densities on $\langle 100 \rangle$ oriented material for the most demanding CCD applications being almost unmeasurably in the low 10^9 cm^{-2} range. Compared to this system it is clear that all other MIS devices fall far short. It is particularly unfortunate that GaAs, perhaps the most highly developed of all the compound semiconductors, would, from the existing data, appear notably poor when considered for MIS applications. Many attempts have been made over the years to develop dielectrics for this semiconductor which will be both insulating and support an acceptably low state density at the surface. In all cases the results have been to a large extent disappointing.

In the light of this, it is often asked whether we should dismiss GaAs from the realm of those semiconductors which may achieve a working MIS status and move on to other apparently more promising compounds such as InP and GaInAs? The answer to this question must be qualified. Certainly, as this chapter is developed, the author's opinions will become apparent on this issue. However, this is not the central question we would wish to address here. Rather, the intent is to present a summary of what has been accomplished and what explanations have been proposed to account for the observed phenomena in what, it must be admitted, is a highly complex physical situation. An attempt will be made to be objective, with the hope being that a presentation of the facts, hopefully largely untainted by the opinions of the author, will allow the reader to draw his own conclusions.

Certainly the development of good insulators and interfaces on GaAs would be a noteworthy and highly significant technical development. The question is, how likely is such a development and what are the most promising avenues of approach to achieve it? These issues will be addressed in this chapter.

First it would be well to consider what sets an insulated gate device technology apart from all others. Stated another way we might ask, why should we spend time on trying to develop an MIS approach on GaAs when perfectly good MESFET devices are available? Examination of the design trade-offs for Si devices shows that MOS transistors are simple to make (far simpler, for example, than bipolar structures) and further that they are smaller, occupying less volume of semiconductor. Moreover, MOS devices can be made to operate in both the depletion as well as enhancement modes and thus are directly applicable to low power and simpler direct coupled logic (DCTL) designs for digital circuit applications. As is discussed elsewhere in this volume, this is certainly also possible with MESFETs on GaAs; however, operating voltage limitations ($\lesssim 0.6$ V) imposed by enhancement FET forward bias gate leakage place severe manufacturing tolerances on the range of threshold voltages, V_{TH}, acceptable for circuit operation. Opinions vary, but 30 mV might be a reasonable estimate for the permitted deviation in this parameter across the working area of the wafer for enhancement–depletion circuitry. Although many laboratories in both the USA as well as Japan and Europe are pursuing this approach, opinion seems divided over whether it can ever be economically feasible. MIS circuitry would eliminate this problem.

Insulated gate devices and circuits are also inherently passivated. Granted, with appropriate dielectrics, MESFET and junction designs can be overcoated to eliminate leakage and stabilize performance against the adverse effects of the environment; however, such layers cannot prevent the problem of chemical interactions between the metal and semiconductor. This effect, although far less pronounced in general than the classic 'purple plague' reaction on Si, is nevertheless worrisome when such performance factors as shelf-life and long-term device stability are concerned or where high-power, high-voltage, or high-temperature operation is at a premium. Finally the possibility of driving a depletion MISFET with input signals of both polarities means that improved linearity with analogue operation around zero bias becomes possible and that larger power levels can be achieved because of the enhanced drive signals permitted.

All this should not, of course, be taken to mean that there are no inherent disadvantages with insulated gate approaches. Radiation tolerance, for one, is notoriously poor in thermal Si/SiO_2 devices (10^4 rad total dose limit) when compared to the GaAs MESFET, for example ($> 10^8$ rad tolerance), and although it is not clear that the same low threshold will occur in other materials, it is cause for some concern.

Our remarks, so far, have primarily addressed the MISFET because this is the device that has attracted both the most interest as well as most study. Other devices can, however, also benefit from the availability of good insulators, including their use to enhance the efficiency of solar cells as well as to enable more flexibility in the design of active devices such as CCDs; and over and against all of this it must be appreciated that the development of a good passivating layer, whether it be for use with MESFETs, JFETs, or any other device configuration, would be a very important and significant development for the GaAs technology as a whole. In any case, it seems clear that the availability of a working MIS system on GaAs would be of considerable benefit in a number of areas. With this as introduction we will now proceed, in turn, to a consideration of evaluation procedures for such insulated gate structures, to a discussion of preparation methods leading, finally, to a presentation of the current state of the art and a review of our present understanding and future prospects for this approach.

7.2 THE MIS SYSTEM

The general properties of metal–insulator–semiconductor structures, albeit primarily on Si, have been discussed, reviewed, and analysed on numerous occasions in the past and it is not our intent to reproduce that material here. The reader who would like a general background in the physics of MIS systems is referred to a number of excellent articles in the open literature.[2,3] Our emphasis here is on GaAs and we shall confine our attention in large measure to the problems, characteristics, and approaches to be employed in analysing such structures based on this semiconductor. It should be realized at the outset, however, that although we will attempt to consider the characteristics of this particular semiconductor and in many respects compare it to Si, we are really comparing GaAs as a representative of a wide class of compounds with an elemental semiconductor. In consequence we will, in many cases, draw conclusions for GaAs based on measurements made on other materials of the class of III–V compounds of which GaAs is but one example.

7.2.1 Basic considerations

The next chapter will consider the fabrication details involved in preparing MIS devices on GaAs. For the moment we assume that such structures can be made and that they are available, and proceed to an examination of their characteristics as these pertain to device operation.

An MIS structure in its ideal form is quite simple, being nothing more than a parallel-plate capacitor with one electrode of sufficiently low carrier density ($\lesssim 5 \times 10^{17}$ cm^{-3}) to permit a significant ($\gtrsim 250$ Å) penetration of the electric field, resulting from any applied bias voltage, into its interior. Practical considerations on convenient voltage levels mandate dielectric thicknesses usually less than a few thousand angstroms and breakdown associated with flaws, such as pinholes, set a lower limit usually around 250–500 Å. This latter value, of course, depends greatly on the degree of development of the dielectric and surface preparation technologies as well as cleanliness.

This field penetration into the semiconductor surface creates a space charge region which is the heart of all MIS semiconductor devices. Changes in the input bias result in changes in the space charge, which in turn result in changes in some output signal.

If this were all that occurred, of course, there would be little further to discuss. In practice, perturbations in the form of electrically active centres, both in the semiconductor as well as in the dielectric and interface, create effects that cause the characteristics of the MIS structures to deviate from the ideal. This deviation in Si structures is now minimal, indicating the high state of perfection of the technology. In GaAs and other compound semiconductors, the non-idealities can be significant if not dominant.

Figure 7.1 compares the ideal MIS case with the general situation likely to be met in practice. In this figure a schematic cross sectional energy diagram together with a charge distribution profile are shown. In a simplistic way it can be said that all effort on these structures is aimed at obtaining something approaching closer to (a) than (b). Failure to do so can result in drastic degradation in device performance, including a reduction in transconductance of FETs, a loss of charge transfer efficiency and high-speed capability of CCDs, and a severe dispersion in device low-frequency characteristics.[4]

The origin of the electrically active centres responsible for trapping of both electrons and holes is, in general, unknown although both defect and material inhomogeneities as well as contaminants or impurities are suspect. Despite considerable theorizing and some experiment, it is certainly clear that no consensus exists as yet as to what causes the non-idealities in the GaAs MIS system. In Si it is well known that Na and K ion contamination lead to severe long-term instabilities owing to ion drift in the thermal SiO_2 oxide and that, by improvements in cleanliness during handling and fabrication and by subsequent treatments such as annealing in hydrogen, this problem has been all but eliminated. Why this should be so and why anneals in H_2 at 900°C or in the presence of Al metallization should also reduce the interfacial trap states is still not clear. In the light of this, it is perhaps not surprising that the compound semiconductor MIS surface is also far from understood.

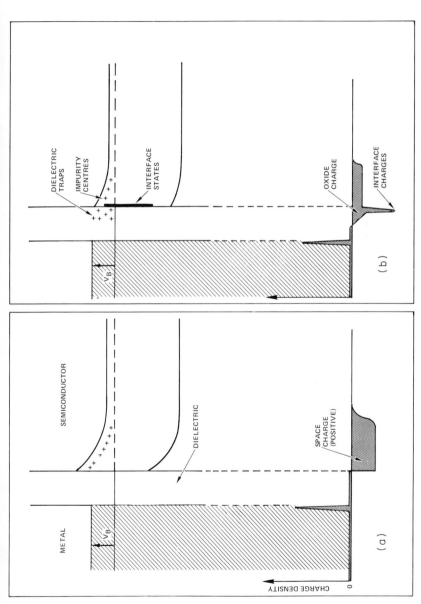

Figure 7.1 Schematic band diagram (upper) and charge distribution (lower) for an (a) ideal and (b) actual MIS diode in the presence of negative bias on the metal. The figure is drawn for n-type material and assumes non-ideality in the form of interface and bulk dielectric states

7.2.2 Measurement procedures

7.2.2.1 Two-terminal structures

In this section we will give consideration to the techniques and resulting data that have been generated from measurements on two-terminal GaAs MIS devices aimed at elucidating the physics of the interface in these structures.

Probably because of its highly successful application to the Si/SiO_2 surface as well as the ease with which the measurement can be made, it is the capacitance–voltage $(C-V)$ technique that has been most widely employed to analyse the interface properties of MIS devices on GaAs. The reader unfamiliar with this technique is referred to the review articles in the general literature for background information.[5] In its simplest form the technique consists of measuring the differential capacitance across a two-terminal MIS diode while at the same time the bias across the structure is varied. If the semiconductor is not too highly doped (net impurities $\lesssim 5 \times 10^{17}$ cm^{-3}) and if the dielectric is not too thick ($\lesssim 2000$ Å), then most of the measured capacitance will be due to the capacitance of the space charge region in the surface of the semiconductor and the contribution corresponding to charge filling of surface states N_{SS}. At high frequency this latter contribution will tend to zero whereas at low frequencies the trap levels can have a significant effect. Varying the bias will result, in general, in both a change in the surface space charge and in the occupancy of surface traps, and by varying the measurement frequency it is possible to generate a series of $C-V$ curves which, by comparison with theory, can be used to deduce information on the trap density and distribution. In practice it has become usual to make such measurements at a frequency sufficiently high that no surface states can respond, in which case an analysis published originally by Terman[6] allows the deduction of surface state densities. Such a high-frequency measurement is, however, fraught with problems, not the least of which is ensuring that no surface traps are responding at the measurement frequency chosen. Moreover, this technique is subject to large uncertainties in accumulation where small errors in deducing the oxide capacitance, C_{ox}, can result in large errors in N_{SS}. Certainly in this regime also, the potentially fast response time of surface states close to the band edge makes the attainment of true high-frequency conditions in general unlikely.

Less ambiguous would seem to be the quasi-static technique of Berglund[7] whereby the $C-V$ data are generated under conditions close to d.c. In this case all the surface states other than those with extremely long time constants contribute to the measurement, reducing to some extent the ambiguity of the deduced numbers. A potential difficulty with this method, however, when used on other than high-quality SiO_2/Si samples, is that any significant gate leakage when compared to the charging current can severely distort the data. Typically, values for resistivity $\rho > 10^{16}$ Ωcm are required to eliminate problems from this source.[5] Analysis of such data also requires knowledge of the carrier concentration near the sample surface, which in practice can often be obtained from a deep depletion C–V curve.

Both of these techniques have been used individually on Si samples with great success and, within the resolution of the C–V method, considerable confidence seems to exist in N_{ss} values deduced by these methods. As we shall see, this is far from the case on the III–V compounds, however, with GaAs MIS C–V data, in particular, having resulted in some controversial interpretations.

Figure 7.2 shows typical C–V data taken on n- and p-type material as has been frequently reported in the open literature. Early interpretations[8] of such curves, because of their often near-ideal appearance (e.g. 5 kHz curve in Figure 7.2a), suggested a low surface state density for the GaAs–dielectric interface, especially for the case of p-type material.[9] Such interpretations, however, were based on the assumption that such a curve, often being measured at frequencies no higher than 1 MHz, was indeed high-frequency in the sense that any surface states present were unable to follow the signal measurement frequency. This assumption, based in large part perhaps on experience with Si and motivated by the convenience of these frequencies of measurement because of available commercial equipment, was subsequently shown to be unfounded. Figure 7.3 shows data published by Meiners[10] where, by measurement to frequencies as high as 150 MHz using an admittance bridge and strip line techniques, he was able to show that significant densities of surface traps were still able to follow signals well in excess of 1 MHz and that the conclusions as previously published based on erroneous assumptions of high-frequency behaviour must be viewed with concern. As is clear from the data in Figure 7.3, the highest-frequency data depart significantly from ideal and, in fact, show very little variation of capacitance with bias. Such a C–V curve would imply an appreciable level of surface trapping, which would also be consistent with the significant dispersion evident in Figure 7.2. Meiners has in fact concluded that the range of surface potential accessible is limited to no more than 0.45 eV. As the frequency of measurements is reduced, the number of surface states able to follow the measurement signal increases and constitutes a component of surface capacitance which adds to that of the space charge. This effect is most pronounced towards stronger accumulation where the speed of surface state response is greater. These arguments are consistent in a qualitative manner with the family of curves shown in the figure. Based on such arguments, an analysis of C–V data has been used to arrive at surface state distributions for both p- and n-type GaAs as are shown in Figure 7.4.[11] Such values, which are extremely large and which would be expected to degrade MIS device performance severely, are consistent with what has also been obtained using conductance versus frequency measurements.[12] In contrast Figure 7.5 shows G–ω data[13] from such measurements together with deduced N_{ss} values which indicate a reasonably low density of states. Whether this is real or due to some artifact of the measurement is unclear. As can be seen in Figure 7.5, the G–ω technique provides information over only a small region of the band gap owing to the practical limitations on the range of

Gallium arsenide

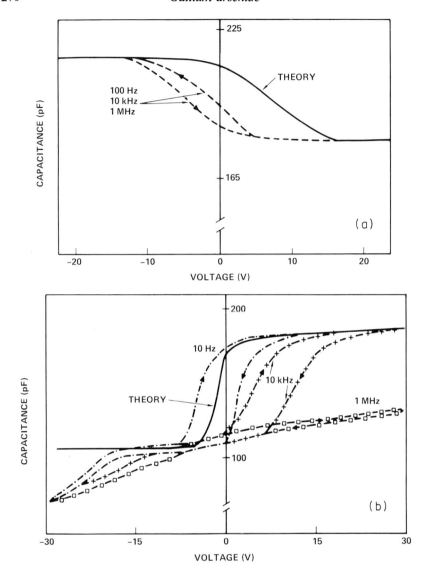

Figure 7.2 Typical $C-V$ data on (a) n-type and (b) p-type (100) oriented GaAs MIS structures. Particular data shown were obtained using plasma anodic oxide for the insulator. (Taken from ref. 9, by permission)

frequencies which may be used. Worse than that, however, for these surfaces, is the problem of drift. This will be discussed more fully later; however, we might note that it is commonly observed on all compound semiconductors where, at other than low temperatures, the characteristics including the surface trap filling

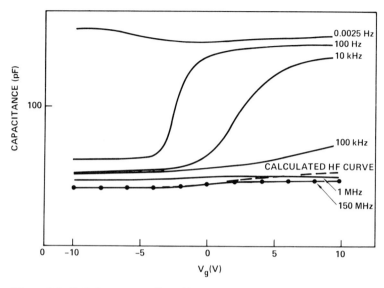

Figure 7.3 C–V data on anodic oxide MIS diode on n-type GaAs. (Taken from ref. 10, by permission)

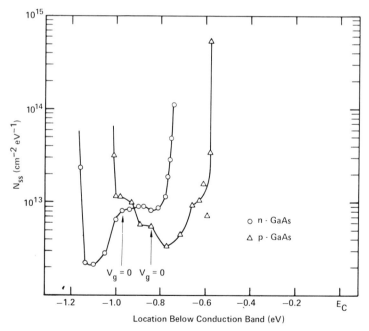

Figure 7.4 Surface state distributions on n- and p-type samples of GaAs as deduced from C-V data. (Taken from ref. 11, by permission)

tend to change with time. For GaAs the problem is especially severe in that looping in $C-V$ data remains, albeit somewhat changed, even at 77 K.[14] In consequence in a $G-\omega$ measurement, where the surface must be scanned in surface potential for various frequencies of measurement, the problem of irreproducibility and inconsistency between each measurement can be severe. This problem can be relieved to some extent by cooling, but this also reduces the time constant of the surface traps further exacerbating the problem of the limited range of surface potentials accessible with available measuring frequencies. This is not to imply that the conductance method cannot be used successfully, however; a number of authors have reported seemingly reasonable results using this technique.[14] What must be kept in mind in any surface measurement technique on such structures as we are considering here is that there are potential sources of error and misinterpretation which must be taken into consideration. Most of the techniques that have been applied to the GaAs MIS system were originally

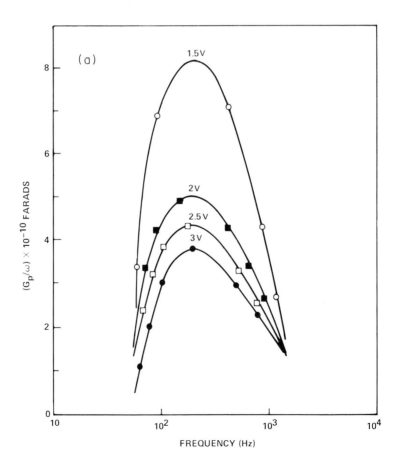

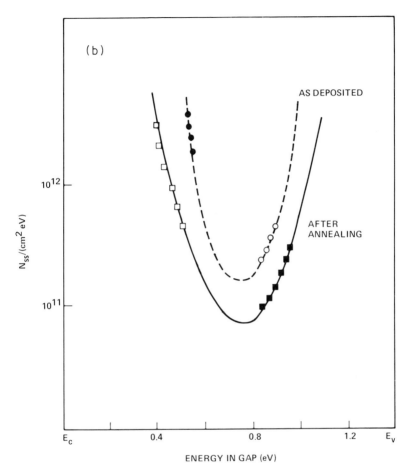

Figure 7.5 (a) Surface state conductance curves plotted versus frequency for various depletion biases on GaN/p-GaAs MIS capacitors. (b) Interface state density distribution before and after annealing in N_2 at 400°C for 30 min obtained using the conductance technique for the same structures as in (a): o, ■ , data obtained on p-type material; ●, □ , data obtained on n-type GaAs. (Taken from ref. 13, by permission)

devised for Si MOS devices and unthinking application to less ideal structures can lead to problems of misinterpretation. Kohn and Hartnagel[15] have discussed this issue and, although their conclusions might be considered extreme, it is certainly a valid comment that care must be taken to include the non-idealities of the MIS structure in data interpretation.

In addition to the $C-V$ and $G-\omega$ techniques described above, more recent analyses of the GaAs–insulator interface have included the use of transient capacitance measurements both in the presence and absence of illumination.

Deep-level transient spectroscopy, for example, has been used by Kassing et al.[16] on anodically prepared MIS diodes of n-type GaAs to elucidate information on the dynamics of the trapping and detrapping process. This technique consists of measuring the device capacitance at two specified instants in time, t_1 and t_2, during the capacitance transient that follows cessation of the trap filling pulse to the gate. ($V_G > 0$ for n-type material.) The DLTS signal, given by the difference of these two capacitance values, then shows a maximum as the temperature is varied, at which point the emission time constant of the filled trap equals the time window, i.e. following Lang[17]

$$\tau_e = \frac{t_1 - t_2}{\ln(t_1/t_2)}$$

Optical measurements, involving the use of monochromatic radiation to depopulate surface trapping centres, have been reported by a number of workers including Alam and Hartnagel[18] and Lagowski et al.[19] Although the details of such measurements vary between groups, a representative technique would be to fill the traps by means of a large accumulating bias to the metal gate ($V_G > 0$ for n-type GaAs) and then to study the discharge current versus wavelength of illumination during subsequent depletion. Such measurements have shown exceptionally large values for the photoionization rate near the band edge which have been interpreted as resulting from an Auger-like process involving donor–acceptor pairs.

Information on the energy band structure at the surface can also be obtained by similar techniques whereby internal photoemission, or the current due to carriers photostimulated into the conduction band of the dielectric, is measured versus wavelength of excitation.

The details of the results of some of these measurements will be discussed in the final section. Suffice it to say that the data, in general, appear consistent with the general understanding that has emerged from the C–V results, with the main ambiguity, at present, seeming to devolve on whether the data are more consistent with discrete or distributed trap distributions.

7.2.2.2 Three-terminal structures

Because of the difficulties inherent in interpreting C–V data on non-ideal MIS structures as well as because the ultimate objective of MIS studies on GaAs is presumably to be able to make good active devices, it is results on three-terminal structures that provide by far the most unambiguous and direct evidence for the state of the art on this as well as any other semiconductor. This is not to say that the C–V data have not provided insights into interfacial effects. Most certainly much of the interpretation of such data is consistent with the three-terminal results reported in the open literature. Where disagreements exist, however, it would seem prudent to weight opinion in favour of the device data, and furthermore where drastic claims are made based on $C-V$ results alone to postpone responses until confirmatory active device data are available.

Non-idealities in MIS gated three-terminal transistor structures manifest themselves in a variety of ways, not least of which is to cause a reduction of device transconductance g_m and a restriction in the region of the band gap over which the surface potential can be varied.[5]

Early results on GaAs MIS transistor structures were encouraging with values of g_m and microwave gain being reported comparable to what could be achieved with the MESFET.[20] Subsequently it was observed that at lower frequencies (below $\sim$ 100 Hz) severe degradation or dispersion in device gain occurred owing presumably to interfacial trapping.[21-24] These results, reproduced in Figure 7.6, were recognized as being consistent with the $C-V$ dispersion previously observed at low frequencies and indicated that the previously reported curve tracer displays of output characteristics and good microwave performance were due not to the good quality of the GaAs/insulator interface but rather to an inability of the extant (and numerous!) surface traps to respond at the frequencies of the measurement. This low-frequency roll-off is obviously of great disadvantage in many potential applications, resulting as it does in an inability to set d.c. bias levels. Furthermore all attempts to improve the low-frequency response by changes in interface preparation and techniques of dielectric deposition appear to have led to little success.

As has been discussed already, one of the main advantages of an MIS technology is that it permits, at least in principle, the fabrication of enhancement mode devices (normally off) of large dynamic range, limited by breakdown or charge injection into the dielectric, based on the formation of inversion or

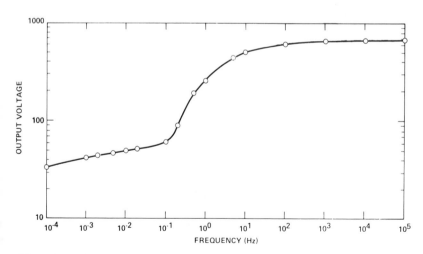

Figure 7.6 Dispersion in GaAs MISFET output with frequency under constant-amplitude input signal conditions

accumulation layers at the dielectric–semiconductor interface. The first report of surface inversion on GaAs appeared in 1974 when Ito and Sakai[25] and Miyazaki et al.[26] published results on enhancement MISFETs fabricated using a variety of insulators. The best results indicated an effective (field-effect) channel mobility for the electrons as high as 2240 $cm^2 V^{-1} s^{-1}$. Despite these seemingly encouraging results this work apparently was not continued and attempts by other groups to duplicate the data appeared to have met with only limited success.[27] These observations together with the C–V data discussed earlier would seem to have led most workers to believe that inversion, at least that of any significant magnitude, cannot be achieved on GaAs presumably due to both the high trap density at the surface as well as to the somewhat undesirable large quiescent surface potential of about 0.8 eV observed on this semiconductor, which places the Fermi level well away from its desired position for inversion near the conduction band edge.

7.3 DEVICE TECHNOLOGY

Although much of the device technology employed with GaAs is an outgrowth of the Si process line, there have been many adaptations and quite a few completely novel developments which have been driven by the unique needs of this material. We shall begin our discussion with a consideration of methods of dielectric growth that have been employed with GaAs. Our treatment will be brief, with those requiring more detailed information on this extensive area of research being referred to the literature.[28–30]

Following this, we shall review, in turn, surface preparation techniques, post-fabrication treatments such as annealing, and general questions of lithography including etches, polishes, and cleaning procedures.

7.3.1 Techniques of dielectric growth

7.3.1.1 Native oxides

In contrast with Si, whose native oxide, SiO_2, possesses extremely good bulk properties as well as near-perfect interfacial characteristics, the native oxide of GaAs suffers from some distinct limitations. Although there are exceptions,[31] many workers have noted such problems as low resistivity and excessive etch rates resulting in uncontrollable lithography. Despite this, working devices have been made using a variety of native oxidation procedures. Perhaps the most simple is thermal oxidation[32] where the sample is raised in temperature in an air or wet or dry oxygen environment. Alternatives include plasma oxidation[33] where the chemical reactivity of the oxidation process is enhanced by locating the sample close to or even inside an oxygen plasma discharge. Both d.c. as well as a.c. methods have been used successfully to produce layers by this means.

Metal–insulator–GaAs structures   277

In all cases such oxides appear to be primarily composed of Ga_2O_3, which is essentially amorphous at temperatures below about 500°C but can be converted to the crystalline β phase by either growth or subsequent annealing above 600°C. The As contributes to the layer, in part, in the form of As_2O_3 but also through the possible presence of metallic excess As at the interface. This question of the presence of excess As and its effect on electrical interfacial properties has been much discussed and would seem still to remain as an unresolved and somewhat controversial issue with statements both for and against its presence being made. Most certainly the high vapour pressure of the column V element mitigates against homogeneous growth. Attempts to overcome this by oxidation in an As-rich ambient have been tried with some success.

More extensively studied and applied to device fabrication has been the process of anodic oxidation whereby the GaAs is positively biased with respect to the oxygen source. This technique has the advantage of flexibility and also results in layers with electrical properties, in general, superior to their thermal counterparts. Wet chemical anodization, involving the transport of oxygen ions to the semiconductor surface from an electrolyte, is the simplest of all these techniques and has been implemented using a variety of solutions. Most widely employed has been the AGW mixture introduced by Hasegawa and Hartnagel[34] consisting of a buffered aqueous solution of tartaric acid and propylene glycol. Although convenient in many respects, other solutions have been applied with similar success.

Such anodically formed layers have been used to make a variety of device structures and undeniably offer a simple, low-cost technique for the generation of an insulator. Like its thermal counterpart, however, this method results in layers that suffer from compositional inhomogeneities and a less-than-ideal resistivity.

Plasma anodization has also been used with some success to form insulating layers for device use. This method resembles closely the plasma oxidation process except that the sample is biased positively with reference to the plasma.

Although varying in their details, all of these methods suffer from the fact that two dissimilar elements with differing oxidation rates are being combined into the one dielectric and that, in general, severe inhomogeneities and non-stoichiometries are likely to result. A variant of the oxidation processes described above that tends to overcome this problem involves first depositing a metal layer on the surface, which is then oxidized to completion either thermally or by anodization.[35] Al layers, for example, can, by such means, be used to form layers of Al_2O_3 and, in principle, any metal that can be oxidized is suitable.

7.3.1.2 Deposited layers

Because the native oxide on GaAs has properties that appear less than ideal for device applications, many workers have investigated the possibility of depositing layers of various insulating compounds using a variety of methods.

278   *Gallium arsenide*

Sputtering   Sputtering is perhaps the most flexible of all deposition procedures allowing, as it does, the formation of layers of almost any vacuum-compatible material. This technique usually involves, however, a fairly large kinetic energy input to the surface of the sample and, as will be discussed later, this seems to be quite undesirable. If sputtered layers are employed, efforts must be taken to minimize disruption of the semiconductor surface by particle bombardment, for example, by using a low-energy ion beam to generate the sputtered species.[36]

Evaporation   Evaporation is perhaps the simplest of all deposition procedures and historically is most certainly the original method used for layer preparation. It has not been used widely in preparing dielectric layers on GaAs, however, despite the fact that some success using thermal evaporation of SiO_x has been achieved. More recently molecular beam epitaxy,[37] which essentially is evaporation under UHV conditions, has been examined for dielectric preparation and offers an attractive possibility for the preparation of the semiconductor–dielectric interfaces *in situ*.

Chemical vapour deposition   Of all deposition techniques, that of CVD has been most widely used for dielectric growth on the III–V compounds in general and on GaAs in particular. The most straightforward method of this family is thermal CVD or pyrolysis where reactive gases, such as silane and oxygen for SiO_2 preparation, are introduced into a heated chamber in which the substrate is located. A primary problem with this technique as applied to compound semiconductors concerns the temperature required to maintain the reaction, which often can be high enough to cause dissociation of the substrate. Two alternatives, plasma-enhanced and photoenhanced CVD,[38] have thus been considered to minimize the thermal energy input required. Both methods rely upon excitation of the gas species to enhance the reaction and thus permit lower temperatures of growth. Some evidence, however, exists to indicate that the plasma itself can exert a derogatory effect on the sample surface and thus, in any such deposition, it would be desirable to separate the growth zone from the immediate vicinity of the ionized gas. Meiners[39] has presented one such approach where a plasma is used to generate an excited but seemingly un-ionized species of oxygen which is then introduced, along with the Si_3H_4, into the reaction zone. This indirect plasma-enhanced CVD technique, a schematic of which is shown in Figure 7.7, has been shown to be very convenient for use with InP and may well be appropriate for the low-temperature, low-energy growth of a variety of dielectrics on other semiconductors also,

Langmuir films   An interesting, if somewhat exotic, and technologically impracticable dielectric preparation procedure that has been used to prepare MIS structures on both the GaAs-related material, GaP, and InP is that of Langmuir and Blodgett.[40] In this technique the sample is repeatedly drawn through a layer

Metal–insulator–GaAs structures

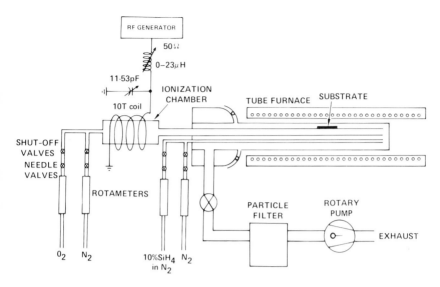

Figure 7.7 Schematic diagram of indirect plasma-enhanced CVD system. (Taken from Ref. 39, by permission)

of diacetylene monomer floating on a water surface, which deposits, atomic layer by atomic layer, an insulating film of apparent high perfection. The method would seem to be distinguished in two primary respects. First, it would appear to be of limited technological utility at least as presently prepared and, secondly, it would seem to be energetically a very gentle process. This latter attribute may be of some significance in elucidating the effects of damage on interfacial properties.

7.3.2 Surface preparation procedures

Other than for the dielectric growth process itself, the pretreatment accorded the semiconductor surface prior to insulator preparation would appear to be the most important aspect of MIS device preparation on GaAs as well as all other III–V compounds. To this end, it is perhaps not surprising that much effort has been devoted to understanding and optimizing the various possible treatments that might be applied. Presumably the starting point for a good surface is a high-quality polish. Wafers as sawn from the boule or following grinding are laden with very high densities of defects. Chemo-mechanical polishing using, for example, a bromine–methanol etchant on a pellon lap is then typically used to remove a few mils (a few thousandths of an inch) of material from one face until a flat mirror-like and hopefully damage-free surface is obtained. This aspect of the processing has become somewhat standard, being a direct carryover from the MESFET development programmes on GaAs. Beyond this point, however, considerable uncertainty and disagreement has been generated as to what type of treatment would best facilitate the formation of low defect trap-free surfaces. One controversy concerns the role of the native oxide layer that inevitably resides on the surface of the GaAs and is retained as an oxide interfacial layer underlying any deposited dielectric. As will be discussed at length later, when we consider the various models that have been suggested to account for the electrical characteristics of the GaAs/insulator interface, this native oxide has been proposed as a source for the high interfacial trapping densities that have been observed. In contrast, no unambiguous evidence would seem to exist to prove the case conclusively and thus other voices have been raised which lay the blame elsewhere. Depending on one's opinion regarding the role of the native oxide, both reducing treatments as well as oxidation steps have been proposed. These can take the form of pretreatments of the surface prior to loading the sample into the reactor or of gaseous *in situ* processes immediately prior to dielectric deposition. In the latter category gaseous HCl pre-etching has perhaps received the widest attention, having been proposed as a means whereby the possibly undesirable native oxide can be reduced.

7.3.3 Postgrowth treatments

It is well known that a postgrowth anneal step is mandatory in the thermal SiO_2/Si technology to achieve low trap density interfaces. Presumably based on this it has been claimed, on a number of occasions, that post dielectric growth heat treatments may be beneficial for the GaAs interface also. In fact, early results using wet chemical anodic layers did show significant improvements in the general appearance of $C-V$ data following annealing, which was interpreted to result from a reduction in surface state densities.[41,42] More recent reports,[10] which have called into question the early data analysis, cloud the issue to some extent. Despite this, it seems that most workers feel, perhaps, justifiably so in the light of its demonstrated efficacy on Si, that annealing should be good. Any anneal process used would, of course, need to be far more moderate than the approximately 1000°C used with Si because of the high volatility of the As. Most work to date would seem to have centred in the range of 350 to 500°C, although higher temperatures, such as the roughly 800°C used for post-implant activations, might be feasible provided suitable precautionary steps were taken to preserve the integrity and stoichiometry of the surface. It is certainly difficult at the moment to argue convincingly either for or against the benefits of annealing when the mechanism of trapping effective in these structures is so little understood. Because of this the same approach as occurred in Si, which was the empirical process of 'try it and see', presumably must be advocated. It is interesting, however, that the model of Sawada and Hasegawa[43] for surface states, which proposes as their source a disordered interface region, would lead one to conclude perhaps that a properly performed anneal, though crystal reconstruction, might indeed lead to a reduced surface state density.

7.3.4 Processing

Much of the processing employed with III–V compounds in general and GaAs in particular is a direct carryover from the Si technology. Moreover, GaAs MIS device procedures have benefited greatly from the considerable process technology development that has occurred on the GaAs MESFET. Contact preparation, for example, is crucial to good device performance and the procedures developed for the GaAs MESFET technology based on alloyed Au–Ge are similarly applicable for GaAs MIS devices. Ion implantation is also a powerful tool that has enjoyed much success in applications to device fabrication on GaAs and certainly would be expected to be the technology of choice for both contact and channel definition in a GaAs MIS structure because of both its controllability and its uniformity. Quite reasonably, the main distinguishing processes for an insulated gate technology are those which affect the dielectric. Our own experience has shown that dry plasma etching in CF_4 works very well for removing the CVD SiO_2 either uniformly or selectively using a photoresist mask. Moreover, the dielectrics are usually compatible with standard lithographic processing including solvent washes. One exception concerns some anodic layers which, presumably because of porosity, adsorb from their ambient with a consequent change in properties. As a practical matter, correct hermetic sealing can accommodate and prevent such sensitivity from affecting device performance although such considerations must be a part of any overall design.

To illustrate the processes involved we will describe a typical fabrication sequence for a small-signal planar depletion mode MIS-gated FET. The values given are meant only to be representative as are the chemical processing steps. It should be understood that variations in the described process may be perfectly acceptable or even beneficial in some cases, including, for example, the use of epitaxial layers for channel and contact definition. Similarly the device geometry shown is only one of a large number of potentially acceptable designs which might be employed for such a device.

With reference to Figure 7.8, the process begins with a wafer of Cr-doped semi-insulating GaAs chemo-mechanically polished on one face. Using standard lithographic procedures, the source and drain contact regions are then defined as apertures in a photoresist layer following which the GaAs is lightly etched (~ 200 Å) using, for example, tartaric acid, which does not attack the photoresist but does provide recesses in the semiconductor for subsequent alignment.

Following this the sample is subjected to a heavy implantation dose of about 5×10^{14} Si^+ ions/cm^2 at roughly 200 keV. The photoresist is then removed and a new pattern generated to define the channel using the recessed source–drain areas and their associated discoloration from ion damage for registration. The channel implant is then performed with a light Si dose of approximately 10^{12} cm^{-2}. The photoresist is removed and the post-implant anneal performed using either a capping layer or an AsH_3 overpressure to maintain the integrity of the surface at

282   *Gallium arsenide*

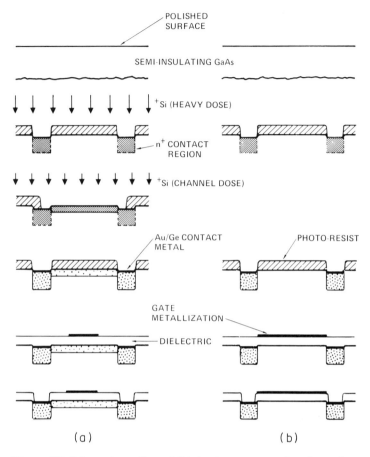

Figure 7.8 Schematic outline of fabrication sequence for planar ion-implanted (a) depletion and (b) enhancement MOSFETs on GaAs

the 800°C temperature required for activation. Following the anneal, contact pads of Au/Ge are defined using standard liftoff of evaporated material. These contacts are then alloyed and the sample is placed in the reactor for dielectric deposition. At this point it should be emphasized that a series of *ex* or *in situ* cleaning procedures might be performed on the sample depending upon their perceived efficacy in improving the surface properties. Similarly a variant is to prepare the dielectric before depositing the contact metallization. This has the advantage of removing a potential source of metallic contamination of the dielectric from the chamber but at the price of subjecting the dielectric and interface to the contact anneal process. Following dielectric growth, to 600–800 Å for a typical small-signal device, the gate pattern is defined in, for example, Al or Au, by liftoff. The final step involves opening contact aperture holes in the oxide using dry plasma etching.

Typically for a 1 μm gate length depletion structure the source–drain gap might be about 3 μm. More recently in MESFET studies this traditional rule of thumb has been changing somewhat as researchers have observed increased microwave performance as the source–gate and drain–gate separations are reduced via self-aligned or 'close-spaced' transistor designs.[44]

It is to be understood that this fabrication sequence is essentially identical to that employed in fabricating MESFETs on GaAs and differs only insofar as a dielectric is deposited prior to gate metallization. A more pronounced difference exists for enhancement devices, of course, where the channel implant dose is omitted and the gate extends across the entire source–drain gap as shown in Figure 7.8b. In this case a self-aligning technique to register automatically the gate and source–drain gap to minimize parasitic interelectrode capacitances would be desirable. At present such an approach has not been demonstrated and will require either an improvement in the ability of the dielectric/semiconductor interface to withstand high-temperature processing or a novel method of alignment to circumvent this thermal restriction.[45]

7.4 DEVICE PERFORMANCE

Of all possible devices, the FET is most certainly the simplest and probably the most useful configuration and lends itself directly to a reasonably unambiguous demonstration of the virtues of any given technological approach. For these reasons, it is perhaps not surprising that by far the largest effort to date on MIS devices has addressed the transistor.

The first report of a MIS-gated FET on GaAs appeared in 1965 when Becke, Hall, and White[22] reported on n-channel depletion mode devices using CVD SiO_2 for the dielectric. These devices, with 7 μm long diffused channels, exhibited g_m values of about $16\,mS\,mm^{-1}$ of gate width and effective (field-effect) channel mobilities of about $1780\,cm^2\,V^{-1}\,s^{-1}$. An inability to invert the surfaces of the GaAs, which was ascribed to a high density of surface states $N_{SS} > 10^{12}\,cm^{-2}\,eV^{-1}$, prevented operation of enhancement devices and also led to an increasing value of g_m with frequency in the depletion structures. These first results were quickly followed by work on an insulated gate FET prepared on polycrystalline GaAs[46] but further efforts to develop MIS devices were somewhat eclipsed by the spectacular success of the MESFET which was proposed in the same year.

Charlson and Weng[47] were the next to report a thin-film MIS transistor in 1968 and then in 1974 Ito and Sakai[25] published a paper describing the development of an inversion mode FET exhibiting a channel mobility of about $2240\,cm^2\,V^{-1}\,s^{-1}$. Since that time numerous results have appeared on both the normally 'on' and normally 'off' structures using various dielectric preparation techniques for gate insulation,[48] including wet chemical anodization[27,41,49–53] and plasma anodization[54] and oxidation,[55,56] plasma anodization of Al,[57] pyrolysis of SiO_2,[22]

284   *Gallium arsenide*

Al_2O_3,[58] $Si_xO_yN_z$,[59] and Ge_3N_4,[60] and thermal oxidation.[23] Without exception, the results have shown that depletion MISFETs may be made on GaAs which exhibit attractive output characteristics as might, for example, be displayed on a curve tracer. By way of illustration the output characteristics shown in Figure 7.9

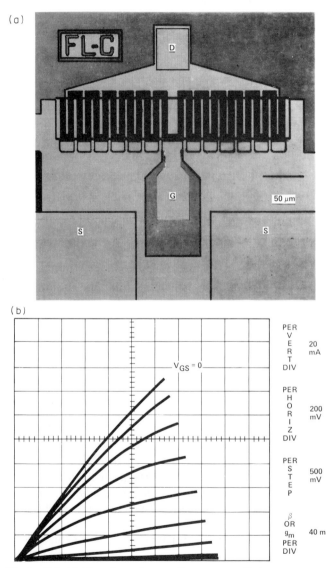

Figure 7.9 (a) Drawing of a medium-power GaAs MISFET. (b) Typical output characteristics for this 1.8 μm gate length depletion mode device. (Taken from ref. 48, by permission)

were obtained on such a GaAs device made using a plasma anodic dielectric. By means of such measurements g_m values as high as 55 mS mm^{-1} of gate width with corresponding values of $\mu_{FE} \sim 3200$ cm^2 V^{-1} s^{-1} have been reported.[48] As has been already discussed, results such as those in Figure 7.9 are only possible because the frequency of measurement (100 or 120 Hz) is beyond that at which the majority of the surface traps can respond. In fact, in some cases it has apparently been found necessary to use pulsed measurements of the output characteristics to overcome the debilitating effects of the slow trap centres.[52] This is clearly evidenced in the dispersion of both $C-V$ characteristics[21] as well as transconductance reported by a number of authors on reducing signal frequency below about 100 Hz.[21-24] As was discussed earlier, the consequences of such a low-frequency characteristic for these devices is quite severe. Particularly it implies a lack of d.c. and hence bias control and would seem to preclude the use of these structures in anything other than a dynamic mode where d.c. signal levels do not have to be held for more than perhaps 10 ms. One such application has, in fact, been demonstrated where a ring oscillator, consisting of a continuously switching series connection of 13 inverter stages, has been demonstrated successfully with a propagation delay of 110 ps at a power level of 2 pJ for a 2 μm length gate.[61] Although dynamic memories and logic approaches where continuous signal refresh is employed are conceptually possible, experience with such logic designs on Si has demonstrated some severe disadvantages including high power consumption and more complex circuitry.

Such debilitating densities of long time constant surface traps would not be expected, however, to impede or degrade the high-frequency performance of such devices and that, in fact, has been shown to be the case.[20,54,62-65] Figure 7.10 demonstrates such microwave response where maximum stable gain, maximum available gain, and deduced unilateral gain for a 1.0 μm gate length depletion GaAs MISFET are shown. Mimura et al.[65] have also reported on a 2.0 μm gate length enhancement structure exhibiting a maximum frequency of oscillation f_u of 13 GHz. A 1.8 μm depletion device of comparable geometry had an f_u of 22 GHz, which is somewhat larger than that to be expected from a MESFET of the same gate length. This improvement was ascribed to be reduced gate parasitic capacitances of the MOSFET structure and appears to be consistent with the subsequent analysis and conclusions of Yamaguchi and Takahashi[66] and Yokoyama et al.[67] Such performance is certainly comparable to what would be expected on a GaAs MESFET of similar geometry, although without stable low-frequency response the practical merit of such devices must be seriously called into question.

Most certainly, one of the major potential advantages of an MIS technology is the possibility it affords of fabricating devices based on the control of surface inversion charge. Unfortunately, in the case of GaAs, with the one exception of the early work of Ito and Sakai[25] and Miyazaki et al.[26] and despite numerous more recent efforts, no convincing demonstration of other than minimal inversion layer

Gallium arsenide

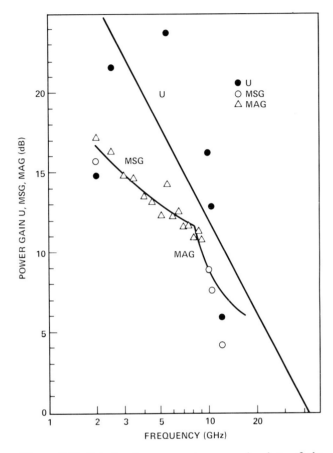

Figure 7.10 Calculated curves and measured values of the maximum stable gain (MSG), maximum available gain (MAG), and unilateral gain (U) for a 1.0 μm gate length depletion MISFET on GaAs. (Taken from ref. 54, by permission)

conduction on GaAs has been forthcoming.[41] This would seem to be consistent with the surface studies reported elsewhere, showing that the n-type GaAs surface is possessed of a quiescent surface potential of approximately 0.85 eV, and a very high density of surface states $N_{SS} > 10^{12}$ cm^{-2} eV^{-1}. By operating at sufficiently high frequencies ($>$ 100 Hz), adequate surface modulation is possible to permit good depletion mode operation. Any attempt to drive the surface into inversion, however, is thwarted not only by the large value of N_{SS} but also because the surface must be initially moved through 0.8 eV even to achieve the onset of inversion. The use of high frequencies to incapacitate the surface traps is precluded in the inversion range of operation because of the decreased response time of traps close to the conduction band edge. Despite the distinctive lack of success in recent attempts to produce good inversion layer performance on GaAs, the early results remain. Although definitive explanations are not possible it is tempting to conjecture that the results observed were due to some phenomenon other than inversion. A primary candidate for GaAs is the inadvertent formation of an n-type conduction layer on the surface due to thermal degradation during processing. Such a layer might be interpreted from FET performance as inversion conduction and is consistent with the historically well documented observation of surface thermal-type conversion in this semiconductor.

Enhancement mode devices, relying upon reduction of the depletion region in a thin epitaxial n-type conduction layer on GaAs, have been demonstrated[51,63,65] with microwave gain being reported to 8 GHz for a 2.0 μm gate length device. These transistors, however, would presumably be expected to suffer from the same low-frequency response restrictions as the depletion structure. It would seem that unless a way can be found either to reduce or otherwise to incapacitate the effects of what is now generally conceded to be a high trap density, then the likelihood for the implementation of such structures in practice is slim. We shall discuss these options in the final section of this chapter. Alternative approaches that attempt to circumvent the surface problem by using novel circuit or device modes have been suggested. Andrade and Braslau,[68] for example, have reported on MIS devices on GaAs in which the dielectric was made intentionally 'conductive' in an attempt to 'draw off' the surface charges. Similarly Schuermeyer[69] has proposed a novel circuit design using what he terms 'electrically setable IGFETs'. In these circuits, depletion MISFET devices can be made to operate in a manner somewhat analogous to enhancement devices by means of a positive 'activation' pulse applied to the gate prior to the actual signal of interest. In this way a variety of integrated circuits including inverters, flip-flops (latches), ring oscillators, and divide-by-two circuits were reported. The only speed or power data given were for the RO, where for 4.5 μm design rules, a gate delay of 150 ps and a speed × power product of 2.5 pJ was observed.

To overcome the restriction of poor surface properties, a number of attempts have been made to engineer a buried channel-type MISFET device where the dielectric is either a high-resistivity region of proton-bombarded GaAs[70,71] or an O_2-doped layer of GaAlAs.[72] In addition to avoiding the surface problems of more conventional MIS designs, such devices may also exhibit enhanced radiation tolerance.[73] Although limited at present, further effort in this direction including perhaps the use of an epitaxially grown high-resistivity layer of GaAs or other wide gap semiconductor such as ZnSe[74] may prove successful. Of much recent interest has been the use of modulation doped superlattice structures between GaAs and GaAlAs, first proposed by Dingle et al.,[75] and implemented into a transistor structure by Mimura et al.[76] These high electron mobility devices (HEMT), which rely for their good performance on the diffusion of electrons from the highly doped large band gap GaAlAs into the high-purity, high-mobility channel layer of GaAs, have conventionally been implemented using Schottky gates. Hotta et al.,[77] however, have recently reported on such devices fabricated with MIS control electrodes. Their results indicate effective mobilities in these structures at 77 K of 27 000 cm^2 V^{-1} s^{-1} using MBE techniques for the semiconductor and Al_2O_3 for the insulator. These metal–insulator–semiconductor (MISS) FETs are perhaps the most promising of all the insulator-based devices to appear to date on GaAs.

In addition to the FET, the application of insulating layers to solar cell efficiency enhancement on GaAs has received some attention,[78] as has their proposed use for MOS light-emitting diode fabrication.[79] In fact, because of the relaxed interfacial dielectric requirements placed on solar cell applications, successful improvements in conversion efficiency and open-circuit voltage V_{oc} have been achieved in this area using MIS configurations on this semiconductor. Increases in V_{oc} of 62% using thin ($\sim$ 20 Å) evaporated layers of SnO_2 and short-circuit current improvements by as much as 60% have been reported by Stirn and Yeh.[80]

Polycrystalline material would have the same advantage in GaAs solar cells as in Si,[81] namely lower cost, and attempts have been made to implement such devices. Pande et al.,[82] in particular, have reported an anodic surface passivation technique for such GaAs devices that eliminates shunting effects of the grain boundaries resulting in a 5 to 6 orders of magnitude improvement in reverse leakage.

Interestingly, the severe trapping effects in GaAs MIS structures have been proposed as a possible source of non-volatile memory. Bayraktaroglu et al.[83] have suggested such an application and, although the data to date are extremely limited, it is clear that effects are present in these structrues that allow for the long-term retention of information.

7.5 STATUS

We have seen in the previous sections how attempts to produce insulating layers on GaAs invariably would appear to result in a large density of surface or interface defects or traps which act to distort and introduce frequency dispersion in C–V curves and FET performance characteristics. Attempts to circumvent these effects have been tried, such as those we described, of using 'leaky' dielectrics or novel circuit modes. However, these would appear to be, at best, no more than stopgap measures, with the real hope, if GaAs MIS circuitry is ever to become a practical reality, being to reduce the actual effective density of states at the surface below the level of perhaps 10^{11} cm^{-2} eV^{-1}.

In this final section we shall discuss what measures are being taken to attempt a reduction of the state trap density on GaAs. To appreciate the signifcance of these endeavours, we shall begin with a brief discussion of the understanding that would appear to exist, at present, on the origin of the surface trap states.

7.5.1 Models for N_{SS}

It should be stated at the outset that, although various explanations have been proposed for the origin of surface states on compounds in general, and on GaAs in particular, no consensus at present would seem to exist on which of these models, if any, is correct. The problem, in part, arises from the extreme complexity of the structure being dealt with, although divergent conclusions based on different experiments also add little to clarifying our understanding. Figure 7.11 is an attempt to illustrate this by displaying schematically the various possible sources of non-ideality in an MIS structure on GaAs.

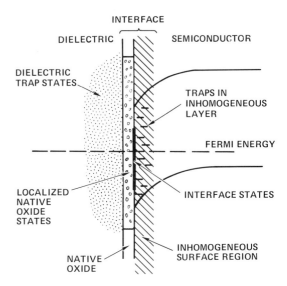

Figure 7.11 Schematic band diagram illustrating the potential sources of surface and interfacial trapping to be expected in an MIS device

Most certainly one obvious source is in the interfacial region of the semiconductor itself where, owing to the differing volatilities of the constituent components Ga and As, a loss of stoichiometry is perhaps likely. Distinct from this, but perhaps related, is the proposal of Hasegawa et al.[43,84] that disorder of the semiconductor crystal lattice, perhaps due to lattice mismatch strains, can cause the introduction of additional unwanted states near the surface. More extreme disorder in the form of vacancy defects has also been proposed by Spicer and coworkers as a source of trapping at the interface.[85] In fact, such defects, in the form of site and antisite vacancies, seem to have engendered much support as a likely source of the traps on GaAs. This so-called 'unified defect model' is primarily based on the observation, from photoemission measurements on the

UHV-cleaved surface of single-crystal samples, of surface potential shifts accompanying exposure of the surfaces to various contaminants including oxygen and various metals. Such deposits might be expected to permit the generation of deep vacancies at the surface that would allow for carrier trapping with the concomitant generation of space charge in the GaAs. Certainly the degrading effect of oxygen on the surface properties of GaAs would seem to be clear from the results of Suzuki and Ogawa,[86] where significant reductions of photoluminescence from samples were observed during oxidation. These results were attributed to enhanced recombination, and thus by inference to increased trap densities, at the oxide–semiconductor interface. Such an indictment of oxygen as the cause of the problems on GaAs is the basis of work such as that of Capasso and Williams[87] whereby a reduction of the surface in a hydrogen atmosphere followed by the deposition of an oxygen-free dielectric such as a nitride is proposed for surface passivation. Whatever the details, it seems clear from the work of Spicer,[88] for example, that, in contrast to Si where oxidation acts to remove intrinsic states present on the semiconductor surface, in the case of GaAs oxidation would seem to introduce levels on an otherwise trap-free clean surface. It might be added that it seems to be accepted that the vacuum-cleaved surface of GaAs is free of states in the gap owing to the effect of surface relaxation in moving the intrinsic centres beyond the band edges.

An alternative explanation ascribes the trapping centres to the ubiquitous native oxide itself, which presumably exists on the surface of GaAs and which, in any practical MIS structure, underlies any deposited or grown dielectric. Of course the deposited insulator itself is a potential source of difficulties and both trapping centres as well as mobile ionic contamination (as was the case in the early SiO_2 results on Si) might well be expected in any given dielectric overlayer.

Intrinsic surface states, resulting from the ideal unterminated or 'dangling' bonds at the metallurgical GaAs/insulator interface have to date received little attention. This would seem realistic for at least two reasons. First, as far as we know from Auger data, for example, with the present state of the art, the interface in these MIS structures is far from abrupt and thus the notion of a GaAs/insulator surface must be replaced by that of an extended interfacial region where the ideal of Tamm–Schockley states would seem to lose much of its meaning. Secondly, no matter what the details, it is clear that the interface region on such compound semiconductor MIS systems is extremely complex and so far from ideal to exclude any real possibility in most people's minds that 'intrinsic' effects are going to exert any real effect.

No matter what the differences between these various possible sources of trapping, it is clear that the compounds are distinguished from Si in one very important respect and that concerns their stoichiometry. Oxidation of Si is relatively straightforward; oxidation of GaAs results in a mix of any of a number of oxides of Ga and of As with the final outcome resulting from the

Metal–insulator–GaAs structures 291

thermodynamics of the system including the relative oxidation rates of the components. The extreme example of this is that some workers have reported the build-up of an elemental metallic layer of As at the dielectric–semiconductor interface following oxidation.[90] Such a layer is itself a severe source of non-ideality in an MIS structure and emphasizes the necessity of processing techniques compatible with maintaining stoichiometry.

From the literature it would seem that the issue of whether the surface states are distributed in energy across the band gap, as would be a result of Hasegawa's interface state band (ISB) disorder model,[43,84] or are localized at one or more discrete levels, as has been proposed by Spicer,[85,88] has been the source of some disagreement. The arguments in favour of a discrete level model for the interface would seem to be largely based on the observed similar 0.8–0.9 eV barrier height on both MIS and Schottky barrier structures. A high density of traps at energy roughly 0.8 eV from the conduction band edge would, for example, then tend to 'pin' or hold immobile the Fermi level in their vicinity, leading to the observed barrier. Of course, to accomplish this on both p- and n-type material, as is observed in practice, somehow requires both donor and acceptor centres. Nedoluha[91] has considered the detils of where such levels might reside and finds that certain limited ranges in energy for one donor and one acceptor level are consistent with the photoemission spectra reported by Spicer et al.[85,88]

It would seem that the issue of the source of the electrical imperfections at the GaAs surface must devolve into a consideration of the chemistry of the interface. A number of papers have identified the problem as resulting from the build-up of excess elemental As together with its oxide As_2O_3 at the interface during oxidation. Capasso and Williams[87] have identified this as a likely source of the problem and have proposed its elimination via a hydrogenation/nitridization passivation process. A similar approach has been adopted by Clark and Anderson[92] where in both cases a hydrogen plasma has been employed. Quite apart from the possible merits of their approach, it is clear that the plasma itself is a definite source of possible trouble owing to its energetically disruptive effect on the surface of the GaAs. It must also be asked whether it is, in practice, possible to generate a truly oxygen-free interface considering the propensity of most materials to oxidation.

Damage to the GaAs surface resulting from the energetics of the deposition process itself as well as from the chemical energy of the deposit arriving at the surface are both matters for concern.[93] Bayraktaroglu et al.[94] have suggested that germanium nitride might, in this respect, be preferable because of its smaller heat of formation and Spicer and coworkers [95] have identified the oxidation of a deposited metal as an attractive possibility.

Ruthenium chemisorbed on the GaAs has also been observed to produce improved surface properties as evidenced by reduced recombination rates. Heller[96] has argued that strong (i.e. short) surface bonds are desirable for low rates of recombination and that the function of the Ru is to bond tightly to the surface. What effect, if any, such treatment might have on the trapping characteristcs of the surface is open to conjecture.

Lucovsky and Bauer[97] have discussed the effects of incomplete termination of surface bonds on the expected trapping behaviour of the surface of III–V compounds from which they have concluded that P-based materials are more likely to complete bonds than are As- or Sb-based materials. Nevertheless Ahrenkiel et al.[98] have examined the possibility that such incomplete termination of the As may be responsible for the large N_{SS} values observed and that oxyfluoride glass compounds of the form $As_xO_yF_z$ may act to terminate and stabilize the surface more completely. Their experiment consisted of comparing C–V data on p-type GaAs MIS structures prepared with plasma oxides with those obtained with plasma-grown dielectrics based on oxygen/fluorine mixtures (from CF_4). In contrast to pure oxide dielectrics, the oxyfluoride structures showed no dispersion in their C–V curves within experimental error over the measurement range from 1 kHz to 5 MHz. From these results it was concluded, in agreement with the model of Lucovsky and Bauer,[97] that fluorine acts to reduce the density of interfacial surface traps due to an increased completion of surface bonding. Related perhaps are the claims of Sher et al.[99] whereby an order of magnitude improvement in N_{SS} was reported from the use of LaF_3.

7.5.2 State of the art

It is interesting to note that the statements and observations reported in the preceding sections of this chapter are drawn, in large part, from two distinct, albeit presumably related, bodies of data. On the one hand FETs have been made and demonstrated with good high-frequency behaviour and poor low-frequency characteristics. On the other, C–V results have been reported showing a variety of forms of behaviour. As stated earlier, it is the author's belief that although C–V (and G–ω, etc.) data are not to be dismissed, at the same time it is the characteristics of the active devices that give a far more direct and unambiguous body of data from which to draw conclusions. It is not without some significance, I think, that it is the device data that have been used to indicate the poor prospects of success for GaAs MIS structures whereas most recent and not so recent claims for promise in this system have been based on C–V results.[100] The lesson seems to be that if indeed some of these new interfacial processing methods are leading to improvement in the trap density at the GaAs–insulator interface, then active devices should be made based on these new methods and be shown to have either improved low-frequency stability and/or significant and unambiguous inversion channel charge controllable by the gate potential. Until such demonstrations are forthcoming, I believe that we must continue to hold in obeyance any enthusiasm for success in this endeavour. The question might be asked, of course and rightly so, as to why indeed many authors have claimed to see and, in fact, have reported apparent improvements in interfacial characteristics as deduced from C–V results without any apparent resulting improvements in three-terminal devices. It would, I think, be foolhardy to dismiss all such data out of hand, but something that seems always to be ignored and can severely distort the interpretation of any results is the time response of the surface state responsible for the non-ideality of the GaAs system. Most, if not all, claims for improvements in the GaAs surface over the years have been based on changes observed in measured $C-V$ characteristics tending hopefully more towards what might ideally be expected. Such changes are then interpreted in terms of reduced values of N_{SS}. The question of these changes perhaps being due to increased time constants for the traps seems always to be ignored, although a changed response time for such traps will most certainly alter the dispersion character of $C-V$ data, for example, just as effectively as will true changes in surface state densities.

It is apparent from the results discussed in this chapter that during the last decade or so a vast variety of dielectrics and interfacial treatments have been applied to GaAs in the hope and expectation of developing a III-V analogue of the Si/SiO_2 system. In large measure these efforts would appear to have led to somewhat disappointing results, suggesting to many groups that the likelihood of low trap density MIS structures on GaAs was, at least in the forseeable future, remote. This fact is perhaps reflected in the level of interest and effort in this area of development, which seems to have peaked in both the USA as well as Europe and Japan around 1977 to 1980. Most certainly relatively little work appears to be currently being expended in this area. Exceptions exist[101] but in large measure it seems that most of these groups with direct experience in this area as well as the majority of onlookers have concluded that, at least in the near term, GaAs MIS is unlikely. Most certainly this is a reasonable position when many other more encouraging areas of research are available and vying for limited research money. The only exception to this option would appear to be the matter of passivation for GaAs, which seemingly will be mandatory if GaAs circuitry of any kind (MESFET, JFET, etc.) is to succeed.

REFERENCES

1. S. M. Sze, VLSI technology overview and trends, *Proc. 14th Int. Conf. on Solid State Devices*, Tokyo, August 24–26, 1982, to be published.
2. A. Many, Y. Goldstein, and N. B. Grover, *Semiconductor Surfaces*, New York and Amsterdam, North-Holland (1965).
3. J. R. Davis, *Instabilities in MOS Devices*, New York, Gordon and Breach (1981).
4. For a review of the consequences of surface imperfections on MIS device performance see: D. L. Lile, Interfacial constraints on III–V compound MIS devices, *Physics and Chemistry of III–V Compound Semiconductor Interfaces*, ed. C. Wilmsen, New York, Plenum, (1983).
5. L. G. Meiners, Electrical properties of insulator–semiconductor interfaces on III–V compounds, *Physics and Chemistry of III–V Compound Semiconductor Interfaces*, ed. C. Wilmsen, New York, Plenum, (1983).

6. L. M. Terman, An investigation of surface states at a silicon/silicon oxide interface employing metal-oxide-silicon diodes, *Solid State Electronics*, **5**, 285-99 (1962).
7. C. N. Berglund, Surface states at steam-grown silicon-silicon dioxide interfaces, *IEEE Trans. Electron Devices*, **ED-13**, 701-5 (1966).
8. R. P. H. Chang and J. J. Coleman, A new method of fabricating gallium arsenide MOS devices, *Appl. Phys. Lett.*, **32**, 332-3 (1978).
9. L. A. Chesler and G. Y. Robinson, Plasma anodization of GaAs in a d.c. discharge, *J. Vac. Sci. Technol.*, **15**, 1525-9 (1978).
10. L. G. Meiners, Electrical properties of the gallium arsenide-insulator interface, *J. Vac. Sci. Technol.*, **15**, 1402-7 (1978).
11. L. G. Meiners, Ph.D. Thesis, Colorado State University (1979).
12. G. Weimann, Oxide and interface properties of anodic oxide MOS structures on III-V compound semiconductors, *Thin Solid Films*, **56**, 173-82 (1979).
13. E. Lakshmi and A. B. Bhattacharyya, Interface state characteristics of GaN/GaAs MIS capacitors, *Solid State Electronics*, **25**, 811-15 (1982).
14. T. Ikoma, H. Yokomizo, and H. Takuda, GaAs passivation and MOS devices, *Jap. J. Appl. Phys.*, **18**, Suppl. 18-1, 131-43 (1979).
15. E. Kohn and H. L. Hartnagel, On the interpretation of electrical measurements on the GaAs-MOS system, *Solid State Electronics*, **21**, 409-16 (1978).
16. R. Kassing, U. Kelberlau, and P. Van Staa, Interface properties of anodically oxidized GaAs MIS capacitors, *Int. J. Electronics*, **52**, 43-55 (1982).
17. D. V. Lang, Fast capacitance transient apparatus: application to ZnO and O centers in GaP p-n junctions, *J. Appl. Phys.*, **45**, 3014-23 (1974).
18. M. S. Alam and H. L. Hartnagel, Pulsed-laser illumination of GaAs MOS structures to study charge trapping, *Int. J. Electronics*, **52**, 61-75 (1982).
19. J. Lagowski, W. Walukiewicz, T. E. Kazior, H. C. Gatos, and J. Siejka, GaAs oxide interface states: a gigantic photoionization effect and its implications to the origin of these states, *Appl. Phys. Lett.*, **39**, 240-2 (1981).
20. D. L. Lile, D. A. Collins, L. Messick, and A. R. Clawson, A microwave GaAs insulated gate FET, *Appl. Phys. Lett.*, **32**, 247-8 (1978).
21. D. L. Lile, The effect of surface states on the characteristics of MIS field-effect transistors, *Solid State Electronics*, **21**, 1199-207 (1978).
22. H. Becke, R. Hall, and J. White, Gallium arsenide MOS transistors, *Solid State Electronics*, **8**, 813-23 (1965).
23. H. Takagi, G. Kono, and I. Teromoto, Thermal oxide gate GaAs MOSFETs, *IEEE Trans. Electron. Devices*, **ED-25**, 551-2 (1978).
24. N. Yokoyama, T. Mimura, and M. Fukuta, Surface states in an n-GaAs/plasma grown native oxide—a modified deep level transient spectroscopy measurement, *Sur. Sci.*, **86**, 826-34 (1979).
25. T. Ito and Y. Sakai, The GaAs inversion-type MIS transistors, *Solid State Electronics*, **17**, 751-9 (1974).
26. T. Miyazaki, N. Nakamura, A. Doi, and T. Takayama, Electrical properties of gallium arsenide-insulator interface, *Jap. J. Appl. Phys.*, Suppl. 2, 441-3 (1974)
27. B. Bayraktaroglu, E. Kohn, and H. L. Hartnagel, First anodic-oxide GaAs MOSFETs based on easy technological processes, *Electronics Lett.*, **12**, 53-4 (1976).
28. W. F. Croydon and E. H. C. Parker, *Dielectric Films on Gallium Arsenide*, New York, Gordon and Breach (1981).
29. B. L. Sharma, Inorganic dielectric films for III-V compounds, *Solid State Technol.*, Pt I, Feb; Pt II, April (1978).
30. C. W. Wilmsen and S. Szpak, MOS processing for III-V compound semiconductors: overview and bibliography, *Thin Solid Films*, **46**, 17-45 (1977).

31. D. H. Laughlin and C. W. Wilmsen, An improved anodic oxide insulator for InP metal–insulator–semiconductor field-effect transistors, *Appl. Phys. Lett.*, **37**, 915–16 (1980).
32. C. W. Wilmsen, Chemical composition and formation of thermal and anodic oxide/III–V compound semiconductor interfaces, *J. Vac. Sci. Technol.*, **19**, 279–89 (1981).
33. R. P. H. Chang, Some properties of plasma-grown GaAs oxides, *Thin Solid Films*, **56**, 89–106 (1979).
34. H. Hasegawa and H. L. Hartnagel, Anodic oxidation of GaAs in mixed solution of glycol and water, *J. Electrochem. Soc.*, **123**, 713–23 (1976).
35. R. K. Smeltzer and C. C. Chen, Oxidized metal film dielectrics for III–V devices, *Thin Solid Films*, **56**, 75–80 (1979).
36. K. Tsubaki, S. Ando, K. Oe, and K. Sugiyama, Surface damage in InP induced during SiO_2 deposition by RF sputtering, *Jap. J. Appl. Phys.*, **18**, 1191–2 (1979).
37. K. Ploog, A. Fischer, R. Trommer, and M. Hirose, MBE-grown insulating oxide films on GaAs, *J. Vac. Sci. Technol.*, **16**, 290–4 (1979).
38. J. W. Peters, Low temperature photo-CVD oxide processing for semiconductor device applications, *Proc. IEDM*, Washington, DC, pp. 240–3 (1981).
39. L. G. Meiners, Indirect plasma deposition of silicon dioxide, *J. Vac. Sci. Technol.*, **21**, 655–8 (1982).
40. K. K. Kan, M. C. Petty, and G. G. Roberts, Polymerized Langmuir film MIS structures, *The Physics of MOS Insulators*, eds G. Lucovsky, S. T. Pantelides, and F. L. Galeener, Oxford, Pergamon Press, pp. 344–8 (1980).
41. B. Weiss, E. Kohn, B. Bayraktaroglu, and H. L. Hartnagel, Native oxides on GaAs for MOSFETs: annealing effects and inversion-layer mobilities, *Gallium Arsenide and Related Compounds 1976*, Inst. Phys. Conf. Ser. 33, London and Bristol, Institute of Physics, pp. 168–76 (1977).
42. B. M. Arora and A. M. Narsale, Electrical instabilities of Al–Anodic-Oxide–n-GaAs MOS structures and the effect of annealing, *Thin Solid Films*, **56**, 153–61 (1979).
43. T. Sawada and H. Hasegawa, Interface state band between GaAs and its anodic native oxide, *Thin Solid Films*, **56**, 183–200 (1979).
44. K. Yamasaki, K. Asai, and K. Kurumada, N^+ self-aligned MESFET for GaAs LSIs, *Proc. 14th Int. Conf. on Solid State Devices*, Tokyo, 1982, to be published.
45. T. Itoh and K. Ohata, X-band self-aligned gate enhancement-mode InP MISFETs, *Proc. IEDM*, San Francisco, 1982, to be published.
46. D. Darmagna and J. Reynaud, A GaAs thin-flim transistor, *Proc. IEEE*, **54**, 2020 (1966).
47. E. J. Charlson and T. H. Weng, Gallium arsenide thin film transistors, *Proc. 20th Annu. Southwestern IEEE Conf. and Exhibition*, 6A1–5, April (1968).
48. T. Mimura and M. Fukuta, Status of the GaAs metal–oxide–semiconductor technology, *IEEE Trans. Electron Devices*, **ED-27**, 1147–55 (1980).
49. D. L. Lile, A. R. Clawson, and D. A. Collins, Depletion-mode GaAs MOSFET, *Appl. Phys. Lett.*, **29**, 207–8 (1976).
50. E. Kohn and A. Colquhoun, Enhancement-mode GaAs MOSFET on semi-insulating substrate using a self-aligned gate technique, *Electronics Lett.*, **13**, 73–4 (1977).
51. E. Kohn, A. Colquhoun, and H. L. Hartnagel, GaAs enhancement/depletion n-channel MOSFET, *Solid State Electronics*, **21**, 877–86 (1978).
52. A. Colquhoun, E. Kohn, and H. L. Hartnagel, Improved enhancement/depletion GaAs MOSFET using anodic oxide as the gate insulator, *IEEE Trans. Electron Devices*, **ED-25**, 375–6 (1978).

53. H. L. Hartnagel, MOS-gate technology on GaAs and other III–V compounds, *J. Vac. Sci. Technol.*, **13**, 860–6 (1976).
54. T. Sugano, F. Koshiga, K. Yamasaki, and S. Takahashi, Application of anodization in oxygen plasma to fabrication of GaAs, IGFETs, *IEEE Trans. Electron Devices*, **ED-27**, 449–55 (1980).
55. T. Mimura, N. Yokoyama, Y. Nakayama, and M. Fukuta, Plasma-grown oxide gate GaAs deep depletion MOSFET, *Jap. J. Appl. Phys.*, **17**, Suppl. 17-1, 153-7 (1978).
56. N. Yokoyama, T. Mimura, K. Odani, and M. Fukuta, Low-temperature plasma oxidation of GaAs, *Appl. Phys. Lett.*, **32**, 58–60 (1978).
57. W. S. Lee and J. G. Swanson, Switching behavior of Al_2O_3–n-GaAs MISFETs, *Electronics Lett.*, **18**, 1049–51 (1982).
58. K. Kakimura and Y. Sakai, The properties of GaAs–Al_2O_3 and InP–Al_2O_3 interfaces and the fabrication of MIS field-effect transistors, *Thin Solid Films*, **56**, 215–23 (1979).
59. L. Messick, A GaAs/$Si_xO_yN_z$ MIS FET, *J. Appl. Phys.*, **47**, 5474–5 (1976).
60. G. D. Bagratishvili, R. B. Dzhanelidze, N. I. Kurdiani, Yu. I. Pashintsev, O. V. Saksaganski, and V. A. Skarikov, GaAs/Ge_3N_4/Al structures and MIS field-effect transistors based on them, *Thin Solid Films*, **56**, 209–13 (1979).
61. N. Yokoyama, T. Mimura, H. Kusakawa, K. Suyama, and M. Fukuta, GaAs MOSFET high-speed logic, *IEEE Trans. Microwave Theor. Tech.*, **MTT-28**, 483–6 (1980).
62. H. Tokuda, Y. Adachi, and T. Ikoma, Microwave capability of 1.5 µm-gate GaAs MOSFET, *Electronics Lett.*, **13**, 761–2 (1977).
63. T. Mimura, K. Odani, N. Yokoyama, and M. Fukuta, New structure of enhancement-mode GaAs microwave MOSFET, *Electronics Lett.*, **14**, 500–2 (1978).
64. L. Messick, Power gain and noise of InP and GaAs insulated gate microwave FETs, *Solid State Electronics*, **22**, 71–6 (1979).
65. T. Mimura, K. Odani, N. Yokoyama, Y. Nakayama, and M. Fukuta, GaAs microwave MOSFETs, *IEEE Trans. Electron Devices*, **ED-25**, 573–9 (1978).
66. K. Yamaguchi and S. Takahashi, Theoretical characterization and high speed performance evaluation of GaAs IGFETs, *IEEE Trans. Electron Devices*, **ED-28**, 581–7 (1981).
67. N. Yokoyama, T. Mimura, and M. Fukuta, Planar GaAs MOSFET integrated logic, *IEEE Trans. Electron Devices*, **ED-27**, 1124–8 (1980).
68. T. L. Andrade and N. Braslau, GaAs lossy gate dielectric FET, presented at *Device Research Conf.*, Santa Barbara, CA, June (1981).
69. F. L. Schuermeyer, GaAs IGFET digital integrated circuits, *IEEE Trans. Electron Devices*, **ED-28**, 541–5 (1981).
70. B. R. Pruniaux, J. C. North, and A. V. Payer, A semi-insulated gate gallium-arsenide field-effect transistor, *IEEE Trans. Electron Devices*, **ED-19**, 672–4 (1972).
71. H. M. Macksey, D. W. Shaw, and W. R. Wisseman, GaAs power FETs with semi-insulated gates, *Electronics Lett.*, **12**, 192–3 (1976).
72. H. C. Casey, Jr, A. Y. Cho, D. V. Lang, E. H. Nicollian, and P. W. Foy, Investigation of heterojunctions for MIS devices with oxygen-doped $Al_xGa_{1-x}As$ on n-type GaAs, *J. Appl. Phys.*, **50**, 3484–91 (1979).
73. P. L. Fleming, A. Meulenberg, Jr, and H. E. Carlson, High-performance GaAs metal insulator semiconductor transistor, *IEEE Electron Devices Lett.*, **EDL-3**, 104–5 (1982).
74. E. J. Bawolek and B. W. Wessels, ZnSe/GaAs heterojunctions for MIS devices,

presented at *Workshop on Dielectric Systems for the III–V Compounds*, San Diego, Ca, June (1982).
75. R. Dingle, H. L. Stormer, A. C. Gossard, and W. Wiegmann, Electron mobilities in modulation-doped semiconductor heterojunction superlatives, *Appl. Phys. Lett.*, **33**, 665–7 (1978).
76. T. Mimura, K. Joshin, S. Hiyamizu, K. Hikosaka, and M. Abe, High electron mobility transistor logic, *Jap. J. Appl. Phys.*, **20**, L598–600 (1981).
77. T. Hotta, H. Sakaki, and H. Ohno, A new GaAs/AlGaAs heterojunction FET with insulated gate structure (MISSFET), *Jap. J. Appl. Phys.* (1982).
78. H. Yamamoto, M. Moniwa, T. Sawada, and H. Hasegawa, High efficiency MOS solar cells by anodic oxidation process, *Jap. J. Appl. Phys.*, **20**, Suppl. 20-2, 87–91 (1981).
79. L. Forbes, J. R. Yeargan, D. L. Keune, and M. G. Craford, Characteristics and potential applications of $GaAs_{1-x}P_x$ MIS structures, *Solid State Electronics*, **17**, 25–9 (1974).
80. R. J. Stirn and Y. C. M. Yeh, Technology of GaAs metal–oxide–semiconductor solar cells, *IEEE Trans. Electron. Devices*, **ED-24**, 476–83 (1977).
81. W. A. Anderson, G. Rajeswaran, V. J. Rao, and M. Thayer, Cr–MIS solar cells using thin epitaxial silicon grown on poly-silicon substrates, *Electron Devices Lett.*, **EDL-2**, 271–4 (1981).
82. K. P. Pande, Y.-S. Hsu, J. M. Bawrego, and S. K. Ghandhi, Grain boundary edge passivation of GaAs films by selective anodization, *Appl. Phys. Lett.*, **33**, 717–19 (1978).
83. B. Bayraktaroglu, S. J. Hannah, and H. L. Hartnagel, Stable charge storage of MAOS diodes on GaAs by new anodic oxidation, *Electronics Lett.*, **13**, 45–6 (1977).
84. H. Hasegawa, T. Sawada, and T. Sakai, Interface-state band model for GaAs and GaP anodic MOS structures, *Surf. Sci.*, **86**, 819–25 (1979).
85. W. E. Spicer, I. Lindau, P. Skeath, and C. Y. Su, Unified defect model and beyond, *J. Vac. Sci. Technol.*, **17**, 1019–27 (1980).
86. T. Suzuki and M. Ogawa, Degradation of photoluminescence intensity caused by excitation-enhanced oxidation of GaAs surfaces, *Appl. Phys. Lett.*, **31**, 473–5 (1977).
87. F. Capasso and G. F. Williams, A proposed hydrogenation/nitridization passivation mechanism for GaAs and other III–V semiconductor devices, including InGaAs long wavelength photodetectors, *J. Electrochem. Soc.*, **129**, 821–4 (1982).
88. W. E. Spicer, P. W. Chye, P. R. Skeath, C. Y. Su, and I. Lindau, New and unified model for Schottky barrier and III–V insulator interface state formation, *J. Vac. Sci. Technol.*, **16**, 1422–33 (1979).
89. W. E. Spicer, I. Lindau, P. E. Gregory, C. M. Garner, P. Pianetta, and P. W. Chye, Synchrotron radiation studies of electronic structure and surface chemistry of GaAs, GaSb and InP, *J. Vac. Sci. Technol.*, **13**, 780–5 (1976).
90. K. Watanabe, M. Hashiba, Y. Hirohata, M. Nishino, and T. Yamashino, Oxide layers on GaAs prepared by thermal, anodic and plasma oxidation: in-depth profiles and annealing effects, *Thin Solid Films*, **56**, 63–73 (1979).
91. A. K. Nedoluha, Surface donors and acceptors on GaAs and InP exposed to oxygen, *J. Vac. Sci. Technol.*, **21**, 429–33 (1982).
92. M. D. Clark and C. L. Anderson, Improvements in GaAs/plasma-deposited silicon nitride interface quality by predeposition GaAs surface treatment and postdeposition annealing, *J. Vac. Sci. Technol.*, **21**, 453–6 (1982).
93. P. Mark and W. F. Creighton, The effect of surface index and atomic order on the GaAs–oxygen interaction, *Thin Solid Films*, **56**, 19–38 (1979).

94. B. Bayraktaroglu, R. L. Johnson, D. W. Langer, and M. G. Mier, Germanium (oxy)nitride based surface passivation technique as applied to GaAs and InP, *The Physics of MOS Insulators*, eds G. Lucovsky, S. T. Pantelides, and F. L. Galeener, Oxford, Pergamon Press, pp. 207–11 (1980).
95. W. E. Spicer, I. Lindau, P. Pianetta, P. W. Chye, and C. M. Garner, Fundamental studies of III–V surfaces and the (III–V)–oxide interface, *Thin Solid Films*, **56,** 1–18 (1979).
96. A. Heller, Chemical control of recombination at grain boundaries and liquid interfaces: electrical power and hydrogen generating photoelectrochemical cells, *J. Vac. Sci. Technol.*, **21,** 559–61 (1982).
97. G. Lucovsky and R. S. Bauer, Local atomic order in native III–V oxides, *J. Vac. Sci. Technol.*, **17,** 946–51 (1980).
98. R. K. Ahrenkiel, L. L. Kazmerski, P. J. Ireland, O. Jamjoum, P. E. Russell, D. Dunlavy, R. S. Wagner, S. Pattillo, and T. Jervis, Reduction of surface states on GaAs by the plasma growth of oxyfluorides, *J. Vac. Sci. Technol.*, **21,** 434–7 (1982).
99. A. Sher, Y. H. Tsuo, J. E. Chern, and W. E. Miller, Interface states in GaAs/LaF$_3$ configurations, *The Physics of MOS Insulators*, eds G. Lucovsky, S. T. Pantelides, and F. L. Galeener, Oxford, Pergamon Press, pp. 280–4 (1980).
100. H. Hayashi, K. Kikuchi, and T. Yamaguchi, Capacitance–voltage characteristics of Al/Al$_2$O$_3$/p-GaAs metal–oxide–semiconductor diodes, *Appl. Phys. Lett.*, **37,** 404–6 (1980).
101. H. C. Gatos, J. Lagowski, and T. E. Kazior, GaAs MIS structures—promising or hopeless? *Proc. 14th Int. Conf. on Solid State Devices*, Tokyo, 1982, to be published.

Gallium Arsenide
Edited by M. J. Howes and D. V. Morgan
© 1985 John Wiley & Sons Ltd

CHAPTER 8

Transferred Electron Devices

M. J. HOWES

8.1 THE TRANSFERRED ELECTRON DEVICE

In 1963, while investigating the noise properties of GaAs, J. B. Gunn[1] observed that at particular bias voltages the current through the device became very noisy. This effect, which now bears his name, was shown by Kroemer,[2] in 1964, to be the first experimental evidence of the prediction of Ridley and Watkins[3] that bulk negative conductance could be achieved in certain III–V semiconductors.

Gunn observed that current oscillations occurred when the applied voltage exceeded a certain critical threshold voltage, V_T. The frequency of these oscillations could be made to lie in the microwave region of the spectrum by suitable choice of the sample thickness. Further measurements revealed a region of high electric field (domain) which, in n-type material, moved through the sample in the direction of electron flow. The domain drift velocity, which remained constant even when the applied voltage was increased above threshold level, was measured to be about 10^7 cm s^{-1}. These measurements showed that, as the voltage was increased past threshold, a high field domain formed near the cathode, consequently reducing the field in the rest of the device and causing the current to drop to about two-thirds of its maximum value. The domain would then drift with the carrier stream across the sample and disappear at the anode contact. As the domain collapsed at the anode, the electric field behind it increased until the threshold field was reached and the current had increased to the threshold value. Subsequently, a new domain would form at the cathode, the current would drop, and the cycle would be repeated.

The theory of Ridley and Watkins[3] and Hilsum[4] showed that the negative differential mobility of electrons in GaAs, as shown in Figure 8.1, arises from the particular form of the band structure[5] in this semiconductor. The shape of the v–E characteristic of GaAs indicates that, if the sample were biased in the region of negative differential mobility and connected to a suitable external circuit, a.c. power would be generated. This mode of operation, limited space-charge

Gallium arsenide

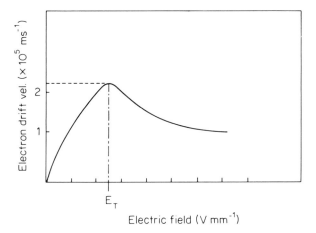

Figure 8.1 The velocity–field (v–E) characteristic of electrons in GaAs

accumulation (LSA), is indeed possible but, owing to the internal domain instabilities, it is difficult to use the device as a pure negative resistor and in practice certain special conditions must be fulfilled in order to obtain LSA operation.

A complete solution to the problem of domain formation and growth requires the non-linear solution of Poisson's equation and the current continuity equation subject to the correct v–E characteristics and also, for more accurate results, the variation of the diffusion constant with electrical field.[7] Unfortunately, domain instabilities present extreme difficulty in measuring the v–E relationship and, furthermore, its theoretical calculation is impeded by the statistical nature of electron scattering by polar lattice vibrations. Nevertheless, Ruch and Fawcett[6] used Monte Carlo computer techniques to calculate the dependence of the v–E characteristic on temperature and impurity concentration, as shown in Figure 8.2.

At first sight it appears that the oscillation frequency is limited to the natural domain transit time frequency, which seems to be a function of temperature and not of voltage. This transit time is given as

$$t_D = L/v_d \tag{8.1}$$

where v_d is the domain drift velocity (10^5 m s^{-1}) and L is the length of the active region. So, for example, an active layer of approximately 10 μm is needed for X-band operation ($\sim$ 10 GHz). In spite of this transit time limitation, Carroll[8] was able to obtain microwave oscillations covering the range 4–31 GHz from a single diode operating in a number of microwave cavities, and Robson and Mahrous[9] obtained frequency tuning in a coaxial cavity by the adjustment of a short-circuit element. These cavity modes of oscillation are a result of the modification of the

Transferred electron devices 301

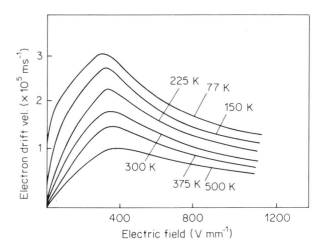

Figure 8.2 Temperature dependence of the v–E characteristic of GaAs. (From Ruch and Fawcett[6])

nucleation and extinction of the domains by the RF voltage and can be categorized into four main types: transit time, delayed domain, quenched domain, and accumulation layer modes. The first three of these modes are illustrated in Figures 8.3a, b, and c.

In the transit time mode, the circuit is heavily loaded so that the RF voltage amplitude is small and does not fall below the threshold voltage V_T. The disadvantage of this mode is its low efficiency and the restriction of the frequency oscillation to the natural domain transit time frequency, as discussed previously.

The delayed domain mode of operation is established if the circuit Q-factor is large enough to sustain the RF voltage, which is of sufficiently large amplitude to cause the total instantaneous voltage across the device to fall below threshold over a portion of the cycle. If the domain transit time is less than the period of the voltage waveform, then another domain will not be nucleated until the voltage once more rises above the threshold voltage. An advantage of this mode is that the frequency of oscillation can be varied by adjustment of the resonant circuit, between the approximate limits

$$2t_D > t_R > t_D$$

where t_D is the natural domain transit time and t_R is the period of oscillation.

The quenched domain mode occurs if the circuit loading is reduced even further, so that the instantaneous voltage falls below V_S, the voltage required to sustain a domain. In this case the domain is quenched in transit and another is not nucleated until the voltage rises above V_T. Theoretically, frequencies above the transit time frequency can be attained and the upper limit is restricted by the time

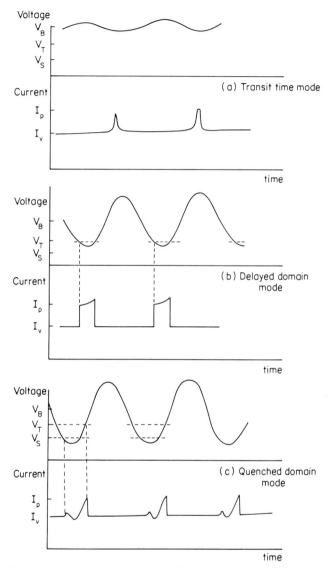

Figure 8.3 Cavity modes of oscillation: (a) transit time mode; (b) delayed domain mode; (c) quenched domain mode

t_S, which is required for domain nucleation and extinction. Thus, the frequency is constrained to lie within the approximate limits

$$2t_D > t_R > t_S$$

and indicates greater circuit tunability than any other mode.

Mechanical tuning of transferred electron effect oscillators is therefore possible over an octave or more. The precise nature of this tuning in any particular cavity is dependent on the interaction between the microwave circuit and the active device, i.e. the impedance presented by the circuit[10-12] to the diode terminals at the frequency of operation and harmonics.[13]

Before a transferred electron effect device can be mounted in a waveguide cavity, it must be encapsulated in some form of package. This facilitates handling, and prevents the bare chip from mechanical damage or contamination. The encapsulation also provides a low-resistance thermal path to the external heat sink as well as enabling easy application of d.c. bias. A disadvantage of the encapsulation is that the electrical performance of the device is restricted by the addition of spurious circuit elements, such as the inductance of the bonding wires and the capacitance of the ceramic stand-off and pedestal (Figure 8.4).

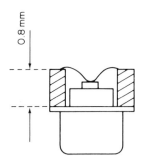

E6 package

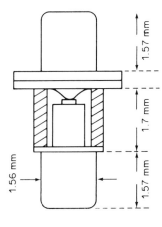

S4 package

Figure 8.4 Typical microwave diode packages

304 Gallium arsenide

A common method of coupling into a waveguide cavity is to mount the encapsulated device under a vertical conducting post as shown in Figure 8.5, a configuration that provides high stability, low frequency-modulated (FM) noise and, by means of a movable short circuit, mechanical tunability. Unless carefully designed, several defects are apparent in the operation of a circuit of this type, such as sudden changes in frequency of oscillation, power output, and spectral purity as the short-circuit position is changed. A typical tuning characteristic is shown in Figure 8.6, which indicates areas of frequency saturation, where the position of the circuit has negligible effect on the frequency of oscillation, and an effect known as tuning hysteresis.

The modelling of transferred electron effect oscillators can be approached from a number of viewpoints. For example, one can construct a time domain computer simulation[15,16] of the device, coupled with a time domain representation of the microwave package and circuit. Alternatively, the use of a Fourier analysis[18,19]

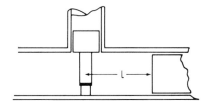

Figure 8.5 Post-mounted oscillator

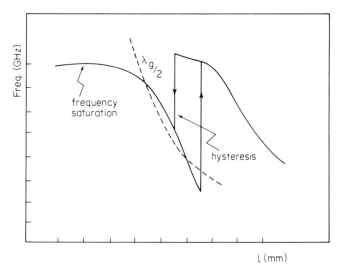

Figure 8.6 Frequency of oscillation as a function of cavity length for a post-mounted diode oscillator

of the voltage and current waveforms will enable an equivalent circuit of the device to be obtained. By considering a particular RF voltage amplitude and bias voltage, the current waveform can be deduced by considering the various modes of operation previously discussed.

Apart from these more detailed approaches, a suitable precursor to a complete study of sinusoidal negative resistance oscillators was provided by Kurokawa,[20] who considered the active device as an amplitude-dependent time-averaged admittance given by

$$Y_D(A) = -G_D(A) + jB_D(A) \tag{8.2}$$

separated from the load by multiple resonant circuits. The total circuit impedance seen by the device is of the form

$$Y_L(\omega) = G_L(\omega) + jB_L(\omega) \tag{8.3}$$

and the condition for steady-state oscillation at ω_0 is given by

$$Y_L(\omega) + Y_D(A) = 0. \tag{8.4}$$

As well as explaining many of the observed phenomena, Kurokawa's analysis also includes conditions for the stability of steady-state oscillations, a treatment of oscillation noise, and methods of broadbanding negative resistance oscillators.

The complete model derived from phenomenological considerations[21,22] is rather complicated to incorporate into most analyses, and the simpler equivalent circuits of Figure 8.7 are more commonly used. In these models, R_0 is the low-field resistance, C_0 the low-field capacitance, and R_D and C_D the domain negative resistance and capacitance.

Theoretical treatises of transferred electron effect oscillators have been conducted using simplified linear models such as those detailed above, but in general the active device impedance is highly non-linear[23] so that Kurokawa's device line must be extended to the concept of a device plane. In this case the impedance of the device is a function of amplitude of oscillation A, frequency ω, and bias voltage V:

$$Y_D(\omega, A, V) = -G_D(\omega, A, V) + B_D(\omega, A, V) \tag{8.5}$$

Experimentally, it is impossible to measure the dynamic impedance of a transferred electron effect device using the usual reflection-type measurement because of the inherent instability of domain mode devices which disturbs the system. Alternatively, a technique can be applied[24-27] that utilizes the fact that for steady-state oscillation the admittance looking into the cavity from the diode terminals is balanced by the admittance of the diode, i.e.

$$G_D(\omega, A, V) = G_L(\omega) \tag{8.6}$$

$$B_D(\omega, A, V) + B_L(\omega) = 0 \tag{8.7}$$

If the circuit admittance $Y_L(\omega)$ is measured at various frequencies, then the actual

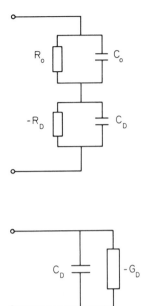

Figure 8.7 Simple equivalent circuits for a transferred electron effect device

frequencies of the free-running oscillator can be used to find the diode admittance. Approximations to the admittance $Y_L(\omega)$ presented to the active device by the circuit have been made by various authors using lumped constant elements[33-35] and more rigorous mathematical treatments such as the use of the dyadic Green's function.[34] These approaches have been useful in identifying some of the gross features of transferred electron effect device oscillator tuning curves, such as frequency saturation and mode jumping. An experimental observation[27] has shown that local resonances arising from the nature of the device encapsulation, mount, microwave choke, and cavity can cause discontinuous changes in power and frequency as the circuit is tuned. This latter technique is useful in the removal of unwanted spurious resonances from the frequency band of interest to enable broadband tuning to be achieved.

As well as mechanical tuning, a certain amount of tuning can be provided by adjustment of the applied voltage,[36] known as 'pushing', the direction and frequency bandwidth being dependent on the particular circuit in which the device is mounted. The effect can be considered as resulting from the domain capacitance dependence on bias voltage[36,37,39] or from the change in device admittance calculated assuming a simple device model.[10,38,40] Only positive df/dV are predicted by these considerations and a better treatment by Tang and Lomax[42] considers a more complete device simulation using a computer analysis. Their results confirm the unpredictability of this type of tuning due to the changes in the device admittance–circuit interaction. The other drawbacks of bias voltage tuning are the large variations in output power that occur and the fact that any practical modulator would have to supply all of the device d.c. bias current.

8.2 ELECTRONICALLY TUNED OSCILLATORS: YIG TUNING

Electronic tuning of transferred electron effect oscillators can also be accomplished by the use of a yttrium iron garnet (YIG) sphere.[42-44] Highly polished spheres of single-crystal YIG when placed in an RF structure under the influence of a d.c. magnetic field exhibit a high Q resonance at a frequency proportional to the d.c. magnetic field (a phenomenon known as ferromagnetic resonance). Broadband tuning of a CW oscillator can be achieved by magnetically coupling the RF current of an oscillating diode to the YIG sphere, so that the frequency is locked to the ferromagnetic resonance. A typical YIG tuned transferred electron effect oscillator circuit is shown in Figure 8.8, which shows the method of coupling the diode of the YIG sphere. The tuning characteristics of the structure illustrated in Figure 8.8 are shown in Figure 8.9, which indicates the good frequency linearity that can be achieved in YIG tuned oscillators.

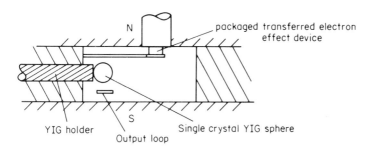

Figure 8.8 YIG tuned oscillator

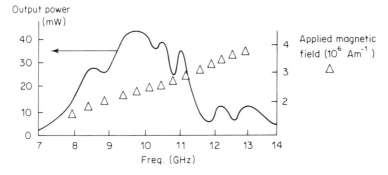

Figure 8.9 Frequency and output power of YIG tuned oscillator

A linear relationship between magnetic field and tuning current is desirable for accurate, direct frequency control, and this requirement determines the choice of magnetic materials and oscillator structure. Non-linear dynamic tuning characteristics are usually the result of both hysteresis and eddy currents in the pole pieces and the RF circuit. Eddy currents in the pole pieces can be further reduced with laminated structures, although they are more difficult to compensate for against thermal expansion and may be more prone to the effects of RF interference and vibration. Oscillators can therefore be constructed with solid cores for applications requiring ruggedness and frequency stability (low df/dT), and with laminated cores for higher sweep rates.

Since the magnetic core exhibits hysteresis, a YIG tuned oscillator is usually swept in only one direction. Care must also be taken in the design of the magnetic circuit in order to minimize the effects of stray magnetic fields so that, if an isolator is to be used, the performance of either unit will not be affected. The requirements of low noise and broad bandwidth can be achieved using a YIG tuned oscillator since the Q of the YIG sphere is about 2000 to 30 000. However, the power supply must be carefully designed to enable the low-noise potential to be achieved, since, with for example a tuning sensitivity of 17 MHz mA^{-1}, a noise current of 1 μA would contribute 17 kHz of frequency deviation.

The rate of frequency tuning is limited by the relatively large inductance of the magnetic coils to about 1 MHz μs^{-1} and consequently YIG tuned oscillators are not suitable for applications requiring rapid frequency changes and high modulation frequencies. The loss of frequency agility can be somewhat alleviated in narrow bandwidth cases by the use of a small ancillary coil in close proximity to the YIG sphere. This coil can also be used as a means of providing automatic frequency control when stable control of frequency is required.

Both the transferred electron effect device and the YIG resonance are temperature sensitive. However, if the YIG sphere is orientated with a suitable crystallographic direction parallel to the magnetic field, its resonance is independent of temperature.[44] The temperature sensitivity (df/dT) of the transferred effect device can then be approximately compensated for by slightly reorientating the sphere, enabling a df/dT of the order of ± 0.1 MHz deg^{-1} to be achieved.

8.3 VARACTOR TUNED TEOS

Electronic tuning of transferred electron oscillators (TEOs) was first reported in 1965 by Kuru[46] who mounted the active device and the varactor in series in a coaxial cavity. Electronic tuning of 200 MHz at 2 GHz was obtained. Electronic tuning bandwidths of over 1 GHz at 13 GHz were achieved[47] by coupling the coaxial configuration of Kuru into a waveguide cavity using an E-plane stub. A notable feature of this circuit is the almost frequency-independent power output

Cawsey analysed the tuning range of single tuned lumped element circuits[48]

and showed that a three-way compromise exists between tuning range, power output, and oscillator stability; improvement in any one of these parameters must degrade the others for a given active device and varactor. The stability/tuning compromise was qualified by Paik.[49]

Wide tuning range is accomplished by coupling the varactor to the active device as closely as possible. In an attempt to realize this, Smith and Crane[50] mounted a TED and varactor in parallel, with both devices in series with the centre conductor of a coaxial cavity. At a centre frequency of 10 GHz, 2 GHz electronic tuning was predicted by a circuit model and 1.1 GHz achieved in practice. This discrepancy was explained by Cawsey[51] who applied his analysis of a single tuned circuit to the oscillator and predicted the experimentally obtained 1.1 GHz. Octave tuned oscillators were first reported by Large[52] who mounted the TED varactor in series in a short coaxial line. A computer analysis of the complete circuit was used to optimize it for best performance. The undesirable package capacitance of the varactors was minimized by using miniature pill packages, and, by mounting two varactors in series, the effective package capacitance was reduced still further. The frequency bands covered were 4–8 GHz and 8–12.4 GHz.

The analysis of a single tuned circuit[48] shows that the tuning range is inversely proportional to the circuit susceptance slope or, since the load is fixed by the active device conductance, inversely proportional to the circuit Q-factor. In Figure 8.10 the circuit diagram for a single tuned lumped element oscillator is shown. Also plotted is a graph of the varactor susceptance and the negative of the circuit susceptance; the intersection of these two quantities gives the oscillation frequency. From this graph it may be seen that a reduction in the circuit susceptance slope will give an increase in the tuning range. This leads to the idea of using reactance compensation to increase the tuning range of an oscillator by introducing a second resonant circuit of the opposite type between the original circuit and the load. A parallel tuned oscillator has a series tuned compensating circuit introduced and vice versa. Figure 8.11 illustrates the technique by showing how the compensating circuit alters the total circuit susceptance in a parallel tuned oscillator. The compensating circuit is arranged to be resonant at the centre of the tuning range of the oscillator and also to cause the circuit to have zero (or near-zero) reactance slope at that frequency. This technique has several disadvantages, which were emphasized by Kurokawa.[53] These are:

(a) the output becomes very noisy as $dB_c/d\omega \to 0$;
(b) tuning characteristic is highly non-linear about the centre frequency; and
(c) the output power may change rapidly as the frequency is tuned away from centre.

Reports have been made of realizations of the technique in both coaxial lines[54] and microstrips[55] giving useful increases in the tuning range of 100% and 70% respectively, but the disadvantages outlined above are apparent in the results. It

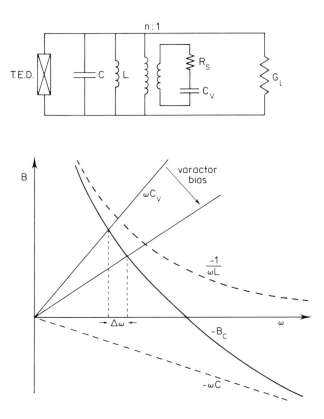

Figure 8.10 The single tuned lumped element circuit

would seem more desirable to use a low-Q single tuned circuit to give the required tuning range, with its constant noise performance and more linear tuning characteristic.

Broadband circuits have also been realized in a reduced height waveguide[56] with 20% tuning in X-band, but a systematic study of circuit configurations and the effects of package strays is easier in coaxial line. Figure 8.12 shows a selection of the structures that have been studied. Tuning of 30% has been obtained by using a parallel structure and by embedding the varactor package into the end of the cavity in order to minimize the effect of the package stray reactances. The degradation in performance due to the stray reactances of packages is widely recognized[57] and is caused by the physical size of the package becoming a significant proportion of the whole circuit at high frequencies. The only way to avoid this problem is to use miniature packages or even to dispense with the package altogether. In addition to this problem, as the frequency increases it becomes more difficult to make lumped element circuits, as the short wavelengths

Transferred electron devices

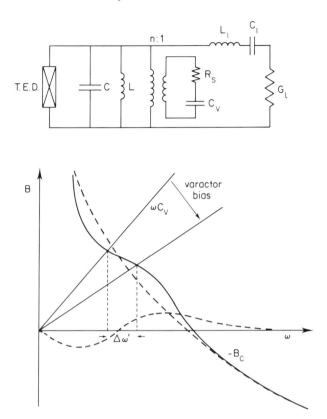

Figure 8.11 An illustration of reactance compensation

cause even small packages to appear distributed. It may be shown that distributed realization of the single tuned lumped element circuit must degrade the tuning performance.[58] To avoid the problems of package strays and distributed circuits, several circuits have been constructed on microstrip or as thick film hybrid integrated circuits.[59,60] This type of circuit lends itself to high-frequency operation (up to 100 GHz) and has the advantages of being simple, cheap to construct, small, and light. The major disadvantage is the low Q since lossy dielectric material is used as the transmission medium.

The noise performance of wide band oscillators is rarely published owing to the low Q circuits resulting in a poor noise performance and because FM noise is difficult to measure. Waveguide circuits offer the possibility of a high Q-factor, and a large number of structures have been studied, some of which are illustrated in Figure 8.13. The devices are normally mounted in inductive posts which carry the bias currents. Close coupling of the varactor and TED is required to produce

312 Gallium arsenide

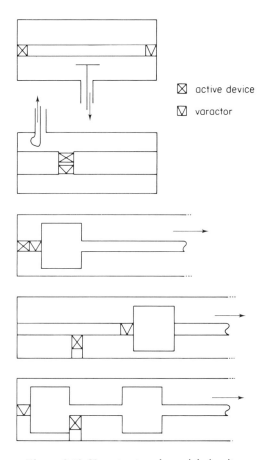

Figure 8.12 Varactor tuned coaxial circuits

wide band tuning. Dean and Howes[62] studied the effect of reducing the spacing between the devices and predicted maximum tuning range with the TED and varactor mounted in the same plane; 15% tuning was obtained in J-band.

Although equivalent circuits may be devised for particular circuit configurations, a general analysis to help the circuit designer is more difficult to create. Joshi and Cornick[63] have made an attempt to relate empirically the electronic tuning range with the mechanical characteristics of the cavity, with a view to presenting a correlation between the two. An understanding of the general nature of the tuning characteristics of electrically and mechanically tuned oscillators is derived from this approach. They concluded that the electrical and mechanical tuning depend upon the propagation characteristics of the cavity and tuning limitations are caused by the effect of a distributed circuit on a point source.

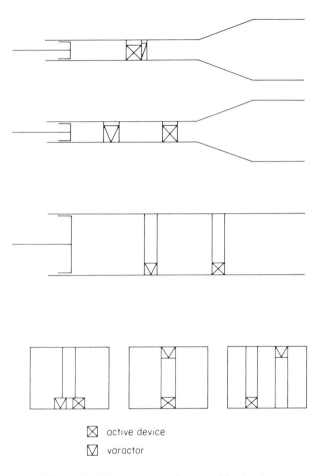

☒ active device
☑ varactor

Figure 8.13 Varactor tuned waveguide circuits

8.3.1 Physical structures for tunable oscillators

The design and performance of a wide variety of varactor tuned oscillators have been reported. These circuits range from narrow band but very stable oscillators to octave tunable miniature oscillators. In general the stability of the source is related directly to the circuit Q-factor and therefore waveguide circuits or large cavity resonators are normally used for low-noise oscillators and microstrip circuits or hybrid circuits constructed on microstrip substrates are used for broadband oscillators. This is an example of the way in which the physical structure of an oscillator may determine its performance and use. The three main types of oscillator are microstrip, coaxial, and waveguide, and Table 8.1 summarizes their electrical and mechanical characteristics.

314   Gallium arsenide

Table 8.1 Characteristics of transmission media for oscillators

	Microstrip	Coaxial line	Waveguide
Q-factor	low: conduction, dielectric, and radiation losses	medium	high: low loss and high impedance levels
Tuning range	high ranges possible using miniature packages or bare chips; lumped elements feasible	good in theory; spoilt by package strays and distributed effects	low if advantage is taken of high Q
Characterization and modelling	difficult to measure circuit elements	easy, well established measurements and modelling techniques	difficult to mount electrical configuration
Mechanical stability	good	good with care in construction	difficult to make satisfactory
Other comments	cheap		prone to mode jumping and spurious resonances

Microstrip is normally operated in a TEM mode and the fields are confined to a thin, lossy dielectric. Although improvements in these dielectrics have been made, the loss results in low Q-factors. In addition to dielectric loss, radiation from the top of the substrate contributes to a low Q. Another disadvantage is the difficulty found in the electrical characterization of microstrip circuits, which makes oscillator design difficult and has led to a 'black art' in adjusting circuits, using foil or metallized paint, to make them operate correctly. The advantages of microstrip over other transmission media are small size, light weight, good mechanical stability, and low cost.

Coaxial line also propagates a TEM mode but has several advantages over microstrip. Q-factors are generally higher than microstrip as airline can be used in order to avoid dielectric losses and, since the fields are completely enclosed, there are no losses from unwanted radiation. The electrical characterization of coaxial line is relatively easy and well defined circuit elements may be constructed. A disadvantage over microstrip is that physically robust device packages must be used as they may be called upon to support the inner conductor and they are generally associated with high stray reactances.

The high impedance levels in waveguides give good matching to low-power TEDs, and the ease with which the guide may be reduced in height to lower the impedance provides a simple method of optimizing the loading with higher-power TEDs. The low losses combined with a large volume of stored energy give high Q-

Transferred electron devices 315

factors, which should enable oscillators with good stability and low noise to be built. Waveguide has several disadvantages over coaxial line and microstrip which make it less popular for many applications. It is physically bulky and therefore heavy, it is prone to mode jumping and spurious resonances, the tuning characteristics may have very small discontinuities, and it is difficult to mount the devices in a known electrical configuration because the posts used to supply bias to, and support, the devices interact with evanescent modes.

Many other types of circuit have been tried and they are usually developed to overcome some of the disadvantages of the above three main types. Lumped element circuits are built on microstrip substrates to avoid the tuning limitations imposed by distributed components and to avoid the dielectric losses. Suspended stripline may have good potential for light, miniature circuits since it has most of the advantages of microstrip, but does not suffer from dielectric losses, as the bulk of the fields are in air, and it does not suffer from radiation losses, as it is totally enclosed. Variations on coaxial line such as trough line[45] are sometimes used to overcome problems of mechanical inaccessibility. Evanescent wave resonators, built like a very short piece of coaxial line, store very little electric energy and have been used for varactor tuned oscillators since the varactor may then dominate the storage of electric energy.

8.4 OPERATION OF PULSED TEOS

The characteristics of a typical crystal detected RF pulse from a pulsed transferred electron oscillator are illustrated in Figure 8.14. The *risetime* of the pulsed output power is defined as the time taken for the leading edge of the RF pulse to increase from 10% to 90% of its maximum amplitude. The *delay time* is the time between the application of the bias pulse and the oscillator switching on. The *jitter width* is defined as the peak-to-peak fluctuation in time of the leading edges of the crystal detected RF pulse waveform when displayed on an oscilloscope. Major problems associated with early pulsed transferred electron oscillators were unacceptably large values of risetime, delay time, and jitter width. The long risetime and delay times have been attributed to temperature-dependent resistive elements at the metal contact to the layer of the device[64-68] (see Figure 8.15). Devices with excessive contact resistance have exhibited delay times of up to several microseconds.[66] The contact resistance may be reduced by growing an n^+ layer between the active layer and metal cathode contact. This fabrication process results in devices that display much shorter risetime and delay times.[64,66] However, for high-power, high-efficiency pulsed devices it is advantageous not to use the n^+ contact layer and special care is needed during the fabrication of the device to keep the contact resistance down to a reasonable level.[69] Many applications of pulsed transferred electron oscillators require fast risetimes of less than 5 ns and delay times of the same order. If an oscillator fails to meet this specification then the

316 *Gallium arsenide*

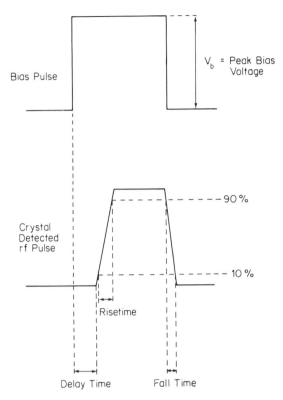

Figure 8.14 Schematic diagram showing characteristics of a typical crystal detected RF pulse from a pulsed transferred electron oscillator

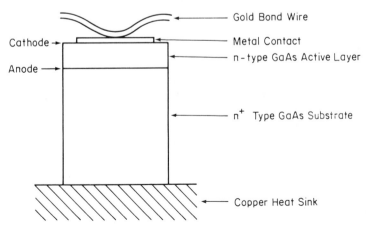

Figure 8.15 Structure of a typical transferred electron device

risetime, delay time, and jitter width can all be reduced to an acceptable level by injecting a relatively small CW signal into the oscillator cavity.[70-72]

Other major problems encountered in the operation of pulsed transferred electron oscillators result from the thermal heating of the device. The temperature of the active layer increases by an amount $|\Delta T|$ during the bias pulse and decreases in between the pulse as shown in Figure 8.16a. If the duty cycle (mark/space ratio of the bias pulse waveform) is sufficiently small, then the active layer will cool down to the ambient temperature in between pulses. However, for higher duty cycle operation the active layer may not cool down to the ambient temperature but to some temperature level T_{min}. Problems in the operation of the oscillator arise as variations in the active layer temperature result in changes of the oscillation frequency, output power, and efficiency.[73-75]

The variation of the output power and efficiency with temperature can be explained by reference to Figure 8.17, which shows the dependence of the v–E characteristic on the active layer temperature (T). It is seen that as T increases, the electron velocity peak-to-valley ratio of the characteristic decreases and this results in the decrease of the efficiency and output power.[76] Therefore the temperature increase ΔT of the active layer during a pulse will result in amplitude modulation (AM) of the crystal detected RF pulse. Generally it is expected that increases in the temperature T_{min} will also result in a decrease in the output power. One of the reasons why a pulse transferred electron oscillator can produce a higher efficiency and peak output power than a CW oscillator is that the pulsed oscillator operates with the active layer at a much lower temperature. For example, a GaAs device used as the basis of a pulsed X-band waveguide oscillator may produce an efficiency of typically 20% whereas the same device operated under CW conditions would produce a maximum efficiency of typically 10%. The difference in the maximum peak output powers would be greater in practice since the pulsed device can operate at much higher bias voltages than the CW device. The efficiency can also be increased by using a device constructed out of a material, such as InP, that has a larger peak-to-valley velocity ratio than GaAs. However, these devices have a much smaller low-field resistance than GaAs devices and are consequently considerably more difficult to use in practice. To obtain the high efficiencies available from a pulsed transferred electron oscillator, it is necessary that the device is well matched into the microwave circuit. For example, by using suitable second harmonic load terminations, pulsed conversion efficiencies of 32% for coaxial (re-entrant cavity) resonators and 27% for coupled TEM bar oscillators have been achieved in L-band.[77] Some of the highest peak powers reported to date for pulsed transferred electron oscillators are listed in Table 8.2. To obtain larger powers, it is possible to place several devices in series or parallel arrays.[78-82]

It is clear, therefore, that the pulsed transferred electron oscillator possesses attractive peak power generating characteristics. However, variations in the temperature of the active layer of the device can result in significant changes of the

318   Gallium arsenide

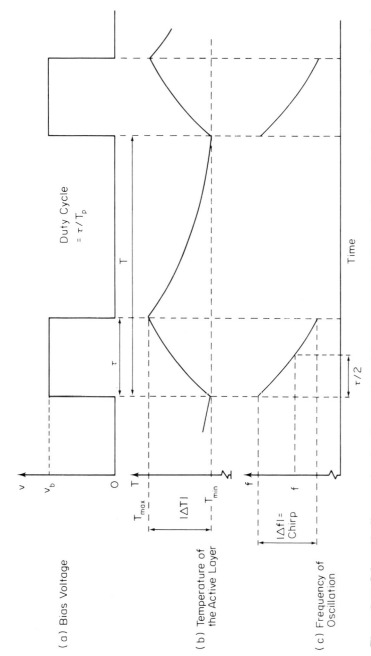

Figure 8.16 Schematic diagram showing the variation of the temperature in the active layer and the frequency of oscillation with time

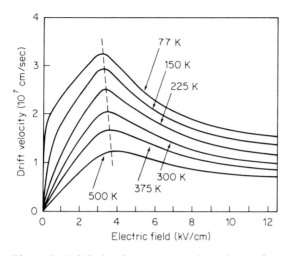

Figure 8.17 Calculated temperature dependence of the electron velocity–electric field characteristic for GaAs as predicted by Ruch and Fawcett[86]

Table 8.2 Some of the highest peak powers reported for pulsed TEDs

Frequency (GHz)	Peak power (W)
2–4	500
4–8	120
8–12	70
12–18	37
18–26	10
26–40	5
40–60	3

peak output power and consequently the RF pulse will be amplitude modulated. However, if these AM effects are found to be unacceptable for a particular application, they can be reduced by using a power limiter. Another, often more serious, problem that also arises from the temperature changes in the active layer of the device is variation in the oscillation frequency. There are two different types of temperature-dependent frequency change encountered in the operation of pulsed transferred electron oscillators. First, there is the rapid change of instantaneous oscillation frequency (Δf) during the pulse, which is known as *chirp*. This frequency change results from the temperature increases (ΔT) in the active layer of the device during a bias pulse. A schematic diagram showing the typical variation of frequency during a pulse is illustrated in Figure 8.16c. The other type of frequency change is a relatively slow drift of the central frequency of oscillation (f_c), which results from changes in the temperature T_{min} as defined in Figure 8.16b. (The central frequency of oscillation is the frequency at the time $\tau/2$ after the start of a pulse, where τ is the pulse width.) Drifts in the central frequency of oscillation result from changes in the bias power, duty cycle, and ambient temperature. A typical pulsed waveguide oscillator operating in X-band exhibits a chirp of approximately 20 MHz per microsecond and a drift in the central frequency of oscillation of approximately -1 MHz/°C. Such characteristics have restricted the use of the pulsed transferred electron oscillator. For example, specifications imposed on the oscillator when used in systems such as portable search radars require chirps of less than 2 MHz over a 0.5 µs pulse and a negligible drift of the central frequency of oscillation in the ambient temperature range of -40 to $+70°$C. To achieve such specifications it is necessary to use frequency compensation techniques to reduce the chirp and the drift of the central frequency of oscillation.

Of the two temperature-dependent changes, chirp is the most difficult to compensate because it occurs over very small pulse widths of typically less than 1 µs. One method of chirp compensation is to couple a fraction of the oscillator output power into a high Q-factor cavity. This technique has been reported[83] to give a 70 % reduction in chirp for only a 1 dB drop in output power. However, the method is narrow band and a small change in the bias power or ambient temperature is sufficient to shift the frequency of oscillation outside the stabilizing bandwidth. A more acceptable wide band method of chirp compensation is electronically to couple a varactor into the oscillator cavity and bias it with pulses shaped from the transferred electron device bias pulses.[82-84] The shaping of the varactor bias pulses is mainly 'trial and error' and a change in the conditions of operation or device properties reduces its effectiveness. Chirp can be completely eliminated by injecting a frequency stable CW signal into the oscillator cavity. This method is known as injection locking and unless large CW locking powers are used it is only effective over a narrow band of frequencies. Large values of chirp can be eliminated by using a combination of varactor chirp compensation and injection locking.[83]

Changes in the central frequency of oscillation resulting from changes in the ambient temperature are more easily compensated as they occur over relatively large time intervals. One method of conpensation is to place an invar stub in the broad wall of the waveguide cavity. The penetration of the stub into the cavity varies with temperature and the resonant frequency of the cavity is changed so as to counteract the drift in the central frequency. The method is only effective for small frequency drifts and is reported[85] to have reduced a 30 MHz frequency drift in the ambient temperature range of -40 to $+70°$C to less than 1 MHz. A method used to compensate for larger frequency drifts involves coating the side walls of the waveguide cavity with metallized perspex. The perspex expands with temperature and reduces the broad wall width, which results in an increase of the waveguide cut-off frequency. This effect increases the resonant frequency of the oscillator and so compensates for the central frequency drift ($\partial f_c/\partial T$), which is usually negative. This technique has been reported[85] to have reduced a drift in the central frequency of 100 MHz over the ambient temperature range of -40 to $+70°$C to less than 2 MHz. Varactor chirp compensated pulsed oscillators use thermistors in the varactor bias circuit which detect changes in the ambient temperature and convert them into bias pulse changes, thereby compensating for the drift in the central frequency of oscillation.[82-84]

8.5 TRANSFERRED ELECTRON DEVICE AMPLIFIERS

A transferred electron device (TED) consists of a sample of material, such as GaAs, which exhibits the transferred electron effect, sandwiched between contacts which are used for connection to an external circuit. It can be utilized as a microwave amplifier in two basic ways. First, the travelling domains in an oscillating TED present a negative resistance over a wide range of frequencies.[87] It is therefore possible to use a TED oscillator as a negative resistance element in a reflection amplifier where filters are required to isolate the oscillation frequency from the amplifier signals. Secondly, it is possible to stabilize a TED such that oscillations are suppressed without totally eliminating the negative resistance. This has advantages over the oscillating amplifier because the circuit is generally simpler and the resulting stable device admittance is easier to characterize.

8.5.1 Stabilization mechanisms is TEAs

A stabilized TED is one in which the electric field and carrier density across the device remain constant with time. In a previous section it has been shown that space-charge fluctuations grow in a region of negative differential mobility and that this can lead to Gunn-type oscillations. However, the device can also be stable, and various stabilization mechanisms have been identified. Early work relied on analytic solutions but more recently greater insight into TED stabilization has been gained by using numerical techniques to solve the basic device equations. The results of different workers are difficult to compare without detailed knowledge of the numerical procedures adopted in each case, but qualitative comparisons can be made.

The stabilization mechanisms fall into four basic categories, which are useful for descriptive purposes, although in practice stabilization is rarely due to any one process. The basic categories are: (i) subcritical $n_0 l$, (ii) circuit stabilization, (iii) diffusion stabilization, and (iv) injection limiting cathodes.

8.5.1.1 Subcritical $n_0 l$

In 1964 Kroemer[2] deduced that space-charge growth is proportional to the differential mobility, the carrier density, and the active length. Later work by McCumber and Chynoweth[88] confirmed this and, by using the Nyquist stability test, it was concluded that a device is short-circuit stable if the space-charge growth can be limited to below a critical value. This argument yields the stability criterion

$$n_0 l < \frac{2.09 \varepsilon r}{q|\mu|} \tag{8.8}$$

Substituting typical values for GaAs ($v \sim 7 \times 10^4$ m s^{-1}; $\mu = -0.05$ m^2 V^{-1} s^{-1}) gives

$$n_0 l < 2 \times 10^{15} \text{ m}^{-2} \tag{8.9}$$

This is known as the $n_0 l$ stability criterion and devices satisfying this relation are said to be subcritical and those which do not are termed supercritical. In practice this is only a rough guide because the internal electric field is non-uniform and the differential mobility and velocity depend on the applied field.[89] Thus the approximate relation, equation (8.9), is only useful as a means of device classification. On the other hand, as the differential mobility and velocity both decrease with increasing temperature and bias voltage, equation (8.8) does show that a nominally supercritical device can actually be subcritical at a high enough bias voltage or temperature.

8.5.1.2 Circuit stabilization

Nyquist stability analysis has shown that all TEDs are unconditionally stable under constant current conditions.[88] It is not, however, practicable to bias a TED in the negative differential mobility region using a constant current source. The device always switches to the low-energy positive mobility operating point. Nevertheless, this does mean that devices that are unstable under constant voltage conditions can be stabilized using a suitable series positive resistor.[90] This technique is limited at microwave frequencies by stray reactance, which is inevitably associated with any resistive element and is only useful over a narrow band. Another circuit stabilization technique is based on the limited space-charge accumulation mode.[91] The circuit required is very complicated and is similar in principle to the oscillating amplifier discussed previously.

8.5.1.3 Diffusion stabilization

Early analysis did not consider diffusion, but more recently its significance has become recognized.[92] Basically, diffusion tends to oppose local concentration gradients, and so it limits the growth of space-charge waves. Thus the carrier concentration can be stabilized by balancing the accumulation of carriers due to the space-charge growth with the outflow of carriers due to diffusion. The resulting stable configuration consisting of a high field domain near the node contact has been predicted by computer simulation[88, 93] and verified experimentally using capacitive probes.[94]

The effect of diffusion can be enhanced by constricting the active area or by loading the surface with a high dielectric layer.[95] This technique is not practicable for layered structures such as TEDs but has been used extensively in travelling wave amplifiers.[96]

8.5.1.4 Injection limiting cathode

In 1971 Thim[97] deduced that the optimum electric field inside a TED is uniform from cathode to anode and since then various methods have been proposed to achieve this optimum. Essentially, the stable solution can be adjusted to the optimum by tailoring the doping of the active region. One common technique is to introduce a doping notch close to the cathode contact. The width and depth of the notch are selected to give a uniform electric field distribution.[98] Such a structure can be fabricated by controlled epitaxial growth or by ion implantation. Another technique employs a graded active region with a doping gradient from anode to cathode and may or may not include a notch. There are, however, conflicting results in this respect with Magarshak and Mircea[93] advocating a doping decrease towards the anode and Hobson et al.[99] recommending a doping increase. Similar confusion exists over the stabilizing influence of temperature since in practice a temperature gradient exists when the device has only a heat sink at one end.[100, 101]

Finally, the possibility of injection limiting using p–n junctions at the cathode or a low height Schottky barrier has been proposed, but neither of these have been successfully demonstrated in practice.[102, 103]

8.5.2 Negative resistance amplifiers

A stabilized TED can be characterized by an admittance with a negative real part over a particular frequency band. If, for simplicity, we assume that, within the band of interest, the device admittance can be represented by a parallel combination of a capacitor and a negative resistor, then with the aid of simple circuit analysis we can illustrate how such an element can be utilized as an amplifier. The usual definitions of amplifier performance such as transducer gain, 3 dB bandwidth, and available gain are used in this section in this context. But note that unlike more conventional active elements a TED has only two terminals and so both the input and the output signals appear at the same port. This leads to certain interesting properties common to all negative resistance amplifiers.

There are two basic classes of negative resistance amplifiers: the transmission type and the reflection type. These are illustrated by Figure 8.18 where in each case

Gallium arsenide

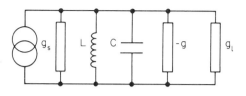

(a) Transmission amplifier

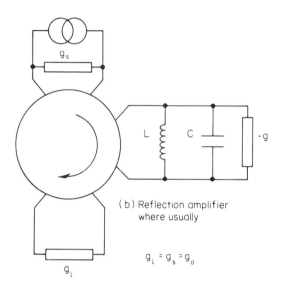

(b) Reflection amplifier where usually

$g_l = g_s = g_o$

Figure 8.18 Examples of negative resistance amplifier circuits

the TED is represented by a parallel RC combination and an inductance (L) has been introduced to resonate the capacitor at some frequency (ω_0) within the band of interest. In the transmission amplifier circuit (Figure 8.18a) power gain is obtained by using the negative resistor to lower the source resistance and thus increase the available power. In the reflection amplifier circuit (Figure 8.18b) a power wave incident on the active element is reflected with an increased magnitude. The power waves must be split in order to isolate the input and the output and the most common method using a circulator is shown as an example.

The transducer gain and 3 dB bandwidth of these circuits have been calculated by Scanlan[104] and the results are summarized in Table 8.3. The gain at midband can be set to any arbitrary value by the suitable choice of source and load conductances. This is an important feature which distinguishes negative resistance amplifiers from more conventional two-port amplifiers. But it should be noted that in very high gain circuits the stability is poor because the sum of the

Table 8.3 Basic circuit relations for the two types of reflection amplifiers as derived by Scanlan[104]

	Transmission circuit (Figure 8.18a)	Reflection circuit (Figure 8.18b)
Transducer grain G_T (at ω_0)	$\dfrac{4g_s g_2}{(g_s + g_1 - g)^2}$	$\left(\dfrac{g_0 + g}{g_0 - g}\right)^2$
Stability	$g < g_s + g_1$	$g < g_0$
Gain bandwidth relation	$(G_T^{1/2} - 1)B = \dfrac{g}{2\pi C}$	$(G_T^{1/2} - 1)\left(1 - \dfrac{2}{G_T}\right)^{1/2} B = \dfrac{g}{\pi C}$

conductances is very nearly zero. Thus any small variation in TED or circuit parameters causes a large change in the circuit gain.

The bandwidth of both amplifiers for the simple single tuned case is also given in Table 8.3. The transducer gain can be chosen as desired, but for a given device specified by the values of g and C the bandwidth is limited. Thus the value of $\omega C/g$, called the negative Q, represents a figure of merit for these devices and the lower the negative Q the larger the gain bandwidth relation. Typical values for the negative Q of a TED chip are between 1 and 5. Thus for a device with $Q = 2$ and a design requiring 10 dB gain the fractional bandwidth for a transmission amplifier is 0.23 and for the reflection amplifier 0.52. In fact the reflection circuit is generally superior to the transmission case. For example, assuming G_T is large then the gain bandwidth relation for the reflection amplifier reduces to

$$(G_T^{1/2} - 1)B = g/\pi C \tag{8.10}$$

which represents double the bandwidth for a given gain or 3 dB extra gain for a given handwidth compared to the transmission amplifier. Thus transmission circuits are limited in use but in wide band applications they may be preferable because of the high cost of circulators. That is, several low gain transmission stages could be more cost effective than a single reflection stage with a circulator.

REFERENCES

1. J. B. Gunn, Microwave oscillations of current in III–V semi-conductors, *Solid State Commun.*, **1**, 88–91 (Sept. 1963).
2. H. Kroemer, Theory of the Gunn-effect, *Proc. IEEE (Corres.)*, **52**, 1736 (Dec. 1964).
3. B. K. Ridley and T. B. Watkins, The possibility of negative resistance in semi-conductors, *Proc. Phys. Soc.*, **78**, Pt 3, 293–304 (Aug. 1961).
4. C. Hilsum, Transferred electron amplifiers and oscillators, *Proc. IRE*, **50**, 185–9 (Feb. 1962).

326 Gallium arsenide

5. J. E. Carroll, *Hot Electron Microwave Generators*, London, Edward Arnold (1970).
6. J. G. Ruch and W. Fawcett, Temperature dependence of the transport properties of gallium arsenide determined by a Monte Carlo method, *J. Appl. Phys.*, **41**, 3843–50 (Aug. 1970).
7. P. N. Butcher, W. Fawcett, and N. R. Ogg, Effect of field dependent diffusion on stable domain propagation in the Gunn effect, *Br. J. Appl. Phys.*, **18**, 755–9 (1967).
8. J. Carroll, Oscillations covering 4–31 GHz from a single Gunn diode, *Electron. Lett.*, **2**(4), 141 (April 1966).
9. P. Robson and S. M. Mahrous, Some aspects of Gunn-effect oscillators, *Radio Electronic Eng.*, **50**, 345–52 (1965).
10. H. Pollman, R. Engelmann, W. Frey, and B. G. Bosch, Load dependence of Gunn-oscillator performance, *IEEE Trans. Microwave Theor. Tech.*, **MTT-18**(11), 817–27 (Nov. 1970).
11. G. C. Dalman and W. T. Chen, Effect of circuit load on the microwave properties of Gunn diodes, *Eur. Microwave Conf. Proc.*, pp 319–23 (1983).
12. T. Ikoma and H. Yanai, Effect of external circuit on Gunn oscillation, *Int. J. Solid State Circuits*, **2**, 108–13 (Sept. 1967).
13. E. M. Bastida and G. Conciauro, Influence of the harmonics on the power generated by waveguide tuneable Gunn oscillators, *IEEE Trans. Microwave Theor. Tech.*, **MTT-22**(8), 796–8 (Aug. 1974).
14. B. Van der Pol, Forced oscillations in a circuit with non-linear resistance, *Phil. Mag.*, **3**, 65–80 (Jan. 1927).
15. P. N. Butcher, Theory of stable domain propagation in the Gunn-effect, *Phys. Lett.*, **19**, 546–7 (1965).
16. J. A. Copeland, Theoretical study of a Gunn diode in a resonant circuit, *IEEE Trans. Electron. Devices*, **ED-14**, 59–62 (1967).
17. G. D. Hobson, Frequency temperature relationships of X-band Gunn oscillators, *Solid State Electronics*, **15**, 431–41 (1972).
18. R. B. Robrock, Analysis and simulation of domain propagation in non-uniformly doped bulk GaAs, *IEEE Trans. Electron Devices*, **ED-16**, 647–53 (1969).
19. E. L. Warner, Extension of the Gunn-effect theory given by Robson and Mahrous, *Electronics Lett.*, **2**(7), 260–1 (July 1966).
20. K. Kurokawa, Some basic characteristics of broadband negative resistance oscillator circuits, *Bell System Tech. J.*, 1937–55 (July 1969).
21. G. S. Hobson, Small signal admittance of a Gunn effect device, *Electron. Lett.*, **2**(6), 207–8 (June 1966).
22. W. Heinle, On the equivalent circuit of a Gunn diode, *Int. J. Electronics*, **23**, 541–6 (1967).
23. I. W. Pence and P. J. Khan, Broadband equivalent circuit determination of Gunn diodes, *IEEE Trans. Microwave Theor. Tech.*, **MTT-18**(11), 784–90 (Nov. 1970).
24. P. W. Dorman, Gunn diode impedance measurements using a single tuned oscillator, *Int. Microwave Symp. Digest*, pp. 150–1 (May 1971).
25. Y. Ito, H. Komizo, T. Meguro, Y. Daido, and I. Umebu, Experimental and computer simulation analysis of a Gunn-diode, *Int. Microwave Symp. Digest*, pp. 152–3 (May 1971).
26. P. J. De Waard, Measurement of admittance of Gunn-diodes in passive and active regions of bias voltage, *Electronic Lett.*, **9**(3), 59–60 (Feb. 1973).
27. M. J. Howes and M. L. Jeremy, Large signal characterisation of solid state microwave diodes, *IEEE Trans. Electron. Devices*, **ED-21**(8), 488–99 (Aug. 1974).
28. M. Kawashima and H. Hartnage, New measurement method of Gunn-diode impedance, *Electronic Lett.*, **8**(12), 305–6 (1972).

Transferred electron devices

29. J. C. T. Young and I. M. Stephenson, Measurement of the large signal characteristics of microwave solid state devices using an injection locking technique, *IEEE Trans. Microwave Theor. Tech.*, **MTT-22**(12), 1320–3 (Dec. 1974).
30. J. McBretney and M. J. Howes, Electrical characterisation of transferred-electron device, *Electronic Lett.*, **12**(20), 532–3 (Sept. 1976).
31. W. C. Tsai, E. J. Rosenbaum, and L. A. MacKenzie, Circuit analysis of waveguide cavity Gunn-effect oscillator, *IEEE Trans. Microwave Theor. Tech.*, **MTT-18**(11), 808–16 (Nov. 1970).
32. C. P. Jethwa and R. L. Gunshor, Circuit characterisation of waveguide-mounted Gunn-effect oscillators, *Electron. Lett.*, **7**(15), 433–6 (July 1971).
33. M. Dean and M. J. Howes, J-Band transferred electron oscillators, *IEEE Trans. Microwave Theor. Tech.*, **MTT-21**(3), 121–6 (March 1973).
34. R. L. Eisenhart and P. J. Kahn, Theoretical and experimental analysis of a waveguide mounting structure, *IEEE Trans. Microwave Theor. Tech.*, **MTT-19**(8), 706–19 (Aug. 1971).
35. J. McBretney and M. J. Howes, Electrical characterisation of transferred-electron device, *Electronic Lett.*, **12**(20), 532–3 (Sept. 1976).
36. B. C. Taylor, S. J. Fray, and S. Gibbs, Frequency saturation effects in transferred electron oscillators, *IEEE Trans. Microwave Theor. Tech.*, **MTT-18**(11), 799–807 (Nov. 1970).
37. M. Shoji, A voltage tuneable Gunn effect oscillator, *Proc. IEEE (Lett.)*, **55**, 130–1 (Jan. 1967).
38. G. King and M. P. Wasse, Frequency modulation of Gunn-effect oscillators, *IEEE Trans. Electron Devices*, **ED-14**, 717–18 (Oct. 1967).
39. W. C. Tsai and F. J. Rosenbaum, Amplitude and frequency modulation of a waveguide cavity C. W. Gunn oscillator, *IEEE Trans. Microwave Theor. Tech.*, **MTT-18**(11), 877–84 (Nov. 1970).
40. W. E. Wilson, Domain capacitance tuning of Gunn oscillators, *Proc. IEEE (Lett.)*, **57**, 688–90 (Sept. 1969).
41. D. D. Khandelwal and W. R. Curtice, A study of the single frequency quenched domain mode Gunn-effect oscillators, *IEEE Trans. Microwave Theor. Tech.*, **MTT-18**(4), 178–87 (April 1970).
42. D. D. Tang and R. J. Lomax, Bias tuning and modulation characteristics of transferred electron oscillators, *IEEE Trans. Microwave Theor. Tech.*, **MTT-23**(9), 748–53 (Sept. 1975).
43. N. S. Chang, T. Hayamizu, and Y. Matsuo, YIG-tuned Gunn effect oscillator, *Proc. IEEE (Lett.)*, **55**, 1621 (Nov. 1967).
44. M. Omori, Octave tuning of a C. W. Gunn diode using a YIG sphere, *Proc. IEEE (Lett.)*, **57**, 97 (Jan. 1969).
45. P. W. Braddock and R. Hodges, High power 10% tuning bandwidth varactor controlled Impatt oscillator/amplifier, *Electronics Lett.*, **13**, 179 (March 1977).
46. I. Kuru, Frequency modulation of the Gunn oscillator, *Proc. IEEE*, **53**, 1642 (1965)
47. B. K. Lee and M. S. Hodgart, Microwave Gunn oscillator tuned electronically over 1 GHz, *Electronics Lett.*, **4**, 240 (June 1968).
48. D. Cawsey, Wide range tuning of solid-state microwave oscillators, *IEEE J. Solid State Circuits*, **SC-5**, 82 (April 1970).
49. S. F. Paik, Q-degradation in varactor tuned oscillators, *IEEE Trans. Microwave Theor. Tech.*, **MTT-22**, 578 (May 1974).
50. R. B. Smith and P. W. Crane, Varactor tuned Gunn effect oscillator, *Electronics Lett.*, **6**, 139 (March 1970).

51. D. Cawsey, Varactor tuned Gunn effect oscillators, *Electronics Lett.*, **6**, 246 (April 1970).
52. D. Large, Octave band varactor tuned Gunn diode sources, *Microwave J.*, 49 (Oct. 1970).
53. K. Kurokawa, Some basic characteristics of broadband negative resistance oscillator circuits, *Bell Syst. Tech. J.*, **48**, 1937 (Aug. 1969).
54. C. S. Aitchison and R. V. Celsthorpe, A circuit technique for broadbanding the electronic tuning range of Gunn oscillators, *IEEE J. Solid State Circuits*, **SC-12**, 21 (Feb. 1977).
55. C. S. Aitchison, Broadband varactor tuned oscillators, *IEE Colloq. on Tunable Microwave Oscillators*, Digest No. 1977/52 (Nov. 1977).
56. B. J. Downing and F. A. Myers, Broadband (1.95 GHz) varactor tuned X-band Gunn oscillator, *Electronics Lett.*, **7**, 407 (July 1971).
57. C. D. Corbey et al., Wide band varactor tuned coaxial oscillators, *IEEE Trans. Microwave Theor. Tech.*, **MTT-24**, 31 (Jan. 1976).
58. M. J. Howes, The application of variable impedance devices in electronically tunable microwave oscillator modules, *Variable Impedance Devices*, eds. M. J. Howes and D. V. Morgan, New York, Wiley Interscience (1978).
59. R. Gough and B. Newton, An integrated wide band varactor tuned Gunn oscillator, *IEEE Trans. Electron Devices*, **ED-20**, 863 (Oct. 1973).
60. B. S. Glance and M. V. Schneider, Millimetre wave microstrip oscillators, *IEEE Trans. Microwave Theor. Tech.*, **MTT-22**, 1281 (Dec. 1974).
61. B. S. Glance, Microstrip varactor tuned mm-wave transmitter, *IEEE Trans. Microwave Theor. Tech.*, **MTT-24**, 156 (March 1976).
62. M. Dean and M. J. Howes, Electronic tuning of stable transferred electron oscillators, *IEEE Trans. Electron Devices*, **ED-21**, 563 (Sept. 1974).
63. J. S. Joshi and J. Cornick, Some general observations on the tuning characteristics of electromechanically tuned Gunn oscillators, *IEEE Trans. Microwave Theor. Tech.*, **MTT-21**, 582 (Sept. 1973).
64. J. Bird, R. M. G. Bolton, A. L. Edridge, B. A. E. De Sa, and G. S. Hobson, Gunn diodes with improved frequency stability/temperature variations, *Electronic. Lett.*, **7**(11), 299–301 (1971).
65. G. S. Hobson, Frequency temperature relationships of X-band Gunn oscillators, *Solid State Electronics*, **15**, 431–41 (1972).
66. D. Edwards, G. Kellet, P. Turner, and F. Myers, Effect of contact resistance on Gunn diode risetimes, *Electronic Lett.*, **8**(24), 596–7 (1972).
67. Z. U. Kocabiyikoglu, G. S. Hobson, and B. A. E. De Sa, Relationship of the starting delay time and frequency/temperature characteristics of X-band Gunn oscillators, *Electronic Lett.*, **7**(18), 550–2 (1971).
68. W. S. C. Gurney and J. W. Orton, New techniques for the study of Gunn diode contacts, *Solid State Electronics*, **17**, 743–50 (1974).
69. F. A. Myers, Gunn-effect technology, *Microwave System News*, 67–71 (April 1975).
70. S. Sugimoto and T. Sugivra, Nanosecond-pulse generation at 11 GHz with Gunn effect diodes, *IEEE Proc.*, **56**, 1215–17 (1968).
71. B. A. Prew, Measurements of the close to carrier FM noise of pulsed InP TEDs primed by an external oscillator, *Electronic Lett.*, **13**(12), 344–6 (1977).
72. H. Pollmann and B. G. Bosch, Injection priming of pulsed Gunn oscillators, *IEEE Trans. Electron Devices*, **ED-14**, 609–10 (1967).
73. J. S. Bravman and L. F. Eastman, Thermal effects of the operation of high average power Gunn devices, *IEEE Trans. Electron Devices*, **ED-17**(9), 744–50 (1970).

74. I. B. Bott and H. R. Holliday, The performance of X-band Gunn oscillators over the temperature range 30°C to 120°C, *IEEE Trans. Electron Devices*, **ED-14**(9), 522–5 (1967).
75. R. J. Clarke, Frequency temperature in Gunn oscillators, *Design Electronics*, **2**(2), 31–2 (Nov. 1971).
76. S. Y. Narayan and J. P. Paczkowski, Integral heat sink transferred electron oscillators, *RCA Rev.*, **33**, 752–65 (1972).
77. J. F. Reynolds, B. E. Bertrand, and R. E. Enstron, Microwave circuits for high-efficiency operation of transferred electron oscillators, *IEEE Trans. Microwave Theor. Tech.*, **MTT-18**(11), 827–34 (1970).
78. R. Stevens, D. Tarrant, and F. A. Myers, Pulsed Gunn-diode oscillator: 40 W at 16 GHz, *Electronic Lett.*, **10**(25/26), 531–3 (1974).
79. R. E. Cooke, J. J. Crisp, R. F. B. Conlon, and J. S. Heeks, Transferred electron oscillators for phased array radar, *IEE Conf. on Radar, Present and Future*, p. 118, Publication No. 105 (1973).
80. K. M. Baughan, F. A. Myers, and G. Kellett, High power pulsed Gunn effect oscillators using series and parallel arrays of devices, *Microwave and Optical Group Association Conf.*, pp. 11–15 (Sept. 1970).
81. J. Gelbwachs and S. Mao, Phase locking of pulsed Gunn oscillators, Proc. *IEEE*, **54**, 1591–2 (1966).
82. M. P. Wasse and E. Denison, An array of pulsed X-band microstrip Gunn diode transmitters with temperature stabilisation, *IEEE Trans. Microwave Theor. Tech.*, **MTT-19**(7), 616–22 (1971).
83. E. D. Bullimore, B. J. Downing, and F. A. Myers, Frequency stable pulsed Gunn oscillators, *Electronics Lett.*, **10**(11), 220–2 (1974).
84. B. J. Downing, Electronic tuning of pulsed Gunn oscillators, Chapter 5, PhD Thesis, Sheffield University (1973).
85. R. Stevens and F. A. Myers, Temperature compensation of Gunn oscillators, *Electronics Lett.*, **10**(22), 463–4 (1974).
86. J. G. Ruch and W. Fawcett, Temperature dependence of the transport properties of gallium arsenide determined by a Monte Carlo method, *J. Appl. Phys.*, **41**(9), 3843–9 (1970).
87. H. W. Thim, Linear negative conductance amplification with Gunn oscillators, *IEEE Trans. Electron Devices*, **ED-14**, 517–22 (Sept. 1967).
88. D. E. McCumber and A. G. Chynoweth, Theory of negative-conductance amplification and of Gunn instabilities in 'two-valley' semiconductors, *IEEE Trans. Electron Devices*, **ED-13**, 4–21 (Jan. 1966).
89. B. W. Hakki and S. Knight, Microwave phenomena in bulk GaAs, *IEEE Trans. Electron Devices*, **ED-13**, 94–105 (Jan. 1966).
90. P. Stertzer, Stabilization of supercritical transferred electron amplifiers, *Proc. IEEE Lett.*, **57**, 1781–3 (Oct. 1969).
91. F. J. Lidgey and K. W. H. Foulds, An X band LSA amplifier, *IEEE Trans. Microwave Theor. Tech.*, **MTT-21**, 736–8 (Nov. 1973)
92. P. Jeppesson and B. I. Jeppsson, The influence of diffusion on the stability of the supercritical transferred electron amplifier, *Proc. IEEE*, **60**, 452 (1972).
93. J. Magarshak and A. Mircea, Stabilisation and wide band amplification using over critically doped transferred electron diodes, *Proc. Int. Conf. Microwave and Optical Generation and Amplification*, Amsterdam, pp. 16.19–23. (1970).
94. H. W. Thim and S. Knight, Carrier generation and switching phenomena in n-GaAs devices, *Appl. Phys. Lett.*, **11**, 85–7 (1967).

95. S. Kataoka, H. Tateno and M. Kawashima, Suppression of travelling high field domain mode oscillations in GaAs by dielectric surface loading, *Electron. Lett.*, **5**, 48–50 (1969).
96. P. N. Robson, G. S. Kino, and B. Fay, Two port microwave amplification in long samples of GaAs, *IEEE Trans. Electron Devices*, **ED-14**, 612–15 (Sept. 1967).
97. H. W. Thim, Noise reduction in bulk negative resistance amplifiers, *Electron. Lett.*, **7**, 106–8 (1971).
98. J. Magarshak, A. Rabier, and R. Spitalnik, Optimum design of transferred electron amplifier devices in GaAs, *IEEE Trans. Electron Devices*, **ED-21**, 652–4 (Oct. 1974).
99. R. Charlton, K. R. Freeman and G. S. Hobson, A stabilisation mechanism for supercritical transferred electron amplifiers, *Electron. Lett.*, **7**, 575–7 (1971).
100. A. K. Talwar and W. R. Curtice, Effect of donor density and temperature on the performance of stabilised transferred electron devices, *IEEE Trans. Electron Devices*, **ED-20**, 544–50 (June 1973).
101. B. Kallback, Stabilising effects for the supercritical $n^+ nn^+$ transferred electron amplifier, *Solid State Electronics*, **18**, 257–62 (1975).
102. M. M. Atalla and J. L. Moll, Emitter controlled negative resistance in GaAs, *Solid State Electronics*, **12**, 119–29 (1969).
103. T. Hariu, S. Ono, and Y. Shibata, Wideband performance of the injection limited Gunn diode, *Electron Lett.*, **6**, 666–7 (1970).
104. J. O. Scanlan, Analysis and synthesis of tunnel diode circuits, London, John Wiley & Sons, p. 76ff (1966).

Gallium Arsenide
Edited by M. J. Howes and D. V. Morgan
© 1985 John Wiley & Sons Ltd

CHAPTER 9

Gaas IMPATT Diodes

J. SMITH

9.1 INTRODUCTION

The IMPATT device, in its various forms, is now widely used as the source of prime or amplified power over a wide frequency range (1–300 GHz). The essential mechanisms involved in the operation of these devices are contained in the expanded title, i.e. IMPact Avalanche and Transit Time diode. The basic device model uses two concepts: the generation of carriers within a semiconductor by impact ionization, and the transport of these carriers across a defined distance, within the device, at the saturated drift velocity.

IMPATT oscillation was predicted by Read[1] almost a quarter of a century ago. In 1965 the first experimental observation[2] of this phenomenon was seen in a silicon p^+–n junction device. Since then a number of semiconductors (Si, GaAs, Ge, GaAlAs, InP), along with various device structures, including n^+–p and symmetrical p–n junctions, Schottky contact devices, and devices with modified doping profiles in both n and p regions for enhanced efficiency, have been the subject of much research and development.

Before concentrating on the development of GaAs devices it will be useful to introduce the concepts of IMPATT operation.

9.2 BASIC MODEL

The diode structure considered is an abrupt p^+–n junction grown on a low-resistivity n^+ substrate. The doping profile is shown in Figure 9.1a. Figure 9.1b indicates the electric field profile under reverse bias. If $p^+ \gg n$, the depletion region in the p-type material can be neglected. While the reverse voltage remains below breakdown, only a small reverse leakage current flows. At breakdown the field is sufficiently large that carriers are accelerated and acquire the required energy to create hole–electron pairs by impact ionization. The holes and electrons

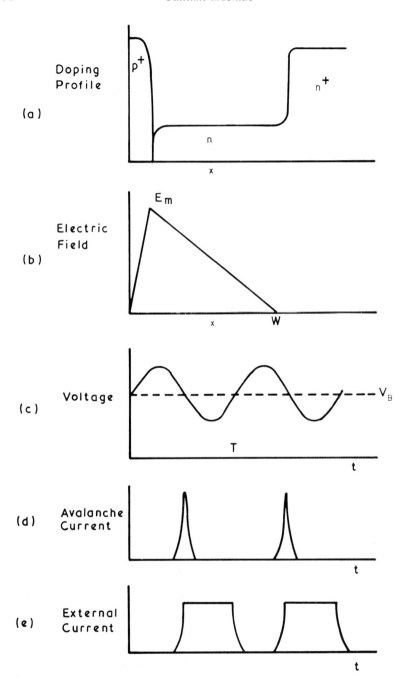

Figure 9.1 Basic model of IMPATT operation

GaAs IMPATT diodes

are separated in the electric field and flow towards the p^+ and n^+ contacts respectively.

The ionization rates for electrons and holes are denoted $\alpha(E)$ and $\beta(E)$ respectively, and as indicated are electric-field-dependent. The coefficients $\alpha(E)$ and $\beta(E)$ have dimensions cm^{-1}. If n electrons traverse a small distance dx within a semiconductor where the electric field is E, the number of extra electrons created by impact ionization of these n electrons is $n\alpha(E)\,dx$. The number of extra holes created by this process is of course the same. A similar situation occurs for a number of holes p entering this region dx and travelling in the same electric field but in the opposite direction. The increase in the number of holes and the number of electrons from impact ionization is $p\beta(E)\,dx$.

It can be shown[3] that avalanche breakdown occurs when

$$\int_0^W \alpha \left[\exp\left(-\int_0^x (\alpha - \beta)\,dx' \right) \right] dx = 1$$

where W is the depletion region width. In GaAs $\alpha = \beta$, thus simplifying the above to

$$\int_0^W \alpha\,dx = 1$$

The ionization rates can be expressed as

$$\alpha(E) = \beta(E) = A \exp[-(b/E)^2]$$

where E is the electric field (V cm^{-1}), $b = 5.55 \times 10^5$ V cm^{-1} at room temperature, and $A = 1.18 \times 10^5$ cm^{-1}, also at room temperature. The maximum field occurs at the metallurgical junction and is given by

$$E_m = \frac{N_0 q W}{\varepsilon} \tag{9.1}$$

The field throughout the depletion region is given by

$$E(x) = E_m - \frac{N_0 q x}{\varepsilon} \tag{9.2}$$

The ionization rates are strongly field-dependent. For example, at a value of electric field of 500 kV cm^{-1}, a typical maximum field, the rates α and β are proportional to $E^{2.5}$. At lower fields, say 300 kV cm^{-1}, the rates are proportional to E^6.

With reference to Figure 9.1b, as the field decreases the probability of a carrier achieving sufficient energy to create a hole–electron pair by impact ionization decreases rapidly. The strong dependence of α and β on electric field enables the depletion region to be considered in two parts: an avalanche region, where the impact ionization process is occurring, and the remainder of the depletion region, where the field is low enough to be able to neglect multiplication and yet is high enough to ensure that carriers travel at the saturated drift velocity. The second region is often referred to as the drift region.

A commonly found, if somewhat arbitrary, definition of the avalanche region width x_a is

$$\int_0^{x_a} \alpha \, dx = 0.95$$

As an example, for GaAs and a uniform doping level of $N_D = 1 \times 10^{16}$ cm^{-3}, the avalanche region is one-third of the total depletion region.

In Figure 9.1c the device is biased to just below breakdown, and a small a.c. voltage superimposed. When the voltage across the diode is above V_B, avalanche multiplication takes place and the number of carriers continues to increase until the voltage falls below V_B. At this point in time the carriers generated by this process can be regarded as being launched into the drift region. Each of these electrons will induce a current to flow in the external circuit. The magnitude of this current is derived using the Ramo–Schockley theorem[5,6] as qv_s/W.

If the device is designed such that the injected carriers traverse the drift region $\sim (W - x_a)$ in a time $T/2$, the external current will cease to flow at the end of the RF cycle; this is shown in Figure 9.1e. Considering the Fourier components of this waveform, it can be seen that there is a sinusoidal current flowing at the same frequency as the RF voltage, but 180° out of phase.

At the transit time frequency the device can be thought of as a negative resistance. So far only the conduction current has been considered. There is also a displacement current flowing which can be accounted for by introducing a capacitance (the diode depletion capacitance) in shunt with the negative resistance. When the device is connected to an electrical circuit such that the sum of the total device–circuit impedance is zero, oscillation will take place.

This is a simple schematic of the device operation. It is possible to arrive at an analytical description of the impedance under small-signal conditions.

9.2.1 Small-signal analysis

In this analysis a slightly modified diode structure is considered to simplify the calculations. Figure 9.2a shows a p^{++}–n$^+$–i–n^{++} doping profile, and the associated reverse field profile. The electric field in the intrinsic region is sufficiently low to neglect the probability of ionization; in other words the avalanche is completely confined within a width x_a.

Figure 9.2b shows a thin section dx within the avalanche region $0 < x < x_a$; n and p are the numbers of electrons and holes, and v_n and v_p are the saturated drift velocities.

GaAs IMPATT diodes

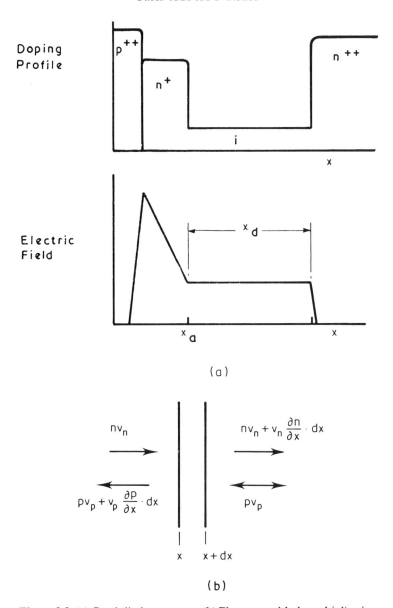

Figure 9.2 (a) Read diode structure. (b) Electron and hole multiplication

The increase in the number of electrons within the region dx is given by

$$\frac{dn}{dt} = \alpha n v_n + \beta p v_p - \frac{d}{dx}(nv_n) \tag{9.3}$$

Introducing the electron current density J_n, where

$$J_n = nqv_n$$

(similarly

$$J_p = pqv_p$$

for holes) we obtain

$$\frac{1}{v_n}\frac{dJ_n}{dt} = \alpha J_n + \beta J_p - \frac{dJ_n}{dx} \tag{9.4}$$

A similar analysis for holes yields

$$\frac{1}{v_p}\frac{dJ_p}{dt} = \beta J_p + \alpha J_n + \frac{dJ_p}{dx} \tag{9.5}$$

Let the combined hole and electron current density be defined as

$$J = J_n + J_p \tag{9.6}$$

Adding (9.4) and (9.5) gives

$$\frac{1}{v}\frac{dJ}{dt} = 2\alpha J_n + 2\beta J_p + \frac{d}{dx}(J_p - J_n) \tag{9.7}$$

For GaAs $\alpha = \beta$ and assuming $v_n = v_p = v$ this simplifies to

$$\frac{1}{v}\frac{dJ}{dt} = 2\alpha J + \frac{d}{dx}(J_p - J_n) \tag{9.8}$$

Integrating over the avalanche region from $x = 0$ to $x = x_a$ we get

$$\frac{x_a}{v_n}\frac{dJ}{dt} = 2J \int_0^{x_a} \alpha\, dx + [J_p - J_n]_0^{x_a} \tag{9.9}$$

At $x = 0$, $J \sim J_{p0}$, i.e. $J_{p0} \gg J_{n0}$; and at $x = x_a$, $J \sim J_n(x_a)$, i.e. $J_n(x_a) \gg J_p(x_a)$; so we obtain

$$\frac{x_a}{v_n}\frac{dJ}{dt} = 2J\left(\int_0^{x_a} \alpha\, dx - 1\right)$$

Now let $\tau_a = x_a/v_n$, i.e. time to traverse the avalanche region, then

$$\frac{dJ}{dt} = \frac{2}{\tau_a}J\left(\int_0^{x_a} \alpha\, dx - 1\right) \tag{9.10}$$

This equation is often referred to as Read's equation and shows the time dependence of the avalanche current. When the voltage is above the breakdown voltage, i.e. $\int \alpha\, dx > 1$, the avalanche current increases. Below breakdown dJ/dt is negative and the current decreases.

GaAs IMPATT diodes

The above analysis has ignored the space charge effect of the generated carriers within the avalanche region; a more rigorous treatment yields the same features as above, except that the factor 2 is replaced by 3 in Read's equation:

$$\frac{dJ}{dt} = \frac{3}{\tau_a} J \left(\int_0^{x_a} \alpha \, dx - 1 \right) \tag{9.11}$$

Under small-signal conditions, that is small sinusoidal perturbations about a mean value,

$$J = J_0 + J_{ac}$$
$$E = E_0 + E_{ac}$$
$$V = V_0 + V_{ac} \quad \text{(voltage across device)}$$
$$\alpha = \alpha_0 + \alpha' E_{ac}$$

The Read equation then reduces to

$$\frac{dJ_{ac}}{dt} = \frac{3}{\tau_a} J_0 x_a \alpha' E_{ac}$$

$x_a E_{ac}$ is the small-signal RF voltage across the avalanche region, and therefore by comparing the previous equation with

$$|V(t)| = L \frac{dI}{dt}$$

this region can be regarded as an inductance

$$L_a = \tau_a / 3 J_0 \alpha' A$$

where A is the diode area.

The avalanche region also sustains a displacement current due to its capacitance

$$C_a = \varepsilon A / x_a$$

These two elements are in shunt and lead to the concept of an avalanche resonant frequency ω_a given by

$$\omega_a^2 = (L_a C_a)^{-1} = 3 J \alpha' v_s / \varepsilon \tag{9.12}$$

Consider now the drift region where the injected carriers travel at the saturated drift velocity v. Using the same small-signal notation, the approximate electron current density within the drift region is $J_c(x, t)$. As the carriers travel at constant velocity, the density at $x = x_1$ is the same as at $x = 0$ only at time $t - x_1/v$, i.e.

$$J_c(x_1, t) = J_c(0, t - x_1/v) \tag{9.13}$$

Let the electron density at $x = 0$ be $J_c \cos[\omega(t - x/v)]$, i.e.

$$J_c(x, t) = J_c \cos[\omega(t - x/v)] \tag{9.14}$$

Gallium arsenide

We have seen earlier that a sheet of electrons of density n travelling at velocity v across a depletion region x_d induce an external current nqv/x_d. The total external current therefore is given by

$$J_e = \int_0^{x_d} \frac{nqv}{x_d} dx = \frac{1}{x_d} \int_0^{x_d} J(x,t) dx \qquad (9.15)$$

$$= \frac{1}{x_d} \int_0^{x_d} J_c \cos[\omega(t - x/v)] dx \qquad (9.16)$$

$$= J_c \cos(\omega t - \theta)(\sin \theta)/\theta \qquad (9.17)$$

where

$$\theta = \tfrac{1}{2} \omega x_d / v$$

Inspection of equation (9.17) shows that there is a phase delay of θ between the external induced current and the injected particle current, and θ is often denoted the transit angle. When this transit angle θ is 90°, the total phase delay between the applied RF voltage and the induced current is 180°. This is the combined phase delay due first to the inductive nature of the avalanche process and secondly to the transit time delay.

This type of small-signal analysis has been used by Gilden and Hines[7] to arrive at the small-signal impedance of the device:

$$Z = \frac{x_d^2}{v\varepsilon A} \frac{1}{1 - \omega^2/\omega_a^2} \frac{1 - \cos\phi}{\theta^2/2}$$

$$+ \frac{1}{j\omega C_d} \left(1 - \frac{\sin\theta}{\theta} + \frac{(\sin\theta)/\theta + x_a/x_d}{1 - \omega_a^2/\omega^2} \right) \qquad (9.18)$$

where as before θ is the transit angle, ω_a is the avalanche resonance frequency, and x_a and x_d are avalanche and drift lengths.

Both the real and imaginary components show a resonance at an angular frequency $\omega = \omega_a$, the avalanche resonance frequency derived earlier in equation (9.12). Above this frequency the device has negative resistance and is capacitive.

This type of small-signal analysis indicates the role of the device parameters on the diode impedance. However, the device is seldom used under small-signal conditions. In order to investigate the large-signal behaviour we shall return to the uniformly doped device.

9.2.2 Sharp pulse approximation

Figure 9.3a shows a uniformly doped n-type Schottky contact diode. At reverse breakdown the voltage across the diode is given by $V_B = E_B W_D$, where E_B is the

GaAs IMPATT diodes 339

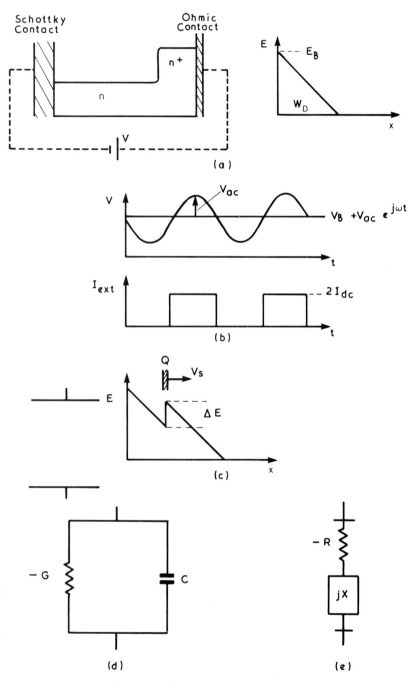

Figure 9.3 Sharp pulse approximation

peak electric field and W_D is the depletion region width. Figure 9.3b shows the voltage across the diode $V_B + V_{ac}e^{j\omega t}$ and the induced current in the external circuit caused by an assumed sheet of avalanche-generated carriers drifting across the depletion region W at the transit time frequency, i.e.

$$W_D = v_s/2f \quad \text{where } 2\pi f = \omega$$

The peak current is therefore $2I_{dc}$. The Fourier component of this current waveform at frequency f has an amplitude

$$i_{ac} = \frac{4}{\pi} I_{dc}$$

A delta-function of charge Q travelling across the depletion region will modify the field profile by space charge effects as shown in Figure 9.3c. The field step ΔE is given by

$$\Delta E = Q/\varepsilon A$$

where A is the device area. The current density

$$J_{dc} = \frac{Qf}{A} = \varepsilon f \Delta E \tag{9.19}$$

The field step ΔE must not exceed the value at which avalanche breakdown will occur in the drift region. This limit on ΔE fixes the current density limit for a particular frequency of operation.

The RF voltage across the device can be expressed as a fraction of the breakdown voltage V_B:

$$V_{ac} = mV_B$$

where m is the modulation factor. There is a limit to this factor m as the field in the drift region at the peak of the RF cycle must be low enough to prevent avalanche multiplication in the drift region. The efficiency of the device is given simply by

$$z = \text{RF power out/d.c. power in}$$

$$z = \tfrac{1}{2} V_{ac} i_{ac} / V_{dc} I_{dc}$$

$$= m\frac{2}{\pi}$$

(This assumes a square-wave-induced current in Figure 9.1.) The negative conductance generated by this process is

$$G = \frac{i_{ac}}{V_{ac}} = \frac{4}{\pi m} \frac{I_{dc}}{V_B} \tag{9.20}$$

So far only the conduction current has been considered. The displacement current can easily be accounted for by adding a capacitance C across the negative

GaAs IMPATT diodes

resistance as in Figure 9.3d. It is useful now to transform this parallel circuit to a series arrangment as in Figure 9.3e. The equivalent series resistance is given by

$$-R_d = \frac{G}{G^2 + C^2\omega^2}$$

For most practical devices $C\omega \gg G$, so

$$-R_d \sim \frac{G}{C^2\omega^2} \qquad C = \varepsilon A/W_D$$

$$\sim \frac{1}{Af^2} \frac{\Delta E v_s}{\pi^3 m E_B \varepsilon} \qquad (9.21)$$

This shows the device has negative resistance proportional to 1/area.

The condition that the device will oscillate when connected to an external circuit of impedance $R_E + jX$ is that $\Sigma Z = 0$. More specifically the device negative resistance $|-R_D|$ must equal the positive resistance in the circuit R_E. This external resistance comprises the load resistance as well as unwanted resistive elements such as series resistance in the connections and circuit losses. If all the unwanted parasitic resistive components are combined and denoted R_c, then

$$R_E = R_L + R_c = -R_d$$

Only the power dissipated in the load is useful. The fraction of the power generated that can be utilized is therefore

$$\frac{R_L}{R_c + R_L} = \frac{|-R_d| - R_c}{|-R_d|} = 1 - \left|\frac{R_c}{R_d}\right|$$

There is a practical limit to how far series resistance and losses can be reduced. To prevent this factor $(1 - |R_c/R_d|)$ becoming unacceptably small, the device impedance must be larger than a certain value $|R_{min}|$, e.g. $|R_{min}| \geqslant 10 R_c$.

From these considerations it can be seen that there is a maximum device area (minimum impedance) that can be used in order to avoid an unacceptable reduction in overall efficiency.

The (electrical) power generated by the device is given by

$$P_{ge} = \tfrac{1}{2} V_{ac} i_{ac}$$

$$= \frac{1}{2\pi} \varepsilon A m E_B \Delta E v_s$$

The power generated is proportional to device area A. The maximum power that can be generated is determined by the maximum area of device that can be acceptably matched in a circuit:

$$A_{max} = \frac{1}{|R_{min}|f^2} \frac{\Delta E v_s}{\pi^3 m E_B \varepsilon} \qquad (9.22)$$

and thus
$$P_{\text{ge(max)}} = \frac{1}{2\pi^4}(\Delta E v_s)^2 \frac{1}{|R_{\min}|f^2} \quad (9.23)$$

This result is often referred to as the PZf^2 product. As can be seen from the above, this product is constant for a particular material. There is a maximum ΔE (related to E_B) such that avalanching in the drift region is prevented. Material parameters that lead to high output powers are those with high breakdown field (large band gap E_g) and high saturated drift velocity. High power generation also requires low-impedance circuits with minimum loss. For high efficiency the device must be able to withstand a high modulation factor m. This feature will be dealt with later. The limit on power generation, as given above, is due to space charge limits on current density, and limitations of low loss matching circuits.

Equation (9.23) can be used to indicate the approximate power levels that may be achieved. Taking numerical values that may be encountered, e.g. $\Delta E = 250 \text{ kV cm}^{-1}$, $v_s = 5 \times 10^6 \text{ cm s}^{-1}$, and $|R_{\min}| = 2.5 \, \Omega$, this yields $P_{\text{ge}} = 3 \times 10^3 \text{ W GHz}^{-2}$, i.e. 13 W at 15 GHz and 3.3 W at 30 GHz. It must be pointed out that this analysis is somewhat simplified and the frequency scaling is not absolutely exact; however, it does indicate the type of performance observed.

This electronic or impedance limit is not the only limit to be considered. The thermal effects within the device must be taken into account. The d.c. input power not converted to RF power must be removed from the device. The fabrication of IMPATT diodes will be covered later. However, these devices are usually constructed with the active region in close proximity to a metal (gold or silver) heat sink. The resistance against the heat flow causes the temperature at the diode junction to rise above ambient:

$$T_J = T_{\text{ambient}} + \Delta T$$

where
$$\Delta T = \text{power dissipated } (P_{\text{dis}}) \times \text{thermal resistance } (R_{\text{th}}) \quad (9.24)$$

There is a limit to the junction temperature at which the device can be operated, with satisfactory life expectancy. The most common failure mechanism is a diffusion of the heat sink or contact metals into the semiconductor. This process is accelerated at elevated temperatures. For a required life there is a maximum ΔT of

$$\Delta T_{\max} = R_{\text{th}} P_{\text{dis}} = R_{\text{th}} P_{\text{gt}} \left(\frac{1}{z} - 1\right) \quad (9.25)$$

The device thermal impedance is the sum of several components. In general the spreading resistance of the heat sink provides the major contribution, and can be expressed as

$$R_{\text{th}} = k/\sqrt{A} \quad (9.26)$$

where A is the device area.

GaAs IMPATT diodes

The maximum (thermal) power that can be generated is thus given by

$$P_{\text{gt(max)}} = \frac{\Delta T_{\max}}{k} \sqrt{A \frac{z}{1-z}} \qquad (9.27)$$

This equation shows the device parameters to be considered when optimizing output power under thermal limitations, i.e. high-reliability contacts must be used to maximize $\Delta T_{\max}$. The constant k must be low, i.e. the heat sink conductivity must be high, and finally device efficiency must be maximized.

Comparing equations (9.22), (9.23), and (9.27) it can be seen that there is a frequency f_c above which the device is impedance limited, and below which it is thermally limited.

Rearranging these equations yields

$$f_c = \frac{1}{4\pi^3} \frac{k}{\Delta T} \sqrt{\left(\frac{\pi E_B \varepsilon}{|R_{\min}|}\right)} (\Delta E v_s)^{3/2} \frac{\pi - 2m}{\sqrt{m}}$$

For a single drift uniformly doped device, typical constants may be $k = 0.2$, $\Delta E = 2.5 \times 10^5 \,\text{V cm}^{-1}$, $E_B = 600 \,\text{V cm}^{-1}$, $|R_{\min}| = 2.5\,\Omega$, $v_s = 5 \times 10^6 \,\text{cm s}^{-1}$, $\Delta T = 250°\text{C}$, and $m = 30\%$ modulation. This yields a value for $f_c \sim 35\,\text{GHz}$. Below this frequency it is not possible to maximize the device current density and area simultaneously, without exceeding the junction temperature limit.

If the maximum current density is used as in equation (9.19) then the power generated is proportional to diode area. This results in the fact that the maximum area is frequency-independent and hence the output power is constant. If on the other hand the diode area is maintained at its maximum from impedance considerations, i.e. $Af^2 = \text{constant}$, the current density must necessarily be reduced.

Using $R_{\text{th}} = k/\sqrt{A}$ and equation (9.22) yields the thermally limited output power

$$P_{\text{gt(max)}} = \Delta T_{\max} \frac{z}{1-z} \frac{1}{k} \left(\frac{\Delta E v_s}{\pi^3 m E_B \varepsilon R_{\min}}\right)^{1/2} \frac{1}{f} \qquad (9.28)$$

$$\propto \frac{1}{f}$$

This $1/f$ dependence is often quoted as characteristic of the thermal limit. The reduction of current density, necessitated by using the maximum diode area, has other implications. As shown earlier the avalanche resonance frequency is proportional to $\sqrt{J}$. The space charge effects of the particle current density also must be considered in both the avalanche and drift regions. In practice the $1/f$ relationship may not be as simple.

If the IMPATT is operated in a pulsed mode, then it is possible to avoid the thermal limit by adjusting the pulse length and duty cycle. In this case the peak power available is determined by impedance considerations.

It is appropriate at this point to introduce a variant of the IMPATT structure known as the double drift device. It is, as the name suggests, an IMPATT device with two drift regions. In a conventional single structure, such as a p^+–n diode, the holes generated in the avalanche process are collected soon after generation and play no further role in the operation. In the double drift structure these holes are allowed to drift across a depletion region formed in a low-doped p region. The structure is shown in Figure 9.4. The structure is designed such that the transit angles in both the n and p sides are approximately equal. The drift regions share a common avalanche zone. The total depletion width is approximately 1.5 times the equivalent depletion width in a single drift structure. The breakdown voltage of the device is again about 1.5 times larger. The maximum current density is the same, however, as the single drift device, as this is limited as in equation (9.19):

$$J = \varepsilon f \Delta E$$

the limitations on ΔE being the same for both single and double drift devices.

It is still the case that the impedance of the device is dominated by the reactance. It can be shown from these approximate calculations that for the same impedance the double drift device can be about 1.5 times larger, and hence the current is increased by a factor of 1.5. For a device that is impedance limited, the d.c. input power is $(1.5)^2$ times larger than a single drift diode. There is also an improvement in efficiency as the two drift regions share a common avalanche region. The avalanche region occupies a smaller fraction of the total depletion region and this will be shown later to increase the efficiency. The net result is that the output

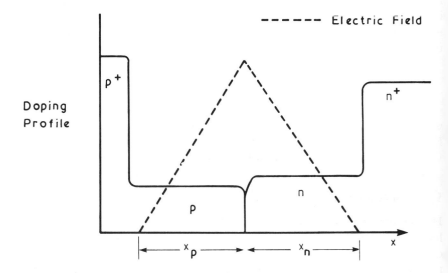

Figure 9.4 Electric field and doping profile of double drift IMPATT structure

GaAs IMPATT diodes

power from a single drift diode can be improved by a factor of about 2.5 by use of a double drift device.

The thermal properties however are degraded. The point of maximum heat generation at the p–n junction is separated from the heat sink beyond the p^+ layer by a thickness of p^+ and p semiconductor. The thermal conductivity of most semiconductors is not high and usually decreases with temperature. The increase in thermal impedance introduced by this region is frequency dependent, i.e. the thickness of p region increases as frequency of operation decreases. At low frequencies these devices tend to be operated pulsed to overcome the thermal problem. At high frequencies full advantage can be taken of the increased power performance, i.e. at a frequency above the changeover from thermal to impedance-limited output power.

So far the performance of an IMPATT diode operating under large signal has been analysed using a model of device operation, often referred to as the sharp pulse approximation, i.e. the avalanche-generated carriers are assumed to be injected simultaneously into the drift region, where they travel together as a pulse of charge. Although this is a gross simplification, and there are many other factors (covered later) that determine the performance, the approach so far is useful in giving an intuitive feel for the parameters that play a role in the operation of the IMPATT.

It is possible to return to the Read equation (9.11) and attempt to find an analytical solution under large-signal operation. An approximate solution was derived by Tager.[8] The expression for the large-signal impedance is similar in form to equation (9.18); the introduction of large-signal parameters manifests itself in the term ω_a, the avalanche resonance frequency. As the signal level increases, the avalanche resonance frequency reduces. It can also be shown that maximum power and efficiency are achieved when the transit angle θ is $0.75\,\pi$ or in other words the device is operated at 0.75 times its transit time frequency.

The efficiency is given by

$$z = k \frac{V_{ac}}{V_B} \frac{1}{1 + x_a/x_d} \qquad (9.29)$$

where x_a is the width of the avalanche region and x_d is the total depletion width. For a device to operate at high efficiency, not only is it necessary to achieve a high voltage modulation $m = V_{ac}/V_B$, but the factor x_a/x_d must be minimized.

9.2.3 Application to GaAs IMPATTs

Having introduced the basic concepts of IMPATT operation, it is now time to look at GaAs IMPATT diodes specifically, and also to introduce factors affecting device performance such as leakage currents, diffusion of carriers, depletion region modulation, velocity–field characteristics, and low-field effects, all of which are too difficult to include in a full analytical solution.

Read's original paper dealt with a diode structure as shown in Figure 9.2a. The first experimental observations in 1965, and in fact the majority of IMPATT developments until the early 1970s, were on uniformly doped silicon p^+–n junction devices. Device performance was improved throughout this period with a move to more reliable contacts, better heat sinking, and improved technology: all leading to higher powers and higher operating frequencies. The maximum efficiency, however, never exceeded about 11%, well below the Read efficiency. From 1969 to 1972 several GaAs results on uniformly doped structures were reported[9–12] with efficiencies typically 12–15%. The higher efficiency was attributed to the reduction of the factor x_a/x_d in equation (9.29) as the avalanche region in GaAs occupies a smaller percentage of the overall depletion width than is the case in silicon p^+–n structures. This is due to the fact that in GaAs the electron and hole ionization coefficients are equal, whereas in Si $\alpha/\beta > 1$. With reference to Figure 9.1, the holes generated by the avalanche process are accelerated towards the high-field region but have a low probability of ionization. If a complementary structure is considered, i.e. an n^+–p junction, the roles are reversed and the ionization process is more efficient, leading to a thinner avalanche region x_a. However, the very low mobility of p-type Si introduces significant series loss which offsets the advantage achieved by the thin avalanche region.

Returning to GaAs devices, in 1972 Salmer and Pribetich[4] predicted that much higher efficiencies could be achieved with GaAs diodes with modified doping profiles. These profiles have become known as hi-lo and lo-hi-lo. Figure 9.4 shows a schematic of these profiles with the associated reverse electric field. In the hi-lo device the electric field towards the junction is increased by a layer of doping $N_1 > N_2$. In the lo-hi-lo structure the hi layer, which is a very thin region of donors, causes a step ΔE in the electric field, increasing its magnitude towards the junction. In both cases the higher electric field increases the probability for ionization. The field profile is designed such that the avalanche process is confined to this region. In this way the ratio of x_a/x_d can be reduced over conventional structures, leading to an increase in efficiency from consideration of the simple model of IMPATT operation. Initial results published at this time were 22% efficiency at 6.5 GHz.

Over the next three years there was a number of experimental results.[13–15] Efficiencies over 35% were achieved at frequencies from 3 to 10 GHz. Above this frequency, efficiencies reduced a little, and at around 15 GHz the efficiency is down to around 25%.

In these high-efficiency structures, the confinement of the avalanche region reduces the factor x_a/x_d, thus increasing the predicted efficiency; however, there are further implications, from using these structures, on IMPATT operation within the device.

For a given frequency of operation f, the depletion region x_d is proportional to $1/f$. The designed x_d is achieved in a uniformly doped device by using the

GaAs IMPATT diodes

appropriate doping level. In the hi-lo or lo-hi-lo device there are more degrees of freedom, but in all cases the doping level in the drift region N_2 (see Figure 9.5) is lower than in the uniformly doped device. The slope of the electric field profile $\partial E/\partial x$ is therefore less. With an RF voltage across the device, the edge of the depletion region will move. The shallow slope of the field profile in the high-efficiency structures will enhance the amplitude and speed of this movement.

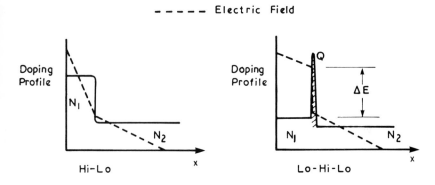

Figure 9.5 Hi-Lo and Lo-Hi-Lo structures

The movement of the depletion region edge has not yet been taken into account. First we return to the Ramo–Schockley theorem and introduce a time-dependent depletion width. The basic theorem states that a charge q moving at velocity v_s between two plates (e.g. the anode and cathode of a parallel-plate capacitor with separation W) will induce a current in the external circuit given by qv_s/W. The same result is given when the region between the plates is a depleted semiconductor. If the analysis is repeated with a time-dependent separation $W(t)$ (Figure 9.6), the induced current is given by

$$\frac{q}{W(t)}\left(v_s - \frac{x}{W(t)}\frac{\mathrm{d}}{\mathrm{d}x}W(t)\right) \qquad (9.30)$$

The first term in this expression is the simple expression with the separation W replaced with the time-dependent value $W(t)$. Remember that the drift process is occurring during the negative half of the RF voltage cycle, i.e. $\mathrm{d}W(t)/\mathrm{d}t$ is negative and $W(t)$ is reducing. The induced external current is increased due to the movement of the depletion region edge. The square-wave current in Figure 9.1 is modified, and the peaking of the induced current at the minimum of the RF voltage is beneficial in increasing the efficiency of the device.

No attempt is made to quantify this effect using this model as there are some cautionary qualifications to be pointed out, such as the fact that in a real device the assumed sheet of charge is of course a distribution of charge. The carriers are

348   Gallium arsenide

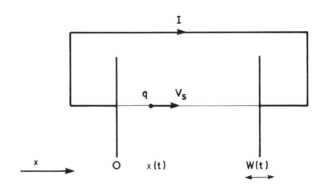

Figure 9.6 Ramo Shockley theorem

generated with a finite distribution in space which spreads due to diffusion effects as the carriers traverse the drift region. The injection and collection of this charge is thus spread in time. The depletion region edge is also a simplification as the field in the so-called undepleted region of the device has a finite non-zero value. The collection process of this charge is difficult to model simply as the field profile itself is modified by the space charge effect of the injected carriers. These factors can only be dealt with using large simulation programs. Reference to this type of work will be made later; however, it is the intention at this point to introduce conceptually the mechanisms that affect device performance and to these ends the depletion region modulation is an important factor and justifies a non-rigorous analysis.

The diffusion of the injected carriers has two effects on device performance. First of all, for low-frequency devices, the spatial spreading of the assumed pulse of carriers leads to an extension of the induced current waveform, such that a current flows during the positive half of the RF voltage cycle, leading to power dissipation. If the device is designed below the transit time frequency to allow these late carriers to be collected, the magnitude of the induced current is reduced, again reducing efficiency. For high-frequency devices, where the transit time is so short that the trailing edge of the diffused pulse is still within the high-field region of the device at the start of the next RF voltage cycle, the avalanche multiplication process starts from a much higher initial carrier density. The large space charge effect from the high density of avalanche-generated carriers can have the effect of prematurely extinguishing the process due to the field drop behind the carriers. This in turn leads to premature injection and a degradation in performance. The diffusion of carriers has been suggested as a frequency-limiting factor for GaAs IMPATT devices.[17] The predicted cut-off frequency of 80 GHz depends directly on the value of the diffusion coefficient, which is not confidently known at the high fields encountered in these devices. Although the exact frequency is in doubt especially as reasonable 80 GHz performance has recently been achieved,[18] the spreading out of the injected carriers will degrade efficiency.

The velocity–field characteristic of GaAs shown in Figure 9.7a may, however, counter this effect to some extent. Figure 9.7b shows the field profile of a hi-lo device with a pulse of carriers of total charge Q travelling across the depletion region, indicating the space charge effect.

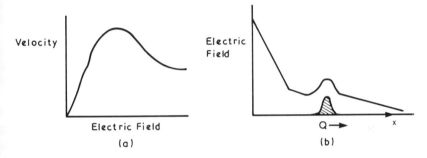

Figure 9.7 Velocity-field and space charge effects

The diffusion process is tending to spread out the charge. When the charge bunch reaches the low-field region of the depletion region, the velocity–field relationship comes into play. Electrons at the back of the bunch, i.e. trailing behind, experience a lower field and hence travel faster. This effect counteracts the diffusion process and tends to keep the bunch together. The other effect from this process is the increase in average velocity of all the electrons in the low-field region. This increases the induced external current and efficiency.

There is another drift region effect that has been proposed as a mechanism increasing efficiency. This is often referred to as the 'surfing mode'.[19] The effect is a combination of the depletion edge modulation and the velocity–field characteristic of electrons in GaAs. As highlighted above, electrons near the low-field edge of the depletion region are accelerated. However, as they venture into lower-field regions, they slow again. There is a sort of trapping action where the charge and the depletion edge move in unison. The picture one forms is of the charge bunch 'surfing' on the depletion edge 'wave'. The interaction of the charge bunch, space charge, and the depletion edge has the effect of increasing the speed of the carriers. Other workers have also proposed transferred electron effects as beneficial in MPATT operation.[20, 21]

There have been, of course, many large-scale computer simulations of the device.[22-25] The agreement with experimental results at frequencies below 20 GHz is good, indicating that the input data in terms of material and device parameters and the models used are valid. It is sometimes difficult, however, to highlight the particular feature responsible for good device performance from this

350 *Gallium arsenide*

type of numerical analysis. At higher frequencies, there are further complications, which will be discussed in a later section.

9.3 TECHNOLOGY

The first step in the fabrication of any GaAs IMPATT diode is the epitaxial growth of the active layers on a low-resistivity substrate, the simplest structure being a uniformly doped single drift n-type structure (using a Schottky contact), i.e. n layer on n^+ substrate. An example of a complex growth cycle is a double drift lo-hi-lo structure with a p^{++} contact layer, i.e. $P^{++}_{contact}-p_2-p_{spike}-p_1-n_{spike}-n_2-n^+_{substrate}$.

The range of doping levels may be between mid 10^{15} cm^{-3} and $(2-3) \times 10^{19}$ cm^{-3}. The thickness of the various layers may vary from 3–4 μm for the drift regions to $<$ 100 Å for the hi region in a lo-hi-lo structure.

Details of the various types of epitaxial growth are not covered here, other than to mention that GaAs IMPATT devices have been fabricated on material grown by LPE (liquid phase epitaxy); VPE (vapour phase epitaxy) including the trichloride, hydride, and metal-organic systems; MBE (molecular beam epitaxy) and ion implantation. Details of these various techniques can be found in Chapter 3.

The IMPATT device is a power device and as such is almost always fabricated with an integral heat sink. A number of techniques have been developed by several workers in this field to fabricate GaAs IMPATT diodes in this form.[13, 15, 26] The exact details of the technology can be found in these references. The common features of most technologies can be highlighted here with reference to Figure 9.8.

(a) Following the growth of the active layers, the initial metallization is deposited. This may be a Schottky contact such as Pt or Ti on n-type surface layer or an ohmic contact formed on a p^+ contact layer. These layers would be deposited by evaporation or sputtering and are only a few thousand angstroms thick.

A final overlay gold layer would also be deposited with an appropriate barrier layer between the Au and the contact metallization.

(b) The heat sink is then formed on this metallized surface. The most commonly used heat sink is plated gold. The heat sink may be around 100μm thick.

(c) The slice is then inverted and the GaAs substrate thinned from an initial thickness likely to be around 400μm to about 10μm or less. The thinning may be achieved by mechanical polishing or etching or a combination of both. It is important to thin the device as much as possible to reduce the series resistance of the substrate material to a minimum.

(d) Metal contacts are made to the n^+ substrate. These can be ohmic contacts such as Au–Ge–Ni or Schottky contacts such as Ti. In device operation this

Gaas IMPATT diodes

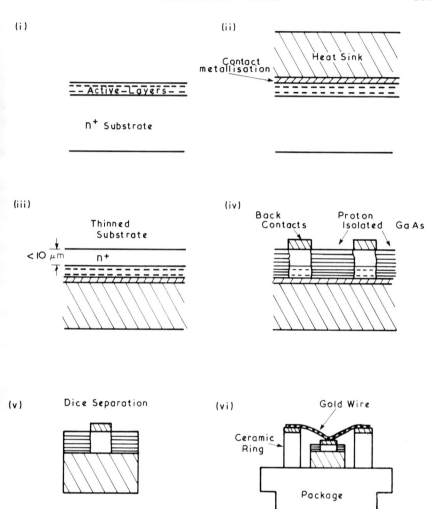

Figure 9.8 Device fabrication schedule

contact, often referred to as the back contact, is forward biased at a high current density and hence the slope resistance is low. There is, however, a voltage drop across the forward-biased contact and hence power dissipation. For these reasons ohmic contacts are used if compatible with the device fabrication schedule.

(e) The individual diodes are formed. One approach is chemically to etch away the semiconductor around the device to form a mesa structure. Another technique is to form a thick plated back contact in gold before the wafer is exposed to high-energy protons. The areas below the plated back contacts are protected from the protons and the GaAs in these areas remains unaltered. The GaAs exposed to the high-energy protons is converted to an insulating form. Resistivities as high as 10^8 Ω cm can be achieved. The technique is referred to as proton isolation.[27]

The range of the protons within GaAs is approximately 1 μm/100 keV. The distribution is spread across about 1 μm. Multiple implants are therefore used, e.g. 0–1 MeV in 100 keV steps for a 10 μm thick layer of GaAs. The dose required is doping-dependent but doses of mid 10^{14} cm^{-2} are typically used.

Using this proton isolation technique the IMPATT diodes are thus embedded within GaAs and there are no exposed mesa edges.

(f) The wafer can now be diced, either by etching or mechanical cutting, to separate the diodes on individual heat sinks.

(g) The devices are packaged, the most common technique being to bond the diode heat sink to a metal carrier. Thermocompression bonding or soldering can be used. The device is bonded inside a ceramic ring or between ceramic pillars, the top of the insulators being metallized. The so-called back contact of the diode is then connected to the ceramic ring using fine gold wires or gold tapes. The device is now in a fairly rugged form and can be mounted in the appropriate microwave circuit.

The schematic of the above technology is shown in Figure 9.8. This type of technology, other than the proton isolation technique, is not specific to GaAs alone, and can be applied to a number of two-terminal semiconductor devices. It is possible to return to a number of features that are unique to GaAs IMPATT devices.

The use of high-efficiency structures such as the hi-lo or the lo-hi-lo profiles as outlined before (Figure 9.5) introduces demanding accuracy on the profile parameters required. Taking the hi-lo structure as an example, the device is designed to confine the avalanche region in the hi region, with a particular depletion width in the lo region to achieve the required operating frequency. The field profile at breakdown depends on the doping levels N_1 and N_2 in the hi and lo regions, and also on thickness of the high region. Denoting this thickness X_1 and the width of the depletion region in the lo region X_2, it can be shown that X_2 is extremely sensitive to variations in X_1. Doping ratios (i.e. $N_1:N_2$) of 20:1 are not uncommon in these devices, and with this type of structure a small change in X_1 of about 2.5% will change X_2 by 10%. Other device parameters will also be altered e.g. breakdown voltage and the degree of avalanche confinement. For a device operating at around 15 GHz, a 2.5% variation in X_1 is in fact only 60 Å. An analysis of the other high-efficiency structure (lo-hi-lo) yields a similar degree of sensitivity of device profile to the depth of the thin region of hi dopant as shown in Figure 9.4.

The challenge of fabricating these precise layers has been tackled in two ways. First of all, a very high degree of control over the growth process has been successfully achieved using both VPE (trichloride) and MBE growth systems in a number of laboratories. The alternative approach is to grow the surface layer thicker than required, and then to etch the layer in a controlled manner until the required thickness is achieved. This requires a removal technique with a fine degree of control. The anodic oxidation process[28] is particularly suited to this (with a removal rate of 13 Å V^{-1}).

Both the above techniques are used to achieve the required material profile. The next problem is to maintain this profile through device fabrication and subsequent device operating life. The problem is particularly severe when Schottky contacts are used to form the reverse junction. A slight interdiffusion of metal and GaAs will have the same effect as reducing the thickness of the surface layer of GaAs. The need is therefore for a highly stable Schottky contact metal compatible with the device technology. A number of metals have been investigated.[29, 30] The Ti-GaAs interface has been shown to be stable up to temperatures around 450°C. The use of a Ti Schottky contact is also compatible with most device technologies. High-efficiency IMPATTS have been operated for over 15 000 with no interaction occurring between the Ti and the GaAs.

Whereas Ti has been found to be comparatively unreactive at elevated temperatures, Pt interdiffuses readily into GaAs at temperatures between 250 and 300°C, forming a number of compounds. Although this may seem somewhat dramatic, Pt has been used successfully as the Schottky contact to high-efficiency GaAs IMPATTS.[31.] The technique is to deposit a thin Pt layer (< 300 Å) followed by a barrier layer such as Ta[31] or W.[32] A sintering process is used to react the thin Pt layer fully with the GaAs, the consumption of GaAs at the surface having been allowed for in the device design. The barrier layer prevents further penetration of metal, while the Pt compounds formed are stable at normal device operating temperatures.

This form of device degradation is only encountered with the high-efficiency structures (hi-lo or lo-hi-lo) when a Schottky contact is used.

If the profile is determined during the epitaxial growth process, i.e. with a p$^+$ junction contact or a double drift structure, the profile of the structure is unlikely to change during device fabrication or operation. The reliability of the device is then determined by more conventional failure mechanisms.

As outlined in the device technology, the IMPATT is fabricated on a metal heat sink, usually Au. A contact metallization is incorporated between the heat sink and the GaAs. This metallization system must fulfil a number of roles, provide good adherence to the semiconductor, act as the Schottky contact in some cases, and prevent the metal from the heat sink reaching the GaAs at elevated operating temperatures. Once the Au reaches the GaAs there is a rapid diffusion usually causing a short-circuit failure of the junction. For GaAs IMPATTS, Ti is favoured as the contact layer. Several barrier metallizations have been used such as Pd, Pt, W. Accelerated life tests have predicted a median time to failure of greater than 5 years for devices operating at junction temperatures as high as 300°C using a Ti/Pt/Au metallization system.[33] Reducing the operating device temperature to 250°C, the predicted life increases to 10^7 h.

354 Gallium arsenide

9.4 PRESENT PERFORMANCE OF GaAs IMPATTS

The output powers achieved, to date, from GaAs IMPATT diodes are shown as a function of frequency in Figure 9.9a.[34-41] CW powers in excess of 20 W are

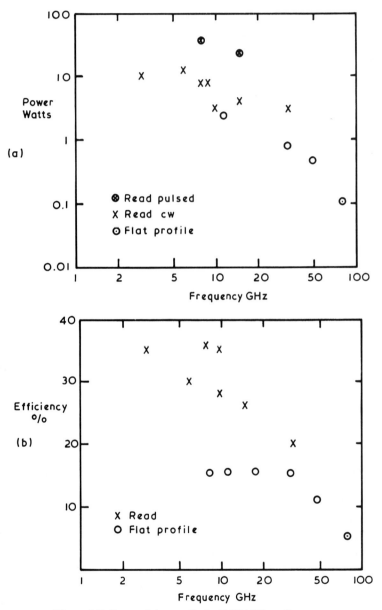

Figure 9.9 State of the art GaAs IMPATT performance

available up to around 10 GHz from high-efficiency structures, either hi-lo or lo-hi-lo. The highest peak power is from a hybrid double drift diode giving 35 W at 8.3 GHz. This device is a hi-lo structure on the n side with a uniformly doped p side. Devices of this type have also given 22 W peak power at 15 GHz. A full double drift Read diode (lo-hi-lo on both sides) has been operated at 33 GHz with output powers approaching 3 W. The output power available decreases as might be expected from pZf^2 considerations. No high-efficiency structure has been operated with significant performance above 40 GHz. Figure 9.9b shows the efficiencies recorded by GaAs IMPATTS, against frequency. Both the Read and flat profile results are shown. The efficiency of the Read-type diodes at frequencies below 10 GHz is around 35%. The efficiency gradually falls to about 20% at 33 GHz.

A number of points can be made. As the design frequency increases, the doping level in the drift region increases. In order to maintain avalanche confinement, the field in the avalanche region must be increased, attempting to keep the ratio x_a/W small. The higher fields and field gradients promote the probability of tunnel currents in the device. A tunnel current, unlike an avalanche current, leads to a component in phase with the RF voltage and thus reduces efficiency. Another feature is that the ionization coefficients α and β for GaAs tend to saturate at high fields ($> 600\,\text{kV cm}^{-1}$). The result of this is that it becomes increasingly difficult to reduce the avalanche region even though the fields are increasing for the higher-frequency operation. There may be other frequency-limiting effects such as the depletion width modulation velocity. The experimental results to date show a gradual decrease in efficiency with frequency.

Figures 9.9a and 9.9b also show the power versus frequency performance of GaAs flat profile devices. Although the results below 20 GHz tend to be eclipsed by the Read diodes, the operating frequency does extend out into the millimetre wave region with efficiencies of 11% at 50 GHz[44] and 5% at 80 GHz.[18] This last result is the highest operating frequency with reasonable efficiency and leads to a brief look at high-frequency limitations.

9.4.1 Frequency limitations

The fall in performance of GaAs IMPATT diodes at high frequencies has been the subject of debate for over a decade. Some of the parameters involved have been mentioned previously.

The role of the diffusion of carriers has been discussed, and the topic of tunnel currents introduced. It was shown earlier that the IMPATT device utilizes the avalanche process to achieve a phase lag between current and voltage. Without this phase lag there would be an increase in the current components in phase with the RF voltage, thus dissipating power and reducing the device efficiency. In the limit when the reverse current is totally tunnel current, the diode is still capable of producing microwave power. The device has been named a 'TUNNETT'.

Experimental results have been achieved at frequencies as high as 248 GHz.[44] The efficiency in this mode is very low.

The transition from avalanche to tunnel breakdown is not sharp, and there is a range of doping levels where the reverse current is composed of a mixture of avalanche and tunnel currents. A method that has been used to differentiate experimentally between the mechanisms is to measure the breakdown voltage as a function of temperature: dV/dT being positive for avalanche breakdown and negative for tunnel breakdown. It is possible to define a doping level N_c such that the reverse current is dominated by avalanche for $n < N_c$, and by tunnelling for $n > N_c$. Calculations of this doping level have been made[45] with $N_c = 1.8 \times 10^{17}$ cm^{-3} for n-type GaAs. This corresponds to a reverse voltage of about 9–10 V.

The effect of this tunnel current is to change the operating mode from IMPATT to TUNNETT as the doping level in the diode increases for higher operating frequencies. There is not a cut-off frequency for operation but a gradual reduction in device efficiency and output power.

Theoretical predictions have been made[46] for TUNNETT operation above 300 GHz with efficiencies of about 1%; and for devices operation in a mixed avalanche and tunnelling mode at frequencies up to 200 GHz with efficiencies of several per cent.

The ionization rates in GaAs have been the most widely discussed topic with reference to the frequency limitations. Experimental determination of ionization rates usually involves the measurement of photocarrier multiplication as a function of d.c. reverse bias.[47, 48] The variation of ionization coefficient with field $\alpha(E)$ is generated in the form

$$\alpha(E) = A \exp[-(b/E)^2]$$

fitting the constants A and b. When plotted as a function of E the coefficients $\alpha(E)$ and $\beta(E)$ [$\alpha = \beta$] can be seen to saturate at high values of electric field. This is an extrapolation as these rates are measured at lower electric fields.

In the region before saturation, as the doping level in a device increases, the breakdown voltage and overall depletion region decreases; the ratio between the avalanche region x_a, as defined earlier, and the drift region x_d stays fairly constant. It was shown previously that the factor x_a/x_d appears in the expression for device efficiency. When the doping level reaches the levels where the ionization rates begin to saturate, the breakdown voltage and depletion region still reduce, but the ratio increases and degrades the efficiency.

The prediction of a fall in device performance from this effect is somewhat tenuous, as it relies on the extrapolation of ionization rates measured at low fields and under d.c. conditions, and their application to a particular model of device operation under RF conditions.

More recently this debate has been superseded by a discussion on the dynamic behaviour of the avalanche process at high frequencies or the so called 'avalanche response time' arguments.

GaAs IMPATT diodes

In essence the concept is that the carrier, either hole or electron, in an electric field takes a finite time to acquire sufficient energy to create a hole–electron pair by impact ionization. If this response time is a significant fraction of the RF period, then the extent of the avalanche multiplication will be limited. At sufficiently high frequencies, or short times, the avalanche process will not be possible.

The debate started in 1977 when an attempt was made to determine the avalanche response time from noise measurements on IMPATT diodes.[49] It appeared that this time was more than three times larger than theoretically predicted. There then followed investigations into the conduction band structure of GaAs and the orientation dependence of the ionization rates. It was predicted that the response time would be anisotropic. Further analysis of noise data on device fabrication in different orientations again gave response times larger than theory, but did not show any anisotropic effects.

Even more recently it has been suggested that in fact there is no discrepancy between the predicted and theoretical values of response time.[51] This postulate arises from the fact that these researchers were able to apply the Read theory for IMPATT operation to predict theoretically the performance of an IMPATT oscillating at 9.5 GHz. The avalanche response time used to match theory and experiment successfully was not anomalously high and fitted the expression

$$\tau_i = \frac{1}{3} x_a / v_s$$

where x_a is the avalanche region width and v_s is the saturated drift velocity.

The measurement of the avalanche response time from the modelling of IMPATT operation at low frequencies, and whether or not this value agrees with theory or not, is a subject that will no doubt receive further attention. However, the question as to the frequency limit on GaAs IMPATT operation that might be determined by avalanche response time still remains.

Inspection of Figure 9.9b shows that the efficiency of GaAs IMPATT devices falls rapidly between 35 and 50 GHz. The efficiency at around 80 GHz has fallen to about 5%. It is not clear at the present time whether the fall in efficiency is due to the slow response of the avalanche mechanism or to other factors such as the increasing probability of tunnel currents.

9.5 FUTURE TRENDS

The development of GaAs IMPATTS over the last decade and a half has been outlined. In this time the IMPATT device has established itself as a high-power solid state source. Silicon devices have been operated at frequencies as high as 220 GHz with peak output powers of 1 W, whereas GaAs devices have achieved efficiencies of 35% at low frequencies and peak output powers of several tens of watts. Can the future be predicted?

It is unlikely that the record efficiencies of around 35% will be exceeded. The GaAs high-efficiency diode has matured from a novel research activity to a developed semiconductor production device. Improvements are likely to come in the area of yield, reproducibility, and reliability for devices up to 20 GHz.

Further improvements in the controlled growth and assessment of thin GaAs epitaxial layers either by molecular beam epitaxy or metal-organic- CVD may extend the operation of these high-efficiency structures to around 35 GHz.

The high-frequency future is more difficult to predict. No doubt the theortical debate will continue. However, the extent of the necessary experimental programmes will depend on the motivation for achieving millimetre wave performance in GaAs, and what advantages there may be over existing milimetre wave sources.

REFERENCES

1. W. T. Read, A proposed high-frequency negative resistance diode, *Bell System Tech. J.*, **37**, 401–46 (March 1958).
2. R. L. Johnston and B. C. DeLoach, A silicon diode microwave oscillator, *Bell System Tech. J.*, **44**, 369–72 (Feb. 1965).
3. S. L. Miller, *Phys. Rev.*, **99**, 1234 (1955).
4. G. Salmer and J. Pribetich, Theoretical and experimental study of GaAs Impatt oscillator efficiency, *J. Appl. Phys.*, **44**, 314–24 (Jan. 1973).
5. S. Ramo, Currents induced by electron motion, *Proc. IRE*, **27**, 584–5 (1939).
6. W. Shockley, *Electrons and Holes in Semiconductors*, New York, Van Nostrand (1950).
7. M. Gilden and M. E. Hines, Electronic tuning effects in the Read microwave avalanche diode, *IEEE Trans. Electron Devices*, **ED-13**(1), 169–75 (1966).
8. A. S. Tager, *Sov. Phys. Usp.*, **9**, 892 (1967).
9. C. H. Kim and L. D. Armstrong, GaAs Schottky barrier avalanche diodes, *Solid State Electronics*, **13**, 53–6 (1970).
10. R. A. Zetter and A. M. Cowley, Batch fabrication of Impatt diodes, *Electron Lett.*, **5**, 693–4 (1969).
11. Y. S. Lee, Two-watt CW GaAs Schottky barrier Impatt diodes, *Proc. IEEE (Lett.)*, **58**, 1153–4 (July 1970).
12. A. Mircea, A. Farrayre, and B. Kramer, X band GaAs diffused Impatt diodes for high efficiency, *Proc. IEEE*, 1376–7 (Sept. 1971).
13. W. R. Wisseman, D. W. Shaw, R. L. Adams, and T. E. Hasty, GaAs Schottky Read diodes for X band operation, *IEEE Trans. Electron Devices*, **ED-21**(6), 317–23 (June 1974).
14. R. E. Goldwasser and F. E. Rosztoczy, High efficiency GaAs lo-hi-lo Impatt devices by liquid phase for X band, *Appl. Phys. Lett.*, **25**(1), 92–3 (July 1974).
15. L. C. Upadhyaynla *et al.*, High efficiency GaAs Impatt structures, *RCA Rev.* **35**, 567–78 (Dec. 1974).
16. C. O. Bozler, J. P. Donnelly, R. A. Murphy, R. W. Laton, and R. W. Sidbury, High efficiency ion implanted lo-hi-lo Impatt diodes, *Appl. Phys. Lett.*, **29**(2), 123–5 (July 1976).
17. B. Culshaw, R. A. Giblin, and P. A. Blakey, Avalanche diode oscillators, *Int. J. Electronics*, **37**(5), 577–632 (1974).
18. J. G. Smith, RSRE Malvern, Unpublished result presented at *Microwave Workshop*, Darmstadt (1981).
19. Y. Hirachi *et al.*, A new concept for high efficiency operation of hi-lo type GaAs

Impatt diodes, *IEEE Trans. Electron Devices*, **ED-25**(6), 666–74 (June 1978).
20. E. Constant, A. Mircea, J. Pribetich, and A. Farrayre, Effect of transferred electron velocity modulation in high efficiency GaAs Impatts, *J. Appl. Phys.*, **46**(9), 3934–40 (Sept. 1975).
21. P. Blakey, B. Culshaw, and R. A. Giblin, Flat field approximation; a model for drift region in high efficiency Impatts, *Solid State Electron Devices*, **1**(2), 57–61 (Jan. 1977).
22. M. S. Gupta and R. J. Lomax, A self consistent large signal analysis of Read type Impatt diode oscillator, *IEEE Trans. Electron Devices*, **ED-18**(8), 544–50 (Aug. 1971).
23. P. Blakey, R. A. Giblin, and A. J. Seeds, Large signal time domain modelling of avalanche diodes, *IEEE Trans. Electron Devices*, **ED-26**(11), 1718–28 (Nov. 1979).
24. J. W. Gewartowski, Progress with CW Impatt diode circuits at microwave frequencies, *IEEE Trans. Microwave Theor. Tech.*, **MTT-27**(5), 434–49 (May 1979).
25. S. O. Scanlan and T. J. Brazil, Large signal computer simulation of Impatt diodes, *IEEE Trans. Electron Devices*, **ED-28**(1), 12–21 (Jan. 1981).
26. P. Brook, J. G. Smith, L. D. Clough, C. A. Tearle, G. Ball, and J. C. H. Birbeck, Design, fabrication and performance of GaAs high efficiency Impatt diodes in J band, *Proc. Cornell Conf.* (1977).
27. A. G. Foyt, W. T. Lindley, C. M. Wolfe, and J. P. Donelly, Isolation of junction devices in GaAs using proton bombardment, *Solid State Electronics*, **12**, 209 (1968).
28. H. Hasegawa, K. E. Forward, and H. Hartenagel, Improved method of anodic oxidation of GaAs, *Electronics Lett.*, **11**(3), 53–4 (Feb. 1975).
29. S. D. Mukherjee, D. V. Morgan, M. J. Howes, J. G. Smith, and P. Brook, Reactions of vacuum deposited thin Schottky barrier metallizations on GaAs, *J. Vac. Sci. Technol.*, **16**(2), 138–40 (March 1979).
30. J. G. Smith and L. D. Clough, The effect of the Schottky contact metallisation on hi-lo GaAs Impatt diodes, *Proc. Cornell Conf.* (1979).
31. J. L. Heaton, R. E. Walline, and J. F. Carroll, Low-hi-low profile GaAs Impatt reliability, *IEEE Trans. Electron Devices*, **ED-26**(1), 96–101 (Jan. 1979).
32. G. E. Mahoney, *Appl. Phys. Lett.*, **27**, 613 (1975).
33. R. A. Murphy, W. T. Lindley, D. F. Peterson, and P. Staecher, Performance and reliability of K_a band GaAs Impatt diodes, *European Microwave Conference Proceedings* (1974).
34. C. O. Bozler *et al.*, High efficiency ion implanted lo-hi-lo GaAs Impatt, *Appl. Phys. Lett.*, **29**(2), 123–5 (July 1976).
35. R. E. Goldwasser and F. E. Rosztoczy, High efficiency GaAs lo-hi-lo Impatt devices by LPE, *Appl. Phys. Lett.*, **25**(1), 92–4 (July 1974).
36. J. R. Grierson *et al.*, High power 11 GHz GaAs hi-lo Impatt diodes with titanium Schottky barriers, *Electronics Lett.*, **15**(1), 13–15 (Jan. 1979).
37. T. L. Hierl, J. J. Berenz, J. Minoshita, and I. V. Zubeck, High efficiency pulsed GaAs Read Impatt diodes, *Electronics Lett.*, **14**(5), 155–7 (March 1978).
38. J. J. Berenz, J. Minoshita, T. Hierl, and F. B. Fank, High power pulsed GaAs double drift hybrid Read Impatt diodes for X band, *Electronics Lett.*, **15**(10), 277–8 (May 1979).
39. M. G. Adlerstein and E. L. Moore, Microwave properties of GaAs Impatt diodes at 33 GHz, *Proc. Cornell Conf.* (1981).
40. K. Nishitani and O. Ishihara, High power high reliability p–n junction GaAs Impatts for J band, *Jap. J. Appl. Phys.*, **16**, Suppl. 16-1, 93–7 (1977).
41. T. Watanabe, H. Kodera, and M. Migitaka, GaAs 50 GHz Schottky barrier Impatt diodes, *Electronics Lett.*, **10**(1), 7–8 (Jan. 1974).

42. J. G. Smith, RSRE Malvern, Millimetre wave GaAs Impatts, Unpublished *Microwave Workshop Conf.*, Darmstadt, Germany (1980).
43. M. Elta and G. Haddad, Large signal performance of microwave transit time devices in mixed tunneling and avalanche breakdown, *IEEE Trans. Electron Devices*, **ED-26**(6), 941–8 (June 1979).
44. J. Nishizawa, K. Motoya, and T. Ohano, GaAs Tunnett diodes, *IEEE MTT 5th Int. Microwave Symp. Digest*, pp. 159–61 (1978).
45. H. Albrect and L. Lerach, Distinction between avalanche and tunneling breakdown in one sided abrupt-junctions, *Appl. Phys.*, **16**, 191–4 (1979).
46. M. Elta, High frequency calculations of Impatt and Tunnett diodes, *Proc. 8th Biennial Cornell Conf.* (1981).
47. R. Hall and J. H. Leck, *Int. J. Electronics*, **25**, 529 (1968).
48. R. A. Logan and S. M. Sze, *J. Phys. Soc. Jap.*, Suppl. 21, 434 (1966).
49. C. A. Lee, J. Berenz, and G. C. Dalman, Determination of GaAs intrinsic response time for noise measurements, *Proc. 6th Biennial Cornell Elect. Eng. Conf.*, p. 233 (1977).
50. J. J. Berenz, Orientation dependence of the n-type GaAs intrinsic response time, *Electronics Lett.*, **15**, 150–2 (March 1979).
51. M. G. Adlerstein, J. W. McClymonds, and H. Statz, Avalanche response time in GaAs as determined from microwave admittance measurements, *IEEE Trans. Electron Devices*, **ED-28**(7), 808–11 (July 1981).
52. J. J. Goedbloed and B. B. Van Iperen, Intrinsic response time of related quantities describing semiconductor avalanches, *Electronics Lett.*, **13**, 448 (July 1977).

Gallium Arsenide
Edited by M. J. Howes and D. V. Morgan
© 1985 John Wiley & Sons Ltd

CHAPTER 10

Gaas MESFETs

B. TURNER

10.1 INTRODUCTION

One of the major influences on microwave technology in recent years has been the rise to maturity of the GaAs MESFET. The GaAs MEtal–Semiconductor Field-Effect Transistor is finding ever-increasing application in microwave amplifiers for civilian and military systems, with amplifiers operating up to 40 GHz now commercially available. The MESFET forms the basis of the GaAs monolithic circuit technologies which are now emerging from research into pilot production. The purpose of this chapter is to give an account of the structure and properties of this device, and to indicate key design features and performance limitations.

Figure 10.1a shows a plan view of a device with dimensions appropriate to operation at about 10 GHz; Figure 10.1b shows a cross section through the FET with typical electrode voltages, referred to the source electrode which is taken as ground. The basic GaAs MESFET material structure consists of an active layer, in this case 0.3 μm thick, with a doping density of 10^{23} net donors/m^3, which has been epitaxially grown either directly on a semi-insulating substrate or on an intervening high-resistivity buffer layer as shown here. Two ohmic contact regions, separated by about 5 μm, form the source and drain electrodes of the device, and a Schottky barrier gate electrode is located between them, forming the control electrode. Conventionally the gate dimension measured parallel to the direction of electron flow is the gate length; that normal to the electron flow is the gate width. Thus the device of Figure 10.1 has a gate length of 1 μm and a gate width of 300 μm. When a positive potential V_{DS} is applied to the drain electrode, electrons flow from source to drain, giving a current I_{DS} from drain to source. The gate is at a negative potential relative to the semiconductor beneath it, and a part of this semiconductor is depleted of carriers, restricting the current path to the 'channel' between the depletion region and the high-resistivity buffer layer. The result is that the drain–source current I_{DS} is a function not only of the drain–source voltage V_{DS} but also of the gate–source voltage V_{GS}. At zero

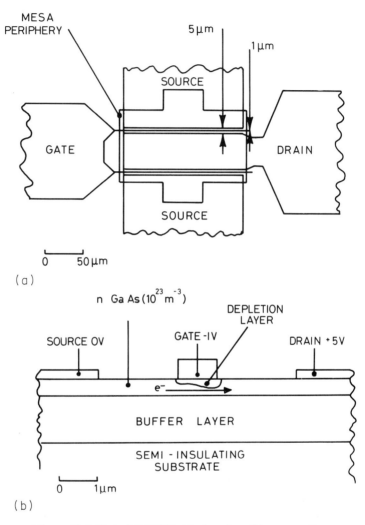

Figure 10.1 GaAs MESFET: (a) plan view; (b) cross section

drain–source bias, the depletion layer beneath the gate is symmetric, and of a depth dependent on the gate–source potential V_{GS} (Figure 10.2a). At zero gate bias, the depletion depth is that of the built-in potential of the Schottky barrier. As V_{DS} is increased from zero, the depletion layer becomes asymmetric since the potential difference between the gate and the active semiconductor beneath the gate is greater at the drain end of the gate than at the source end. The channel is more constricted at the drain end of the gate (Figure 10.2b) and so the field is higher in that region. As the drain–source voltage is increased still further, the

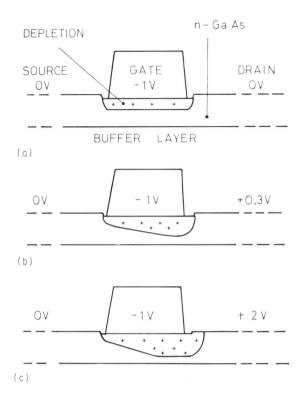

Figure 10.2 Depletion layer profiles (schematic): (a) $V_{DS} = 0$; (b) $V_{DS} = 0.3\,\text{V}$ (below velocity saturation); (c) $V_{DS} = 2\,\text{V}$ (above velocity saturation)

field at the drain end of the gate approaches the value at which the electron velocity saturates. Beyond this value, which corresponds to the 'knee voltage' on the d.c. characteristics, further increases in drain–source voltage do not substantially increase the drain–source current. The length of the region of the channel over which the electrons are in velocity saturation increases (Figure 10.2c) and, as shown in Section 10.3, there is some carrier accumulation within the channel. The depletion layer edge at the drain end of the gate moves closer to the drain as the drain–source voltage increases. At a sufficients negative gate potential, the depletion layer punches through to the high-resistivity buffer layer, and the source and drain electrodes are connected only by leakage paths within the buffer layer and substrate. These paths are located not only directly beneath the channel but also between any contact pads formed on the buffer layer.

Figure 10.3 shows typical d.c. characteristics for a device with a total gate width of 400 μm. At low drain–source voltage V_{DS}, the channel acts as a resistor with a resistance dependent on the gate–source voltage. Above the knee voltage, the

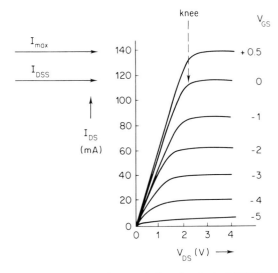

Figure 10.3 D.c. characteristics of a GaAs MESFET (gate width 400 μm)

drain–source current I_{DS} is essentially independent of V_{DS}: the value of this current at zero gate–source bias is referred to as the saturated drain–source current I_{DSS}. Since the Schottky barrier has a built-in negative potential of between 0.6 and 0.8 V, the gate–source potential may be taken somewhat positive before the gate electrode takes significant forward current. The drain–source current corresponding to this condition is denoted I_{max}. The gate–source voltage V_{GS} necessary to reduce the drain–source current to a very small value (typically less than 1 % I_{DSS}) is the 'pinch-off' voltage V_P. The mutual conductance g_m is the rate of change of drain–source current with gate–source voltage, i.e.

$$g_m = \left(\frac{\partial I_{DS}}{\partial V_{GS}} \right)_{V_{DS}}$$

and the drain conductance g_D is defined by

$$g_D = \left(\frac{\partial I_{DS}}{\partial V_{DS}} \right)_{V_{GS}}$$

A typical value for g_m at zero gate bias is 100 mA V^{-1} per millimetre gate width or 100 mS mm^{-1}. Although the device of figure 10.3 requires a negative potential applied to the gate to reduce I_{DS} to zero, some MESFETs used as 'normally off' logic gates are 'pinched off' at $V_{GS} = 0$ by the built-in potential of the Schottky gate.

The high-frequency performance of the MESFET is largely governed by the time taken for carriers to traverse the region of the device beneath the gate, and is

GaAs MESFETs

promoted by a high average electron velocity through this region. By comparison with silicon, GaAs offers a higher electron mobility (typically $0.4 \text{ m}^2 \text{ V}^{-1} \text{ s}^{-1}$ at 10^{23} net donors/m^3 as opposed to $0.08 \text{ m}^2 \text{ V}^{-1} \text{ s}^{-1}$), the opportunity in short gate devices for enhanced electron velocity in non-steady-state conditions, and a semi-insulating substrate of resistivity typically 10^6 Ω m, which is particularly significant in integrated circuit applications. By taking advantage of these properties, allied to submicrometre lithographic techniques and close control of epitaxial material growth, GaAs MESFETs have demonstrated stable amplification at 60 GHz,[1] oscillator characteristics up to 69 GHz,[2] low-noise amplifier performance (5.5 dB noise figure at 6.7 dB associated gain) at 38 GHz,[3] noise figures from commercial devices less than 1 dB at 4 GHz, and output powers up to 1 W at 18 GHz[4] and 18.5 W at 4 GHz.[5]

In this introductory section, a semiquantitative account of GaAs MESFET action will be given, based on a simple model of the device. Fabrication techniques and device parameters are covered in Section 10.2. Section 10.3 is concerned with the analytical theories of the GaAs MESFET, which emphasize velocity-saturation effects in limiting I_{DS}, by contrast with the account of silicon FETs given by Shockley.[6] A proper description of electron dynamics requires computer simulation of the device, and this is considered in Section 10.4. Sections 10.5 and 10.6 consider low-noise and high-power amplification, respectively, and Section 10.7 presents some alternative MESFET configurations. Inevitably, a single chapter can only cover these topics superficially. The ever-expanding literature on GaAs MESFETs contains further details. Two recent books cover in detail many aspects of GaAs field-effect transistors: *Microwave Field-Effect Transistors—Theory, Design and Applications* by Pengelly[7] and *GaAs FET Principles and Technology* edited by DiLorenzo and Khandelwal.[8]

An estimate of some parameters of the intrinsic MESFET can be made using a very simple model applicable above the knee voltage. The velocity–field characteristic of GaAs, which in the steady state shows a negative differential mobility region, is represented by a two-piecewise linear approximation as shown in Figure 10.4. This approximation is very widely made in simple device models and is justified principally by simplicity. The electron dynamics in a microwave FET are in reality more complex than deduced from the steady-state velocity–field characteristic of GaAs, since the transit time of electrons through the high-field region of the device is comparable with the time constants characterizing relaxation to the steady-state velocity–field characteristic. Consider a MESFET formed on an active layer having thickness a, doping density n_0 donors/m^3, and width W. The maximum, or 'saturation', current that can be carried by this layer, I_0, is given by

$$I_0 = n_0 q W v_s a \qquad (10.1)$$

where q is the electronic charge and v_s the saturated velocity. If a metal gate forms a Schottky barrier contact to this layer, and $V_{DS} = 0$, the depth of the depletion

Gallium arsenide

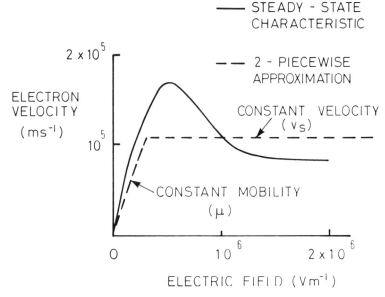

Figure 10.4 Two-piecewise representation of the GaAs steady-state velocity–field characteristic

layer, h, obeys

$$-V_{GS} + \phi = n_0 q h^2 / 2\varepsilon\varepsilon_0 \quad (10.2)$$

where ϕ is the magnitude of the built-in Schottky barrier potential and ε and ε_0 are the relative permittivities of GaAs and free space. The potential V_0 required to deplete the channel is given by

$$V_0 = n_0 q a^2 / 2\varepsilon\varepsilon_0 \quad (10.3)$$

An initial estimate may be made of the transconductance of the MESFET at and above the knee voltage by making the assumptions that the electrons travel at their saturated velocity over the whole length of the gate and that the depletion layer depth remains approximately constant over the length of the gate (Figure 10.5a). Then, if V_{ch} is the potential at the source end of the gate,

$$-V_{GS} + \phi + V_{ch} = n_0 q h^2 / 2\varepsilon\varepsilon_0 \quad (10.4)$$

and

$$I_{DS} = I_0 b/a \quad (10.5)$$

where b is the channel height $(a - h)$. Beyond the source end of the velocity-saturated region of the channel, the channel potential continues to rise. A consequence of carrier velocity saturation is that the depletion depth increases only gradually, current continuity being maintained by carrier accumulation, as

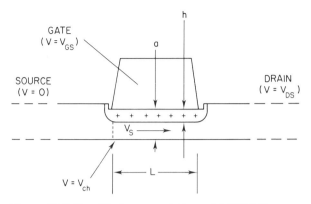

Figure 10.5 Simplified representation of MESFET cross section and dimensions for analysis

discussed in Section 10.3. From equations (10.4) and (10.5) we have

$$I_{DS} = I_0 \left[1 - \left(\frac{-V_{GS} + \phi + V_{ch}}{V_0} \right)^{1/2} \right] \quad (10.6)$$

and

$$g_{m(i)} = \frac{I_0}{2V_0} \left(\frac{V_0}{-V_{GS} + \phi + V_{ch}} \right)^{1/2} \quad (10.7)$$

where $g_{m(i)}$ is the mutual conductance of the intrinsic device.

Pursuing the approximation that the depletion depth is approximately constant over the length of the gate, the charge Q_{dep} in the gate depletion layer is given by

$$Q_{dep} \approx n_0 q h L W \quad (10.8)$$

where L is the gate length.

Since the charge in the depletion layer is balanced by a negative charge $Q_G = -Q_{dep}$ on the gate, the rate of change of gate charge with V_{GS} is given by

$$\left(\frac{\partial Q_G}{\partial V_{GS}} \right)_{V_{DS}} \approx LW \left(\frac{\varepsilon \varepsilon_0 n_0 q}{2(-V_{GS} + \phi + V_{ch})} \right)^{1/2} \quad (10.9)$$

from equations (10.4) and (10.8). For the purposes of this section, $(\partial Q_G / \partial V_{GS})$ may legitimately be associated with the gate–source capacitance C_{GS}, above the knee voltage as the drain–source current does not depend markedly on drain–source voltage. Thus

$$C_{GS} = \left(\frac{\partial Q_G}{\partial V_{GS}} \right)_{V_{DS}} = \varepsilon \varepsilon_0 LW / h \quad (10.10)$$

The current gain β in common source configuration is given by

$$\beta = i_{DS}/i_{GS} = g_m / j\omega C_{GS} \quad (10.11)$$

where the use of lower-case symbols i_{DS} and i_{GS} denotes small-amplitude a.c. signals at angular frequency ω. The current gain drops to unity at frequency f_β where;

$$f_\beta = g_m/2\pi C_{GS} \qquad (10.12)$$

From equations (10.7) and (10.10) it may be seen that both g_m and C_{GS} are functions of gate–source bias; specifically, both decrease as the gate bias is made more negative, while their ratio remains constant, and

$$|\beta| = v_s/\omega L = 1/\omega\tau \qquad (10.13)$$

where τ is the time L/v_s taken for electrons to travel the length of the gate at their saturated velocity. The GaAs MESFET is a straightforward example of a charge-controlled device of the form described by Johnson and Rose,[9] wherein the current gain is limited by the transit time through the control region of the device. The transit time cut-off frequency f_T at which the current gain drops to unity is given, according to this description, by

$$f_T = 1/2\pi\tau \qquad (10.14)$$

In practice, although f_T and f_β are often equated in initial estimates of device performance, not all the charge imposed on the gate terminal is translated into mobile charge in the channel: gate fringing capacities and interelectrode capacities reduce the practical unity current gain cut-off frequency below the value given by equation (10.14). Nonetheless, equation (10.14) indicates the importance of a low transit time in maintaining high-frequency operation of the GaAs MESFET.

Experimental measurements (e.g. ref. 10) show that the ratio g_m/C_{GS} is a function of V_{GS} and V_{DS} even above the knee voltage. In practice, velocity saturation is not attained at the source end of the gate, and the properties of the constant mobility region of the channel must be taken into account. This causes g_m to diminish more rapidly with V_{GS} than equation (10.7) implies. Close to the active/substrate interface, there can be a diminution in carrier density or carrier mobility which reduces g_m. Conduction in the substrate or buffer layer reduces g_m close to pinch-off. The gate–source capacitance C_{GS} involves not only the depletion capacitance below the gate but also side-wall capacities due to fringing charges at the source and drain end of the gate. These do not diminish with increasing negative gate–source potential. The combined effect of these deviations from the simple model is to cause the current gain to diminish as the gate potential approaches pinch-off and the channel current approaches zero.

The potential V_{ch} at the source end of the gate is given by

$$V_{ch} = I_{DS} r_S \qquad (10.15)$$

where r_S is an effective source resistance, attributable to the source metallization and contact resistances and the resistance of the GaAs between the source contact and the gate. The mutual conductance g_m, measured between the source gate and drain terminals, is degraded relative to the intrinsic mutual conductance $g_{m(i)}$ according to

$$g_m = g_{m(i)}/(1 + r_S g_{m(i)}) \qquad (10.16)$$

There has recently been significant effort devoted to reducing the source resistance of GaAs MESFETs, particularly in the context of logic gates, to reduce this degradation in g_m. Further degradation of the mutual conductance in the common source configuration arises from impedance in the source grounding connection caused, for example, by an inductance L_S of a bond wire between source and ground. Reduction of such inductance is an important feature of device design and mounting, particularly in GaAs power FETs where many elemental devices are connected in parallel.

Figure 10.6 shows a rudimentary common source equivalent circuit for the MESFET, based on the account of device operation given above. Simplifications include neglect of any output admittance and gate-to-drain feedback admittance.

The input impedance Z_{in} is given by

$$Z_{in} = r_G + 1/j\omega C_{GS} + (r_S + j\omega L_S)(1 + g_{m(i)}/j\omega C_{GS}) \qquad (10.17)$$

where r_G is the gate electrode resistance.

The power gain G_P for a load resistance R_L is given by

$$G_P = \frac{\left(\dfrac{g_{m(i)}}{\omega C_{GS}}\right)^2 R_L}{r_G + r_S + g_{m(i)} L_S/C_{GS}} \qquad (10.18)$$

In practice, output conductance limits the load resistance that can be imposed on the output, and useful gain is not achieved at frequencies much in excess of the

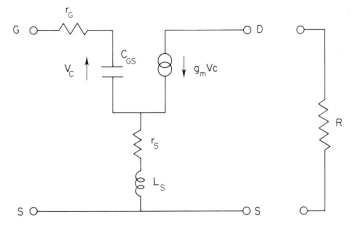

Figure 10.6 Rudimentary equivalent circuit for GaAs MESFET in common source

cut-off frequency f_β. Equation (10.18) shows the importance of reducing the 'parasitic' resistances r_S and r_G to reduce the power dissipated in the input circuit. Note also that common source inductance is transformed into an equivalent source resistance.

Common source operation is the configuration giving the highest stable microwave gain, although other operating configurations are used for specific applications. Microwave oscillators may be realized[11,12] by introducing suitable feedback, either externally or by operation in common drain. In addition, the control of the drain–source conductance by means of the gate electrode allows the GaAs MESFET to be used as a switch both in logic circuits and in microwave circuits: the versatility of the device is particularly valuable in monolithic integrated circuits, enabling a wide range of functions to be realized on a single semiconductor chip.

10.2 FABRICATION TECHNOLOGY, DEVICE PARAMETERS, AND EQUIVALENT CIRCUIT

Typical material structures for the active region of the device are shown in Figure 10.7. These are grown on the (100) face of a wafer, or on a face orientated a few degrees from the [100] direction to enhance nucleation. Epitaxial growth on semi-insulating GaAs usually involves initial growth of a high-resistivity or semi-insulating buffer layer, typically 3–5 μm thick, in order that the FET channel region shall be well separated from the disordered and defected surface of the starting wafer, followed by growth of the active n-type layer, which may be up to 0.7 μm thick and doped to give between 5×10^{22} and 4×10^{23} net donors/m^3 (Figure 10.7a). Material for use in short gate (< 0.5 μm) devices tends towards the higher end of the doping density range. In some cases an n$^+$ contact layer of thickness typically 0.1 μm and doping density 10^{24} donors/m^3 is grown on top of the active layer (Figure 10.7b) to reduce parasitic resistances. Similar structures, albeit with a rather thinner active layer, may also be realized by ion implantation. Selective implantation enables an active channel region to be realized with a doping density of the order of 10^{23} donors/m^3, and highly doped contact regions to be located immediately adjacent to the channel to minimize contact resistance (Figure 10.7c). In this last case, implantation is restricted to those areas necessary for fabrication of the device, leaving the rest of the wafer in its semi-insulating state.

Carrier confinement in the vertical direction is realized at the GaAs free surface with its built-in depletion, the gate depletion layer, and the interface between the active layer and substrate or buffer layer. If the Fermi level is pinned at mid-band in the buffer or substrate, confinement is achieved by a potential barrier of about 0.7 V. A low doped (e.g. 10^{20} donors/m^3) buffer layer would be expected to give less efficient confinement for energetic electrons. Carriers must also be confined in the horizontal direction, and this is realized by a mesa etch, by the use of selective

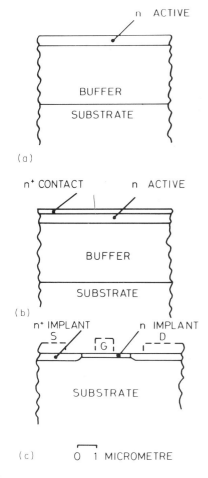

Figure 10.7 Some material structures for GaAs MESFET fabrication: (a) single layer on buffer; (b) contact and active layer on buffer; (c) selectively ion implanted, showing eventual electrode locations

area implantation, or by use of damage implants. It is necessary to ensure that bonding areas and connections to the active area of the device are located on high-resistivity semiconductor, i.e. the buffer layer or substrate.

Chemical etches for GaAs normally consist of an oxidizing agent, such as hydrogen peroxide, mixed with an acid (e.g. sulphuric acid) or base (e.g. ammonium hydroxide) that reacts with oxides of Ga and As. Figure 10.8 shows schematically etch profiles beneath a photoresist mask for three orientations in the (100) plane. This anisotropy—which is found with most GaAs etches—arises[13] because (111) crystal planes revealed by etching beneath a (100) surface are of two kinds: A planes (Ga atoms) and B planes (As atoms). The former planes etch more slowly and are thus revealed by the etching process. The [011] directions lie in natural cleavage planes, and slopes that are either overhung or

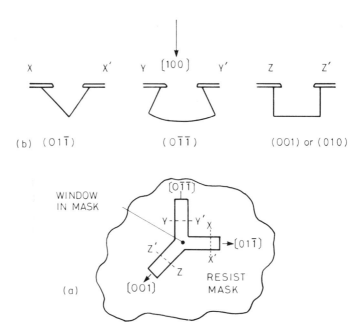

Figure 10.8 Schematic etch profiles in GaAs: (a) directions in (100) plane, showing etch resist mask and window; (b) cross sections in planes shown, normal to (100) plane

ramped are obtained along orthogonal directions parallel to natural cleavage planes. To achieve a uniform drain–source current across a wafer, the thickness of the active layer may be controlled by the use of a chemical etching process involving successive anodic oxidation and stripping. This process automatically references the active layer thickness to the depletion depth in the active layer and is thus particularly useful for 'normally off' MESFETs for logic applications.[14]

Proton isolation[15] is a technique for rendering semi-insulating a doped GaAs layer, by creating damage centres that trap electrons and holes. A typical proton implantation fluence is of the order of 5×10^{18} m^{-2}, with an implant energy of 90 kV at the GaAs surface, giving isolation to a depth of approximately 0.9 μm. It is necessary to mask off the active region of the device from the proton beam, and a thick (~ 1 μm) gold layer supported by a dielectric has been used. The mask may be removed by etching away the supporting insulator after use. While the use of selective ion implantation in principle replaces a separate isolation technology, proton isolation has been used in conjunction with selective ion implantation to reduce interaction ('back-gating') between closely spaced active regions.[15]

The metallizations required to complete the basic MESFET structure are the source and drain ohmic contacts and the gate, though overlays are frequently used to reduce metallization resistance and to thicken up bonding pads. The gate length in microwave FETs is normally between 1.5 and 0.25 μm. Control of such dimensions by wet chemical etching is not straightforward and it has been usual to define not only the gate but also the source and drain contacts by the process of 'float off' or ('lift off'), using optical or electron beam lithography. Figure 10.9 indicates the list-off technique by contact photolithography and electron beam lithography, applied to definition of the source and drain electrodes and using positive resists in each case. Contact photolithography is normally restricted by diffraction to dimensions of 0.5 μm or above, comparable with the wavelength of the light used (0.4 μm for a conventional source and 0.2–0.26 μm for a deep UV source). The dimensions that can be realized in electron beam lithography are limited by scattering in the resist, as each electron undergoes many scattering events both within the resist and at the resist/substrate interface. As a consequence, while conventional photoresist profiles after development tend to have vertical or ramped edges, electron resist profiles can be undercut (Figure

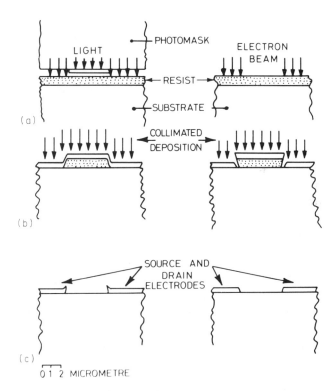

Figure 10.9 Metallization using the 'lift-off' process: (a) exposure by optical or electron beam lithography; (b) collimated deposition of metal after development of resist pattern; (c) electrodes remaining after dissolution of resist and removal of excess metal

10.9b). When metal is evaporated onto the wafer, it adheres in each case to the substrate and the top layer of the resist, but the side-walls of the photoresist are also coated. The undercut portion of the electron resist edge is not coated and the electron resist can be dissolved by immersion in suitable solvent (e.g. acetone for polymethylmethacrylate, a common electron resist), whereupon the unwanted metal floats off into the solvent (Figure 10.9c). Using photoresist, the thinly coated side-wall needs to be broken or torn and somewhat ragged edges can result, although techniques have been developed to give clean edges to patterns defined by float-off using optical lithography.

The most commonly used ohmic contacts are based on the formation of a germanium-rich layer at the surface of the semiconductor, and require that the metallization (e.g. In/Ge/Au or Ge/Au/Ni multilayers) be heated to a temperature of about 450°C for a short period of time. This is not compatible with the integrity of many Schottky barrier contacts to GaAs, in particular those including a layer of Au, and ohmic contacts are usually formed before the gate is deposited, preceded by a light chemical etch of the ohmic contact area. A source–drain gap of approximately four times the gate length (i.e. 4 μm for 1 μm long gate) is used to relax constraints on subsequent gate alignment.

Using optical or electron beam lithography techniques, the resist window for the gate evaporation is located between the source and drain pads, with an offset towards the source in certain processes to reduce source resistance. The gate resist window may be used (Figure 10.10a) as a mask for a self-aligned etch to generate a recess in which the gate is subsequently located. Thr recess profile depends on the etch used and the orientation of the gate. There are a number of motivations for using a recessed gate technology:

(a) The gate recess etch may be used to remove an n^+ contact layer.

(b) The GaAs immediately below a free surface is depleted by the presence of surface states having a density of about 10^{16} m^{-2}, giving a surface depletion depth of ~ 0.1 μm in GaAs with 10^{23} donors/m^3. For the channel current to be dominated by the action of the gate rather than surface depletion, the gate should be recessed by approximately this amount below the free surface.[16]

(c) A gate recess has been claimed[17] to give a high gate–drain breakdown particularly if the recess has smoothly ramped sides.

(d) Recessing the gate enables material between the gate and the ohmic electrodes to be thicker than that below the gate, giving a low-resistance structure, and alleviating any potential field concentration at the drain electrode (see Section 10.6).

There are disadvantages accruing to the gate recess etch:

(a) A chemical etching process may be a cause of non-uniformities over a wafer or within a multi-device integrated circuit.

Gaאs MESFETs 375

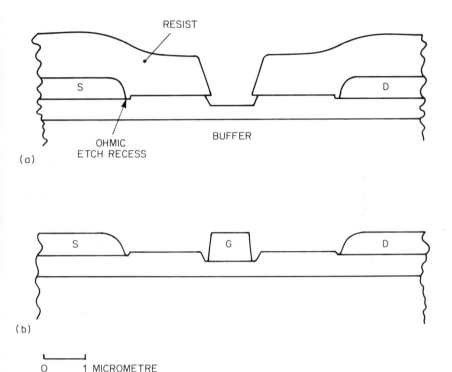

Figure 10.10 Gate electrode definition: (a) self-aligned gate recess; (b) gate metallization defined by lift-off. The etch profile shown is appropriate to the use of an anisotropic chemical etch and orientation of the gate along the [01$\bar{1}$] direction

(b) Calculations[18] have suggested that too short a recess may decrease the breakdown voltage of power FETs by constraining the extent to which the depletion layer may move towards the drain.

(c) A deep recess etch was shown to increase gate–drain capacitance and drain conductance in the power MESFET studies of Macksey *et al.*,[19] although the mechanism is not fully resolved.

There is still considerable discussion in the literature concerning the recess etch; most discrete devices now have such a recess, but this trend may not be followed by GaAs integrated circuits in the future.

Following any etching of a gate recess, the gate metallization is deposited (Figure 10.10b). Typical gate metallizations include aluminium or a multilayer structure such as Ti/Pt/Au. After floating off the excess metal, overlays and bonding pads are defined using a low-resistance overlay metallization scheme based on gold. Figure 10.11a shows a photomicrograph of a low-noise MESFET (Plessey GAT VI). Figure 10.11b is a scanning electron micrograph of the channel

Gallium arsenide

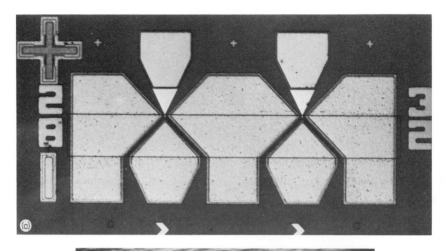

Figure 10.11 (a) Plessey GAT VI low-noise MESFET. (Courtesy of Plessey Research (Caswell) Ltd.) (b) Scanning electron micrograph of the channel region of a GaAs MESFET fabricated at RSRE

region of an experimental GaAs MESFET made at RSRE. Both of these devices embody fabrication technology broadly of the form described above.

The conventional processing schedule requires a good alignment of the gate between the source and drain contacts. To overcome this requirement and to realize a small source–drain spacing with associated low parasitic resistances, a number of self-aligned processes have been developed. Two of these are briefly illustrated below.

Figure 10.12a illustrates a self-aligned gate technology of the form used commercially by Microwave Semiconductor Corporation,[20,75] whereby an ohmic contact is deposited and alloyed and a gap, equal to the eventual gate length, is defined in this metallization. This is followed by a chemical etch that undercuts the source and drain contacts and enables a subsequent gate

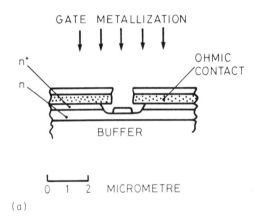

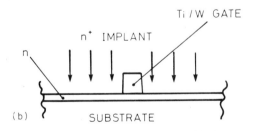

Figure 10.12 Two self-alignment techniques for MESFET fabrication. (a) Self-aligned gate using an undercut etched recess. (From ref. 75. Reproduced by permission of *Microwave Journal*.) (b) Contact implant self-aligned about gate metallization. (Based on figure 1 in M. Abe *et al.*, New technology towards GaAs LSI/VLSI for computer applications, *IEEE Trans. Microwave Theor. Tech.*, **MTT-30**, 992–8, July (1982). Copyright © 1982 IEEE. Reprinted with permission.)

metallization to be defined without any further alignment steps. Fujitsu[21] have reported a self-aligned implantation process, illustrated schematically in Figure 10.12b, whereby n$^+$ contact regions were aligned to a Ti/W gate. Activation of this implant at 800°C did not degrade the quality of the Schottky barrier contact. Self-alignment of the n$^+$ implant eases constraints on the subsequent source–drain metallization alignment.

Reduction of the gate length of a MESFET carries, in a conventional structure, the penalty of increasing the gate resistance and inductance. This may be overcome by limiting the gate width of each elemental FET or 'unit cell', and joining many such cells in parallel. Alternatively, T-shaped gate technologies have been developed, which use a double-layer gate metallization, the lower layer of which may be undercut by chemical plasma etching to give a short gate with a conducting cap.[22] Huang et al.[23] have reported the use of plating to achieve a mushroom-shaped gate and so reduce gate resistance in GaAs MESFETs operating up to 40 GHz.

In addition to the evolutionary development of GaAs MESFET fabrication, major steps forward in the processing technology of GaAs MESFETs may be anticipated. These include a widespread use of ion implantation and probably molecular beam epitaxy to achieve more uniform material parameters over a wafer, exploitation in GaAs of implantation-based self-aligned technologies[21,24] which are common in Si processing, and increased use of dry processing to enhance dimensional control and uniformity compared with chemical etching processes. For very short gate (less than 0.25 μm) devices, x-ray lithography becomes an attractive technology, permitting very fine features with vertical sidewalls to be defined in resist. More speculatively, non-alloyed ohmic contacts realized for example by molecular beam epitaxy (MBE) growth techniques[25] may be used to eliminate lateral diffusion effects during alloying and possibly lead to enhanced reliability.

The choice of basic device parameters (e.g. gate length, doping density) is governed principally by the operating frequency. The gate length L has the most influence on the maximum useful operating frequency of the GaAs MESFET. This dimension constrains the active layer thickness beneath the gate, as shown schematically in Figure 10.13. A thin active layer increases the source resistance, and also the fraction of the channel close to the interface with the buffer layer. (However, 'normally off' MESFET logic gates use a thin active layer, depleted at zero gate bias.) Too thick an epitaxial layer brings a number of problems including a severe drop in mutual conductance between the open channel and pinched-off conditions. The gate–source capacitance includes side-wall effects that increase with depletion depth (Figure 10.13b) and parasitic interpad capacitances, and so the drop in g_m is not balanced by an equivalent drop in C_{GS} and the gain falls. In an extreme case of a thick active layer the device may not be pinched off before the vertical field in the depletion layer exceeds the value appropriate to avalanche breakdown within the depletion region. Typically an

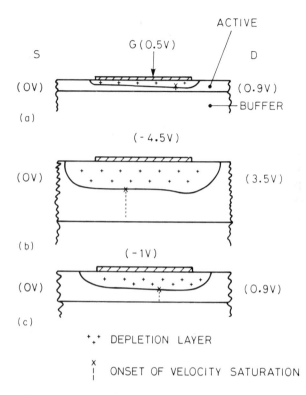

Figure 10.13 Effect of active layer thickness: depletion region profiles at $I_{DSS} = I_0/2$. (a) Thin active layer, showing long, high-resistance channel in constant mobility regime. (b) Thick active layer, showing excessive depletion layer spreading. (c) Compromise value, $L/a = 3$

active layer thickness of approximately $L/3$ (Figure 10.13c) is chosen as a compromise between these constraints. For a given L and active layer thickness a below the gate, the channel current rises with doping density n_0. At a given operating current, the mutual conductance rises with doping density and parasitic resistances fall as the doping density increases. However, the breakdown voltage of a Schottky barrier decreases with increase in doping density,[26] and a compromise value is found, in the region of 10^{23} donors/m³. Doping densities above this value are generally associated with small-signal MESFETs at high frequencies, where gain is at a premium, and lower values are reported in, for example, 4 GHz power MESFETs.[27]

The gate–source input circuit may be considered to be a lossy transmission line, with capacitance per unit length dominated by C_{GS} and inductance appropriate to the source and gate metallizations. Loss occurs by dissipation in the gate metal resistance and the charging resistance for C_{GS} through the semiconductor and ohmic contact. According to Fukui[28] the optimum unit gate width is attained when the gate metallization resistance equals the source series resistance, and diminishes with reduction in gate length at a given gate metallization height. Experimental studies have indicated that for GaAs power MESFETs, where it is desirable to maximize the unit cell gate width to decrease the number of cell interconnections and improve the heat dissipation properties, unit gate widths up to 500 μm can be used at 4 GHz,[5] 200 μm gate width at 8 GHz,[29] and 150 μm gate width at 15 GHz.[19] Low-noise microwave MESFETs tend to have rather smaller unit gate widths, since there is less emphasis on power handling capability, with 150 μm at X band and 75 μm for 26–40 GHz[30] being typical values.

From the above discussion of the device configuration, a simple equivalent circuit for the GaAs MESFET may be built up, and is shown in Figure 10.14a. In this equivalent circuit, r_G is the gate metallization resistance, r_i is the 'intrinsic

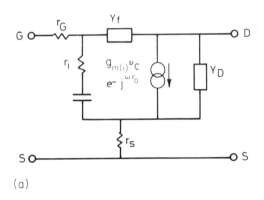

(a)

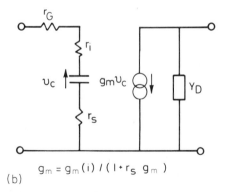

$g_m = g_m(i) / (1 + r_S g_m)$

(b)

Figure 10.14 Equivalent circuit representations of the GaAs MESFET: (a) including feedback components, Y_F and r_S; (b) unilateralized simplification of (a)

resistance' and represents the resistance of the semiconductor below the gate, C_{GS} is the gate–source capacitance, identified with the depletion layer capacitance, r_S is the sum of the source contact resistance and resistance of the semiconductor between the source contact and the gate, Y_F is the feedback admittance (principally gate–drain capacitance), and Y_D is the output admittance of the MESFET, which consists partly of a capacitive component between electrodes and partly of output conductance g_D. This latter arises not only from the nature of electron transport in the channel but also from current injected into the buffer layer. At high drain voltages, incipient avalanche breakdown also contributes to Y_D.

The current generator in the output circuit i_0 has magnitude $g_m v_c$, where v_c is the potential drop across the gate–source capacitance, and a phaseshift $\exp(-j\omega\tau_0)$ representing the transit time through the velocity-saturated region of the channel.

This equivalent circuit represents in simplified form the MESFET unit cell; further allowance needs to be made, for example for bonding pad capacitances, bond wire inductances, etc. Typical values for a 1 μm gate MESFET at I_{DSS}, adapted from data given by Pengelly in chapter 2 of ref. 7 and normalized to 1 mm gate width, are $r_G = 0.61\ \Omega$, $r_i = 0.9\ \Omega$, $C_{GS} = 1.33$ pF, $r_S = 1.5\ \Omega$, $Y_F = j\omega \times 0.03$ pF, $Y_D = 6.6$ mmho $+ j\omega \times 0.23$ pF $g_m = 100$ mS, and $\tau_0 = 3$ ps.

A full analysis of the equivalent circuit is beyond the province of this section. However, a simple unilateralized equivalent circuit, valid well below f_T, is shown in Figure 10.14b where the feedback effects of source resistance and gate–drain capacity have been neglected. From the circuit of Figure 10.14b, the power gain G_P is given by

$$G_P = \tfrac{1}{4}(g_m/\omega C_{GS})^2 \{1/[g_D(r_G + r_i + r_S)]\} \quad (10.19)$$

where the real part R_L of the load impedance has been optimized such that $g_D R_L = 1$. The principal consequence of neglecting the effect of feedback in deriving the equivalent circuit of Figure 10.14b is that no account is taken of the potential low-frequency instability of the GaAs MESFET. This is a feature that may be exacerbated by elements outside the unit cell, such as source bond wires, and is an important element in microwave amplifier design. Further details are to be found in, for example, ref. 7.

The equivalent circuit element values given above were derived at zero gate–source bias (i.e. at I_{DSS}). The bias dependence of equivalent circuit parameters has been investigated by, for example, Willing et al.[10] and Figure 10.15 shows the bias dependence of the instrinsic gate–source capacitance C_{GS}, g_m and r_i for a GaAs MESFET with a 1.7 μm gate, a pinch-off voltage of -5.1 V, and $I_{DSS} = 225$ mA mm^{-1}. The values are taken from the data of ref. 10, and normalized to 1 mm gate width. The values of V_{DS} correspond approximately to the knee voltage ($V_{DS} = 2$ V) and 2 V increments above that value. The reduction in both g_m and C_{GS} with increasing negative gate bias is explained in general terms by the deepening of the gate depletion layer; similarly the rise in r_i as the gate is

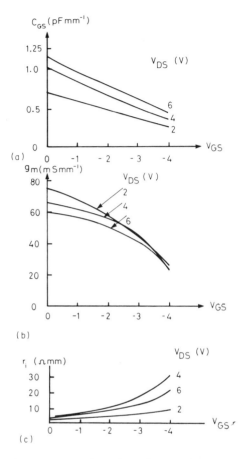

Figure 10.15 Bias dependence of GaAs MESFET equivalent circuit parameters: (a) intrinsic gate–source capacitance; (b) mutual conductance; (c) voltage-dependent component r_i of input resistance

made more negative is a consequence of the reduction in the height of the undepleted channel and hence a rise in the resistance of the non-velocity-saturated region of the channel. As the drain–source bias is increased, the gate depletion layer protrudes further beyond the gate, the control region of the device becomes longer, and the charge it contains increases. Hence C_{GS} increases with V_{DS} as shown. The reduction in g_m with V_{DS} at high currents is also attributable to the increase in the length of the charge control region and an increase in the transit time; at lower currents this is counterbalanced by changes in the degree of carrier accumulation within the channel. The overall effect of the changes in C_{GS}

GaAs MESFETs

and g_m is to reduce the unity current gain cut-off frequency f_β (equation (10.12)) as the drain–source bias is increased.

The strong bias dependence of equivalent circuit elements is significant in determining d.c. operating conditions for small-signal devices and the terminal impedances required for optimized large-signal operation.

10.3 GaAs MESFET THEORY—ANALYTICAL METHODS

In this section, and the next, an account is given of some of the theoretical methods used to explain and predict the performance of GaAs MESFETs. These methods fall broadly into two categories: the analytical or semi-analytical methods derived from studies of long gate silicon FETs, and computer simulation techniques particularly appropriate to short (< 1 μm long) gate devices. The latter are briefly considered in Section 10.4.

Analytical descriptions of the MESFET action work on the basis of simplifying assumptions concerning the electric field distribution within the device, the carrier dynamics, and the profile of the channel. It turns out that, when the drain–source potential V_{DS} is below the knee voltage and carriers travel with constant mobility, a straightforward account of device operation can be given and the d.c. characteristics can be derived. Operation above the knee voltage is considered by dividing in channel into three regions, as shown in Figure 10.16, determining the properties of each region, and using boundary conditions between regions I, II, and III to evaluate unknown parameters and coefficients. In region I, carriers are transported with constant mobility between the source end of the gate and the velocity saturation point $x = L_1$ where the electric field rises to the saturation field E_S. In regions II and III, the carriers travel at their saturated velocity, the joining point between regions II and III coming at $x = L$, the drain end of the gate. Region III terminates when the electric field in the channel falls to the saturation field. Before considering the details of the analysis of the three-region model of the GaAs MESFET, a summary of the model and its description of the device will be given.

The velocity–field characteristic of GaAs is represented by the two-piecewise linear approximation of Figure 10.4. Figure 10.16 includes a schematic representation of the longitudinal electric field E_x as a function of position within the channel. There is an initial increase in electric field as carriers enter the constricted region beneath the gate, and the electric field continues to rise beneath the gate until the field in the channel reaches the saturation field E_S. At the knee voltage, the velocity saturation point is located at the drain end of the gate. At higher drain–source potentials, the velocity saturation point moves towards the source end of the gate, and the electric field in regions II and III increases to accommodate the drain–source potential.

In region II, carriers travel at their saturated velocity v_s. In order to satisfy Gauss's law and current continuity within the channel, the increase in the

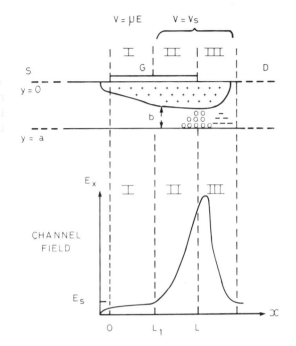

Figure 10.16 Three-region model of the GaAs MESFET for drain–source potential above the knee voltage, showing regions of carrier depletion and accumulation, and the electric field profile within the channel

longitudinal electric field E_x must be accompanied by a reduction in the channel height and by carrier accumulation beyond the charge-neutral value of region I. These arguments predict that the electric field in the channel shall peak close to the drain end of the gate. Beyond the drain end of the gate, in region III, the carrier density falls as the channel height increases and, as the channel height rises above that at the velocity saturation point, the mobile carrier density drops below the charge-neutral value. This transition from carrier accumulation to carrier depletion is accompanied by a sharp drop in the electric field in the channel. Charge neutrality is restored when the electric field falls to the saturation field and electrons re-enter the constant mobility regime.

The strong saturation in drain–source current above the knee voltage is a consequence of most of the drain–source potential V_{DS} being dropped across the accumulation/depletion regions in the channel, sometimes called the 'stationary domain'. Increasing V_{DS} emphasizes these regions, and also increases the length of the velocity-saturated region of the channel.

Further detail, and some quantification, of this description of the MESFET above the knee voltage is given later in this section. It is appropriate first to consider the analysis of the FET channel below the knee voltage, i.e. where the electric field is below that required for velocity saturation. This analysis is based on the 'gradual channel' model used by Shockley[6] in his original analysis of a silicon FET. Figure 10.17 shows the symmetrical structure conventionally used for analysis. Note that the boundary condition at the plane of symmetry is that there shall be no transverse field, in contrast to carrier confinement by a potential barrier between, for example, n-type and semi-insulating GaAs. In using the gradual channel approximation to analyse the GaAs MESFET, the following assumptions are generally made:

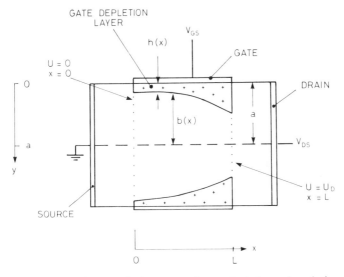

Figure 10.17 Symmetrical structure for gradual channel analysis of MESFET

(a) The depletion layer edge is abrupt, i.e. all dimensions are significantly greater than the Debye length (0.014 μm in GaAs with 10^{23} donors/m³ at 300 K).

(b) The transverse electric field E_y in the undepleted semiconductor is very much less than the longitudinal field E_x so that at any point x the potential in the channel is independent of y, whereas the longitudinal electric field in the depletion layer may be neglected relative to the transverse electric field.

(c) The electron velocity at each point is directed longitudinally, consistent with (b) and implying that the angle made by the depletion layer edge with the x-axis is small.

Then, for a channel current I_{DS} per half-device in Figure 10.17, the channel potential $V(x)$ obeys

$$I_{DS} = G_0 L\,(1 - h/a)\,dV/dx \tag{10.20}$$

G_0, the conductance of a fully open channel, is given by

$$G_0 = n_0 q \mu\, Wa/L \tag{10.21}$$

From a one-dimensional integration of Poisson's equation between the depletion layer edge and the gate, justified by assumption (b) above,

$$V = V_{GS} - \phi + V_0 h^2/a^2 \tag{10.22}$$

If the source and drain potentials are assumed to be translated to the planes $x = 0$ and $x = L$ respectively (i.e. source and drain resistances are neglected), then

$$I_{DS} = G_0 [V_{DS} - \tfrac{2}{3}(V_{DS} - V_{GS} + \phi)^{3/2} + \tfrac{2}{3}(-V_{GS} + \phi)^{3/2}] \tag{10.23}$$

There are two limits to the applicability of equation (10.23). First, from equation (10.20), at a sufficiently high normalized drain potential, $h = a$ at the drain end of the gate. The channel is completely 'pinched down', i.e. it has a zero height. This occurs for

$$V_{DS} = V_0 + V_{GS} - \phi \tag{10.24}$$

In the absence of velocity saturation this would require a finite current to be carried by an infinitesimally thin sheet of carriers moving with infinite velocity in infinite longitudinal field. This condition has been analysed by, for example, Lehovec and Miller.[31]

A second limitation to the applicability of equation (10.23) arises from the phenomenon of velocity saturation. If (Figure 10.4) it is assumed that the electron velocity saturates sharply at v_s at electric field E_s, then from equations (10.20) and (10.22) velocity saturation occurs first at the drain end of the gate at a drain potential given by

$$V_{DS} = V_{GS} - \phi + V_0[(I_0 - I_{DS})/I_0]^2 \tag{10.25}$$

A measure[32] of the relative importance of the roles of velocity saturation and pinch-down is the 'saturation index' ξ, the ratio of the potential drop along the gate at the saturation field to the potential V_0 required totally to deplete the channel, i.e. $\xi = E_s L/V_0$. The smaller ξ, the greater the importance of velocity saturation in limiting the source–drain current. To illustrate the calculations made using analytical descriptions of the MESFET, a model GaAs MESFET structure is considered, with parameters listed in Table 10.1.

Figure 10.18a shows the d.c. characteristics of a GaAs MESFET up to pinch-

Table 10.1 Model GaAs MESFET

Epitaxial layer thickness	a	0.3 μm
Doping density	n_0	$10^{23}/m^3$
Low-field mobility	μ	$0.35\ m^2\ V^{-1}\ s^{-1}$
Saturation field	E_s	$3 \times 10^5\ V\ m^{-1}$
Saturated velocity	v_s	$1.05 \times 10^5\ m\ s^{-1}$
Built-in voltage	ϕ	0.6 V
Channel depletion voltage	V_0	6.26 V
Saturation current	I_0	$504\ A\ m^{-1}$
Gate length	L	1 μm
Saturation index	ξ	0.048
Saturated drain–source current	I_{DSS}	$325\ A\ m^{-1}$

down as calculated from equation (10.23). The solid lines indicate the characteristics generated in the absence of velocity saturation; the dotted line indicates the region of the characteristics in which the channel field is below E_s and in which equation (10.23) may legitimately be applied. The strong effect of velocity saturation in limiting I_{DS} may clearly be seen.

Figure 10.18b shows the transfer characteristic (I_{DS} as a function of V_{GS}) for the GaAs MESFET of Table 10.1, both at the channel pinch-down condition according to equations (10.23) and (10.24) and at velocity saturation, which effectively corresponds to the limiting value of I_{DS}. A similar comparison for a material having channel mobility $0.08\ m^2\ V^{-1}\ s^{-1}$, saturation field $1.3 \times 10^6\ V\ m^{-1}$, and saturated velocity $1.05 \times 10^5\ m\ s^{-1}$ is also shown. The effect of reducing the mobility to a value representative of a Si MESFET is to reduce strongly the role of velocity saturation in determining the channel current. Note that, even at equal saturated velocity, I_{DS} and g_m increase with channel mobility; for the lower mobility material the potential rise along the constant mobility region of the channel is greater, and hence the channel is further pinched down, at the higher saturation field. To determine the drain potential and drain–source current at the knee voltage, equations (10.23) and (10.25) must be solved simultaneously. Turner and Wilson[33] have analysed the d.c. characteristics, mutual conductance and intrinsic gate–source capacitance at the onset of velocity saturation, i.e. along the dotted line of Figure 10.18a. They were able to derive closed-form expressions for the mutual conductance and for the gate–source capacitance and unity current gain cut-off frequency at $I_{DS} = I_{max}$. However, equations (10.23) and (10.25) do not yield explicit general solutions for I_{DS} and V_{DS} at the knee voltage and need to be solved numerically or graphically. Thus, the onset of velocity saturation marks the boundary of truly analytical theories for GaAs MESFET operation. Semi-analytical models for the GaAs MESFET, valid at drain–source potentials above the knee voltage, have been developed by Pucel et al.[32] and by Sone and Takayama.[34,35] Although such models necessarily involve some lack of self-consistency and considerable simplification of the device

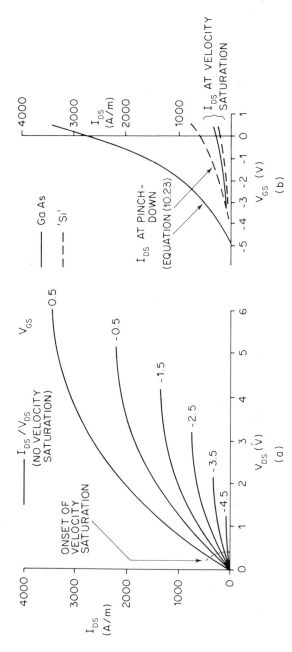

Figure 10.18 The effects of channel pinch-down and velocity saturation in limiting the drain–source current. (a) L_{DS}/V_{DS} characteristics for a GaAs MESFET according to equation (10.23), showing the small region of validity of that equation. (b) A comparison of the transfer characteristics of a GaAs and a 'Si' MESFET, showing the reduced importance of velocity saturation in lower mobility material

physics, they have been shown[32,35] to give good accounts of the d.c. and microwave properties of GaAs MESFETs with gate lengths in excess of 1 μm. The three-region model of Sone and Takayama, which forms the basis for the discussion of this section, considers the effects of carrier accumulation and depletion in the channel.

Consider a small section of the channel, bounded by the depletion layer edge and a notional plane of symmetry at $y = a$ (Figure 10.16). The electric field normal to these boundaries is zero. Thus, application of Gauss's law to the channel gives

$$\frac{1}{b}\frac{db}{dx} + \frac{1}{E_x}\frac{dE_x}{dx} = \frac{n-n_0}{|E_x|}\frac{q}{\varepsilon\varepsilon_0} \quad (10.26)$$

where E_x is the longitudinal electric field, which is assumed constant over the channel height $b(x)$, and n is the mobile electron density. Note that $E_x < 0$. Equation (10.26) and the requirement for current continuity are satisfied in region I of the channel with $n = n_0$. It may be seen that, if db/dx is positive, or negative and sufficiently small, an increase in the electric field in the channel is associated with carrier accumulation ($n > n_0$). To a good approximation, this represents the situation in region II of the channel. Beyond the drain end of the gate, according to the model of Sone and Takayama, the protrusion of the depletion layer into the gate–drain space takes the form of a quarter of an ellipse. As the channel height increases in region III, the carrier density falls. The electric field rises to a maximum value shortly beyond the drain end of the gate, then falls rapidly with the transition from mobile carrier accumulation to carrier depletion in the channel, when the channel height beyond the drain end of the gate exceeds that at the velocity saturation point. As the electric field drops to the saturation field, charge neutrality is restored and region III is considered to terminate. The description advanced above is illustrated later by Figures 10.20 to 10.22, based on data output from application of the three-region model of Sone and Takayama to the model MESFET of Table 10.1. A summary of the analysis of the three-region model is given immediately below.

Figure 10.19 shows the three-region MESFET, with dimensions used in the analysis. It is convenient to assume initially that the gate depletion layer in region II has a rectangular cross section, leaving a uniform channel height b_2 in region II. To preserve current continuity between the velocity saturation point and the drain end of the gate in the presence of carrier accumulation, the channel height b_1 at the velocity saturation point is different from b_2 as shown. Having chosen a value for L_1, a gradual channel solution for the channel potential in region I may be found, terminating at the condition $E_x = -E_s$ at $x = L_1$. The channel potential at the velocity saturation point and drain–source current corresponds to the dotted line in Figure 10.18a, with $G_0 \equiv n_0 \mu q\, Wa/L_1$.

The potential in the depletion layer of region II is that appropriate to a solution of Poisson's equation in a rectangular box, with uniform charge density $n_0 q$. Such

Gallium arsenide

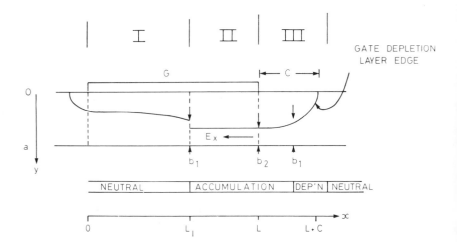

Figure 10.19 Dimensions for analysis of the three-region GaAs MESFET model

a solution may be written as

$$V(x, y) = V_{GS} - \phi + V_0 \left(\frac{(a-b_2)^2 - (a-b_2-y)^2}{a^2} \right) + \psi(x, y) \quad (10.27)$$

The first three terms represent a one-dimensional integration of Poisson's equation

$$\nabla^2 V = -n_0 q/\varepsilon\varepsilon_0$$

with $\partial V/\partial y = 0$ at the depletion layer edge and $V = (V_{GS} - \phi)$ at the gate electrode. $\psi(x, y)$ is a solution of Laplace's equation

$$\nabla^2 \psi = 0$$

within the depletion layer in region II. The potential $\psi(x, y)$ must be zero at the gate electrode and in the plane $x = L_1$ of the velocity saturation point. The electric field E_y normal to the depletion layer edge at $y = a - b_2$ must also be zero. At the velocity saturation point, $x = L_1$, $y = a - b_2$,

$$\left. \frac{\partial \psi}{\partial x} \right|_{L_1, a-b_2} = E_s \quad (10.28)$$

The increase in the channel potential over the length of region II is $\psi(L, a - b_2)$. A suitable form for $\psi(x, y)$ is[32,34,35]

$$\psi = \sum_{m=0}^{\infty} A_m \sinh\left(\frac{(2m+1)\pi(x - L_1)}{2(a - b_2)} \right) \sin\left(\frac{(2m+1)\pi y}{2(a - b_2)} \right) \quad (10.29)$$

The electric field lines associated with this solution are generated on positive charges in the depletion layer in region III beyond the drain end of the gate, and terminate at negative charges on the gate electrode in regions I and II. The electric field in the channel is given by differentiation of equation (10.29). Pucel et al.[32] consider only the first terms in equation (10.29) and take a solution for ψ appropriate to $b_2 = 0$. They do not consider explicitly accumulation effects in the channel. A_0 is found from (10.28) and V_{DS} evaluated (apart from a parasitic resistive drop) at $x = L$. By contrast, Sone and Takayama[34,35] take the first two terms in (10.29) and consider the protrusion of the depletion layer into region III to be a quarter of an ellipse of semi-axes $(a - b_2)$ and c. There are thus four unknown parameters to be evaluated, A_0, A_1, b_2 and c. The boundary conditions between regions I, II, and III are used for this purpose. Equation (10.28) represents one such boundary condition. The others come from the requirement for current continuity between the charge-neutral channel at $x = L_1$ and the carrier-accumulated channel at $x = L$; from the maintenance of continuity in the second differential of the electric field at $x = L$; and from the requirement that the integral of E_x across the plane $x = L$ in the depletion layer (from (10.29)) shall be related by Gauss's law to the charge contained in the depletion layer of region III. Finding a set of values satisfying these conditions is an iterative process, suitable for a small desktop computer.

Having determined the geometric parameters b_1, b_2 and c of the three-region model, the potential rise in region III may be determined by applying Gauss's law to the channel in region III, with the local carrier density $n(x)$ determined by current continuity. Current continuity is not satisfied in region II except at $x = L$, and in Figures 10.20 and 10.21 the channel height has been drawn as a continuously varying quantity, to satisfy current continuity. This invalidates, strictly, the potential function in (10.29), but enables more realistic steady-state velocity–field curves to be used if necessary. The difficulty in satisfying both current continuity and a consistent channel potential solution highlights a major weakness of analytical MESFET methods. Nonetheless, it may be seen that comparatively simple physical arguments may be used to build up a semi-analytical account of the MESFET which explains many of the features revealed by more sophisticated numerical modelling techniques. The model of Sone and Takayama has been applied to the model MESFET, with parameters shown in Table 10.1, and the results are presented below.

Figure 10.20 shows the depletion layer, channel potential, and carrier density profiles for the model MESFET, using the two-piecewise approximation to the velocity–field characteristic, at a moderate drain–source voltage (0.9 V). The protrusion of the gate depletion layer beyond the source end of the gate has been taken as a quadrant. In Figure 10.20a, the depletion layer profiles show a rather small variation in channel height over the length of the gate, and the depletion layer extensions at each end of the gate increase with depletion layer depth. Figures 10.20b and 10.20c show that most of the drain–source potential is taken up about the drain end of the gate. As the gate voltage is made more negative, the channel potential profile becomes smoother, and the accumulation region is less

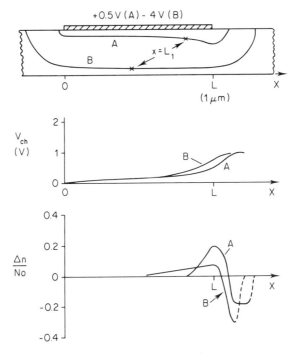

Figure 10.20 Analysis of the model MESFET for $V_{DS} = 0.9$ V with $V_{GS} = +0.5$ V, $I_{DS} = 388$ A m^{-1} (case A) and $V_{GS} = -4$ V, $I_{DS} = 65$ A m^{-1} (case B): (a) depletion layer profiles; (b) potential distribution along the channel; (c) carrier density profile along the channel

pronounced. For the open-channel case (case A), carrier depletion in the channel extends beyond the protrusion of the gate depletion layer to satisfy current continuity, whereas, close to pinch-off (case B), the electric field drops to E_s below the depletion layer protrusion. Figure 10.21 presents the same information for $V_{DS} = 2.5$ V. Increasing the drain–source voltage increases the fraction of the gate length over which velocity saturation is present, the extent of the depletion layer protrusion beyond the drain end of the gate, and the magnitude of the carrier accumulation and depletion regions in the channel. It is clear from Figures 10.20 and 10.21 that the extent of the region of the device controlled by the gate electrode, i.e. the charge control region, extends not only beyond the physical length of the gate but, under some bias conditions, beyond the extent of the gate depletion layer (e.g. Figures 10.20 and 10.21, case A). This contributes to a reduction in the unity current gain cut-off frequency at high drain–source potentials (see Section 10.2).

Figure 10.22 shows electric field profiles for $V_{DS} = 2.5$ V at $V_{GS} = 0.5$ V and -4 V, i.e. the conditions of Figure 10.21. The maximum electric field occurs just

GaAs MESFETs

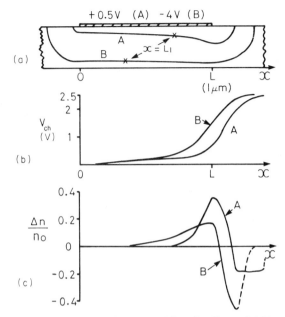

Figure 10.21 As Figure 10.20 but for $V_{DS} = 2.5$ V

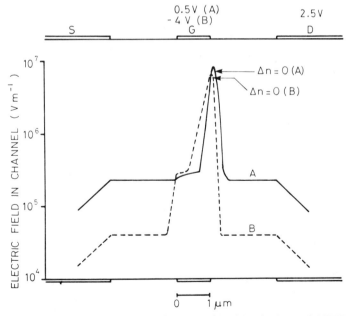

Figure 10.22 Electric field as a function of position in the model FET for the conditions of Figure 10.21. The figure shows the crossover from carrier accumulation to depletion ($\Delta n = 0$)

beyond the drain end of the gate, and falls to a low value (which, for the open-channel case, is nonetheless comparable with the saturation field E_s) in the semiconductor between the contacts. A source–gate gap of 2 μm is assumed and the electric field decay beneath the ohmic contacts is that appropriate to a transfer length of 1 μm. It may be inferred that the potential drop between the gate and source electrodes is of the order of 1 V. The maximum electric field in the channel is of the order of V_{DS} divided by the depletion depth. Since avalanche multiplication occurs at an electric field between 3×10^7 and 6×10^7 V m^{-1}, depending on the field profile, a limit is set to the maximum V_{DS} that may be tolerated.

The calculations presented above have, for simplicity, neglected any negative differential mobility regime in the velocity–field characteristic. Such may readily be encompassed,[34,35] by extending the constant mobility region to a higher electric field, defining a peak velocity at the velocity saturation point $x = L_1$, and demanding that the velocity at the drain end of the gate be a lower value, appropriate to V_s. Constraining the peak carrier accumulation to the drain end of the gate precludes any study of transferred electron instabilities in MESFET channels. However, these can occur at high channel currents, close to the knee voltage.[36] Oscillations are promoted by a thick active layer, and inhibited by the use of a recessed gate. The latter prevents the accumulation/depletion layer from propagating by constraining the electric field to drop rapidly beyond the drain end of the gate.

A further simplification, the neglect of surface depletion, is also significant at high channel currents. Surface depletion reduces the effective thickness of the active layer in the interelectrode spaces to below the value a, and, at a sufficiently high channel current, carrier accumulation needs to be maintained in the gate–drain spacing to satisfy current continuity. This condition can lead to catastrophic breakdown, since the channel electric field does not drop beyond the drain end of the gate, but it can be prevented by use of a recessed gate technology (see Section 10.6).

A major deficiency in analytical or semi-analytical models for the GaAs MESFET lies in the neglect of non-steady-state effects in the relationship between carrier velocity and electric field. The use of a steady-state velocity–field characteristic can produce significant inaccuracies when considering devices having a gate length less than 1 μm. Computer simulation of the operation of the device is then required, and this is considered in the next section.

10.4 GaAs MESFET THEORY—SHORT GATES AND COMPUTER MODELLING

The most immediate consequence of reducing the gate length of a GaAs MESFET is to reduce the transit time τ through the control region (i.e. the channel) and hence to increase the maximum operating frequency. Estimates of f_T

may be made on the basis of a simplified velocity–field characteristics (Figure 10.4), assuming that the electron velocity is that appropriate to the local electric field; this postulate becomes increasingly questionable as the gate length is reduced below about 1 μm and the transit time τ becomes comparable with the times taken for the electron energy and momentum distributions to relax to their steady-state values. Under such conditions, dynamic effects become significant and the transit time is predicted to decrease more rapidly than geometric scaling as $1/L$ would imply. Investigation of this regime of device operation is not an appropriate area for analytical techniques. Computer modelling is essential for this purpose by, for example, extending Monte Carlo techniques[37] to model the particle behaviour of the electrons in the device, or using a fluid flow model[38] in which the processes of energy and momentum relaxation to their steady-state values are described by characteristic time constants that are themselves deduced from Monte Carlo simulations of electron transport. Curtice and Yun[39] have used a GaAs MESFET model based on computing the electron temperature distribution within the device. Rather than review the growing field of the numerical modelling of devices, this section will highlight areas where the study of short gate devices, and the computational methods needed for this purpose, reveal significant differences from the steady-state quasi-analytical method described in Section 10.3.

Monte Carlo simulations,[40–42] showing the evolution of the average drift velocity of electrons as a function of time after applying a constant electric field, form a convenient introduction to dynamic effects. Figure 10.23, from ref. 42, shows that the transient response of the electron distribution leads to velocities significantly above the steady-state value, reflecting the finite rate of electron transfer between the central and satellite valleys. Similar results were obtained by Hill et al.,[41] and the effects of ionized impurity scattering were showed to be significant at fields near 3×10^5 V m^{-1}. Relaxation times associated with this transient response are of the order of 6 ps at 4×10^5 V m^{-1}, falling to about 1 ps above 2×10^6 V m^{-1}.[42] Hence, departures from the steady-state velocity–field characteristic are significant in microwave MESFETs, particularly for gate lengths less than 1 μm.

Relaxation effects were shown by Jones and Rees[43] to limit the high-frequency performance of the LSA mode of transferred electron oscillators. By contrast, in GaAs MESFETs, relaxation effects serve to increase the average velocity beneath the gate, and hence the drain–source current, mutual conductance, and current gain, above the values deduced from steady-state parameters. Wang and Hseih[44] have accommodated the velocity overshoot effect within a quasi-analytical model of the GaAs MESFET by extracting equivalent piecewise velocity–field characteristics which have a saturation field E_s and saturation velocity v_s increasing with decreasing gate length. Figure 10.24, based on data from ref. 44, shows the saturation field and saturated velocity as a function of gate length for the two-piecewise linear approximation for the velocity–field characteristic of GaAs. This

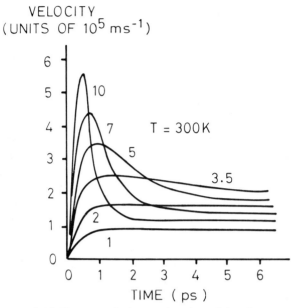

Figure 10.23 Electron velocity as a function of time for a range of electric fields. Electric field is shown in units of 10^5 V m^{-1}. (From ref. 42 Reproduced by permission of IEE)

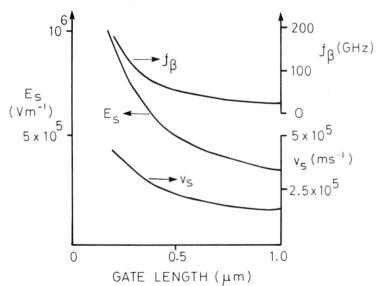

Figure 10.24 Saturation electric field E_s, saturated electron velocity v_s and unity current gain cut-off frequency f_β as a function of gate length, according to ref. 44

figure also shows the increase in the unity current gain cut-off frequency of the GaAs MESFET modelled by Wang and Hseih as the gate length is reduced. It can be seen that f_β rises more rapidly than $1/L$. The calculations were performed for an active layer thickness of 0.2 μm and a doping density of $10^{23}/m^3$.

The rigour of the approach of Wang and Hseih may be called into question; nonetheless the general features that they predict are supported in general terms by more detailed simulations. Carnez et al.[45] have compared the $I-V$ characteristics of GaAs MESFETs having gate lengths of 0.2 and 1.0 μm, contrasting predictions made using a steady-state velocity–field curve with those made using the fluid-flow model of Shur.[38] Including transient electron dynamic effects increased the drain–source current and unity current gain cut-off frequency by factors up to 1.2 for the 1.0 μm device and 1.75 for a gate length of 0.2 μm, relative to the steady-state calculation. This application of the fluid-flow model predicts electron accumulation at the drain end of the gate for the 1.0 μm gate length MESFET, consistent with the model of Sone and Takayama. By contrast, Monte Carlo simulation by Warriner[46] of a 1.0 μm gate length MESFET showed transfer to the satellite valleys occurring principally in the space between the gate and drain. The consequent reduction in electron velocity at constant layer thickness gives rise to carrier accumulation in the gate–drain region, as shown in Figure 10.25, where the electron content of the active layer, normalized to the $n_0 a$ product, is plotted as a function of distance.

The most desirable situation for a high-frequency MESFET occurs when the

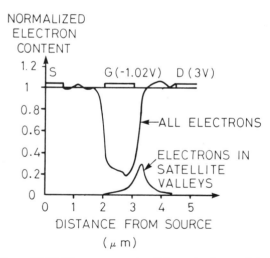

Figure 10.25 Electron content of the active layer of the GaAs MESFET, showing the gate depletion region, and carrier accumulation between the gate and drain. (From ref. 46. Reproduced by permission of IEE)

transit time between source and drain electrodes is sufficiently short that transfer to the satellite valleys has not time to occur during transit. Awano et al.[47] have modelled a 0.25 μm gate length MESFET with short (0.1 μm) gate–source and gate–drain spacings, and find no evidence for electron accumulation, or significant population of the satellite valleys apart from electrons back-scattered from the n^+ region below the drain contact. This effectively represents the condition where the constant mobility regime applies over the whole channel of a short gate MESFET, albeit with electron velocities up to 7×10^5 ms^{-1}. The device is characterized by a very high mutual conductance (643 mS mm^{-1}, compared with about 100 mS mm^{-1} for the model MESFET of Section 10.3), but also by a rather high output conductance; this latter is to be expected as a general feature of MESFETs in the absence of velocity saturation. The short gate–drain spacing is seen as a necessary condition for the inhibition of velocity saturation, to reduce the time spent at electric fields where transitions may be made to the satellite valleys.

The extreme case of device operation taking advantage of velocity overshoot is the use of 'ballistic transport' in semiconductors. Truly ballistic transport occurs when a carrier suffers no scattering events in transport between electrodes. In the case of the GaAs MESFET, this would involve transport from source to drain without interaction with phonons, other electrons, or impurity atoms. The latter two requirements lead to the fabrication of the device in low doping density material and the first to cryogenic operation to reduce the extent of phonon interaction. The 'Cold FET' has been suggested[48] as a suitable device for fast logic operation, and involves MESFET fabrication on material with a doping density n_0 of about 10^{21} donors/m^3. In operation at 77 K, the velocity–field characteristic is predicted to show a saturated carrier velocity of $(1 \text{ to } 2) \times 10^5$ ms^{-1}, with velocity saturation due to optical phonon emission at a field above about 10^4 Vm^{-1}. A logic gate delay of 40 ps at 25 μW dissipation was predicted, representing a very low-power semiconductor logic gate.

There has been much debate in the literature about the possibilities for ballistic transport in GaAs devices. Hess[49] has concluded that truly ballistic transport should be observed over distances of several micrometres for energies below 0.035 eV, the threshold for polar optical phonon emission. At higher energies, but below 0.3 eV, ballistic transport takes place over shorter distances, 0.01–0.04 μm, although velocity overshoot phenomena due to the finite scattering rate can occur over distances up to approximately 0.3 μm. It is not, at the time of writing, clear to what extent ballistic electron transport will be used in practical devices; it is clear that there is scope for reduction in the effect of phonon and impurity scattering below those appropriate to conventional MESFET operation. Predictions of velocity overshoot effects in GaAs MESFETs, along with some experimental verification,[50] and the successful operation of high electron mobility transistors (HEMTs) (Section 10.7) support this view. Exploitation of ballistic transport would lead to the development of GaAs MESFETs operating at frequencies in excess of 100 GHz, given suitably short (< 0.3 μm) gate lengths, precise control of material parameters, and short source–gate and gate–drain gaps.

The GaAs MESFET model of Curtice and Yun[39] evaluates the local electron temperature T_e within the device by equating the excess electron kinetic energy over that ascribed to the drift velocity with $3k_B T_e/2$ where k_B is Boltzmann's constant. The electron temperature is of the order of 1000 K in the active regions of the GaAs MESFET and the velocities associated with such an electron temperature dominate the drift velocity under most conditions of operation. The model shows, in addition to increased carrier velocity in the channel relative to values predicted by the steady-state velocity–field characteristic, evidence for a distribution of longitudinal electron velocity across the height of the channel, with the peak velocity located—for a fully open channel—well above the bottom of the channel (see also ref. 51). A consequence of the high electron temperature in GaAs MESFETs is that carrier confinement by the gate depletion layer and the n/substrate or n/buffer interface are not perfect; refs 39 and 45 show significant penetration of a high-resistivity buffer layer close to pinch-off, where the confining potential is simply that due to the space charge of carriers injected into the buffer layer.

While the analytical approaches of Section 10.3 assume zero transverse channel field, the mechanism of carrier confinement by surface depletion or a Schottky barrier at the upper surface and the interface at the lower surface of the device demand that the electron be located in a potential well defined by these interfaces. The transverse fields associated with this well mean that the longitudinal velocity–field relationship is not necessarily that deduced from one-dimensional Monte Carlo simulation. This topic has been studied recently by Maxfield et al.[52] for InP, albeit with general conclusions applicable to GaAs. The transverse field suppresses to some extent the velocity overshoot, steady-state velocities, and the negative differential mobility.

Circuit aspects of the GaAs MESFET may be encompassed by extracting y parameters from Fourier analysis of the terminal current response to an applied potential step.[48] An alternative approach specifies the impedance of the microwave circuit connected to the device terminals, and maintains a voltage waveform consistent with this impedance and the instantaneous terminal currents.[53] This latter approach is particularly appropriate to large-signal, non-linear operation.

While quasi-analytica analyses of GaAs MESFET operation are sufficient to predict the principal features of the performance of long gate devices (greater than about 1 μm), a full appreciation of the competing physical processes within the transistor and meaningful predictions of high-frequency limitations to device operation require a full computer simulation. It should, however, be noted that there are many features of the physical structure of the device that are not sufficiently well understood to permit an accurate model to be incorporated within a simulation. Specifically, the GaAs surface, contact regions, and the nature of the real n/semi-insulating interface are not yet well understood (particularly in relation to dynamic effects), and most simulations model significantly simplified versions of the device topology. There is, therefore, a need for some caution in translating the fine detail of computer simulation into expectations of the performance of real devices.

10.5 LOW-NOISE GaAs MESFETS

Amplification of an electronic signal degrades the signal-to-noise ratio, because an amplifier contributes added noise to the output circuit. This degradation is expressed in the noise figure F of an amplifier, defined as the signal-to-noise ratio at the amplifier input, when the input signal is associated with a noise source at 290 K, divided by the signal-to-noise ratio at the amplifier output. Figure 10.26 shows the state of the art (1982) for noise figure and associated gain that can be realized with GaAs MESFETs up to 40 GHz. The attraction of MESFETs as

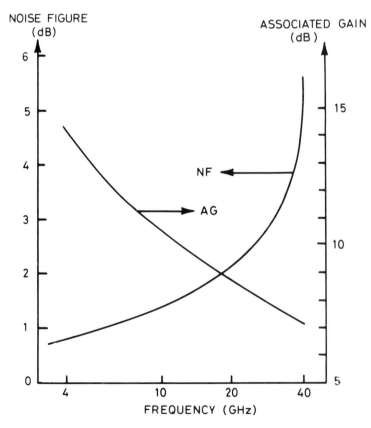

Figure 10.26 Noise figure and associated gain for GaAs MESFETs (1982)

low-noise preamplifiers for communication and radar systems is apparent. These low noise figures are attained as a consequence in part of the high current gain of GaAs MESFETs at microwave frequencies and in part of the low intrinsic and parasitic dissipation of the device. Improvements in noise figure are still being reported over the whole frequency range covered in Figure 10.26, arising from reduced gate length in commercial devices up to about 20 GHz and advances in the performance of research devices up to 40 GHz. In this section, a brief account is presented of the noise performance of GaAs MESFETs following closely the work of Pucel and coworkers[32] and Brewitt-Taylor et al.[54] Device implications for low-noise operation are also discussed.

The principal sources of added noise at microwave frequencies in the GaAs MESFET are associated with dissipative processes in the intrinsic device and in the parasitic resistances, although at the lower microwave frequencies generation–recombination noise associated with traps may also be significant.[55] In the constant mobility region of the channel, local fluctuations in the channel current generate noise in the drain–source circuit. The magnitude of these fluctuations can be related to the Johnson noise associated with the resistance of the channel. A variation in the channel current represents a fluctuation in the charge in the channel, and this is coupled via the gate–source capacitance to the gate electrode, giving rise to a noise source in the gate–source circuit that is partly correlated with the noise in the drain–source circuit. According to the treatment of Pucel et al.,[32] 'dipole noise' in the velocity-saturated region of the channel, caused by the spontaneous creation of dipoles, also contributes to the noise in the drain–source circuit and induces noise in the gate–source circuit. Parasitic resistances in the gate–source circuit contribute Johnson noise in the usual manner. Typically, the low-noise operating point for a GaAs MESFET has a drain–source potential slightly above the knee voltage, and the gate biased such that the drain–source current is approximately one-sixth of I_{DSS}—see, for example, Hasegawa in ref. 8. The lowest noise figure is not, therefore, attained at the d.c. bias conditions for maximum gain, which occurs close to I_{DSS}; nor, as will be shown, are the input microwave matching conditions for minimum noise figure those for a conjugate match.

A circuit analysis of the low-noise MESFET[32] shows how the intrinsic and extrinsic device parameters interact. For this purpose, the noise figure of a two-port amplifier may be defined as the ratio of the total noise power delivered into the output termination, when the noise temperature of the input termination is 290 K, to that portion of the output noise power attributable to the input termination alone. To examine further the noise figure of the GaAs MESFET, consider the noise-equivalent circuit of Figure 10.27, which is derived from Figure 10.14b. Resistances r_S, r_G, and r_{in}, the real part of the input termination, are associated with noise voltage generation e_S, e_G, and e_{in} respectively, where

$$\overline{e_S^2} = 4k_B T_0 \Delta f r_S \qquad (10.30)$$

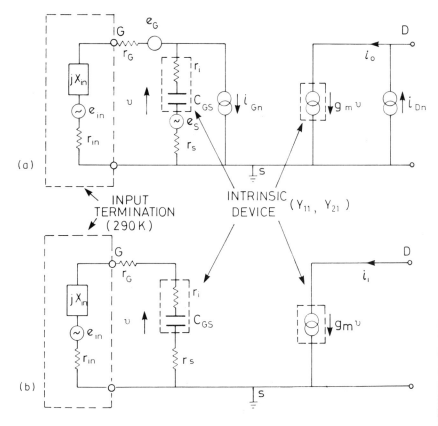

Figure 10.27 Equivalent circuit for MESFET noise figure calculation: (a) including all noise sources; (b) showing only the noise source associated with the input termination

and similarly for e_G and e_{in}. Temperature $T_0 = 290$ K, and Δf is the noise-equivalent bandwidth. Noise in the channel is represented by a current generator i_{Dn} in the drain–source circuit. Dissipation associated with r_i is included in i_{Dn}. Fluctuations in the local drain–source current couple via the gate–source capacitance C_{GS} to the gate and give rise to a gate noise current i_{Gn}. Also i_{Gn} and i_{Dn} are partially correlated. Since the channel is coupled capacitively to the gate, this correlation may be expressed by

$$\overline{i_{Gn} \cdot i_{Dn}} = jC(\overline{|i_{Gn}|^2}\,\overline{|i_{Dn}|^2})^{1/2} \tag{10.31}$$

where $j^2 = -1$ and C is the correlation coefficient. The noise figure of the transistor circuit may be determined by considering the ratio of the total noise current i_{n0} delivered to the output terminals of the device to that portion i_{n1} attributable to the amplification of the input termination noise voltage source e_{in},

whence
$$F = \overline{|i_{n0}|^2} / \overline{|i_{n1}|^2}$$

Pucel et al.[32] relate the gate and drain noise current generators to parameters R and P,

$$\overline{|i_{Gn}|^2} = 4k_B T_0 \Delta f \omega^2 C_{GS}^2 R/g_m \qquad (10.32a)$$

$$\overline{|i_{Dn}|^2} = 4k_B T_0 \Delta f g_m P \qquad (10.32b)$$

and show by analysis of the circuits of Figure 10.27 that

$$F = (1/r_{in})(r_n + g_n |Z_{in} + Z_C|^2) \qquad (10.33)$$

The qualities r_n and g_n are themselves relate to the fundamental noise coefficients,[32] given by

$$K_G = P\{[1 - C\sqrt{(R/P)}]^2 + (1 - C^2)R/P\} \qquad (10.34a)$$

$$K_C = [1 - C\sqrt{(R/P)}]/\{[1 - C\sqrt{(R/P)}]^2 + (1 - C^2)R/P\} \qquad (10.34b)$$

$$K_R = R(1 - C^2)/\{[1 - C\sqrt{(R/P)}]^2 + (1 - C^2)R/P\} \qquad (10.34c)$$

The 'noise resistance' r_n is given by

$$r_n = r_S + r_G + K_R(1 + \omega^2 C_{GS}^2 r_i^2)/g_m \qquad (10.35a)$$

and the 'noise conductance' g_n by

$$g_n = K_G \omega^2 C_{GS}^2 / g_m \qquad (10.35b)$$

Z_C in equation (10.33) is a 'correlation impedance' given by

$$Z_C = r_S + r_G + K_C(1 + j\omega C_{GS} r_i)/j\omega C_{GS} \qquad (10.35c)$$

The fundamental noise coefficients relate to the properties and structure of the intrinsic device, through R, P, and C, and the roles of the intrinsic device and the parasitic elements in governing the FET noise figure may be seen from equation (10.33). From equation (10.33), it may be seen that the minimum noise figure F_{min} is attained for

$$Z_{in} = Z_{opt} = r_{opt} + jX_{opt} \qquad (10.36)$$

where

$$X_{opt} = -\text{Im}(Z_C) \qquad (10.37a)$$

and

$$r_{in}^2 = r_{opt}^2 = (\text{Re}(Z_C))^2 + r_n/g_n \qquad (10.37b)$$

Note that these conditions do not correspond to a matched input.

From equations (10.33) to (10.35), $F_{\min}$ may be expanded as a power series, giving [32]

$$F_{\min} = 1 + 2(\omega C_{GS}/g_m)\{K_G[K_R + g_m(r_S + r_G)]\}^{1/2}$$
$$+ 2(\omega C_{GS}/g_m)^2[K_G g_m(r_S + r_G + K_C r_i)]$$
$$+ \ldots \qquad (10.38)$$

Two limiting cases of this analysis may be examined:

(a) $R \to 0$, i.e. noise induced on the gate may be neglected. Then $K_C \to 1$, and $X_{opt} \to -1/j\omega C_{GS}$, i.e. the reactive part of the input termination corresponds to a perfect match. Similarly, $C \to 0$ leads to the same conclusion. Thus, it is deduced that the reactive input mismatch at the low-noise operating condition is attributable to the noise induced on the gate and partially correlated with the drain noise generator. At high values of r_{in}, the noise currents in the input circuit are small, and the noise in the output circuit is dominated by the drain–source noise current. At values of r_{in} much less than $r_S + r_G$, noise in the output circuit is dominated by high noise currents in the low-impedance input circuit, driven by voltage generators associated with the parasitic resistances r_s and r_G and amplified by the current gain of the device. The noise figure, therefore rises at high and low values of r_{in}, and there is an optimum value differing from the conjugately matched condition.

(b) $g_m(r_S + r_G) \gg K_R$, i.e. R is small and/or the gate and drain noise sources are highly correlated. The leading term of (10.38) then becomes

$$F_{\min} \approx 1 + 2K_G^{1/2}(\omega C_{GS}/g_m)[g_m(r_S + r_G)]^{1/2} \qquad (10.39)$$

Equation (10.39) has been shown by Fukui[28] to represent well the behaviour of GaAs MESFETs over a wide range of operating conditions with $K_G^{1/2} \approx 2.5$, and predicts a noise figure increasing with increasing drain–source current, as C_{GS} rises more rapidly than $g_m^{1/2}$ with increasing channel height. At very low currents, g_m tends to zero whereas C_{GS} becomes dominated by fringing capacities. Thus at very low currents, $F_{\min}$ is expected to rise again, giving a minimum value of $F_{\min}$ at low I_{DS}. It may be seen that $F_{\min} - 1$ rises as ω.

The general form of equation (10.39) thus indicates the importance of maintaining a high g_m to low values of I_{DS} and of reducing the parasitic resistances r_S and r_G.

The relative contribution of intrinsic and extrinsic noise sources to the MESFET noise figure are not yet fully elucidated by theory. To determine P, R, and C requires a model quantifying noise current fluctuations in the channel and relating them to the total drain noise generator i_{Dn} and the induced gate noise generator i_{Gn}. Much of the framework for the description is based on developments of the 'impedance-field' concept of Shockley et al.,[56] which relates potential fluctuations at the terminals of a semiconductor device to noise current dipoles in the bulk of the material, and assumes that these current dipoles are uncorrelated. Brewitt-Taylor et al.[54] use 'influence functions' relating the noise current dipole to current fluctuations in the drain–source circuit and the gate–source circuit. Consider (Figure 10.28) a current dipole in the drain–source current of the MESFET channel below velocity saturation; the 'current dipole' consists of a local increase ΔI in the channel current over a length Δx of the channel. The total noise current i_{Dn} in the drain–source circuit is given by

$$i_{Dn} = \sum_{\Delta x} \Delta I \, \Delta x f(x)/L \qquad (10.40a)$$

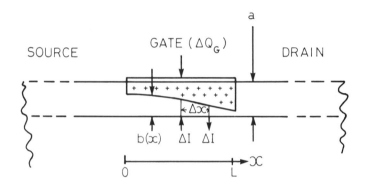

Figure 10.28 Noise dipole $\Delta I \Delta x$ in MESFET channel below velocity saturation

where $f(x)$ is the 'drain influence function' and the factor $1/L$ is introduced so that $f(x)$ is dimensionless. Similarly, a charge ΔQ_G is induced on the gate where

$$\Delta Q_G = \sum_{\Delta x} \Delta I \, \Delta x \, s(x) C_{GS}/g_m L \qquad (10.40b)$$

and the quantity $C_{GS}/g_m L$ ensures that the 'gate influence function' $s(x)$ is dimensionless.[54] The summations are performed over the length of the channel. The noise power generated by ΔI in length Δx of the channel is identified with $4k_B T_0 \Delta f$, which gives using Einstein's relationship

$$\overline{\Delta I^2} = 4q \Delta f Q(x) D(x)/\Delta x \qquad (10.41)$$

where

$$Q(x) = n_0 q W b(x) \qquad (10.42)$$

and $D(x)$ is the local diffusion coefficient.

From equations (10.40a) and (10.40b), assuming that current dipoles in adjacent sections of the channel are uncorrelated, and replacing the summations by integrals over the length of the channel, Brewitt-Taylor et al.[54] derived expressions for i_{Gn} and i_{Dn} and hence P, R, and C.

Noting that the noise figure is not generally a strong function of drain–source potential V_{DS} above the knee voltage, Brewitt-Taylor et al.[54] investigated the noise figure of the MESFET at the condition where velocity saturation is attained at the drain end of the gate. An analytical approach (using the gradual channel approximation) was compared with results from computer modelling (neglecting velocity overshoot) to derive $f(x)$ and $s(x)$ for a device with a 1 μm long gate. Generally good agreement was attained. Simulations performed at drain–source potentials above the knee voltage did not reveal any strong increase in noise figure attributable to the velocity-saturated region. Figure 10.29 shows predictions for the noise figure F of a GaAs MESFET with saturation index $\xi_1 = 0.2$ and parasitic resistances $r_S + r_G = \alpha/G_0$. The predicted minimum in noise figure as a function of I_{DS} may be seen for non-zero parasitic resistance. The general shape of the curve with I_{DS} is typical for practical devices. The influence of bias conditions and parasitic resistances is apparent.

While Brewitt-Taylor et al.[54] analysed only the constant mobility regime, Pucel and coworkers[32] have analysed a two-region model of the GaAs MESFET. They have concluded that the principal source of noise is the spontaneous formation of current dipoles in the velocity-saturated region of the channel, but that the effect of this noise is markedly reduced by strong correlation with the gate noise

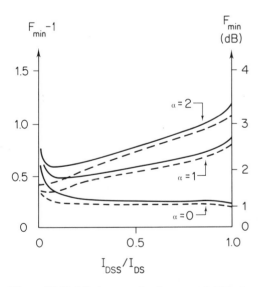

Figure 10.29 Minimum noise figure at 10 GHz for a GaAs MESFET with a 1.0 μm gate, and parasitic resistance α/G_0. Solid lines, $\xi = 0.2$; dotted lines, $\xi = 0.05$. (From ref. 54. Reproduced by permission of IEE.)

generator. Simulation of 1.0 and 0.4 μm gate length devices, including the effects of velocity overshoot by Shur's method,[38] have been reported by Carnez et al.[57] They used the model to calculate the noise parameters P, R, and C, and concluded that noise contributions from the whole channel were significant. Good agreement with the work of Pucel et al.[32] was reported for a 2 μm long gate device and with equation (10.39) for a 0.4 μm gate FET, except in the limit of low parasitic resistances as one might expect. Equation (10.39) has been used by Fukui[28] to develop an account of the noise figure of GaAs MESFETs, showing excellent agreement with experimental results by using detailed semi-empirical expressions for the parasitic elements in the device. This general approach has been used to optimize device configurations[58,59] for low noise, and to predict the performance of millimetre-wave transistors.[60]

The account given above of the theory of the noise performance of GaAs MESFETs has indicated the need not only to reduce the noise figure of the intrinsic transistor but also to minimize the effects of parasitic resistive elements. At low frequencies (below about 20 GHz) this is manifested principally in fine tuning of the device configuration to reduce not only ohmic contact resistances (e.g. by the use of n^+ contact layers) but also the resistance of the semiconductor between the source and drain contacts, and between the edges of any etched gate channel and the gate. Because the lowest noise figure is typically obtained at gate–source voltages comparatively close to pinch-off, it is important that a high mutual conductance be attained when the channel is confined close to the active/semi-insulating or active/buffer interface. The maintenance of a high material mobility in this region, and a wish to minimize trapping and carrier penetration into the semi-insulating material, have been driving forces towards growing an undoped or chromium-doped buffer layer on top of the semi-insulating substrate in order to avoid direct growth of the active layer on the defected surface of a bulk wafer. A further paper by Fukui[61] shows how the bandwidth requirement for a low-noise amplifier can influence the device design parameters; in particular, the use of a thin, heavily doped FET channel is to be preferred for wideband applications. Design rules for devices at high frequencies are not yet well established, and at the time of writing the usefulness of GaAs MESFET LNAs at 40 GHz is limited as much by the low associated gain as by the noise figure. It has been predicted[60] that noise figures down to 3 dB should be attainable at 40 GHz, using a gate length of 0.2 μm, and that operation up to 75 GHz should be possible from a short gate MESFET.

Low-noise operation of GaAs MESFETs has been a major driving force in the research, development, and commercial exploitation of these devices. Among trends for the future will be the search for low-noise amplifiers at millimetre waves and the continuing refinement of transistor structures at lower frequencies, with emphasis on reliability, resistance to burnout, dynamic range, and bandwidth capability, as much as on extracting further improvements in noise figure.

10.6 GaAs POWER MESFETS

There is a growing trend towards the use of solid state power amplifiers in microwave systems. The silicon bipolar transistor is well established in this role at the lower microwave frequencies, and the GaAs MESFET is now becoming a serious contender for many commercial and military applications. Figure 10.30 shows output powers attained as a function of frequency for GaAs MESFETs in research laboratories up to about December 1982, together with an indication of the performance of commercial devices. The research results correspond to a gain of 4 dB or greater, and power added efficiency (the ratio of the microwave added power to d.c. dissipation) typically greater than 30%.

Refs 7 and 8 include chapters on GaAs power MESFETs and ref. 5 is a review by Dilorenzo and Wisseman. In this section, the factors limiting the output power of a GaAs MESFET are discussed and special technology features described.

A high output power from a FET requires high voltage and current swings at the drain–source terminals. Briefly, the output voltage swing is limited by breakdown within the transistor, and the current swing by the total gate width of the device. The latter is in turn limited by the impedance that can be matched into a microwave circuit. The width of each individual gate finger is limited by

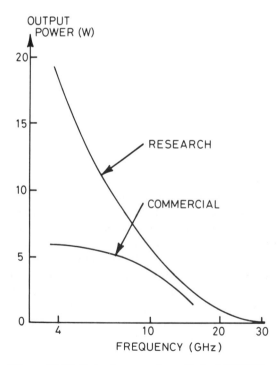

Figure 10.30 Output powers from GaAs MESFETs

attenuation and phase changes down the source–gate transmission line to about 200 μm at X band (Section 10.2). Thus a typical GaAs power MESFET configuration consists of an interdigitated structure with many gate fingers in parallel to give the required gate width. The three-terminal, planar nature of the device demands that one set of electrodes be joined by a non-coplanar interconnection. Since the transistor is normally operated in common source, it has been customary to interconnect the sources in this way, so that interactions with other electrodes do not create feedback elements. Figure 10.31 shows an optical micrograph of two interdigitated GaAs MESFET structures, with source pads to be interconnected.

An estimate of the output power of the GaAs MESFET may be made from the d.c. characteristics (Figure 10.32). It is assumed that the maximum drain–source voltage during the RF cycle is limited by the gate–drain avalanche breakdown voltage V_B, whence the maximum class A output power P_A is given by

$$P_A = I_{max}(V_B - V_P - V_k)/8 \qquad (10.43a)$$

and the power added efficiency η by

$$\eta = 0.5 \frac{V_B - V_P - V_k}{V_B - V_P + V_k}\left(\frac{G-1}{G}\right) \qquad (10.43b)$$

where G is the gain. The quiescent drain–source voltage at the operating point is $(V_B - V_P + V_k)/2$. For example, $V_B = 19$ V, $V_P = 3$ V, $V_k = 2$ V give $V_{DS} = 9$ V, and if $I_{max} = 300$ mA mm^{-1}, $P_A = 0.52$ W per millimetre gate width and at 5 dB gain $\eta = 26\%$. This illustrates the importance of maintaining a high gate–drain breakdown voltage, consistent with a large maximum channel current. It is apparent that the power added efficiency is significantly degraded from the drain efficiency by the low gain, which is not untypical for GaAs power MESFETs. Class AB and class B operation have yielded very high power added efficiencies, e.g. 72% at 2.45 GHz,[62] 68% at 4 GHz,[63] and 48% at 9.5 GHz.[64] A commonly quoted figure of merit for GaAs power MESFETs is the output power per unit gate width. The highest value reported to date[65] is 1.4 W mm^{-1} and values about 1 W mm^{-1} serve as a laboratory benchmark.

While the general configuration of the GaAs power MESFET unit cell is similar to its low-noise counterpart, emphasis is placed on delaying the onset of breakdown phenomena as the drain–source bias is increased to accommodate a high drain–source voltage swing. Figure 10.33 illustrates some of the breakdown paths in a GaAs MESFET at high drain–source voltage. Destructive breakdown effects, ascribed to high fields at the drain contact edge (A in Figure 10.33) and accompanied by light emission,[66] are inhibited by the use of a recessed gate technology[67] together with an n$^+$/n material structure,[68] giving a low-resistance layer beneath the drain contact, or by an inlaid n$^+$ layer beneath the drain contact, grown by selective epitaxy.[69] Wemple et al.[68] present evidence for a catastrophic burn-out process caused by thermal runaway in the buffer layer at high

410   *Gallium arsenide*

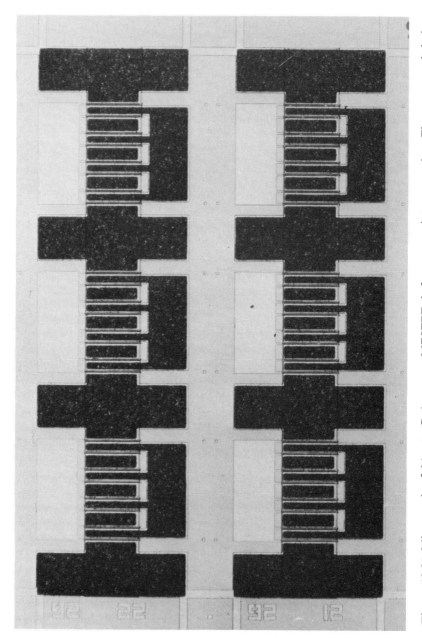

Figure 10.31 Micrograph of 24-gate GaAs power MESFET, before source interconnection. The source and drain electrodes, parts of which are electroplated, are darker than the gate feed. (Courtesy of Plessey Research (Caswell) Ltd.)

GaAs MESFETs

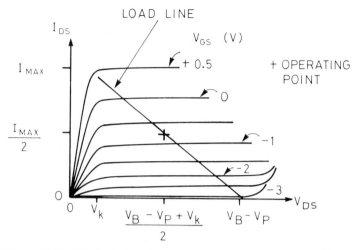

Figure 10.32 Schematic d.c. characteristics of GaAs MESFET, showing the class A operating point, resistive load line, and the effect of gate–drain avalanche breakdown in raising the drain–source current at high V_{DS}

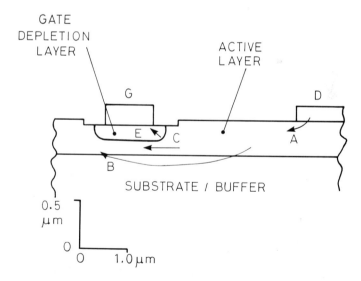

Figure 10.33 Breakdown mechanisms in GaAs MESFETs at high drain–source voltages: A, field concentration at drain contact; B, thermal runaway in buffer or substrate; C, breakdown in the carrier accumulation/depletion layer; E, avalanche breakdown of gate–drain diode

drain–source voltages and drain–source currents and at elevated ambient temperatures (B in Figure 10.33). Eastman and coworkers[70,71] have investigated both theoretically and experimentally the destructive avalanche breakdown (C in Figure 10.33) caused by high electric fields in the carrier accumulation–depletion region (stationary domain) at the drain end of the gate and in the gate–drain spacing (see Figure 10.22). Breakdown phenomena B and C generally occur at rather high drain–source voltages, values well in excess of 30 V being quoted in the literature. By a suitable choice of material structure and channel topology, breakdown at the drain contact (A) may be inhibited, and the limitation to the drain–source voltage swing is set by gate–drain breakdown (E in Figure 10.33). At a sufficiently high drain–source potential, avalanche multiplication occurs in the depletion layer, the drain–source current rises as shown in Figure 10.32, and current flows in the gate–source circuit. The consequence for microwave power amplification is that the drain–source voltage swing can only be increased beyond the value ($V_B - V_P - V_k$) at the expense of a reduced maximum current swing since the FET cannot be completely pinched off at high drain–source voltages. Thus, the output power saturates. Experiment, theory, and simulation[16,18,72] broadly agree that for sufficiently thin active layers the gate–drain breakdown voltage falls as the reciprocal of the charge per unit length beneath the gate $n_0 a$. According to ref. 16, $V_B \approx 50/n_0 a$, where V_B is measured in volts, n_0 in units of $10^{22}/m^3$, and a in micrometres. This relationship holds for $n_0 a < 2.3$. For $n_0 a > 2.6$ the breakdown voltage is that appropriate to a diode on bulk material, implying that at a high drain–source potential the vertical electric field beneath the gate exceeds the breakdown field before the device is pinched off. The increase in breakdown voltage above the values found in bulk GaAs is attributable to the two-dimensional field configuration and to enhanced recombination and the inhibition of avalanche multiplication very close to the surface of the device.[18]

David et al.[18] used computer simulation to model the breakdown of recessed gate MESFETs, and found a strong dependence of breakdown voltage on the $n_0 a$ product, fine detail indicating a preference for thinner, more highly doped layers. Their studies indicated that the spread of the depletion layer edge towards the drain was inhibited by the thickening of the active layer at the gate recess edge, and it was suggested that the optimum recess length is equivalent to the depletion layer extension towards the drain (~ 0.7 μm) at the breakdown of a planar device.

Typically, GaAs power MESFETs operate with $V_{DS} \approx 10$ V, though values up to 24 V[65,73,74] have been reported. From the inverse relationship between V_B and $n_0 a$ (and hence V_B and I_{max}), Wemple et al.[16] predict a maximum power-handling capability of 1.5 W mm^{-1} for GaAs MESFETs, independent of frequency. Assuming that the gate length and channel thickness both scale as (operating frequency)$^{-1}$, the input capacitance per unit gate width would remain frequency-independent. The input circuit of the MESFET is principally capacitive with a Q typically about 5 and thus, assuming there is a minimum circuit impedance that can be realized without excessive loss, the maximum gate width

GaAs MESFETs

and hence maximum output power of a GaAs MESFET should fall as (frequency)$^{-1}$. Figure 10.30 shows a rather more rapid drop with frequency, possibly because of the limited experience with FET devices and circuits at high frequencies.

The limitations to the output power per unit gate width mean that a high-power GaAs MESFET is necessarily a wide gate device. The adverse effects of source impedance on MESFET gain, noted in Section 10.1, mean that it is necessary that a low source impedance interconnection be achieved between the many 'unit cells' of the device. Although wire bonding has been used, this is a yield- and cost-limiting process and can lead to increased parasitic source inductance. Three interconnection approaches are used: flip-chip mounting, overlay, and via hole grounding.

Flip-chip mounting is shown schematically in Figure 10.34a. Plated source pads are bonded to a carrier which forms an integral part of the device packaging. This technique is used by Microwave Semiconductor Corporation (MSC).[20, 75] The device may be heat sunk via these source pads. Recently workers at Mitsubishi[76] have flip-chip bonded all the electrode terminals to a carrier, so reducing still further the requirement for bonding.

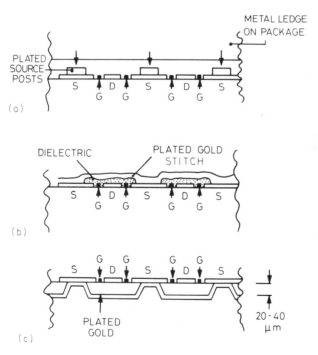

Figure 10.34 Schematic interconnection techniques for GaAs power MESFETs: (a) flip-chip mounting; (b) dielectric overlay (source-over-drain); (c) via hole interconnection

Overlay techniques are widely used for source interconnection in commercial devices. Fujitsu[69] and NEC[77] have reported the use of silicon dioxide to insulate the off-mesa source interconnection from the gate electrode feed. The parasitic capacitances introduced by this technique have been shown to be small when compared with bonding pad capacitances.[78, 79] A thick layer of polyimide, an organic insulator, has also been used[74] in a source-over-drain interconnection, as shown in Figures 10.34b and 10.35. Figure 10.36 shows interconnection of power MESFET source pads using an air bridge, formed over a resist layer that is subsequently dissolved.

Via hole grounding[80] is shown in Figure 10.34c. A hole is etched from the rear face of a thinned GaAs wafer to a source pad on the active surface of the device, and the rear face of the device and the via hole are metallized by sputtering or electroless plating.[80] This metallization is electroplated to form a thick metal support layer, integral heat sink, and low-inductance source interconnection. The technique is used by Bell Laboratories and by Raytheon in commercial devices.

The number of unit cells that can be interconnected is constrained by factors such as the yield of 1 μm gates and the need to maintain uniformity of device parameters and feed length over the unit cells of a chip, together with constraints on the impedance that can be matched by a microwave circuit.

In the presence of large voltage and current swings within the transistor, a reduction in power gain from the small-signal case is to be expected, even in the absence of clipping of the waveforms. For example, a sinusoidal ouput current, while compatible with a sinusoidal input current, requires from the gate–source bias dependence of the input capacitance a non-sinusoidal input voltage waveform.[62] The harmonics so created are reflected into the input circuit or dissipated internally; a similar argument may be advanced with respect to the bias dependence of output conductance in the drain circuit. The result of the bias dependence of the FET parameters is that the gain falls with increasing drive level. The circuit design implications of these non-linearities have been considered by, for example, Rauscher, Willing and coworkers,[10, 81] Tucker,[82, 83] and Tajima et al.[84] It is not yet clear how far power MESFET design can be optimized to reduce the sources of non-linearity, although variations in the vertical doping profile have been proposed or used to control the dependence of mutual conductance on gate–source voltage.[85, 86]

The thermal resistance of GaAs power MESFET structures affects their performance (through the negative temperature dependence of mobility and saturated velocity) and their reliability, since degradation processes generally obey an Arrhenius law and have a rate proportional to $\exp(-E_A/kT)$ where E_A is an activation energy. With the exception of some flip-chip interconnected MESFETs, heat sinking is performed through the chip, which is generally thinned to less than 200 μm. The thermal design of GaAs power MESFETs is reviewed by Wemple and Huang.[87] For the purposes of this section, it is noted that the thermal resistance depends not only on chip thickness but also on the

GaAs MESFETs 415

Figure 10.35 Experimental GaAs power MESFET, made at RSRE, Baldock, and using polyimide (dark areas on the micrograph) in a dielectric overlay source interconnection

packing density on the front face of the device. Heat is generated principally in the gate–drain gap, where the fields are highest, and the heat sinking of an individual MESFET gate would be strongly affected by the spreading resistance from this small region. At a high packing density, heat flow becomes increasingly one-dimensional and the thermal resistance is strongly dependent on chip thickness. It

416 Gallium arsenide

Figure 10.36 Air bridge interconnection of a GaAs power MESFET. (Courtesy of Plessey Research (Caswell) Ltd.)

is desirable to increase so far as is possible the gate width of each unit cell, to reduce the device aspect ratio, and to increase the intergate spacing. Typically GaAs power MESFETs operate with channel temperatures about 100°C above ambient. There have been reports (e.g. ref. 88) of significant improvements in output power and gain by the pulsed operation of GaAs MESFETs, attributed to lowering of the mean device temperature.

Exploitation of GaAs power MESFETs, while well advanced, has not proceeded as rapidly as in the case of low-noise devices. In part this is a consequence of a more difficult device technology, as regards both the development of a source interconnection and the yield of high-resolution lithography features (e.g. a 1 μm long gate with a gate width of perhaps 10 mm per complete device). The low impedance of the input of the power MESFET ($< 5 \Omega$) and difficulties in defining the appropriate input and output matching circuits—which for a large-signal device cannot be predicted by the direct use of small-signal s-parameters—have also hindered advances in circuit and amplifier performance. The reliability of GaAs power MESFETs has received much attention, although recent reports have demonstrated encouraging advances, particularly for passivated devices.[89, 90]

There is still scope for significant advances in GaAs power MESFET performance, particularly above 8 GHz; however, progress will be seen not only in device technology and performance at high frequencies but also in demonstrating highly reliable devices and in circuit techniques, which may include package prematching, push–pull configurations, and non-linear circuit analysis, addressing the problems associated with the use of non-linear, low-impedance transistors.

10.7 SPECIAL MESFET CONFIGURATIONS

The preceding sections have dealt with GaAs MESFETs having a conventional configuration. In this section, some alternative structures are briefly described.

The dual gate GaAs MESFET is shown in a cross section in Figure 10.37. In essence, this can be regarded as two single gate MESFETs in cascade, with their joining point at an undetermined potential, although according to simulations by Allamando et al.,[91] velocity overshoot effects in the intergate spacing are significant when that spacing drops below 0.6 µm. Figure 10.38a shows a representation of a dual gate MESFET as two single gate devices. In Figure 10.38b a construction of the d.c. characteristics is presented, whereby the first FET d.c. characteristics in common source (I_{DS} vs V_{DIS} as a function of V_{G1S}) are shown opposed to the d.c. characteristics of the second FET in common drain (I_{S2D} vs V_{S2D} as a function of V_{G2D}). The point D1 corresponds to the drain of the first device and the source of the second device.[92] It may be seen that by a suitable choice of V_{G1S} and V_{G2D} the change in drain–source current with the potential V_{G2D} (and hence V_{G2S} which in Figure 10.38 is given by $V_{G2S} = 5\text{ V} + V_{G2D}$) can be made large (point A) or small (point B). The ability to alter the mutual conductance of one gate–source circuit using the potential between the other gate and source gives the opportunity for a wide range of electronic functions combined with microwave gain. Such functions include gain control, mixing, and switching; Pengelly[93] describes a number of novel circuit applications of the dual

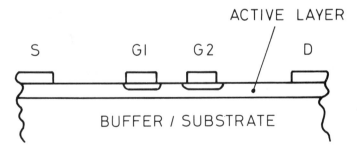

Figure 10.37 Dual gate MESFET cross section (schematic), showing control gates G1 and G2

418   Gallium arsenide

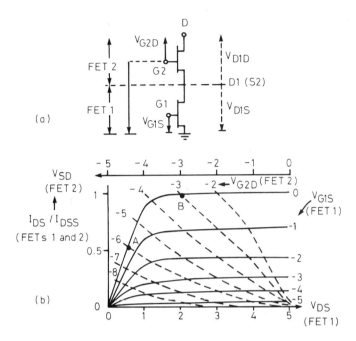

Figure 10.38 D.c. representation of the dual gate MESFET as two single gate devices in cascade. (a) Circuit representation, showing electrode potentials and unknown potential at D1. (b) Interaction of d.c. characteristics of FET 1 in common source (solid lines) and FET 2 in common drain (broken lines) at $V_{DS} = 5$ V. The potential of the point D1 and I_{DS} are given by the intersection of the lines of constant V_{GIS} and V_{G2D}. (Reproduced by permission of IEE.)

gate MESFET. The use of this device for electronic control is particularly relevant to GaAs monolithic circuits, to maintain technology compatibility between control elements and microwave components on a single substrate.

Two device structures aimed at improved MESFET performance are shown in Figure 10.39. The opposed gate–source FET[94] has been proposed as a structure with low source resistance, since the source and gate are separated only by the active layer thickness; at the time of writing details of fabrication have not been published. Vokes and coworkers[95, 96] fabricated GaAs MESFETs on a GaAs active layer separated from the substrate by a thin layer of GaAlAs. The substrate and GaAlAs spacer layer were removed by successive selective chemical etching. Features of this structure included the potential for the control of channel current by electrodes on either side of the active layer and, for low-noise operation, the ability to operate at low currents away from any defected active/buffer interface.

A device that has evolved rapidly in recent years is the High Electron Mobility Transistor (HEMT)[21, 97, 98] or Two-dimensional Electron Gas FET (TEGFET).[99, 100] Such a device is shown in cross section in Figure 10.40,

GaAs MESFETs 419

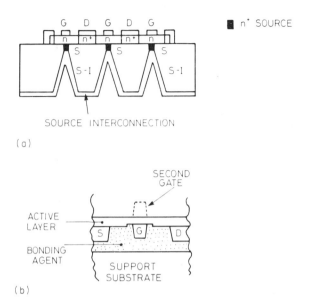

Figure 10.39 Novel GaAs MESFET structures. (a) Opposed gate–source FET. (Reproduced by permission of *Microwaves and RF*. © Hayden Publishing Co. Inc.) (b) Double gated MESFET, fabricated using novel FET technology. The second gate is deposited after the substrate and GaAlAs spacer layer have been removed, the device being bonded to a support layer. (After ref. 95.)

together with a band structure diagram showing the electron confinement well at the GaAs/GaAlAs interface, arising from the difference in electron affinities. Depending on the thickness and doping density of the GaAlAs layer beneath the gate, the structure of Figure 10.40 may be made 'normally off'—such that at zero gate–source bias the built-in depletion punches through the GaAlAs and reaches into the undoped GaAs—or 'normally on'. In the former case a small positive potential applied to the gate enables current to flow. The mechanism for charge control—and hence mutual conductance—is the control of the capacity of the electron confinement well by the potential on the gate.[101] Measurements on device structures show considerably enhanced electron mobility relative to doped MESFET layers, for example $0.62 \text{ m}^2 \text{ V}^{-1} \text{ s}^{-1}$ at 300 K, as opposed to a value of typically $0.4 \text{ m}^2 \text{ V}^{-1} \text{ s}^{-1}$ for GaAs at 10^{23} donors/m^3, and $3.25 \text{ m}^2 \text{ V}^{-1} \text{ s}^{-1}$ at 77 K, compared with $0.58 \text{ m}^2 \text{ V}^{-1} \text{ s}^{-1}$ at 77 K for MESFET layers.[97] This mobility improvement is attributed to a reduction in impurity scattering caused by the separation of electrons from their donor ions, a reduction that may be enhanced by a thin, undoped GaAlAs layer between the n-GaAlAs and the GaAs,[102] and to a modification of the electron–phonon interactions by the two-dimensional nature of the electron gas.[103] HEMTs have demonstrated very high mutual conductance g_m (e.g. 193 mS mm^{-1} at 300 K and 409 mS mm^{-1} at

420   *Gallium arsenide*

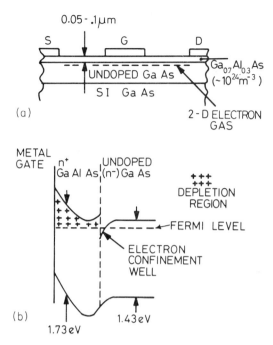

Figure 10.40 High electron mobility transistor: (a) schematic cross section through device; (b) band structure, showing depletion layers due to gate and band gap discontinuity. (Based on figures 3 and 4 in M. Abe *et al.*, New technology towards GaAs LSI/VLSI for computer applications, *IEEE Trans. Microwave Theor. Tech.*, **MTT-30**, 992–8, July (1982). Copyright © 1982 IEEE. Reprinted with permission)

77 K[98]) and this is of particular value in logic gate applications, although promising HEMT microwave results have also been reported.[104] Ring oscillator propagation delays as low as 18.4 ps per gate at room temperature have been achieved.[100] At the time of writing, many aspects of HEMT operation are not well established; for instance the competing roles of high mobility, low source resistance caused by a low surface depletion, and an enhancement in saturated velocity[105,106] are still under discussion. However, the outstanding results already achieved with this structure and the promise of the device, especially in the context of large-scale computer systems, ensure that interest in the physics of the HEMT, and its circuit exploitation, will continue.

The Permeable Base Transistor (PBT) is shown in schematic cross section in Figure 10.41. This novel device demands a range of new device technologies;[107,108] an n-GaAs layer is grown on an n^+ substrate which serves as an emitter, followed by definition of a 30 nm thick tungsten grating with 0.16 μm lines and spaces using x-ray lithography and the float-off process. A further n-layer is then grown on the mixed tungsten/GaAs surface, with growth nucleating

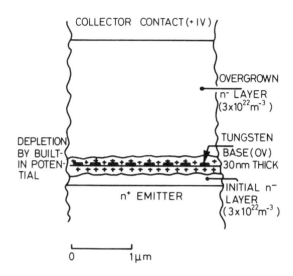

Figure 10.41 Cross section through 'normally-off' permeable base transistor structure

on the GaAs, and spreading through the grid spacing to cover the grid lines. An ohmic contact is then made to the top (collector) surface. The dimensions and n-doping density are chosen such that the semiconductor between the bars of the grating is completely depleted at zero base–emitter bias, and by application of a positive bias to the base, emitter–collector current can flow. (Although the PBT electrodes are conventionally described in bipolar notation, the majority carrier action and operating principle are clearly similar to those of the MESFET.) The principal difficulties encountered to date have arisen from anomalously high resistance in the tungsten base, and interfaces between the base metal and the overgrown GaAs layer. Predictions of f_T above 200 GHz[107] and a logic gate delay of 2 ps[108] have been made; a gain of 16 dB has been reported at 18 GHz from this device, as a consequence of a number of small improvements in the device technology.

The basic GaAs MESFET structure is postulated to give useful small-signal performance to about 100 GHz, by reducing the gate length of device to 0.25 μm and taking advantage of velocity overshoot effects. The novel configurations described above are not yet sufficiently well understood to enable confident predictions to be made as to the limits of their operation; in particular, the trade-off between small intrinsic device dimensions necessary for high-frequency operation and the limiting effects of feedback admittance and low circuit impedance associated with small structures and high-frequency operation need careful study. The GaAs MESFET will, in the future, be challenged not only by the structures described above but also by MESFETs fabricated in alternative

III–V ternary or quaternary semiconductors, such as InGaAsP[109] with higher mobility and saturated velocity.

Recent advances in epitaxial semiconductor growth technology have enabled a range of devices exploiting the properties of semiconductor heterostructures to be fabricated. The HEMT or TEGFET is one example of such a device, which exploits the band gap difference between GaAlAs and GaAs in a majority carrier transport device. The capability for independent control of the transport of electrons and holes across a heterojunction, by a suitable choice of semiconductor composition, offers a wide range of options to the transistor designer, exemplified in the GaAlAs/GaAs heterojunction bipolar transistor, recently the subject of a stimulating review by Kroemer.[110] More speculatively, the exploitation of 'modulation doped' structures,[111] which contain many layers of semiconductor with different band gaps, may lead to new devices either in MESFET form or embodying new concepts in device operation. There is, therefore, every prospect that we shall see in the future not only continued evolution of GaAs MESFET technology and performance but also competing devices based on the properties of semiconductor heterojunctions.

ACKNOWLEDGEMENTS

All figures are 'British Crown Copyright Reserved' unless otherwise stated. The author gratefully acknowledges the contribution made to this chapter by technical discussions over a number of years with colleagues at RSRE, in UK industry, and at the University of Sheffield.

REFERENCES

1. E. T. Watkins, J. M. Schellenberg, L. H. Hackett, H. Yamasaki, and M. Feng, A 60 GHz GaAs FET amplifier, *IEEE Trans. Microwave Theor. Tech.*, **MTT-S**, 145–7 (1983).
2. J. M. Schellenberg, H. Yamasaki, and D. W. Maki, A 69 GHz FET oscillator, *IEEE Trans. Microwave Theor. Tech.*, **MTT-S**, 328–30 (1981).
3. E. T. Watkins, H. Yamasaki, and J. M. Schellenberg, 40 GHz FET amplifiers, *1982 ISSCC Digest of Technical Papers*, pp. 198–200 (1982).
4. T. Noguchi and Y. Aono, K and K_a band power GaAs FETs, *IEEE Trans. Microwave Theor. Tech.*, **MTT-S**, 156–8 (1982).
5. J. V. DiLorenzo and W. R. Wisseman, GaAs power MESFETs, *IEEE Trans. Microwave Theor. Tech.*, **MTT-27**, 367–78 (1979).
6. W. Shockley, A unipolar field-effect transistor, *Proc. IRE*, **40**, 1365–76 (1952).
7. R. S. Pengelly, *Microwave Field-Effect Transistors*, Chichester, John Wiley and Sons (1982).
8. J. V. DiLorenzo and D. D. Khandelwal (eds), *GaAs FET Principles and Technology*, Dedham, MA, Artech House (1982).
9. E. O. Johnson and A. Rose, Simple general analysis of amplifier devices with emitter, control, and collector functions, *Proc. IRE*, **47**, 407–18 (1959).
10. H. A. Willing, C. Rauscher, and P. de Santis, A technique for predicting large-signal

performance of a GaAs MESFET, *IEEE Trans. Microwave Theor. Tech.*, **MTT-26**, 1017–23 (1978).
11. P. C. Wade, Novel FET power oscillator, *Electronics Lett.*, **14**, 672–4 (1978).
12. J. S. Joshi and J. A. Turner, High peripheral power density GaAs FET oscillator, *Electronics Lett.*, **15**, 163–4 (1979).
13. Y. Tarui, Y. Komiya, and Y. Harada, Preferential etching and etched profile of GaAs, *J. Electrochem. Soc.*, **118**, 118–22 (1971).
14. J. Mun, J. A. Phillips, and B. E. Barry, High yield process for GaAs enhancement mode MESFET integrated circuits, *IEE Proc. Pt I*, **128**, 144–7 (1981).
15. D. C. D'Avanzo, Proton isolation for GaAs integrated circuits, *IEEE Trans. Microwave Theor. Tech.*, **MTT-30**, 955–63 (1982).
16. S. H. Wemple, W. C. Niehaus, H. M. Cox, J. V. DiLorenzo, and W. O. Schlosser, Control of gate–drain avalanche in GaAs MESFETs, *IEEE Trans. Electron Devices*, **ED-27**, 1013–18 (1980).
17. T. Furutsuka, A. Higashisaka, Y. Aono, Y. Takayama, and F. Hasegawa, GaAs power MESFETS with a graded recess structure, *Electronics Lett.*, **15**, 417–18 (1979).
18. J. P. R. David, J. E. Sitch, and M. S. Stern, Gate–drain avalanche breakdown in GaAs power MESFETs, *IEEE Trans. Electron Devices*, **ED-29**, 1548–52 (1982).
19. H. M. Macksey, F. H. Doerbeck, and R. C. Vail, Optimisation of GaAs power MESFET device and material parameters for 15 GHz operation, *IEEE Trans. Electron Devices*, **ED-27**, 467–71 (1980).
20. I. Drukier, Power GaAs FETs, in ref. 8, Chapter 4, pp. 202–17.
21. M. Abe, T. Mimura, N. Yokoyama, and H. Ishikawa, New technology towards GaAs LSI/VLSI for computer applications, *IEEE Trans. Microwave Theor. Tech.*, **MTT-30**, 992–8 (1982).
22. S. Takahashi, F. Murai, and H. Kodera, Sub micrometre gate fabrication of GaAs MESFET by plasma etching, *IEEE Trans. Electron Devices*, **ED-25**, 1213–18 (1978).
23. C. Huang, A. Herbig, and R. Anderson, Sub micron GaAs FETs for application through K band, *IEEE Trans. Microwave Theor. Tech.*, **MTT-S**, 25–7 (1981).
24. K. Yamasaki, K. Asai, T. Mizutani, and K. Kurumada, Self-align implantation for n^+ layer technology (SAINT) for high speed GaAs ICs, *Electronics Lett.*, **18**, 119–21 (1982).
25. P. A. Barnes and A. Y. Cho, Non alloyed ohmic contacts to n GaAs by molecular beam epitaxy, *Appl. Phys. Lett.*, **33**, 651–4 (1978).
26. See, for example, S. M. Sze, *Principles of Semiconductor Devices*, New York, Wiley-Interscience (1969).
27. L. Hollan and J. Hallais, Vapor phase epitaxy for GaAs FETs, in ref. 8, Chapter 2, pp. 67–144.
28. H. Fukui, Optimal noise figure of microwave GaAs MESFETs, *IEEE Trans. Electron Devices*, **ED-26**, 1032–7 (1979).
29. A. Higashisaka, Y. Takayama, and F. Hasegawa, A high power GaAs MESFET with an experimentally optimised pattern, *IEEE Trans. Electron Devices*, **ED-27**, 1025–9 (1980).
30. J. Rosenberg, P. Chye, C. Huang, and G. Policky, A 26.5–40.0 GHz GaAs FET amplifier, *IEEE Trans. Microwave Theor. Tech.*, **MTT-S**, 166–8 (1982).
31. K. Lehovec and R. S. Miller, Field distribution in junction field effect transistors at large drain voltages, *IEEE Trans. Electron Devices*, **ED-22**, 273–81 (1975).
32. R. A. Pucel, H. A. Haus, and H. Statz, Signal and noise properties of GaAs microwave FETs, *Adv. Electronics Electron Phys.*, **38**, 195–265 (1975).

33. J. A. Turner and B. L. H. Wilson, Implications of carrier velocity saturation in a GaAs FET, *Proc. 2nd Int. Symp. GaAs*, 1968, Paper 30, pp. 195–204 (1969).
34. J. Sone and Y. Takayama, Analysis of field distributions in a GaAs MESFET at large drain voltages, *Electronics Lett.*, **12**, 622–4 (1976).
35. J. Sone and Y. Takayama, A small-signal analytic theory for GaAs FET's at large drain voltages, *IEEE Trans. Electron Devices*, **ED-25**, 329–37 (1978).
36. H. L. Grubin, D. K. Ferry, and K. R. Gleason, Spontaneous oscillations in GaAs FETs, *Solid State Electronics*, **23**, 157–72 (1980).
37. W. Fawcett, A. D. Boardman, and S. Swain, Monte Carlo determination of electron transport properties in GaAs, *J. Phys. Chem. Solids*, **31**, 1963–90 (1970).
38. M. Shur, Influence of non uniform field distribution on frequency limits of GaAs FETs, *Electronics Lett.*, **12**, 615–16 (1976).
39. W. R. Curtice and Y.-H. Yun, A temperature model for the GaAs MESFET, *IEEE Trans. Electron Devices*, **ED-28**, 954–62 (1981).
40. T. W. Maloney and J. Frey, Transient and steady state electron transport properties of GaAs and InP, *J. Appl. Phys.*, **48**, 781–7 (1977).
41. G. Hill, P. N. Robson, and A. Majerfeld, Effect of ionised impurity scattering on the electron transit time in GaAs and InP FETs, *Electronics Lett.*, **13**, 235–6 (1977).
42. R. A. Warriner, Distribution function relaxation times in GaAs, *IEE J. Solid State Electron Devices*, **1**, 92–6 (1977).
43. D. Jones and H. D. Rees, A reappraisal of instabilities due to the transferred electron effect, *J. Phys. C: Solid State Phys.*, **6**, 1781–93 (1973).
44. Y.-C. Wang and Y.-T. Hseih, Velocity overshoot effects on a short gate microwave MESFET, *Int. J. Electronics*, **47**, 49–66 (1979).
45. B. Carnez, A. Cappy, A. Kaszynski, E. Constant, and G. Salmer, Modelling of a submicrometre gate FET including the effects of non stationary electron dynamics, *J. Appl. Phys.*, **51**, 784–90 (1980).
46. R. A. Warriner, Computer simulation of GaAs FETs using Monte Carlo methods, *IEE J. Solid State Electron Devices*, **1**, 105–10 (1977).
47. Y. Awano, K. Tomizawa, N. Hashizume, and M. Kawashima, Monte Carlo simulation of a GaAs short-channel MESFET, *Electronics Lett.*, **19**, 20–1 (1983).
48. H. D. Rees, G. S. Sanghera, and R. A. Warriner, Low temperature FET for low power high-speed logic, *Electronics Lett.*, **13**, 156–8 (1977).
49. K. Hess, Ballistic electron transport in semiconductors, *IEEE Trans. Electron Devices*, **ED-28**, 937–53 (1981).
50. C. V. Shank, R. L. Fork, B. I. Greene, F. K. Reinhart, and R. A. Logan, Picosecond non-equilibrium carrier transport in GaAs, *Appl. Phys. Lett.*, **38**, 104–5 (1981).
51. M. Deblock, R. Fauquembergue, E. Constant, and B. Boittiaux, Electron dynamics in nearly pinched-off GaAs FET operation, *Appl. Phys. Lett.*, **36**, 756–8 (1980).
52. N. P. Maxfield, J. E. Sitch, and P. N. Robson, Transverse electric field effects on the electron transport properties of InP, *Solid State Electronics*, **25**, 655–63 (1982).
53. C. M. Snowden, M. J. Howes, and D. V. Morgan, Large signal modelling of GaAs MESFET operation, *IEEE Trans. Electron Devices*, **ED-30**, 1817–24 (1983).
54. C. R. Brewitt-Taylor, P. N. Robson, and J. E. Sitch, Noise figure of MESFETs, *IEE Proc. Pt I*, **127**, 1–8 (1980).
55. B. T. Debney, A theory of generation–recombination noise from the velocity-saturated channel of a GaAs MESFET, *Solid State Electronics*, **24**, 703–8 (1981).
56. W. Shockley, J. A. Copeland, and R. P. James, The impedance–field method of noise calculation in active semiconductor devices, in *Quantum Theory of Atoms, Molecules and the Solid State*, ed. P.-O. Lowdin, New York, Academic Press, pp. 537–63 (1966).

57. B. Carnez, A. Cappy, R. Fauquembergue, E. Constant, and G. Salmer, Noise modelling in sub micrometre-gate FETs, *IEEE Trans. Electron Devices*, **ED-28**, 784–9 (1981).
58. H. Fukui, J. V. DiLorenzo, B. S. Hewitt, J. R. Velebir, H. M. Cox, L. C. Luther, and J. A. Seman, Optimisation of low noise GaAs MESFETs, *IEEE Trans. Electron Devices*, **ED-27**, 1034–7 (1980).
59. J. A. Turner, R. S. Butlin, D. Parker, R. Bennett, A. Peake, and A. Hughes, The noise and gain performance of submicron gate length GaAs FETs, in ref. 8, Chapter 3, pp. 151–75.
60. H. Yamasaki, GaAs FET technology: a viable approach to millimetre waves, *Microwave J.*, **25**, 93–105 (1982).
61. H. Fukui, Design of microwave GaAs MESFETs for broadband low noise amplifiers, *IEEE Trans. Microwave Theor. Tech.*, **MTT-27**, 643–50 (1979).
62. F. N. Sechi, High efficiency FET power amplifiers, *Microwave J.*, **24**, 59–66 (1981).
63. H. C. Huang, I. Drukier, R. L. Camisa, S. Y. Narayan, and S. T. Jolly, High efficiency GaAs MESFET amplifiers, *Electronics Lett.*, **11**, 508–9 (1975).
64. M. Cohn, J. E. Degenford, and R. G. Freitag, Class B operation of microwave FETs for array module applications, *IEEE Trans. Microwave Theor. Tech.*, **MTT-S**, 169–74 (1982).
65. H. M. Macksey and F. H. Doerbeck, GaAs FETs having high output power per unit gate width, *IEEE Electron Devices Lett.*, **EDL-2**, 147–8 (1981).
66. R. Yamamoto, A. Higashisaka, and F. Hasegawa, Light emission and burnout characteristics of GaAs Power MESFETs, *IEEE Trans. Electron Devices*, **ED-25**, 567–73 (1978).
67. T. Furutsuka, T. Tsuji, and F. Hasegawa, Improvement of drain breakdown voltage of GaAs power MESFETs by a simple recess structure, *IEEE Trans. Electron Devices*, **ED-25**, 563–7 (1978).
68. S. H. Wemple, W. C. Niehaus, H. Fukui, J. C. Irvin, H. M. Cox, J. C. M. Hwang, J. V. DiLorenzo, and W. O. Schlosser, Long-term and instantaneous burnout in GaAs Power FETs: mechanisms and solutions, *IEEE Trans. Electron Devices*, **ED-28**, 834–40 (1981).
69. M. Fukuta, K. Suyama, H. Suzuki, Y. Nakayama, and H. Ishikawa, Power GaAs MESFET with a high drain–source breakdown voltage, *IEEE Trans. Microwave Theor. Tech.*, **MTT-24**, 312–17 (1976).
70. S. Tiwari, L. F. Eastman, and L. Rathbun, Physical and materials limitations on burnout voltage of GaAs Power MESFETs, *IEEE Trans. Electron Devices*, **ED-27**, 1045–54 (1980).
71. L. F. Eastman, S. Tiwari, and M. S. Shur, Design criteria for GaAs MESFETs related to stationary high field domains, *Solid State Electronics*, **23**, 383–9 (1980).
72. W. Frensley, Power limiting breakdown effects in GaAs MESFETs, *IEEE Trans. Electron Devices*, **ED-28**, 962–70 (1981).
73. S. Wemple, W. C. Niehaus, W. O. Schlosser, J. V. DiLorenzo, and H. M. Cox, Performance of GaAs power MESFETs, *Electronics Lett.*, **14**, 175–6 (1978).
74. B. Turner, W. P. Barr, D. P. Cooper, and D. J. Taylor, GaAs microwave power FET with polyimide overlay interconnection, *Electronics Lett.*, **17**, 185–7 (1981).
75. I. Drukier, The design of a 15 GHz high power GaAs FET, *Microwave J.*, **23**, 59–64 (1980).
76. Y. Mitsui, M. Kobiki, M. Wataze, M. Otsubo, T. Ishi, and S. Mitsui, Flip-chip mounted GaAs power FET with improved performance in X-to-K_u-band, *Electronics Lett.*, **15**, 461–2 (1979).
77. Y. Aono, A. Higashisaka, T. Ogawa, and F. Hasegawa, X and K_u band performance

of submicron gate GaAs power FETs, *Jap. J. Appl. Phys.*, **17**, Suppl. 17-1, 147–52 (1978).
78. Y. Aono, A. Higashisaka, and F. Hasegawa, Effects of capacitance at crossover wirings in power GaAs FETs, *Electronics Lett.*, **16**, 417–18 (1980).
79. A. Higashisaka and F. Hasegawa, Estimation of fringing capacities of electrodes on SI GaAs substrate, *Electronics Lett.*, **16**, 411–12 (1980).
80. L. A. D'Asoro, J. V. DiLorenzo, and H. Fukui, Improved performance of GaAs microwave FETs with low inductance via-connections through the substrate, *IEEE Trans. Electron Devices*, **ED-25**, 1218–21 (1978).
81. C. Rauscher and H. A. Willing, Simulation of non linear microwave FET performance using a quasi-static model, *IEEE Trans. Microwave Theor. Tech.*, **MTT-27**, 834–40 (1979).
82. R. S. Tucker, Third order intermodulation distortion and gain compression in GaAs FETs, *IEEE Trans. Microwave Theor. Tech.*, **MTT-27**, 400–8 (1979).
83. R. S. Tucker, RF characterisation of microwave power FETs, *IEEE Trans. Microwave Theor. Tech.*, **MTT-29**, 776–87 (1981).
84. Y. Tajima, B. Wrona, and K. Mishima, GaAs FET large signal model and its application to circuit designs, *IEEE Trans. Electron Devices*, **ED-28**, 171–5 (1981).
85. R. A. Pucel, Profile design for distortion reduction in microwave FETs, *Electronics Lett.*, **14**, 204–6 (1978).
86. J. A. Higgins and R. L. Kuvas, Analysis and improvement of intermodulation distortion in GaAs power FETs, *IEEE Trans. Microwave Theor. Tech.*, **MTT-28**, 9–17 (1980).
87. S. H. Wemple and H. Huang, Thermal design of power GaAs FETs, in ref. 8, Chapter 5, pp. 309–47.
88. P. C. Wade, D. Rutkowski, and I. Drukier, Pulsed GaAs FET operation for high peak output power, *Electronics Lett.*, **15**, 591–2 (1979).
89. H. Fukui, S. H. Wemple, J. C. Irvin, W. C. Niehaus, J. C. M. Hwang, H. M. Cox, W. O. Schlosser, and J. V. DiLorenzo, Reliability of power GaAs FETs, *IEEE Trans. Electron Devices*, **ED-29**, 395–401 (1982).
90. J. C. Irvin, The reliability of GaAs FETs, in ref. 8, Chapter 6, pp. 351–400.
91. E. Allamando, G. Salmer, M. Bouhess, and E. Constant, Influence of gate interval on the behaviour of submicron dual gate FETs, *Electronics Lett.*, **18**, 791–3 (1982).
92. C. Tsironis and R. Meirer, Equivalent circuit of GaAs dual gate MESFETs, *Electronics Lett.*, **17**, 477–9 (1981).
93. R. S. Pengelly, Novel FET circuits, in ref. 7, Chapter 9, pp. 315–38.
94. J. W. Berenz, G. C. Dalmand, and C. A. Lee, Improved FET design reaches millimetre waves, *Microwaves*, **21**, 67–71 (1982).
95. J. C. Vokes, B. T. Hughes, D. R. Wight, J. R. Dawsey, and S. J. W. Shrubb, Novel microwave GaAs FETs, *Electronics Lett.*, **15**, 627–9 (1979).
96. R. J. M. Griffiths, I. D. Blenkinsop, and D. R. Wight, Preparation and properties of GaAs layers for novel FET structures, *Electronics Lett.*, **15**, 629–30 (1979).
97. T. Mimura, S. Hiyamizu, T. Fujii, and K. Nanbu, A new FET with selectively doped GaAs/n-Al_xGa_{1-x}As heterojunctions, *Jap. J. Appl. Phys.*, **19**, L225–7 (1980).
98. T. Mimura, S. Hiyamizu, K. Joshin, and K. Hikosaka, Enhancement mode high electron mobility transistors for logic applications, *Jap. J. Appl. Phys.*, **20**, L317–19 (1981).
99. P. N. Tung, D. Delagebeaudeuf, M. Laviron, P. Delescluse, J. Chaplart, and N. T. Linh, High speed two-dimensional electron gas FET logic, *Electronics Lett.*, **18**, 109–10 (1982).
100. P. N. Tung, P. Delescluse, D. Delagebeaudeuf, M. Laviron, J. Chaplart, and N. T.

Linh, High speed low power DCFL using planar 2D electron gas FET technology, *Electronics Lett.*, **18**, 517–19 (1982).
101. D. Delagebeaudeuf and N. T. Linh, Metal–(n)AlGaAs–GaAs two dimensional electron gas FET, *IEEE Trans. Electron Devices*, **ED-29**, 955–60 (1982).
102. P. Delescluse, M. Laviron, J. Chaplart, D. Delagebeaudeuf, and N. T. Linh, Transport properties in GaAs–$Al_xGa_{1-x}As$ heterostructures and MESFET applications, *Electronics Lett.*, **17**, 342–4 (1981).
103. A. D. Welbourn, Gigabit logic. A review, *IEE Proc. Pt I*, **129**, 157–72 (1982).
104. M. Laviron, D. Delagebeaudeuf, P. Delescluse, J. Chaplart, and N. T. Linh, Low noise two-dimensional electron gas FET, *Electronics Lett.*, **17**, 536–7 (1981).
105. D. Delagebeaudeuf, M. Laviron, P. Delescluse, J. Chaplart, and N. T. Linh, Planar enhancement mode two-dimensional electron gas FET associated with a low AlGaAs surface potential, *Electronics Lett.*, **18**, 103–5 (1982).
106. S. L. Lu, R. Fischer, T. J. Drummond, W. G. Lyons, R. E. Thorne, W. Kopp, and H. Morkoc, Modulation doped (Al,Ga)As/GaAs FETs with high transconductance and electron velocity, *Electronics Lett.*, **18**, 794–6 (1982).
107. C. O. Bozler and G. D. Alley, Fabrication and numerical simulation of the permeable base transistor, *IEEE Trans. Electron Devices*, **ED-27**, 1128–41 (1980).
108. C. O. Bozler and G. D. Alley, The permeable base transistor and its application to logic circuits, *Proc. IEEE*, **70**, 46–58 (1982).
109. H. Morkoc, T. J. Drummond, and C. M. Stanchak, Schottky barriers and ohmic contacts on n-type InP based compound semiconductors for microwave FETs, *IEEE Trans. Electron Devices*, **ED-28**, 1–5 (1981).
110. H. Kroemer, Heterostructure bipolar transistors and integrated circuits, *Proc. IEEE*, **70**, 13–25 (1982).
111. R. Dingle, H. L. Störmer, A. C. Gossard, and W. Wiegmann, Electron mobilities in modulation-doped semiconductor heterojunction superlattices, *Appl. Phys. Lett.*, **33**, 665–7 (1978).

Gallium Arsenide
Edited by M. J. Howes and D. V. Morgan
© 1985 John Wiley & Sons Ltd

CHAPTER 11

Gaas Optoelectronic Devices

D. H. Newman

11.1 INTRODUCTION

GaAs emerged as an optoelectronic material in the early 1960s and consolidated its position with the development, a decade later, of double heterostructure GaAs/GaAlAs stripe lasers capable of continuous operation at room temperature. With the initial impetus of the prospects for optical fibre transmission, considerable research and development has gone into perfecting designs of GaAs-based lasers and LEDs emitting milliwatts of power and operating stably for hundreds of thousands of hours. These are now being installed with commercial fibre systems worldwide, both in long-distance (inter-city) public telecommunications networks and also in a range of short-distance links. Further applications in fibre optics such as avionics data and control links, military weapons systems and also inter-rack wiring within switching systems and computers are being investigated.

New applications have also emerged in the areas of video discs (as a replacement for the He–Ne laser) and digital audio discs where mass consumer markets are predicted.

The higher power capability being exhibited by some GaAs-based laser configurations is emphasizing the possibilities that exist for semiconductor lasers in optical recording and in printing.

GaAs is being developed by the solar cell industry and may find a substantial niche in concentrator cell arrangements. It is also a possible optical detector material for 600–850 nm optical systems, but, having to compete with silicon, it has seen little development in this direction to date.

Unlike GaAs electronic technology, the optoelectronics applications rely for their major part on the incorporation of GaAlAs/GaAs heterojunctions to achieve efficient operation and this has limited the complexity of devices to date. The development of refined epitaxial growth techniques in the form of MBE and MOCVD (Chapter 3) is now beginning to allow much more precise control of uniformity and a much greater complexity (in terms of number of epilayers that can be controllably grown on a substrate). Thus GaAs-based optoelectronic technology is now coming of age and the prospects are good for novel components, monolithic optoelectronic integration (incorporating electronic and optical functions), and also cheap and reliable components to emerge in the current decade.

11.2 PRINCIPLES OF DEVICE OPERATION

GaAs is a direct gap semiconductor, so that recombination of electron–hole pairs across the energy band gap results in the efficient emission of photons of close to band gap energy, E_g. The wavelength of the emitted photons, λ, is given by

$$\lambda \approx hc/E_g \quad (11.1)$$

where h is Planck's constant and c the velocity of light in vacuum. This wavelength is in the near infrared, close to 900 nm at 300 K. The spectral width of the emission is determined by the density of states functions, transition probabilities, and temperature.

Electron–hole recombination is conveniently achieved by injecting holes and/or electrons across a heavily forward biased p–n junction—a so-called light-emitting diode (LED) structure. This basic structure suffices for a few simple requirements for near-infrared diode emitters (such as in opto-couplers) but improved LED efficiency and also the possibility of efficient laser action can be achieved by incorporating gallium aluminium arsenide ($Ga_{1-x}Al_xAs$) heterojunctions.

In the double heterostructure (DH) arrangement, electrons and/or holes are injected from wider band gap (higher Al content) material into the narrower gap recombination (or active) region (Figure 11.1). The doping scheme adopted determines whether electrons or holes are injected—in the example shown, with a heavily p-type doped active layer, electrons are the injected species. The wider band gap layers ensure efficient confinement of electrons and holes in the narrow band gap active layer when the potential barriers introduced by the heterojunctions are several times kT. In practice, barriers of 0.2–0.3 eV will give good confinement, even well above room temperature. The emission wavelength can be decreased by incorporating Al into the active layer. Active layer thicknesses vary normally, in the range 0.1–2 μm dependent on the application.

An additional aspect of the DH structure is that it forms an optical slab waveguide that is able to guide light in the plane of the p–n junction. This has application in some circumstances, particularly with respect to lasers. The slab waveguide comes about automatically because, as the Al content of the $Ga_{1-x}Al_xAs$ alloy making up the wide gap confining layers is increased, it refractive index, measured at the emission wavelength of the narrow gap active layer, decreases.

GaAs optoelectronic devices 431

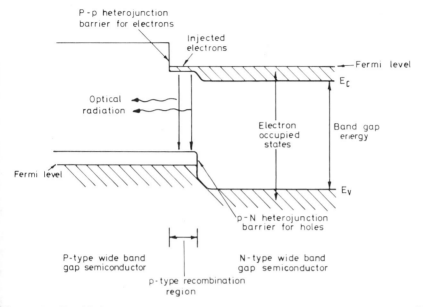

Figure 11.1 Double heterostructure energy band diagram. From Newman and Ritchie.[7] Copyright (1981) John Wiley & Sons Ltd. Reprinted by permission

Although carrier and optical confinement can be achieved using a heterojunction approach, good radiative properties can only be obtained if the heterojunction is free of defects—in practice this means lattice matching. Fortuitously, GaAs and AlAs have similar lattice constants and low interface defect density results.[1]

GaAs structures can also be used for detection of optical photons with an energy greater than the band gap energy (equation (11.1)), i.e. for wavelengths less than about 900 nm at 300 K. Absorption of a photon creates an electron–hole pair and if created in close proximity to a p–n junction can lead to flow of current in an external circuit (Figure 11.2). As both a solar cell and also as a photodetector in fibre communication systems and opto-couplers, GaAs has to compete with silicon, which also absorbs optical radiation in the same wavelength range and has obvious technological advantages.

1.3 LIGHT-EMITTING DIODES

The basic GaAs LED structure is a rectangular die (Figure 11.3) with metallic ohmic contacts to the p and n sides. Because GaAs is a direct band gap material, optical radiation emitted close to the p–n junction is heavily absorbed elsewhere in the structure with resulting low external efficiency. Thus more efficient structures incorporating special geometric features or using wide band gap

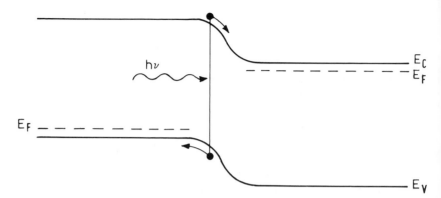

Figure 11.2 Energy band diagram of an illuminated p–n junction with $h\nu > E_g$

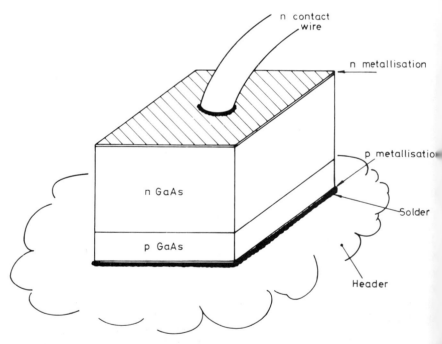

Figure 11.3 Basic GaAs LED structure. From Newman and Ritchie.[7] Copyright (1981) John Wiley & Sons Ltd. Reprinted with permission

heterolayers for both carrier confining and transparency have been developed. This is the case particularly for the fibre optics market, where power launched into the fibre is often the critical aspect determining operating margin or maximum communicating link length before signal regeneration. Two generic types of these so-called high-radiance LEDs (HRLEDs) exist—surface emitting or edge emitting.

The surface-emitting LED (SLED) achieves high radiance (power emitted into unit solid angle) by restricting emission to a small area within a larger chip using a restricted area p contact and a p–n junction close to that contact (1–5 μm distant) so that, at the high current densities employed (1–10 kA cm^{-2}), current spreading is substantially eliminated (Figure 11.4). The optical radiation emitted nearly perpendicular to the p–n junction plane is taken out through a well etched in the otherwise absorbing n-GaAs substrate. Figure 11.4 depicts the double heterostructure version, which is the most efficient, making use of both the current confinement properties and the transparency of the $Ga_{1-x}Al_xAs$ sandwiching layers. Radiation is, of course, emitted isotropically from the active layer, but because of the high refractive indices of GaAs and $Ga_{1-x}Al_xAs$ alloys (about 3.5), only that within a cone of half-angle 16° about the normal to the p–n junction plane can escape from the diode through the well in the substrate.

For fibre optics applications, the emitting area is usually made equal to the fibre core diameter for straightforward butt-fibre coupling; however, microlenses can be employed, either cemented within the LED well or formed on the coupling fibre to increase coupled power. In this latter case, because the lens effectively magnifies the image of the emitting area as seen by the fibre, smaller emitting

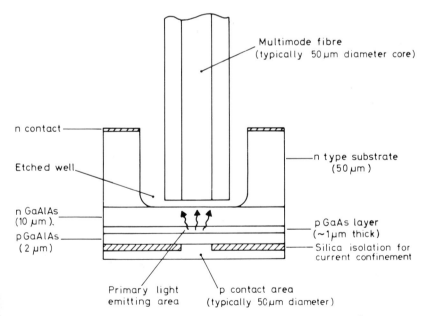

Figure 11.4 Cross section through a high-radiance surface-emitting LED[7] designed for fibre optics applications. From Newman and Ritchie.[7] Copyright (1981) John Wiley & Sons Ltd. Reprinted with permission

areas can be employed. Thus SLEDs with emitting areas between perhaps 5 and 150 μm have been pursued for various applications.[2]

The edge-emitting LED (ELED) structure utilizes the optical radiation emitted in the junction plane and is assisted by the optical guiding effect of the double heterostructure configuration already mentioned. A restricted length stripe p contact electrode is often employed (Figure 11.5) to improve efficiency as self-absorption or loss of radiation emitted deep within the structure can otherwise limit external efficiency. Alternatively, longer stripe regions can be employed to encourage super-radiant emission at high injected carrier densities as a result of single pass optical gain along the length of the active layer. A restricted length stripe contact is necessary in this case also, in order to frustrate the onset of laser action (i.e. round-trip optical gain). ELED output is more directional than SLED output, so that somewhat easier coupling of optical radiation into fibre results, particularly in the super-radiant mode. In addition, because of the optical gain process, this latter device has a narrower spectral width than SLEDs or other ELEDs and this is of benefit in combating the effects of fibre dispersion in some applications. Typical linewidths for SLEDs and ELEDs lie in the range $2kT$ to $3kT$, depending on doping effects (i.e. 30–50 nm), whereas linewidths down to 10–15 nm have been reported for the super-radiant ELED.[3]

An important parameter of a LED is its modulation capability and, although at low injected current levels this may be limited by junction capacitance, in the heavy forward biased operation mode it is fundamentally limited by the lifetime of carriers injected into the recombination region. If the current is modulated at frequency ω, the optical output intensity $I(\omega)$ will vary in the following manner:[4]

$$I(\omega) = I_o[1 + (\omega\tau)^2]^{-1/2} \quad (11.2)$$

where τ is the carrier lifetime. The 3 dB electrical bandwidth will be τ^{-1}. The variation of carrier lifetime with injected current density can be calculated, and, in circumstances where only radiative band-to-band recombination is important, is given by:[5]

$$\tau = \frac{qd}{2J}(p_0 + n_0)\{[1 + 4J/qB_rd(p_0 + n_0)^2]^{1/2} - 1\} \quad (11.3)$$

with p_0 and n_0 the background hole and electron doping, q the electronic charge, d the recombination region width, and B_r the radiative recombination coefficient ($\sim 10^{-10}$ cm^3 s^{-1} for GaAs at 300 K). Thus modulation bandwidth can be increased either by using a highly doped recombination region or by operating at high current densities and small recombination region width (so overcoming the influence of background carrier concentration). With high doping, non-radiative processes begin to dominate so that (radiative) quantum efficiency reduces and is the price that has to be paid for achieving a fast response. Modulation bandwidths up to several hundred megahertz are available from specially designed structures although 10–50 MHz is typical. These lower bandwidth

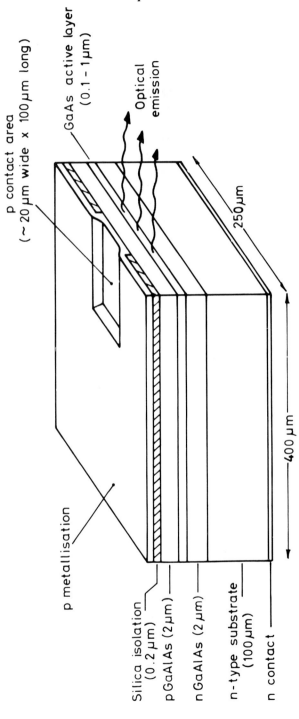

Figure 11.5 Schematic of an edge-emitting LED. From Newman and Ritchie.[7] Copyright (1981) John Wiley & Sons Ltd. Reprinted with permission

devices usually have the best output power performance. The super-radiant ELED can be an exception to this rule, where reduced carrier lifetime is achieved via the influence of stimulated emission and where the resulting single pass optical gain can give enhanced optical output. It is clear, therefore, that devices can be designed for either high speed or output power and that detailed consideration,[6] beyond the scope of this chapter, is necessary in order to optimize device characteristics for a particular application. Output power linearity is another characteristic, particularly important in analogue applications, that has also to be considered along with device thermal properties and the temperature variation of output.

Considerable effort has been expended in the improvement of GaAs/GaAlAs LED reliability which has, in the past, been of extreme concern owing both to localized internal production of non-radiative recombination centres (reducing output power at a fixed injection current) and also to imperfect metallization schemes.[7] However, results of overstress testing predict operating lives of 5 $\times 10^6$ h or more at 300 K and use of devices in fibre systems is also building up confidence that reliability is now satisfactory. Use of low dislocation density substrates, reduction of bonding strain, deposition of impervious metal contact layers, and control of a number of other processing variables have all been essential to the attainment of high reliability for GaAs LEDs.

11.4 LASERS

Although homojunction (i.e. p–n GaAs) laser diodes and single heterostructure GaAs/Ga$_{1-x}$Al$_x$As devices have been developed and the latter still find commercial application as high-power sources, the vast majority of research and development effort over the last decade has been directed towards the double heterostructure laser and its variations. With advances in epitaxial growth techniques achieved (LPE mainly, but also, recently MOCVD and even MBE), it is probable that variations of the basic DH laser can now be more easily and controllably fabricated than single heterostructure versions to meet all types of requirements for GaAs/GaAlAs laser sources.

For a laser structure, the basic DH arrangement is incorporated in an optical cavity (Figure 11.6) that is formed by cleaving the semiconductor wafer perpendicular to the DH layer plane. The high reflectivity ($\sim 30\%$) of a GaAs/air interface gives optical feedback into the cavity so that above a critical value of injected current—the threshold current—round-trip cavity optical gain is achieved, sufficient to overcome the losses, and laser action results. Optical radiation is, of course, amplified in the DH layer plane and guided by the slab waveguide geometry so that, above the lasing threshold, the optical output increases much more rapidly with drive current (Figure 11.7).

It is usually the case that the thickness of the active (recombination) layer (d) is less than the injected carrier diffusion length so that (if the heterobarriers are

GaAs optoelectronic devices

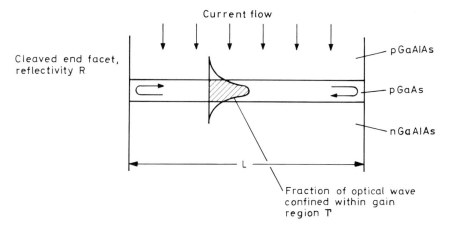

Figure 11.6 Schematic DH laser cavity

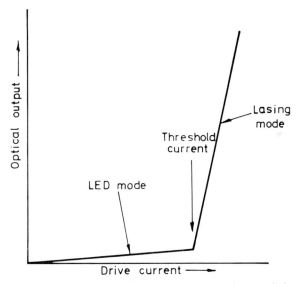

Figure 11.7 Idealized optical output–current characteristic for a laser. From Newman and Ritchie.[7] Copyright (1981) John Wiley & Sons Ltd. Reprinted with permission

sufficient to confine all the injected carriers) it is uniformly excited across its thickness. Under these circumstances, the injected current density J (below threshold) is given by

$$J = nqd/\tau \qquad (11.4)$$

438 *Gallium arsenide*

where n is the number of electron–hole pairs per unit volume and τ the lifetime in the active region, so that small widths are required to achieve a given injected carrier density at a small current density. As d is reduced to less than 1 μm, only the fundamental optical mode can be guided (with the refractive index steps available in the GaAs/GaAlAs system) and only a fraction of this mode energy (Γ) is contained within the active layer (Figure 11.6), a consequence of Maxwell's equations applied to dielectric waveguides.[8]

To achieve laser action in the cavity, it is necessary to ensure a round-trip gain of unity, i.e. at threshold

$$R_1 R_2 \exp\{[g_{th}\Gamma - \alpha_p(1-\Gamma) - \alpha_a\Gamma]2L\} = 1 \qquad (11.5)$$

where R_1, R_2 are mirror reflectivities, g_{th} the threshold gain, α_p and α_a are the (free carrier) absorption coefficients in the passive and active layers respectively, and L is the cavity length. The relationship between gain and current density can often be written in the form[9]

$$g = A(J/d) - B \qquad (11.6)$$

where A and B are constants independent of active region thickness d, so that

$$J_{th} = \frac{d}{A}\left[\alpha_p\left(\frac{1-\Gamma}{\Gamma}\right) + \alpha_a + B + \frac{1}{2L\Gamma}\ln\left(\frac{1}{R_1R_2}\right)\right] \qquad (11.7)$$

For parameters relevant to GaAs DH lasers, this expression gives $J_{th} \sim$ 1 kA cm^{-2} for $d = 0.2$ μm, implying an injected carrier density (from equation (11.4)) of around 10^{18} cm^{-3} at 300 K. There is an implicit temperature dependence through the coefficients A and B. Because of the decrease in Γ as active region thickness decreases, there is an optimum active region thickness that minimizes threshold current density for a given refractive index step (Δn) between the active and passive layers. This occurs for 0.05–0.2 μm active region thickness with aluminium content steps in the range 60 to 20%.

There have been some studies aimed at further reduction of threshold current density by utilization of the quantum size effect. For active region thicknesses of around 0.03 μm or less (equivalent to the de Broglie wavelength for thermalized electrons), the electron and hole energy levels become quantized in a manner given by the bound state energies of a finite square potential well. For deep wells, the allowed energy levels are approximately given by

$$E_n = \frac{\hbar^2}{2m^*}\left(\frac{n\pi}{d}\right)^2 \qquad (11.8)$$

where n is an integer. In GaAs, when $d = 100$ Å, $E_1 = 56$ meV for electrons and 5 meV for heavy holes. The density of states is steplike in a quantum well heterostructure, compared with parabolic in the bulk semiconductor. Figure 11.8 illustrates these points. Thus recombination can occur between two well defined levels in a quantum well laser and calculations indicate that the optical gain is

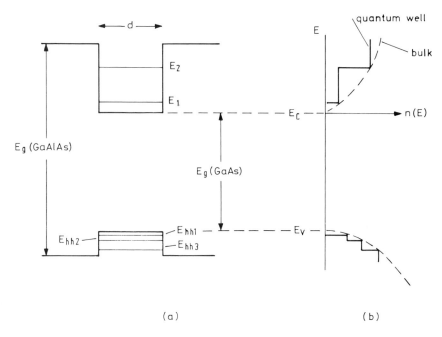

Figure 11.8 (a) Quantum well heterostructure energy levels and (b) density of states functions drawn to approximate scale for the situation with $d = 100$ Å and for $Ga_{0.7}Al_{0.3}As$ confining layers. For simplicity the light hole band is not shown

significantly enhanced at any particular injection level.[10] Thus A in equation (11.6) is increased (B is also changed somewhat) and the net result is that lower threshold current densities are possible in suitable structures incorporating quantum wells.

It is clear that optimization of the performance of a double heterostructure slab laser diode is complex but even this is only part of the overall optimization process for most device structures. For controlled launching of optical radiation into fibres and for focusing to diffraction-limited dimensions with respect to video and audio disc reading, it is necessary to move to stripe geometry laser designs. The restriction of current flow to a narrow stripe also reduces absolute current demand and improves thermal properties so that quasi-continuous (CW) operation is much more easily obtained over a reasonable temperature range around and above normal room temperatures.

The simplest stripe geometry structures are the oxide-isolated version (Figure 11.9a) and also the similar in concept proton-isolated version where the width of the contact stripe is usually of the order of 3–6 μm but can be wider for particular applications in which, for example, output power rather than precise control of the modal behaviour is important. In these simple stripe lasers, the optical field is

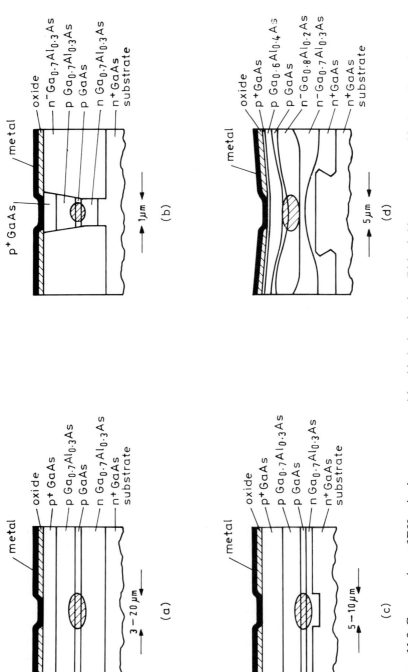

Figure 11.9 Cross sections of DH stripe laser structures: (a) oxide-isolated stripe; (b) buried heterostructure; (c) channelled substrate planar; (d) CDH-LOC structure. (Copyright © 1981 IEEE). Reproduced with permission from Botez[13]

usually controlled in the transverse direction (in the plane of the layer structure perpendicular to the cavity direction) by the injected carrier profile, which peaks in the centre of the stripe, and the laser is then said to be gain-guided. Injected electrons decrease the refractive index so that there is an index antiguiding effect enhancing the diffraction of the optical wave in the transverse direction and so increasing cavity losses (which in turn increases threshold current density through an extension of equation (11.7). In addition, interaction between injected carriers and the optical field can result in changes of transverse mode structure, excess noise, and non-linearities (so-called 'kinks') in the optical output transfer characteristics. However, in spite of these potential hazards, the relative simplicity of the oxide- or proton-isolated stripe structures, incorporating the single stage epitaxial growth of a planar layer structure by LPE (and now also MOCVD and MBE techniques), and the resultant high device yield, has led to the development of stable rather narrow stripe (~ 3 μm wide) versions.[11] The narrow stripe, coupled with good control over layer uniformity, ensures that the optical wave exhibits zero-order transverse mode behaviour and remains located centrally on the stripe. Threshold currents are usually in the range 50–100 mA at 300 K and zero-order transverse mode outputs up to 10 mW are obtained. Gain-guided lasers exhibit a multi-longitudinal mode spectrum due to the high injected current density required to achieve lasing threshold and the variation of carrier density across the stripe in the transverse direction. The longitudinal modes of the laser are the result of superposition of the Fabry–Perot modes of the optical cavity on the gain profile (as a function of wavelength) of the active medium. Because electron–hole recombination is occurring between bands (rather than between discrete levels as in gas lasers for example) the spontaneous LED emission is broad (~ 50 nm in GaAs) and optical gain is achieved near the peak of this spectrum. The modes selected by the lasing cavity are separated by

$$\Delta \lambda = \lambda^2/2n_i L$$

where λ is the free-space wavelength and n_i the group index (~ 4.5 in GaAs). Thus for $L = 300$ μm, $\Delta\lambda$ is about 0.3 nm. The gain-guided structure, which can exhibit up to 40 separate longitudinal modes in its emission, is thus ideally suited to overcome so-called modal noise effects that can arise in multimode optical fibres.[12] However, the optical wavefront in the transverse direction is non-planar because of the lack of refractive index guiding, so that difficulties exist in focusing to the diffraction-limited spot size preferred for disc playback, etc.

To achieve lower operating currents and to improve upon some of the other characteristics of simple stripe lasers, a large range of stripe geometry variants have been developed incorporating positive index guiding of the optical field in the transverse direction. These variants involve growth over non-planar surfaces and also, often, two-stage growth.[2] In the present discussion, it is possible only to mention a small number of these variants, which, however, are representative of the whole class. The extreme example, both in terms of low threshold current and small dimensions, is the buried heterostructure device. By a two-stage process, involving the growth of a sequence of DH planar layers, formation of mesas, and subsequent overgrowth to surround the laser active region with material of both larger band gap and lower refractive index in the transverse as well as in the perpendicular direction (Figure 11.9b), a positive index-guided waveguide structure is formed. The active region width (in the transverse direction) has to be restricted to about 1 μm in order to achieve controlled zero-order transverse mode operation because of the relatively large index steps introduced. Threshold currents around 10 mA are achieved when the structure is fabricated accurately enough to minimize parallel leakage paths. The optical flux density is very high at the output facets because of the very restricted dimensions so that, even with protective Al_2O_3 facet coatings, these devices are not well suited for reliable high-power output performance.

A structure where a somewhat wider transverse dimension can be used (because the transverse index guiding can be made weaker than in the BH case) is the channel substrate variant called the CSP (channel substrate planar) laser. Here, by suitable control of the LPE process, a set of planar DH layers are grown over a shallow groove etched in the GaAs substrate (Figure 11.9c). Outside of the channel, losses are increased, inhibiting laser action there. Because of the change in dimensions inside and outside the channel, transverse waveguiding results, which encourages fundamental transverse mode operation so long as the channel width is less than 8–10 μm. Power handling capability is enhanced above that possible with the BH laser (because of the larger extent of the optical mode) at the expense of an increased threshold current ($\sim$ 50 mA).

An example of a structure that has been developed specifically with the requirement for higher output powers in a stable fundamental transverse mode in view, is the CDH-LOC (constricted double heterostructure–large optical cavity) laser[13] where attention is given to maximizing the optical field cross-sectional area in order to minimize optical power density at the output facets (Figure 11.9d). A single stage of LPE growth over a shallow mesa separating two substrate channels produces the configuration required. In the direction perpendicular to the junction a wide lasing field is achieved through use of the large optical cavity concept,[8] where carriers are restricted to recombine in a narrow active layer but where the optical field spreads into a wider ($\sim$ 1.5 μm thick) region (the LOC) because of the reduced optical confinement barrier between the recombination and adjacent optical guide layer. Fundamental transverse mode (in the junction plane) operation to high output powers is achieved by relatively weak positive refractive index confinement in conjunction with discriminatory higher-order mode loss. The convex active layer in combination with the concave guide layer provide such a situation. The resulting lasing spot is approximately 1.7 μm $\times$ 7 μm and outputs up to 40 mW are reliably emitted. In addition, as with the other index-guided structures discussed, most of the power is contained within a single longitudinal mode (at least under CW operating conditions).

A number of other approaches have been taken to achieve high CW output powers, with particular application to laser printing. Terraced heterostructure large optical cavity (THLOC) lasers emitting 50 mW,[14] coupled multiple stripe (CMS) phased-array lasers emitting 400 mW,[15] stripe buried heterostructure (SBH) lasers emitting 60 mW,[16] and twin ridge substrate (TRS) lasers emitting 100 mW[17] from one facet have all been reported recently.

Major advances have also been made in the fabrication of $Ga_{1-x}Al_xAs$ lasers with shorter emission wavelengths than those required for fibre optics (800–900 nm), with the digital video and audio disc application in view. The shorter wavelength is required in order to reduce the diffraction-limited focused laser beam spot-size to the minimum in order to increase disc information capacity. The shorter wavelength (emission at 680–735 nm has been reported now by a number of laboratories) is obtained by increasing the aluminium content of the active layer in a DH stripe laser up to between 20 and 30% with a corresponding increase in the aluminium content of the passive layers to retain both the carrier and optical confinement effects. Unfortunately, because $Ga_{1-x}Al_xAs$ becomes an indirect gap semiconductor beyond $x = 0.45$, the rate of increase in band gap with increasing x beyond this value is only a fraction of that for $x < 0.45$, so that a natural limit is set close to 680 nm for efficient lasing action in the DH configuration. It is possible that the introduction of quantum well confinement may assist in allowing more controlled device performance to be achieved close to this limit by reducing threshold current density,[18] as already discussed. Also, because the lowest-energy allowed confined-particle states are, in the case of electrons, considerably above the bulk conduction band minimum (see Figure 11.8), somewhat shorter wavelength emission may be possible.

Other factors such as built-in stress may also have an influence on device performance close to 680 nm. It has been shown that lower threshold current densities can be achieved by incorporating a thick (~ 100 μm) $Ga_{1-x}Al_xAs$ ($x = 0.15$) LPE- grown buffer layer which allows the original GaAs substrate to be removed prior to device fabrication.[19] Although GaAlAs alloys can be grown lattice-matched to GaAs substrates over a very wide range of compositions (the lattice constants of GaAs and AlAs are equal to 900°C), stress is introduced on cooling from the growth temperature owing to differential thermal expension between the layers. It is also possible that such effects may influence reliability in a detrimental manner.

These examples discussed give some insight into the complexity involved in the design and fabrication of DH stripe lasers. The majority of work has been carried out using the LPE technique, but, with the control being demonstrated by MOCVD and MBE in achieving good performance of simple planar layer lasers,[20, 21] effort has already begun to investigate the types of non-planar structures that might be possible using these latter techniques where growth conditions are very different from those obtaining during LPE.[22, 23]

One final example of the capability that still exists for improvement in device characteristics is worthy of specific mention. Threshold current densities down to 120 A cm^{-2} have been reported[10, 24] for graded refractive index–separate confinement heterostructure (GRIN-SCH) lasers grown by MOCVD. The structures consist of a 60–100 Å wide GaAs single quantum well, with 250 meV confinement barriers, surrounded by graded regions of varying $Ga_{1-x}Al_xAs$ composition which are thought to enhance carrier confinement in the quantum well active region. This value of threshold current is between a factor of 5 and 10 better than can be achieved with conventional DH structures. No doubt both theoretical understanding of this phenomenon and fabrication technology associated with its realization are still capable of refinement and it is possible that a whole new stage in GaAlAs laser technology will unfold as a result.

Of particular importance to the use of lasers in fibre optics applications in their modulation and transient behaviour. The radiative emission from $Ga_{1-x}Al_xAs$ lasers has been modulated at rates of at least 8 gigabits per second[25] by imposing signals on the laser injection current, but not all types of laser exhibit this capability. The ability of a laser to respond to high modulation rates is limited by both the switch-on delay before the onset of lasing and also by damped oscillations due to coupling between the carrier and photon populations in the laser cavity. The switch-on delay is related to the period required to build up the excess injected population in the required threshold level in the presence of (spontaneous) electron–hole pair recombination. The delay time can be written as[9]

$$t_d = \tau \ln[(I - I_0)/(I - I_{th})] \quad (11.9)$$

where τ is the carrier recombination lifetime (usually of the order of 1–5 ns), I the value of injected current, I_0 is the pre-bias level, and I_{th} the threshold current. Thus, as expected, delay time can be reduced either by high drive currents (usually not permissible) or by pre-biasing very close to (or even above) the lasing threshold. In this way the very rapid turn-on necessary for multigigabit per second modulation can be achieved but at the expense of extinction ratio (i.e. the optical intensity ratio between digital 1s and 0s). Thus it is necessary to control the pre-bias level rather accurately to achieve optimum performance.

Damped relaxation oscillations usually occur in the 0.2–2 GHz range. An approximation to the relaxation frequency v_R is given by[26]

$$v_R = (1/2\pi)[(I - I_{th})/I_{th}\tau_p\tau]^{1/2} \quad (11.10)$$

where τ_p is the photon lifetime in the lasing cavity (of the order of several picoseconds). Again pre-bias close to threshold can minimize relaxation effects since carrier and optical perturbations are reduced. In addition, some stripe laser structures show marked suppression of the relaxation resonance, probably due to effects associated with carrier indiffusion from the edges of the stripes being able to compensate for loss of carriers in the centre where interaction with the optical field is strongest.[2]

GaAs optoelectronic devices

It is also worth noting that the output spectrum of a laser may be altered when it is operating at rates of hundreds of megabits per second. In particular, devices that operate in a single longitudinal mode under d.c. conditions may well emit in a number of modes over periods of nanoseconds during pulse operation due to the transient variations of carrier densities and optical fields. Laser output power may also be 'partitioned' and exist in a different set of longitudinal modes from pulse to pulse. Other transient variations in output may also occur due to external feedback from, for example, fibre ends placed in close proximity that may give a system penalty in a fibre transmission link. On the other hand, the effect of the variation in optical feedback on overall laser diode voltage under constant current drive conditions may be used constructively to detect variations in reflectivity of a surface adjacent to the laser output facet (as in a digital video or audio disc).

Laser reliability has had a long and tortuous history, but now, as in the case of LEDs, with early problems under control, lifetimes for operation at 20°C of 10^5–10^6 h are being extrapolated[7] from temperature overstress testing. As with LEDs, attention to substrate and epitaxial layer quality and morphology, contact metallization, and also the use of protective facet coatings to help stabilize the exposed cavity surfaces has enabled commercial products to appear that meet demands.

11.5 OTHER COMPONENTS

GaAs will probably never be a material seeing large-scale use for photodetectors in fibre systems because of the intense competition that exists from silicon and the very well developed state of silicon technology. However, a number of detector types have been investigated, some of which will find application in monolithically integrated optoelectronic circuits. These detectors include the reverse-biased p–i–n detector, the photo-FET (or optically sensitive FET) and the phototransistor. Other devices, such as avalanche photodiodes (APDs), incorporating graded gap or multilayer structures are not being seriously contemplated as specific detectors for 800–900 nm radiation, but rather as precursors to the development of versions in other materials (when the required growth techniques become available), suitable for 1300–1600 nm detection where there is as yet no equal to the silicon APD in performance.

The p–i–n diode configuration offers improved performance over the simple p–n junction photodiode mentioned earlier. The p–i–n structure incorporates a low-doped (intrinsic) 'i' region between the n- and p-type layers (Figure 11.10). Sufficient reverse voltage is applied to deplete fully the i region and to produce a field in it sufficient for carriers to drift at near their saturated drift velocity. The i region need be only several micrometres thick in the case of a direct gap material such as GaAs, as the high absorption coefficient ensures that the majority of carriers can be absorbed there. This eliminates the slow response due to carriers

Gallium arsenide

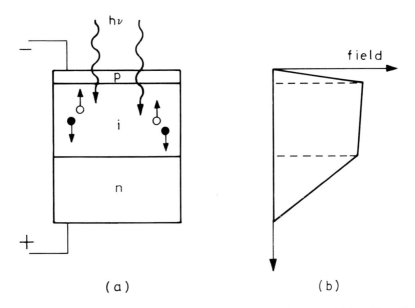

Figure 11.10 (a) Schematic of a p–i–n structure and (b) electric field profile with device reverse-biased

drifting slowly in undepleted material and also assists in achieving high quantum efficiency since the chance of recombination is less. Junction capacitance is reduced because of the wider depleted regions in the p–i–n structure, which is helpful also. Improved efficiency could result by using a $Ga_{1-x}Al_xAs$ window layer for the p layer, thus avoiding absorption in an undepleted p^+-GaAs layer. With this sort of design, quantum efficiencies approaching 100% could be achieved (if an antireflection surface coating were incorporated) and a response time less than 100 ps achieved (i-region transit time $\sim$ 10 ps μm^{-1}).

At higher reverse voltages in a p–i–n structure, avalanche multiplication occurs as a result of impact ionization of carriers across the band gap and use can be made of this effect in the avalanche photodiode (APD), which provides an internal current gain (rather like the photomultiplier tube) dependent on the applied voltage. Avalanche multiplication is a statistical process so that there is an excess noise factor F given by[27]

$$F = M\left[1 - (1-K)\left(\frac{M-1}{M}\right)^2\right] \quad (11.11)$$

where M is the mean avalanche gain and K_W (the ratio of the ionization rates for holes with respect to electrons) equals K for an avalanche process initiated solely by electrons or $K_W = 1/K$ for hole-initiated avalanching.

For applications in optical transmission, this excess noise factor is a critical

GaAs optoelectronic devices 447

parameter determining performance and, whereas K is of the order of 0.02 for Si, it is close to unity for GaAs. This has been one reason for the lack of serious development of GaAs APDs. It is in this context that the study of GaAlAs graded gap and GaAlAs/GaAs heterojunction multilayer structures is attempting to identify possible avenues for obtaining markedly different hole and electron ionization coefficients in GaAs-based materials. Ionization coefficient ratios of greater than 10 have already been reported in such structures.[28,29]

The photo-FET[30] concept involves absorption of optical radiation in the gate regions of a GaAs MESFET (Chapter 10). Generation of carriers in both the depleted regions under the gate and the undepleted channel can affect channel conductance by both variation of channel resistance and variation of space charge region width. Thus the drain current becomes dependent on optical intensity. The phototransistor[31] involves the creation of additional electron–hole pairs in the base of a bipolar transistor (which for ease of illumination and to achieve high gain should be of heterojunction type), again giving rise to an optical intensity dependent emitter–collector current. Both of these devices are thus capable of achieving electrical amplification of the imposed optical signal but efficient collection of the incoming optical flux remains a problem for the short gate length and small area transistors that are required for high-speed detection. The study of these topics is still in its infancy.

GaAs and GaAlAs form an important class of materials with respect to solar energy conversion devices (solar cells). The advantages and potential of GaAs and GaAlAs could only be highlighted by reviewing the whole topic and discussing the capabilities of crystalline and amorphous materials generally and also the device configurations employed,[32] which is beyond the scope of the current discussion. However, the capability that is beginning to exist using MOCVD and MBE techniques to produce up to 10 cm diameter wafers of uniform multilayer configurations is helping to secure a firm place for GaAs-based devices. The design of multilayer structures incorporating several p–n junctions in different band gap materials to extend the fraction of the Sun's spectrum usefully absorbed together with the use of heterojunctions to restrict the diffusion of carriers away from p–n junctions has added a new dimension to solar cell technology.

Possibilities obviously exist for the monolithic integration of various optical devices with electronic functions as GaAs is an established electronic material with the capability for very high speed (> 1 Gbit/s) digital (Chapter 13) and analogue (Chapter 12) operation. The study of such concepts and the development of such components are, however, in their infancy, so that a brief description of a few early examples of possibilities is included as a guide.

To date, the monolithic integration of optical with electronic devices has been applied to fabricate optical sources, optical detectors, and optical repeaters. Both laser diodes[33] and LEDs[34] have been integrated with FETs or bipolar transistors as well as p–i–n diodes with FETs. In addition, the concept of a complete regenerator has been demonstrated (Figure 11.11) using a photo-FET detector

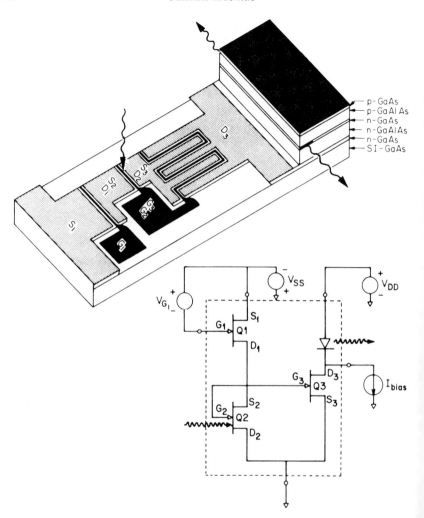

Figure 11.11 Schematic representation of an optoelectronic repeater and its equivalent circuit diagram. Reproduced by permission of the American Institute of Physics from Yust et al.[35]

and FET amplifier and driver stages for a laser with an overall gain of 5 reported.[35] Because of the need to cleave end mirrors for the laser structure, the complexity of the arrangement shown is limited, but recent work using microcleaving[36] or etched facets[37] remove the need for bulk cleaving and will allow a much more flexible arrangement of components with the possibility of greater complexity being engineered. The distributed feedback (DFB) concept,[38] which involves the incorporation of a grating structure within the laser optical cavity, might also be used in this respect.

It is also possible to incorporate passive optical waveguides[39] within the semiconductor layer structure so that optical radiation emitted by a laser can be guided on-chip to another component (possibly an optical modulator or switch in the form of a directional coupler[40] or an optical bistable element[41]) and beyond. Thus the prospect emerges of signal processing and routeing in the optical as well as in the electronic regime, so that a whole range of combinations of optical, optoelectronic, and electronic components may eventually be able to be integrated monolithically in a number of semiconductor systems, including GaAs/GaAlAs. It is predicted by some that the successful development of optoelectronic integrated circuits will be a major breakthrough of the decade because integration is necessary to achieve compactness, and to offer high-speed operation and performance as well as low-cost mass production processes. The growing control of materials parameters and complexity possible with MBE and MOCVD growth techniques is a major step in this direction. Competition exists between the GaAs/GaAlAs system and the InP/InGaAsP system currently in the ascendancy for fibre optics applications. The strong position of GaAs as an electronic component material and the more developed state of the art of GaAs materials technology will, however, ensure it has a continuing future in the optoelectronics arena.

REFERENCES

1. H. C. Casey and M. B. Panish, *Heterostructure Lasers*, Part A, New York, Academic Press (1978).
2. R. C. Goodfellow and R. Davis, Optical source devices, in *Optical Fibre Communications*, eds M. J. Howes and D. V. Morgan, Chichester, Wiley, pp. 27–106 (1980).
3. C. S. Wang, W. H. Cheng, C. J. Hwang, W. K. Burns, and R. P. Moeller, High-power–low-divergence superradiance diode, *Appl. Phys. Lett.*, **41**, 557–89 (1982).
4. Y. S. Liu and D. A. Smith, The frequency response of an amplitude-modulated GaAs luminescent diode, *Proc. IEEE*, **63**, 542–4 (1975).
5. H. Namizaki, H. Kau, M. Ishi, and A. Itoh, Current dependence of spontaneous carrier lifetimes in $GaAs-Ga_{1-x}Al_xAs$ double-heterostructure lasers, *Appl. Phys. Lett.*, **24**, 486–8 (1974).
6. A. C. Carter, Light-emitting diodes for optical fibre systems, *Radio Electr. Eng.*, **51**, 341–8 (1981).
7. D. H. Newman and S. Ritchie, Reliability and degradation of lasers and LEDs, in *Reliability and Degradation: Semiconductor Devices and Circuits*, eds M. J. Howes and D. V. Morgan, Chichester, Wiley, pp. 301–62 (1981).
8. H. Kressel and J. K. Butler, *Semiconductor Lasers and heterojunction LEDs*, New York, Academic Press (1977).
9. H. C. Casey and M. B. Panish, *Heterostructure Lasers*, Part B, New York, Academic Press (1978).
10. D. Kasemset, C. S. Hong, N. B. Patel, and P. D. Dapkus, Very narrow graded-barrier single quantum well lasers grown by metalorganic chemical vapour deposition, *Appl. Phys. Lett.*, **41**, 912–14 (1982).
11. H. D. Wolf, K. Mettler, and K. H. Zschauer, High performance 880 nm

(GaAl)As/GaAs oxide stripe lasers with very low degradation rates at temperatures up to 120°C, *Jap. J. Appl. Phys.*, **20**, L693–6 (1981).
12. A. R. Goodwin, A. W. Davis, P. A. Kirkby, R. E. Epworth, and R. G. Plumb, Narrow stripe semiconductor laser for improved performance of optical communication systems, *Proc. 5th ECOC*, Amsterdam (1979).
13. D. Botez, High-power single-mode semiconductor diode lasers, *Proc. IEDM* (1981).
14. D. Botez, J. C. Connolly, and J. K. Butler, Terraced heterostructure large optical cavity AlGaAs lasers, *Proc. 8th IEEE Int. Semiconductor Laser Conf.* (1982).
15. D. R. Scifres, R. D. Burnham, and W. Streifer, High power phased array lasers, in ref. 14.
16. K. Takahashi, K. Ikeda, J. Ohsawa, and W. Susaki, High efficiency and high power AlGaAs/GaAs laser, in ref. 14.
17. M. Wada, K. Hamada, T. Sugino, H. Shimizu, I. Itoh, G. Kano, and I. Teramoto, A high-power single-mode laser with twin-ridge-substrate structure, in ref. 14.
18. R. D. Burnham, D. R. Scifres, and W. Streifer, Low threshold, high efficiency $Ga_{1-x}Al_xAs$ single quantum well visible diode lasers grown by metalorganic chemical vapour deposition, *Appl. Phys. Lett.*, **41**, 228–30 (1982).
19. S. Yamamoto, H. Hayashi, T. Hayakawa, N. Miyashi, S. Yano, and T. Hijikata, Room-temperature cw operation in the visible spectral range of 680–700 nm by AlGaAs double heterostructure lasers, *Appl. Phys. Lett.*, **41**, 796–8 (1982).
20. R. D. Burnham, D. R. Scifres, and W. Streifer, Current threshold uniformity of shallow proton stripe GaAlAs double heterostructure lasers grown by metalorganic-chemical vapour deposition, *Appl. Phys. Lett.*, **40**, 118–19 (1982).
21. W. T. Tsang, High-throughput, high-yield and highly-reproducible (AlGa)As double-heterostructure laser wafers grown by molecular beam epitaxy, *Appl. Phys. Lett.*, **38**, 587–9 (1981).
22. Y. Mori, O. Matsuda, K. Morizane, and N. Watanabe, V-DH laser: a laser with a V-shaped active region grown by metalorganic CVD, *Electronics Lett.*, **16**, 785–7 (1982).
23. T. P. Lee and A. Y. Cho, Single-transverse-mode injection lasers with embedded stripe layer growth by molecular beam epitaxy, *Appl. Phys. Lett.*, **29**, 164–6 (1976).
24. S. D. Hersee, M. Baldy, P. Assenat, B. de Cremoux, and J. P. Duchemin, Very low threshold GRIN-SCH GaAs/GaAlAs laser structure grown by OM-VPE, *Electronics Lett.*, **18**, 870–1 (1982).
25. R. Tell and S. T. Eng, 8 Gbit/s optical transmission with TJS GaAlAs laser and p–i–n detection, *Electronics Lett.*, **16**, 497–8 (1980).
26. M. J. Adams, Rate equations and transient phenomena in semiconductor lasers, *Opto Electronics*, **5**, 201–15 (1973).
27. P. P. Webb, R. J. McIntyre, and J. Conradi, Properties of avalanche photodiodes, *RCA Rev.*, **35**, 234–78 (1974).
28. F. Capasso, W. T. Tsang, A. L. Hutchinson, and P. W. Foy, The graded bandgap avalanche diode, *Gallium Arsenide and Related Compounds 1981*, Inst. Phys. Conf. Ser. 63, Bristol and London, Institute of Physics (1982).
29. F. Capasso, W. T. Tsang, A. Hutchinson, and G. F. Williams, Enhancement of electron impact ionization in superlattices, *Gallium Arsenide and Related Compounds 1981*, Inst. Phys. Conf. Ser. 63, Bristol and London, Institute of Physics (1982).
30. T. Sugeta and Y. Mizushima, High speed photoresponse mechanism of a GaAs-MESFET, *Jap. J. Appl. Phys.*, **19**, 27–9 (1980).
31. R. A. Milano, T. H. Windhorn, E. R. Anderson, G. E. Stillman, R. D. Dupuis, and P. D. Dapkus, $Al_{0.5}Ga_{0.5}As$–GaAs heterojunction phototransistors grown by metal-organic chemical vapour deposition, *Appl. Phys. Lett.*, **34**, 562–3 (1979).

32. S. M. Sze, *Physics of Semiconductor Devices*, New York, Wiley (1981).
33. I. Ury, K. Y. Lau, N. Bar-Chaim, and A. Yariv, Very high frequency GaAlAs laser field-effect transistor monolithic integrated circuit, *Appl. Phys. Lett.*, **41**, 126–8 (1982).
34. R. Davis, N. Forbes, A. C. Carter, and R. C. Goodfellow, An optoelectronic integrated circuit in the GaAlAs/GaAs system, *Gallium Arsenide and Related Compounds 1981*, Inst. Phys. Conf. Ser. 63, Bristol and London, Institute of Physics, pp. 565–6 (1982).
35. M. Yust, N. Bar-Chaim, S. H. Izadpanah, S. Margalit, I. Ury, D. Wilt, and A. Yariv, A monolithically integrated optical repeater, *Appl. Phys. Lett.*, **35**, 795–7 (1979).
36. H. Blauvelt, N. Bar-Chaim, D. Fekete, S. Margalit, and A. Yariv, AlGaAs lasers with micro-cleaved mirrors suitable for monolithic integration, *Appl. Phys. Lett.*, **40**, 289–90 (1982).
37. N. Bouadma, J. Rion, and J. C. Bouley, Short-cavity GaAlAs laser by wet chemical etching, *Electronics Lett.*, **18**, 879–80 (1982).
38. M. Nakamura, K. Aiki, J. Umeda, and A. Yariv, CW operation of distributed feedback GaAs–GaAlAs diode lasers at temperatures up to 300 K, *Appl. Phys. Lett.*, **27**, 403–5 (1975).
39. J. C. Shelton, F. K. Reinhart, and R. A. Logan, Characteristics of rib waveguides in AlGaAs, *J. Appl. Phys.*, **50**, 6675–87 (1969).
40. J. C. Campbell, F. A. Blum, D. W. Shaw, and K. L. Lawley, GaAs electro-optic directional-coupler switch, *Appl. Phys. Lett.*, **27**, 202–5 (1975).
41. H. M. Gibbs, S. S. Tarng, J. L. Jewell, D. H. Weinberger, K. Tai, A. C. Gossard, S. L. McCall, A. Passner, and W. Wiegmann, Room-temperature excitonic optical bistability in a GaAs–GaAlAs superlattice etalon, *Appl. Phys. Lett.*, **41**, 221–2 (1982).

Gallium Arsenide
Edited by M. J. Howes and D. V. Morgan
© 1985 John Wiley & Sons Ltd

CHAPTER 12

GaAs Microwave Monolithic Circuits

D. MAKI

12.1 INTRODUCTION

Microwave monolithic integrated circuits (MMICs) are composed of active and control devices, transmission lines, and lumped matching and bias elements fabricated on a semi-insulating substrate. The circuits are fabricated in a batch mode with typically hundreds of circuits contained on a single wafer. The production cost of an individual circuit is then roughly proportional to its area and can potentially be very low in the same manner that silicon digital ICs containing over 10^5 active devices can be produced for a dollar or so. This promise of inexpensive microwave circuits in large quantities has dramatically changed the way in which designers approach microwave system configurations and has attracted relatively large amounts of government and commercial funding.

Prominent among the potential applications are active array radar and direct broadcast television receivers. Both systems require millions of near state-of-the-art components at very low costs if they are to be successful.

Work on GaAs monolithic ICs began in the late 1960s at Texas Instruments[1] with the development of a 94 GHz receiver. While fundamentally sound, the programme had serious problems owing to the poorly developed state of material, processing, and device technologies. In 1976 the first monolithic FET amplifier was demonstrated at Plessey,[2] and shortly thereafter major monolithic IC efforts were under way at many large radar system companies. At the point of writing, late 1982, most microwave houses with some processing capabilities are at least experimenting with monolithic ICs. Such devices are being designed into the next generation active array radar, while low-frequency GaAs ICs are commercially available, and a number of companies are developing TVRO systems based on monolithic circuits. Although they have proved to be technically feasible and useful in a wide range of system applications, their ability to compete economically with new hybrid circuit designs has not yet been demonstrated. This test will determine their ultimate viability.

In this chapter we will examine the passive and active elements typically used in monolithic ICs, the processing steps involved in fabricating them, and monolithic circuit design.

12.2 TRANSMISSION LINES

Semi-insulating GaAs, with a dielectric constant of 12.9 and a resistivity of 10^7 Ω cm, forms an excellent medium for low-loss microwave transmission lines. Figure 12.1 illustrates four common transmission lines that are useful for monolithic circuits. Of these, microstrip and coplanar waveguide are the most widely used.

In general, microstrip is lower loss since the current is fairly uniformly distributed across the top conductor, whereas in coplanar waveguide the currents are concentrated on the strip edges. Coplanar circuits also have problems in heat extraction and with parasitic microstrip and slot line modes.

Schneider[3] has given the microstrip effective dielectric constant and characteristic impedance at zero frequency with an air dielectric as

$$\varepsilon_{\text{eff}} = \frac{\varepsilon_r + 1}{2} + \frac{\varepsilon_r - 1}{2} \left(1 + 10\frac{h}{W}\right)^{-1/2}$$

with an accuracy of $\pm 2\%$ and

$$Z_0 = \begin{cases} 60 \ln\left(\dfrac{8h}{W} + \dfrac{W}{4h}\right) & W/h \leqslant 1 \\[2ex] \dfrac{120\pi}{(W/h) + 2.42 - 0.44(h/W) + (1 - h/W)^6} & W/h \geqslant 1 \end{cases}$$

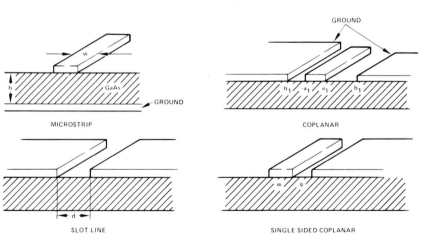

Figure 12.1 Transmission lines

GaAs microwave monolithic circuits

The characteristic impedance of microstrip on a dielectric substrate is given by

$$Z_{0d} = Z_0 \varepsilon_{eff}^{-1/2}$$

Figure 12.2 shows a plot of Z_0 vs W/h for GaAs and alumina. From this we see that for 50 Ω line on GaAs, $W/h \approx 0.75$. Owing to the air–dielectric interface at the top surface of microstrip, a pure TEM wave is impossible, and the impedance and effective dielectric constant will vary with frequency.

Getsinger[4] has derived an expression for the effective dielectric constant of microstrip as a function of frequency as

$$\varepsilon_{eff} = \varepsilon_r - \frac{\varepsilon_r - \varepsilon_0}{1 + G(f/f_p)^2}$$

where

$$f_p = \frac{Z_0}{2\mu_0 h}$$

and

$$G = 0.5 + 0.01 Z_0$$

ε_r is the relative dielectric constant of the substrate, ε_0 and Z_0 are the effective dielectric constant and characteristic impedance of the microstrip at zero frequency, μ_0 is the permeability of free space, and h is the thickness of the substrate.

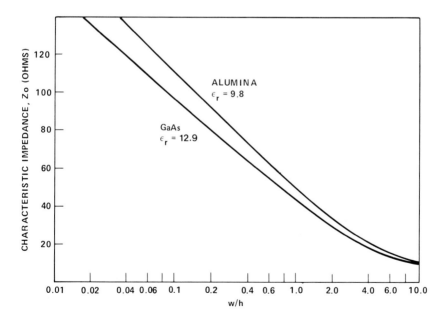

Figure 12.2 Microstrip characteristic impedance

Gallium arsenide

Microstrip loss is due primarily to metallization resistance and is given by Pucel et al.[5] as

$$\alpha_c = \frac{\sqrt{(\pi f u)}}{2Z_0 W} \frac{1}{\sqrt{\sigma_{c1}}} + \frac{1}{\sqrt{\sigma_{c2}}}$$

where σ_c is the conductivity of the metal in mho m^{-1}. Neglecting dielectric losses we see that, for a constant value of Z_0, the attenuation constant is proportional to $f^{1/2} h^{-1}$. For gold conductors on 4 and 8 mil (0.004 and 0.008 inch) GaAs, the theoretical loss is shown in Figure 12.3.

Although the attenuation per unit length rises as $f^{1/2}$, the guide wavelength is inversely proportional to frequency. The unloaded Q of microstrip is given by

$$Q_0 = \frac{20\pi}{\ln 10} \frac{1}{\alpha_0 \lambda_0}$$

and is, therefore, proportional to $f^{1/2}$ as shown in Figure 12.4 and will increase linearly with substrate thickness, neglecting radiation losses. As is apparent from the curve, the loss of microstrip is quite high for thin substrates and to minimize loss we should choose as large a substrate thickness as practicable. However, we desire thin substrates to facilitate heat extraction and they must remain thinner than

$$h_{TE} = \tfrac{1}{4} C_0 f^{-1} (\varepsilon_r - 1)^{-1/2}$$

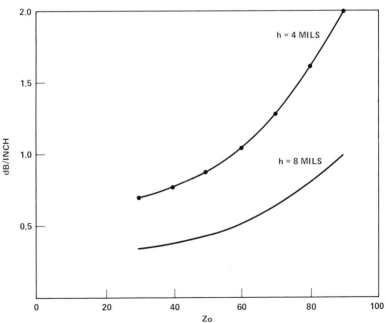

Figure 12.3 Insertion loss versus impedance for 4 and 8 mil microstrip

GaAs microwave monolithic circuits

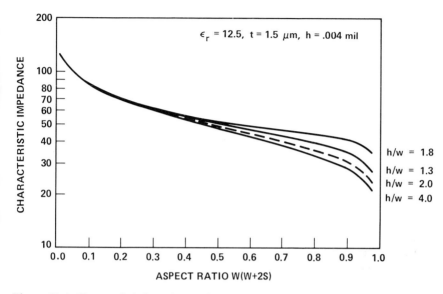

Figure 12.4 Characteristic impedance of coplanar waveguide on infinitely thick GaAs

where h_{TE} is the minimum thickness necessary to support the lowest-order TE surface mode. For power amplifiers, 100 μm thick substrates are a reasonable compromise. This gives a Q of 50 at 10 GHz and also allows via holes to be etched through the substrate for grounding. A thicker substrate would make grounding more difficult. Low-noise microstrip circuits, which do not have the thermal problems of a power amplifier, are generally fabricated on 200 μm thick substrates to reduce loss and to ease the fabrication and handling difficulties encountered with thin, fragile wafers.

Microstrip has several definite advantages over other transmission lines. The presence of the ground plane not only improves the thermal properties of the circuit but also confines the fringing fields to within a couple of ground plane spacings of the transmission lines. This greatly increases the calculability of the circuit performance owing to the wealth of microstrip design information available and decreases crosstalk and parasitic coupling between elements.

The treatment of microstrip discontinuities is important in monolithic design but is rather lengthy to undertake here. Studies of the effects of gaps, crossings, width changes, and open circuits can be found in refs 6–10.

Wen[11] has given an approximation for the coplanar waveguide impedance as

$$Z_0 = \frac{1}{Cv_{ph}}$$

where phase velocity v_{ph} is

$$v_{ph} = \left(\frac{2}{\varepsilon_r + 1}\right)^{1/2} c$$

c is the velocity of light, and the capacitance per unit length, C, is given by

$$C = (\varepsilon_r + 1)\varepsilon_0\, 2a/b$$

where

$$\frac{a}{b} = \frac{K(a_1/b_1)}{K'(a_1/b_1)}$$

a_1 and b_1 are shown in Figure 12.1 and $k(a_1/b_1)$ is the complete elliptic integral of the first kind. Hillberg[12] has given approximations for $K(k)/K(k')$ accurate to better than 3 parts in 10^6 for $k = 0$ to 1. These relations are

$$G = \frac{K(k)}{K(k')} = \frac{K(k)}{K'(k)} = \begin{cases} \dfrac{1}{\pi}\ln\left[2\left(\dfrac{1+k}{1-k}\right)^{1/2}\right] & \dfrac{1}{\sqrt{2}} \leqslant k \leqslant 1 \\[2ex] \pi\bigg/\ln\left[2\left(\dfrac{1+k'}{1-k'}\right)^{1/2}\right] & 0 \leqslant k \leqslant \dfrac{1}{\sqrt{2}} \end{cases}$$

where

$$k' = \sqrt{(1 - k^2)}$$

Using these we can calculate the characteristic impedance for coplanar waveguide on GaAs as shown in Figure 12.5 with varying substrate thickness. As we bring a ground plane close to the coplanar lines, Z_0 decreases owing to a parasitic microstrip mode and gives potential problems in terms of effective grounding of components. This makes heat extraction difficult in a coplanar circuit because of the poor thermal conductivity (0.81 W cm^{-1} K^{-1}) of GaAs. Proposed solutions have involved the mounting of thin coplanar monolithic chips on BeO carriers and the use of thick air bridge structures to extract heat from power FETs.

Care must also be taken when using shunt elements in coplanar lines. If, for example, we connect an inductance, L, from the centre conductor to one ground plane, there will be a potential difference between the grounds and a slot line mode could be generated. This is avoided by connecting between the ground planes at regular intervals using air bridges or wire bonds or by using an inductance of $2L$ to each ground.

Slot line is treated in refs 13 and 14 and is of limited use in monolithic circuits. In coplanar circuits at millimetre wave frequencies, slot transmission line forms useful series elements, and broadband mixers can be formed using slot–microstrip hybrids but require access to both sides of the circuit. Losses in

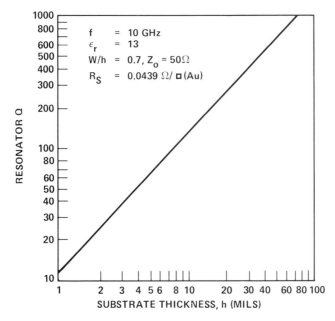

Figure 12.5 Microstrip Q on GaAs substrate

slot line are higher than in microstrip owing to the high concentration of currents near the edge of the conductors.

12.3 LUMPED ELEMENTS

Lumped elements are those elements in a circuit which are small compared to wavelength. Arbitrarily, a tenth of a wavelength is normally picked as the maximum size. In monolithic circuits, however, it is best to treat all circuit elements as distributed so that their behaviour can be more accurately predicted. At frequencies of 20 GHz or higher, most circuit elements will fall into the range of 0.1λ to 0.25λ; at X-band we will avoid long distributed transmission lines whenever possible (since $\lambda/4 \approx 0.090$ inch), and at low frequencies almost all elements will be less than 0.1λ in size.

The types of elements necessary for monolithic circuits can be roughly grouped into two categories: those necessary to provide impedance matching or transforming functions, and those needed for d.c. blocking, RF bypassing, and biasing. We will consider inductors, interdigital and overlay capacitors, distributed transmission lines, and resistors for RF applications, and overlay capacitors, resistors, and occasional quarter-wavelength line for bias and bypassing functions.

12.3.1 Interdigital capacitors

Interdigital capacitors are composed of a single layer of metallization, defined as shown in Figure 12.6, forming a number of interleaved fingers connected on each end by a shorting bar. In analysing this structure, Alley[15] has proposed the low-frequency model also shown in the figure when l and W are short compared to wavelength. C_T is the capacitance of the shorting bar to ground and C_1 is the capacitance of the fingers to ground. A_1 and A_2 are geometry-dependent constants relating interfinger capacitance to X/t and are given in refs 15 and 16. The interfinger capacitance can be calculated exactly for an infinite array of lines on an infinitely thick substrate with metallization thickness equal to 0:

$$C_g = (\varepsilon_r + 1)(\varepsilon_0/2)[F(k)/F(k')]$$

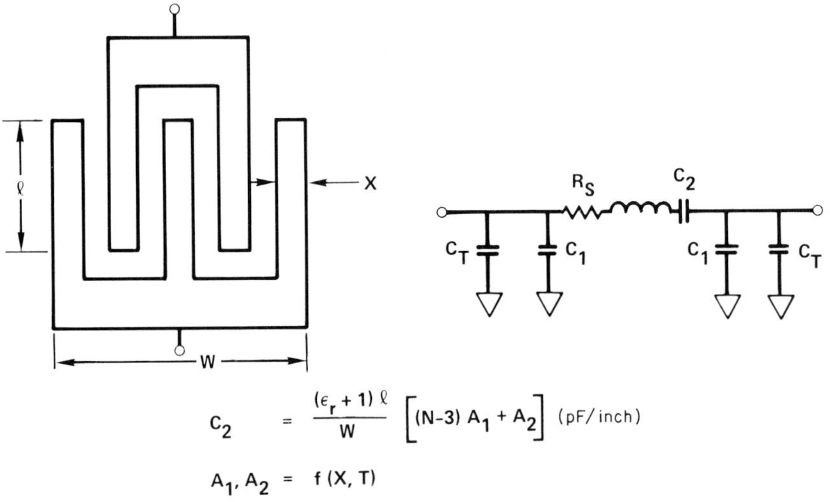

$$C_2 = \frac{(\varepsilon_r + 1)\,l}{W}\left[(N-3)A_1 + A_2\right] \text{ (pF/inch)}$$

$$A_1, A_2 = f(X, T)$$

Figure 12.6 Interdigital capacitor geometry and equivalent circuit

where

$$k = \sin[W/2)(W+g)^{-1}]$$

Using Hillberg's[12] approximation again we get

$$C_g = (\varepsilon_r + 1)0.112\,395\,G \text{ pF/inch of gap}$$

with G given by

n	G	n	G	n	G
0.1	0.485	0.4	0.865	0.7	1.349
0.2	0.619	0.5	1.000	0.8	1.614
0.3	0.741	0.6	1.156	0.9	2.059

GaAs microwave monolithic circuits

where
$$n = W(W+g)^{-1}$$

As the dimensions of the capacitor become large compared to wavelength, we must consider the distributed nature of the structure.[13] Using coupled transmission line theory[14,15] we can represent pairs of coupled fingers at any given frequency by a lumped equivalent model. In Figure 12.7, these models then form a periodic discontinuity along the transmission lines formed by the pair of terminal strips. If the fingers are spaced closely together, we can write distributed transmission line equations for the terminal strips, which are now reactively loaded and coupled by the finger pairs. We now have two different resonant modes in the interdigital capacitor: lengthwise along the fingers or transversely along the terminal strips.

Capacitor Q can be expressed approximately as

$$Q_c = \frac{XN}{WC\frac{4}{3}R_S}$$

for metallization thickness greater than several skin depths, where X is the finger width and R_S is the sheet resistivity. This equation predicts a Q somewhat higher than that measured, and Figure 12.8 gives a curve of capacitance vs number of fingers for two finger lengths and their corresponding Qs at X-band. Note that doubling the finger length and holding capacitance fixed divides Q by 4. Figure 12.9 gives gap capacitance on a 100 μm substrate as a function of finger and gap widths.

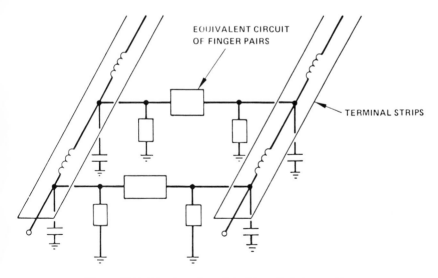

Figure 12.7 Distributed model of interdigital capacitor

462   *Gallium arsenide*

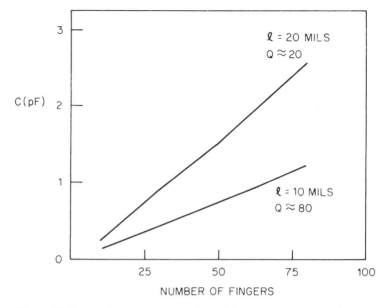

Figure 12.8 Capacitance versus number of fingers for two finger lengths

Since lumped elements are difficult to test accurately individually at microwave frequencies owing to parasitics, measuring the lumped elements at resonance is more convenient. Figure 12.10 shows a circuit containing two inductors and capacitors. By connecting in parallel various combinations of elements, both the elemental reactances and their associated Q can be measured.

Microwave Qs of 75 and 50 have been achieved at 10 and 15 GHz respectively for a 0.3 pF interdigital capacitor with 5 μm gaps, 10 μm lines, and a metallization thickness of 1.5 μm.

The Q calculations have assumed no loss in the dielectric, which is an excellent approximation for semi-insulating GaAs. If, however, these capacitors are formed on a conductive medium, such as buffer layers, which can have resistivities as low as 10^2 Ω cm, their performance will be degraded. Figure 12.11 shows the total Q of a capacitor with a metal Q of 50 as a function of resistivity. If we desire to use interdigital capacitors on wafers with buffer layers, high losses may result unless:

(a) a Cr-doped buffer is used to guarantee high resistivity;

(b) the substrate is bombarded with protons, oxygen or boron to damage the crystal and form a high-resistivity layer;

(c) the buffer is etched through and the capacitor formed on SI material (this poses some difficult fabrication problems if the buffer is 2–3 μm thick).

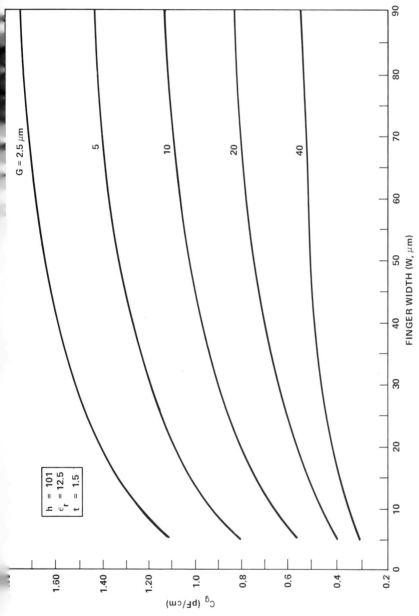

Figure 12.9 Gap capacitance versus finger width for various gaps

Gallium arsenide

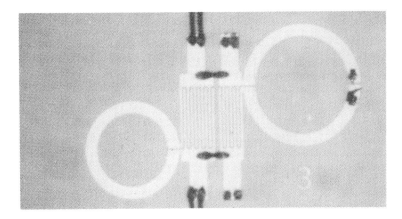

Figure 12.10 Interdigital capacitor test structure

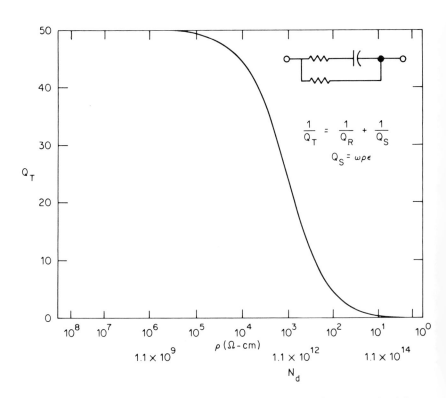

Figure 12.11 Degradation of capacitor Q due to finite substrate conductivity

12.3.2 Overlay capacitors

An overlay capacitor consists of a bottom metal layer, a thin dielectric layer (normally 1000–10 000 Å), and a top metal layer. Capacitance values of 0.025–0.5 pF/mil² can be easily obtained, making them well suited for RF bypass or d.c. blocking applications where 5–20 pF may be required in a small area. They may also be used for RF tuning applications if the dielectric constant and thickness can be accurately controlled and the microwave loss minimized.

The capacitor is analysed as a lossy transmission line, and referring to Figure 12.12 we can write

$$Z_{in} \simeq \frac{Rl}{3} - j\frac{1}{l\omega C}$$

$$C = \frac{\varepsilon_r \varepsilon_0 W}{t}$$

$$R = R_S l W^{-1}$$

assuming the metallization thickness is greater than two skin depths and ignoring, for the present, dielectric losses. We can therefore express Q_m, the Q due to metal losses, as

$$Q_m = \frac{3t}{2 W \varepsilon_r \varepsilon_0 R_S l^2}$$

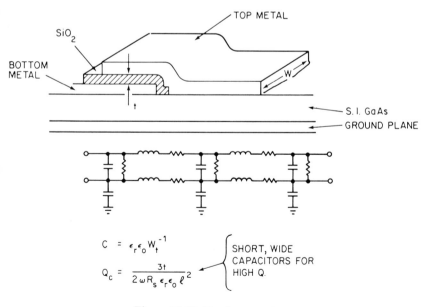

Figure 12.12 Overlay capacitors

To this must be added the dielectric loss

$$Q_D = 1/\tan \delta$$

If we use the loss tangent for bulk SiO_2 or Si_3N_4 ($\sim 10^{-3}$ to 10^{-4}) we would expect the total Q to be dominated by metal losses at microwave frequencies and that Qs of several hundred could be achievable. The first thorough study of overlay capacitors was done by Sobol and Caulton,[19] who determined that the dielectric loss of deposited SiO_2 or Si_3N_4 was much higher than anticipated and that Qs of 30 to 50 were common at X-band. This could be improved to near-theoretical values by heat treating the films. The current state of the art is such that Qs of 30 to 100 can be fabricated on monolithic circuits with good yield, but much work needs to be done to improve the dielectric Q. The dielectrics used to date include the following.

(a) Sputtered SiO_2 and Si_3N_4: Currently the best Q and film integrity.

(b) 'Silox'-deposited SiO_2: Low Q (~ 4) probably due to water trapped in the film.

(c) Polyimide: An organic dielectric that has good step coverage and few pinholes. It has good potential for commercial applications, but may not be useful for military applications owing to possible deterioration of the film. It is difficult to form in thin layers since it is spun-on like photoresist. It is frequently used in thick layers above the bottom metal layer to provide a low dielectric base for a second metal layer.

(d) Al_2O_3: Relatively good, if sputtered. When formed using anodization, the Qs have been unmeasurably low.

(e) Ta_2O_5: Either sputtered or formed using anodization. High capacitance per unit area.

Another parameter of interest is the voltage breakdown of the capacitors. Significant problems were encountered with the capacitor as shown in Figure 12.12, owing to breakdown of the dielectric where it goes over the edge of the bottom metal. The configuration shown in Figure 12.13 solves this problem by using an air bridge to connect to the metal, and breakdown voltages of 75–100 V are achieved.

If the overlay capacitors are used in a microstrip circuit, care must be taken to account for the characteristic impedance of the bottom plate of the capacitor relative to ground. Figure 12.14 shows a common application of an overlay capacitor used in a microstrip circuit with a ground connection over the edge of the substrate. The capacitor can be modelled as an open-circuit low-impedance transmission line of characteristic impedance

$$Z_0 = \frac{120\pi t}{\varepsilon_r^{1/2} W}$$

GaAs microwave monolithic circuits

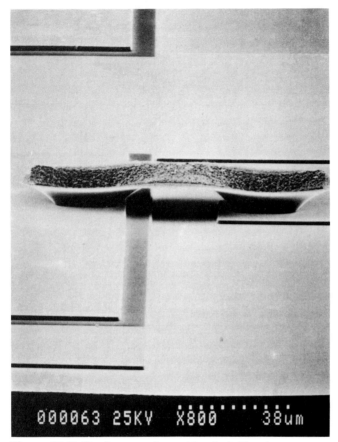

Figure 12.13 Overlay capacitor with air bridge interconnect. (Photo courtesy of Hughes Aircraft Company)

where W is the width of the top plate and t is the thickness of the dielectric. For l much less than a wavelength, the input impedance reduces to the low-frequency value

$$Z_{in} = -j\frac{t}{\varepsilon_r \varepsilon_0 l W \omega}$$

The bottom plate of the capacitor is then modelled as a short-circuit microstrip transmission line. This, as shown in Figure 12.15, presents an effective inductance in series with the capacitor, which must be accounted for in the circuit design.

To summarize, overlay capacitors using low-temperature sputtered SiO_2, Si_3N_4 or Al_2O_3 and an air bridge interconnect can reliably provide high-value capacitors with Qs of 30 at X-band and good breakdown voltage. Capacitors formed using Ta_2O_5 are useful for small-area, high-value capacitors at low frequencies.

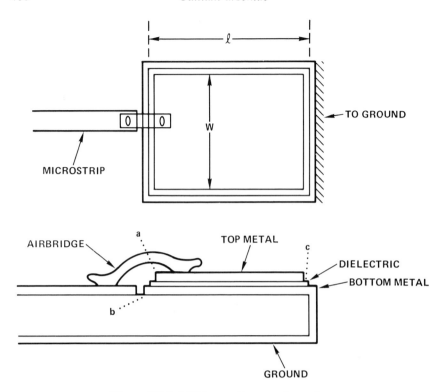

Figure 12.14 MIM capacitor on microstrip

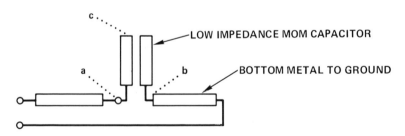

Figure 12.15 Approximate model of MIM capacitor grounded at end

12.3.3 Inductors

Inductors need to be carefully designed in monolithic circuits to keep their number and size within reasonable bounds and to arrange them to a topologically useful manner to facilitate biasing of the active devices. The simplest inductor is a rectangular ribbon with inductance given by Terman[20] as

$$L = 5.08 \times 10^{-3} l \left[\ln\left(\frac{l}{W+h}\right) + 1.193 + 0.2235\left(\frac{W+h}{l}\right) \right]$$

where L is in nH and all dimensions are in mils. The Q of the inductor, assuming thick copper metallization, is

$$Q = 4.81 \times 10^7 f^{1/2} \, l W / k$$

where k is a correction factor, due to current crowding at the edges of the conductor, which varies from 1.3 to 2.0. This equation assumes that the inductor is isolated from ground. If this is not the case, we must estimate the characteristic impedance of the structure and use the transmission line equations to calculate inductance.

If we form a single-turn loop inductor, the inductance (in nH) is calculated from[19]

$$L = 5.08 \times 10^{-3} l \left[\ln\left(\frac{l}{W+t}\right) - 1.76 \right]$$

Comparing this equation with the previous equation for a straight conductor, it is apparent that, for a constant length, we have lost inductance due to mutual coupling from one side of the loop to the other. Again, it is worth noting that this equation is for an isolated loop and the results will be different as we bring a ground near the inductor. As shown in Figure 12.16, the presence of a ground plane can be modelled by an identical image conductor carrying a current opposite to that of the top conductor. The effect of this is to reduce the inductance by the mutual inductance and to increase the loss. For thin substrates, the isolated inductor equations must be used with care, or abandoned in favour of distributed calculations. As an example, many of the monolithic circuits in the literature use microstrip circuitry and loop inductors whose diameter is large compared to the substrate thickness. In this case it is best to analyse it as a length of microstrip and to ignore the fact that it resembles a loop inductor.

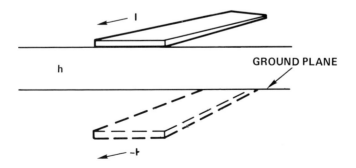

Figure 12.16 Effect of ground plane on conductor

To generate a large amount of inductance in a small area we need to use a spiral inductor. Terman[20] has given a low-frequency approximation:

$$L = \frac{a^2 n^2}{8a + 11c}$$

where L is in nH, and

$$a = (d_o + d_i)/4 \qquad c = (d_o - d_i)/2$$

n is the number of turns, and d_o and d_i are the outer and inner diameter in mils. For maximum Q an optimum value of d_i/d_o of 0.2 was derived and for this condition

$$Q \approx \frac{130\, WL^{1/2}}{d_o^{1/2} K} \left(\frac{f(\text{GHz})}{2}\right)^{1/2} \left(\frac{\rho(\text{Cu})}{\rho}\right)^{1/2}$$

Podell[21] has given approximations for spiral inductors as

$$L(\text{nH}) = 0.0124 D n^2$$

$$Q \simeq D\sqrt{f(\text{GHz})}$$

$$Z_0 \simeq 28 n \ln(16T/D)$$

where all dimensions are in mils, D is the diameter, and T is the metallization thickness.

Care must be taken to keep the total length of the inductor below a quarter wavelength, unless we wish to take advantage of the self-resonance of the structure for bias purposes.

A square spiral may also be used, giving slightly lower Q than a circular spiral, but being easier to generate. The inductance is given as[22]

$$L(\text{nH}) = 6.12 \times 10^{-2} a n^{5/3} \ln(a/c)$$

where a is the mean length of the square (in mils) and c is the depth of the winding.

Qs of 50 to 70 are readily achieved with circular spirals at 10 GHz for inductance values of 1 nH and their Q is proportional to $f^{1/2}$ so they become more attractive at high frequencies.

The analysis of the spiral inductor is based on the generalized transmission line equation

$$Z_n = Z_0 \frac{Z_L \cosh \gamma l + Z_0 \sinh \gamma l}{Z_0 \cosh \gamma l + Z_L \sinh \gamma l}$$

where Z_L is the line terminating impedance, and γ is the complex propagation constant $\alpha + j\beta$ used to account for line loss.

While this equation is strictly true only for uniform TEM transmission lines, it can be applied with reasonable accuracy to less than ideal geometries. For example, a spiral inductor is in reality a non-uniform transmission line with a variable Z_0 along its length. However, by assigning an effective Z_0 to the spiral, it

Gaאs microwave monolithic circuits

can be represented quite accurately by the above equation. For a spiral inductor with a total unwrapped length of less than $\lambda/4$, the effective Z_0 can be calculated as the even mode impedance of the strip above the ground plane.

This model approximately accounts for the distributed capacitance along the inductor spiral due to the ground plane and the adjacent lines. The inductor is then self-resonant when the total electrical length is $\lambda/4$. The self-resonant frequency is given by

$$f_r = \frac{c}{4\varepsilon_r l}$$

where c is the velocity of light and ε_r is the even mode effective dielectric constant of the spiral line. Assuming an upper limit on even mode line impedance of 150 Ω, a spiral inductor having a minimum reactance above 900 Ω over a 20% bandwidth can be realized in X-band making use of this resonance effect. This is adequate for many MMIC RF choke requirements.

Another approach to increasing the inductance possible in an MMIC is to increase the local permeability by locally depositing a ferrite on the substrate. This approach puts a burden on the fabrication technology. The design of RF chokes using such a technology is closely tied to the processing and is based on experiment.

12.3.4 Resistors

Resistors are used in monolithic circuits for biasing, terminations, and lossy elements in impedance matching and feedback networks. They can be formed either by using portions of the active GaAs layer or by depositing thin layers of resistive material. The parameters of interest include:

(a) power handling capability;
(b) temperature coefficient of resistivity;
(c) sheet resistivity;
(d) frequency response due to distributed effects;
(e) reproducibility;
(f) aging effects.

The materials available for resistors include:

(a) resistive metals (Cr, Ta, Ti);
(b) resistive compounds (TaN, NiCr);
(c) active GaAs;
(d) cermet (mixture of SiO and Cr).

Active GaAs is easily implemented in monolithic ICs since it requires no processing steps other than those already employed to fabricate FETs so long as the sheet resistivity of the active channel layer is satisfactory. This resistivity varies, of course, with channel doping density and thickness, but usually falls in the range $100\text{--}400\,\Omega/\square$. The problems with the use of resistors GaAs include a severe positive temperature coefficient and non-linear $I\text{--}V$ characteristics for high electric fields. Thin metal films tend to have low resistivities, which makes their reproducibility more difficult to control. Tantalum has the advantage of being trimmable by anodization.

Diagrams of thin film and GaAs resistors are shown in Figure 12.17. Any resistive line will have a certain inductance and parasitic capacitance to ground as shown in Figure 12.17c, which forms a distributed characteristic impedance and

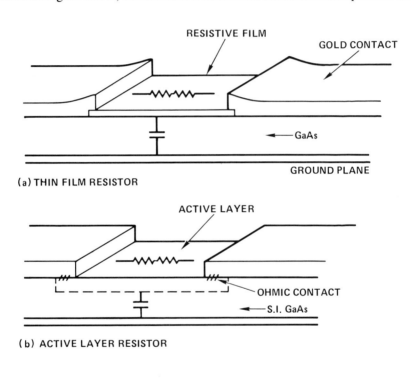

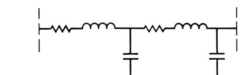

Figure 12.17 Monolithic resistors

propagation constant given by

$$Z_0 = \left(\frac{r+jWl}{jW}\right)^{1/2} \qquad \gamma = [(r+jWl)jWc]^{1/2}$$

where r, l, and c are the resistance, inductance, and capacitance per unit length. The input impedance of a length of lossy line, shorted at the far end, is given by

$$Z_{in} = Z_0 \tanh \gamma L$$

which reduces for lengths that are short compared to wavelength to

$$Z_{in} \approx rL + jWlL$$

The design of a microwave resistor is sometimes complicated by occasionally conflicting requirements. The width of the resistor should match the width of the line feeding it to minimize reflections, the length must be long enough to obtain the desired loss or resistance, yet not so long that distributed effects degrade performance, and the total surface area must be large enough to dissipate the required power.

12.4 ACTIVE DEVICES

The variety of active devices available in a monolithic circuit is rather spartan when compared to those available in hybrid circuits. Single and multiple gate FETs and Schottky barrier diodes and GaAs PIN diodes are available, however are difficult to combine on a common wafer. The FET is a versatile device, however, and can be used as a switch, attenuator, variable resistor, mixer, oscillator, multiplier and phase shifter, as well as an amplifier.

12.4.1 Field-effect transistors

The theory of operation of FETs is covered in detail in Chapter 10 of this book and will not be repeated here. For our purposes we will examine the simplified device cross section and equivalent circuit shown in Figure 12.18. Electrons flow from the source to the drain under the gate. The gate is normally reverse-biased and draws no current. However, a modulating voltage on it varies the depth of the depletion region, thereby controlling the drain current and the drain–source resistance. At $V_g = -V_p$, the pinch-off voltage, the depletion region fills the channel and the d.c. current drops to zero.

The I–V characteristics for a common source mounted device are approximated by Figure 12.19 shown superimposed with a number of bias points. Examining the I–V characteristics of the FET about these bias points, we see the possibility of many modes of operation.

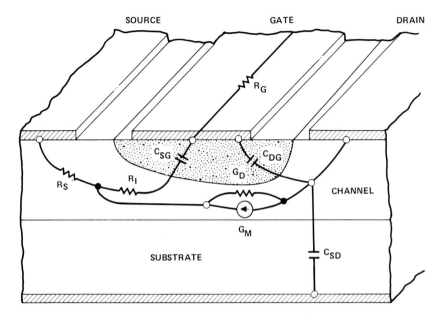

Figure 12.18 Equivalent circuit model of FET

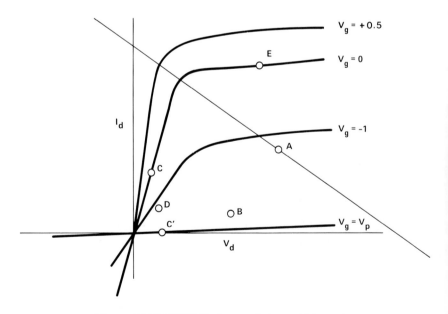

Figure 12.19 FET I–V characteristics and bias points

Amplifier By biasing the device at $V_D > V_{knee}$ and setting $V_p < V_g$ LO, we can form an amplifier. Biasing at point A and establishing a load line as shown allows the delivery of maximum power from the FET. The load line is typically set from the knee of the $V_g = 0.5$ curve to the breakdown voltage at pinch-off. Low-noise FETs are typically biased at 3 V and 15% of I_{dss}, modelled by point B on the curve.

Switch By switching the gate voltage from zero to pinch-off with the source and drain d.c. connected, the FET switches from a low-resistance state with resistance given by

$$R \approx \rho l (Wt)^{-1}$$

where l, W, and t are the length, width, and thickness of the channel, to a high-resistance state, shunted by the drain–source capacitance. This switch requires no holding current in either state.

Variable resistor If $V_D < V_{knee}$ we are operating in the linear region about the origin and can change the slope of the I–V curve by varying V_g.

Constant current source If $V_D > V_{knee}$ and V_g is fixed, as for example at point E, we see that I is relatively independent of V_D, forming a constant current source, shunted by the source–drain capacitance. This is useful for biasing applications but increases the noise figure of the circuit due to noise contributions from current flow through the drain–source channel.

Active matching Although most applications use FETs in the grounded source configuration, common gate and common drain FETs can provide wide band impedance matching using a minimum of reactive elements.

Dual gate FETs allow the gain of the device to be varied by adjusting the second gate bias without substantially affecting the input impedance. This allows the device to be used as a wide band variable gain amplifier or as a mixer with separate ports for the insertion of LO and RF signals.

12.4.2 Schottky barrier diodes

Schottky barrier (SB) diodes are used in monolithic ICs to form mixers, variable capacitors, bias capacitors, and level shifters. The quality of an SB diode is expressed as a cut-off frequency given by

$$f_{co} = 1/(2\pi r_s C_0)$$

where r_s is the series resistance in the diode and C_0 is the zero bias capacitance. Discrete diodes for use at high frequencies have cut-off frequencies of

1000–2000 GHz and such performance is necessary to build high-quality MM wave mixers and analogue phase shifters at X-band with a minimum of amplitude change as a function of phase shift. If a simple SB diode is formed of n-GaAs on SI GaAs, which is used for FETs, the cut-off frequency will typically be 100–200 GHz. This is sufficient for some non-critical applications but inadequate for most. Figure 12.20 shows two basic diode configurations. The first illustrates the problem of fabricating diodes without an n^+ layer. The capacitance is generated between the SB contact and the n layer beneath the depletion region. By varying the voltage, the depletion region changes thickness, thus varying the capacitance. The n layer under the diode can be modelled as an open-circuit R–C transmission line with an equivalent resistance of $r/3$, where r is the total resistance from the centre to the periphery of the depletion region, integrated around the SB contact geometry. Compare this with the geometry in Figure 12.20b. In this case we again have the capacitance across the depletion region but the resistance is now calculated from the bottom of the depletion region down to the n^+ layer. This is easily an order of magnitude improvement in resistance, but requires a completely different active layer structure than those

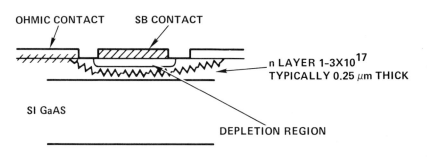

(a) SCHOTTKY BARRIER DIODE ON FET TYPE R_i ACTIVE LAYER

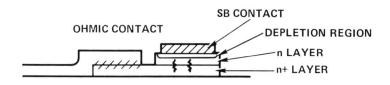

(b) SCHOTTKY BARRIER DIODE USING n ON n^+ LAYERS

Figure 12.20 Monolithic Schottky barrier diodes

Gaas microwave monolithic circuits

used for FETs. This is no problem if only diodes are being fabricated on a wafer, but requires selective implantation or selective epitaxial growth to provide both high-quality FETs and diodes on the same wafer.

12.5 INTEGRATED CIRCUIT PROCESSING

Monolithic IC processing includes all of the processing described earlier to produce state-of-the-art devices, as well as that necessary to produce high-quality passive elements for RF matching and d.c. bias. Owing to the difficulty in trimming ICs, the reproducibility requirements for the processing are substantially higher for monolithic circuits than for discrete devices. Monolithic technology must also be compatible with future high-volume processing techniques, remembering that the main justification for developing monolithic ICs is to provide large quantities of inexpensive circuits.

There are, of course, as many processing schemes as there are companies fabricating circuits and we will not attempt a comprehensive survey of the technology.

Figure 12.21 shows a simplified, fairly representative processing flow chart for a microstrip-based microwave monolithic IC. The process starts by selecting both a semi-insulating substrate and the means of obtaining the active layer. As with silicon processing, it is desirable to work with as large a slice of material as possible, while still preserving yield, and to use round, uniform diameter wafers to facilitate automated processing. Active layers can be formed using ion implantation directly into a semi-insulating substrate or into an undoped buffer layer, or they may be formed epitaxially.

If the wafer is ion implanted it must be annealed to reform the crystal surface after the damage during implantation. The wafer is annealed by heating it to 850–900°C while protecting the surface to prevent loss of arsenic from the crystal. The surface is protected by either coating the wafer with Si_3N_4 or by heating the wafer in a chamber with an arsenic overpressure to maintain equilibrium. At this point we must define the areas in which we are fabricating devices and assure that the rest of the wafer is high resistivity to minimize losses in our passive elements. This isolation process can be accomplished using mesa etching, proton bombardment or selective implantation as shown in Figure 12.22. In mesa etching we simply etch off the active layer, exposing the underlying semi-insulating or buffer. With selective implantation a mask allows heavy doping of the GaAs where exposed, and light doping or none at all under the mask. We can also isolate the wafer using high-energy proton, oxygen or boron bombardment to damage the active layer, thereby significantly increasing its resistivity. We must protect the areas that we wish to remain active by using a thick photoresist, polyimide or gold mask.

FET sources and drains and GaAs resistor contacts are formed with a thin layer of Au–Ge/Ni/Au which is alloyed at 400°C to form ohmic contacts. The gate

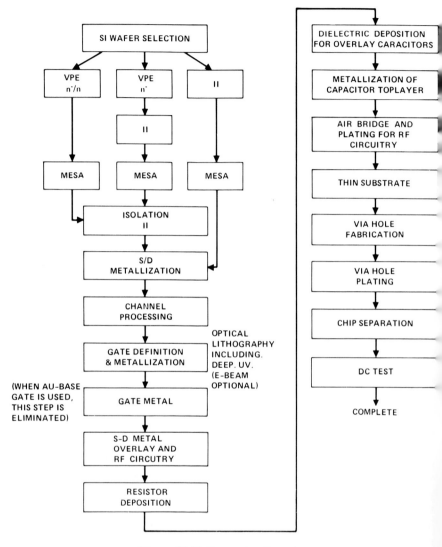

Figure 12.21 Processing flow chart

channel is then etched to get the desired profile and saturated current; then the gate is formed in the channel. The gate metallization is typically aluminium or either Ti–Pt–Au or Ti–Pd–Au. If aluminium is used, a barrier must be formed between the aluminium gate and the gold circuitry. A thin layer of Ti–Pd can be used.

After the device has been formed a source–drain overlay metallization, typically 1.0–1.5 μm of Cr–Au, is deposited through a photoresist mask to reduce

GaAs microwave monolithic circuits

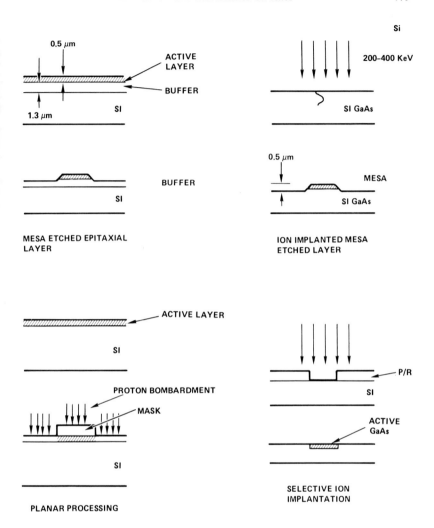

Figure 12.22 Device isolation techniques

the loss of the ohmic contacts and to form the bottom metallization pattern, which includes the transmission lines, interdigital capacitors, and the bottom plates of overlay capacitors. Resistors, if needed, can be formed at the same time as the FETs if GaAs resistors are used or a thin-film resistor can be deposited. The dielectric layer is formed next, which in the example shown consists of 2000 Å of sputtered SiO_2. To improve adhesion a thin layer of chromium is typically deposited before the dielectric. The dielectric layer is follwed by the capacitor top plates and finally, for the top surface, the air bridges. The wafer is then thinned to

480   Gallium arsenide

the desired thickness, typically 100–200 μm, and via holes are etched if required. The final processing step is chip separation either by sawing, scribing and breaking, or chemical etching.

We will now briefly discuss those processing areas that are substantially different than those required for either discrete devices or hybrid circuits. FET device materials and processing considerations have been comprehensively examined in a previous chapter and we will not cover them here.

12.5.1 Lithography

Most of the metal and dielectric layers on monolithic ICs are formed using a lift-off technique. Since they tend to attack GaAs, wet chemical etches, which are commonly used in hybrid circuitry to pattern metal layers, are avoided. The basic lift-off process shown in Figure 12.23 consists of patterning a photoresist layer

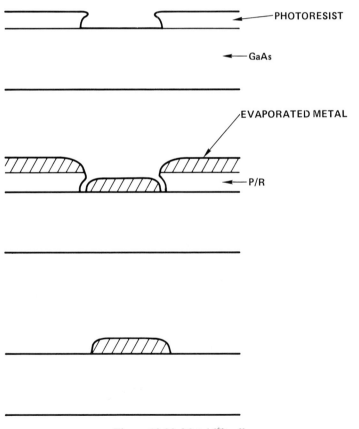

Figure 12.23 Metal lift-off

GaAs microwave monolithic circuits

with the inverse of the desired metallization pattern. Metal is deposited over the entire wafer, landing on top of the photoresist and on the bare GaAs. The photoresist is then dissolved, leaving the desired metal layer. This allows tight control of the metal dimensions and avoids undercutting problems which occur with wet etching techniques. If the layer to be deposited is thin, as with discrete FETs where the thickest layer is typically 8000 Å, the basic lift-off technique provides good resolution and yield. For low loss in monolithic ICs, we desire two skin depths of metal: 1.6 μm at X-band and 3–5 μm at low frequencies. Special techniques must be employed for these thicker layers to provide high-yield, well defined lift-offs. Figure 12.24 shows an SEM of a photoresist profile after treatment with chlorobenzene, followed by normal exposure and development. The pronounced lip effectively shadows the vertical face of the photoresist, preventing the deposited material from bridging the gap between the desired

Figure 12.24 Chlorobenzene-treated photoresist profile. (Photo courtesy of Hughes Aircraft Company)

pattern and the area coated with photoresist. This technique was first developed at IBM[24] and is now used almost universally for thick-film lift-offs. To provide high yield, the deposited material is typically evaporated rather than sputtered or vapour deposited. Evaporation is a line-of-sight process that provides sharp shadowing, thus protecting the inset vertical walls of the resist, and can also be done at low substrate temperatures, preventing excessive heating of the photoresist which makes it difficult or impossible to remove.

Metals or dielectrics may also be patterned using ion milling. An argon ion beam is used to erode the material while the desired pattern is protected with a thick layer of photoresist. This technique provides good definition and high yield, but care must be taken to avoid eroding the GaAs surface which is removed at a much higher rate than the metal layer. A protecting layer, typically polyimide, is used between the GaAs and the metal layer to protect the semiconductor surface.

The photoresist can be exposed using standard optical lithography, deep UV or electron beam lithography. Conventional optical techniques can achieve high-yield, 0.8 μm linewidth, which is satisfactory for low-frequency circuits and power amplifiers through X-band. Low-noise amplifiers at X-band require 0.5 μm or smaller gates to achieve noise figures below 2.0 dB, while FET amplifiers at 30–60 GHz require 0.25 μm or smaller gates to provide adequate performance. To achieve these dimensions, a wide variety of special techniques involving angle evaporation and multiple photoresist layers have evolved using standard optical lithography. A more straightforward approach is to use deep UV to expose the photoresist, taking advantage of the shorter wavelength and greater resolution. These systems require special quartz masks that are transparent at the higher frequency and are capable of 0.5 μm resolution.

The use of an electron beam lithography system combines the benefits of fine line resolution with a great degree of flexibility. The system is basically a computer-driven scanning electron beam, programmed to trace out the desired pattern on either a mask or directly on a wafer. The electron spot exposes a special photoresist (typically PMMA) which, when developed, forms the pattern. This technique produces lines as small as 0.15 μm or less and is regularly used to produce 0.25 μm gates. More importantly, complete circuits can be produced with no glass masks by using the E-beam system for every level since the beam is driven by software. This saves the time required to produce a set of glass masks which can amount to 1 to 3 months if they must be procured externally.

12.5.2 Thin-film capacitors

Capacitors are required in circuits for RF matching, d.c. blocking, and RF bypassing functions. Interdigital capacitors are useful for values below 1 pF, and thin-film capacitors are generally used for larger values. As discussed, the dielectrics commonly used are SiO_2, Si_3N_4, Ta_2O_5, and to a lesser extent polyimide.

Silicon dioxide and silicon nitride are typically formed using sputtering or plasma enhanced chemical vapour deposition (PECVD). Sputtering[25] involves the transport of material from a target to a substrate. Atoms of the target material are ejected due to collisions with heavy ions, typically argon, which are accelerated in an electric field. Silicon dioxide is sputtered using a mixture of argon and oxygen gases to form the plasma, and a quartz target. The presence of the oxygen in the plasma reduces the likelihood of depositing silicon monoxide. Silicon nitride uses a silicon target and a mixture of argon and nitrogen gases. Deposition rates for both of these materials are fairly slow, typically 10–20 Å min^{-1}. The molecules depositing on the substrate give up their energy in the form of heat; slow deposition rates and water cooling keep the substrate temperature below some upper bound. If a lift-off technique is being used, we keep the temperature below 200°C to avoid damage to the photoresist layer. An alternative to lift-off processing is to coat the wafer completely with dielectric, which is removed from unwanted areas by plasma etching. Etch rates of 2500 Å min^{-1} are obtainable when etching silicon nitride in DE-100, a carbon tetrafluoride based etchant. This process allows the film to be deposited at a higher temperature than when formed on photoresist, producing a denser, higher-quality film.

PECVD of silicon nitride uses a mixture of silane (SiH_4) and nitrogen gases excited by an RF discharge in a reactor. The substrate is heated on a grounded plate beneath the discharge and Si_3N_4 grows uniformly on the surface. A lift-off cannot be used to pattern the dielectric since the material is formed at the surface, rather than being deposited from above, and would coat the walls of the photoresist.

Anodization is used to produce thin layers of Ta_2O_5 and Al_2O_3 whose thicknesses are linearly dependent on the anodization voltage. Durschag and Vorhaus[26] have reported a tantalum based monolithic processing scheme which uses thin-film tantalum resistors and anodized Ta_2O_5 capacitors. With an applied voltage of 100 V, a film thickness of 1400 Å and a dielectric constant of 21 are achieved. Uniformity is given as 2% across a wafer and less than 1% variation from run to run, which is significantly better control than can be achieved with other technologies. Anodization also produces high values of capacitance per unit area, 1200 pF mm^{-2}. This is an order of magnitude higher than can be obtained with SiO_2 or Si_3N_4 thin-film capacitors, which becomes important when high-value capacitors are needed for RF bypass at low frequencies. Tantalum can also be sputtered with somewhat less control in thicker layers to avoid pinholes and with a higher reported dielectric constant.[27]

Low-value capacitors can also be formed using polyimide, an organic polymer which is spun onto the wafer like photoresist and cured. It has excellent step coverage owing to the method of application and provides a high-quality, high-yield and extremely tough film. The film has a dielectric constant of 3 to 4 and can be patterned using plasma etching in oxygen. It can be used for capacitors, as a barrier to ion milling, or to separate metal layers on low-frequency circuits, but the thickness of the layer is difficult to control and layers below 1.5 μm are non-reproducible and low-yield.

12.5.3 Via holes

When microstrip transmission lines are used on an IC, frequently low-inductance connections must be formed to ground, which is on the bottom of the chip. The topography of the circuit layout can be arranged such that all points requiring ground fall at a chip edge so that wire bonds or ribbons can be used to attach to ground. This labour intensive procedure normally wastes chip area. A more satisfactory solution is to etch holes through the GaAs substrate which are then metallized.[28] Although a seemingly simple process, much effort has been expended to develop a high-yield, reproducible process. Although laser drilling, ion milling and plasma etching have been used, wet chemical etching is the preferred technology at most companies.

The wafer is completed on the front surface, waxed to a carrier and polished down to the desired thickness, normally 75–100 μm. It is critical at this point that the wafer be flat so that the via holes will etch through the substrate at the same time. The wafer is then removed and cleaned, and photoresist is applied to the back surface. The via hole pattern is aligned to the front surface using an IR aligner since GaAs is transparent to infrared, and the wafer is then etched. As shown in Figure 12.25a, the etched hole undercuts the photoresist mask and tapers towards the front surface. If an anisotropic etch is used, the etched hole tends to follow the crystal planes and a 'hatchet' shaped hole is formed on the front surface. Figure 12.26 shows a cross section of an etched via hole. Note that the hole tapers smoothly from the back to the front allowing a high-yield contact

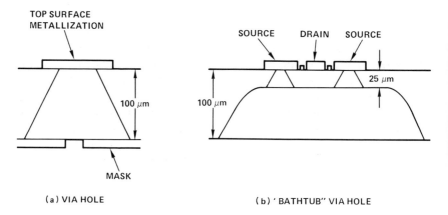

Figure 12.25 Via holes

to be made when metal is deposited from the back. To save space and to reduce parasitic inductance, it is frequently desired to form vias directly beneath overlay capacitors. When a via is etched beneath a metal layer the difference in stress in the metal on GaAs and over the hole tends to buckle the metal slightly into the hole. If a brittle layer of dielectric is above the metal it will crack and short out. Thick plating of the top plate of the capacitor can be used to support the films above the hole.

Another via technique is to etch a large tub via as shown in Figure 12.25b to within 25 μm of the front surface. Smaller vias are then etched through the thinner GaAs providing closer spacing of the vias, allowing them to be used in each source of a power FET if desired.

12.5.4 Air bridges

Air bridges are used to interconnect sources of multi-source FETs, to connect to the top plates of thin film capacitors, and to connect to topologically isolated elements, such as the centre contact of a spiral inductor.

Air bridges are formed by opening holes to form the contacts to the circuit in a thick layer of photoresist. The photoresist may then be heat treated if desired to shape the air bridge 'posts'. A thin layer of metal is evaporated over the entire wafer, a second photoresist layer defines the shape of the air bridge, and a 2–5 μm layer of gold is selectively plated to form the bridge. The photoresist and thin metal layer are then washed away. Figure 12.13 shows a completed air bridge contacting a thin-film capacitor.

12.6 MONOLITHIC CIRCUIT DESIGN

The majority of monolithic circuits that have been designed can be divided into two distinct circuit types: a 'microwave' circuit configuration utilizing lumped or distributed matching elements to effect maximum power transfer type designs; and a direct coupled, low-frequency design, characterized by large numbers of small active devices and few passive matching elements. As an example of the microwave design, consider the two-stage power amplifier[29] shown in Figure 12.27, which consists of a 900 × 0.8 μm driver FET, a 2400 × 0.8 μm output FET, interdigital capacitors and microstrip transmission line for RF matching, overlay capacitors for RF bypassing, air bridges to interconnect sources and to connect to the top plates of the capacitors, and via holes to achieve top surface ground. The chip size is 1.6 × 2.3 × 0.1 mm, of which 17% is devoted to active devices and the rest to passive matching and biasing functions. By using reactive matching we can achieve maximum gain, maximum power or minimum noise figure from the device using standard microwave design techniques. These circuits closely resemble conventional hybrid microwave circuits. The problems inherent in this design are not technical, but rather financial. It has been demonstrated that a wide variety of circuit functions can be built using this technology with results that approach those of hybrid circuits. What has not been demonstrated at this time (late 1982) is that they can be fabricated in volume at attractive prices. Circuits fabricated in this manner occupy relatively large areas of GaAs, which is still an expensive commodity, and require tight controls on the materials and processing technologies to achieve reproducibility from circuit to circuit and from wafer to wafer.

The low-frequency integrated circuit is best typified by the heterodyne IC signal generation chip developed by Hewlett Packard[30] in Figure 12.28. On a chip measuring 0.65 × 0.6 mm, 35% of which is active area, is included a local oscillator, a doubly balanced mixer, an on/off modulator, an RF phase shifter, an

(a)

(b)

Figure 12.26 Etched via holes. (Photo courtesy of Hughes Aircraft Company)

oscillator monitor, and an IF amplifier. As shown by the schematic (Figure 12.29), most of the components are FETs, capacitors are realized using back-biased diodes, and the only inductors used are located off chip. Heavy use is made of balanced circuits, d.c. feedback, level shifting diodes, and FETs as constant current sources. Circuits such as this are typically used from d.c. to 6 GHz in applications where small size and high circuit density are important, but where low noise, high power, and high gain per element are not. The d.c. feedback and balanced circuitry make these circuits more tolerant of material and processing variations, and makes them potentially higher yield and lower cost.

488 Gallium arsenide

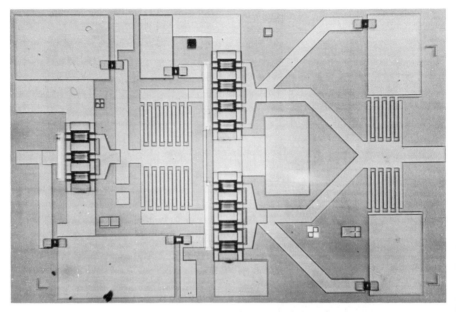

Figure 12.27 Two-stage 10 GHz power amplifier. (Photo courtesy of Hughes Aircraft Company)

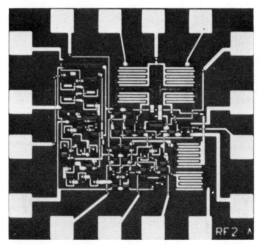

Figure 12.28 Signal generation chip. (Photo courtesy of Hewlett Packard)

12.6.1 Low-frequency circuits

At low microwave frequencies (0–5 GHz), the parasitics associated with small periphery FETs may be neglected and the devices analysed as simple voltage-

GaAs microwave monolithic circuits 489

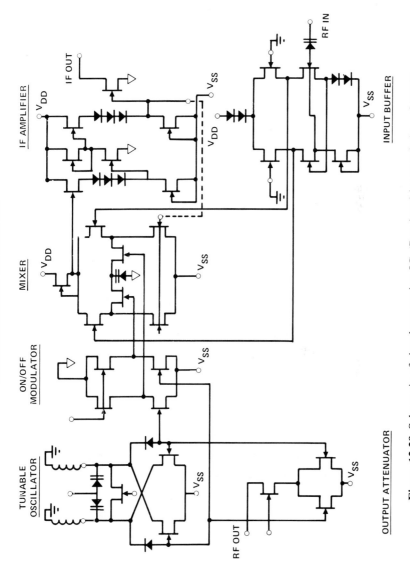

Figure 12.29 Schematic of signal generation IC. (Reprinted by permission of IEEE)

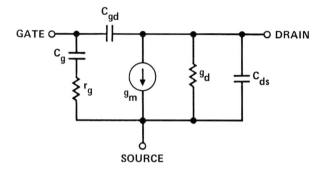

Figure 12.30 Simplified FET model

dependent current sources. Figure 12.30 shows a simplified lumped equivalent circuit model of a FET, neglecting parasitic inductances, resistances, and a capacitance from each of the three external terminals to ground.

If we ground the source, the voltage gain $A = V_D/V_g$ is given by

$$A = \frac{j\omega C_{gd} - g_m(1 + j\omega r_g C_g)^{-1}}{j\omega C_{gd} + Y_L + g_d + j\omega C_{ds}}$$

where Y_L is an external load connected from drain to source. For $\omega \ll (RF_g C_g)^{-1} - 1$ and $Y_L + g_d \gg j\omega(C_{gd} + C_{ds})$ this reduces to

$$A \approx -g_m Z_T$$

where Z_T is the total output impedance given by

$$Z_T = \frac{Z_L}{1 + Z_L(g_d + j\omega C_{ds})}$$

$$Z_L = 1/Y_L$$

The input admittance is given by

$$Y_{in} = Y_g + j\omega C_{gd}(1 - A)$$

where Y_g is the gate admittance

$$Y_g = j\omega C_g (1 + j\omega R_g C_g)^{-1}$$

If we ground the gate and apply a signal at the source, the low-frequency voltage gain is given by

$$A = g_m Z_L$$

and the input impedance given approximately by

$$Z_{in} = \frac{1 + Z_L g_d}{g_m + g_d}$$

where Z_L is the load impedance from drain to ground. For values of $Z_L \ll g_d^{-1}$ the input impedance is equal to g_m^{-1}. By selecting the device periphery appropriately over a wide frequency range, in practice, a 200 × 0.5 µm gate FET with a g_m of 20 mS can demonstrate a value of S_{11} below 0.1 from 0 to 14 GHz.

A common drain FET has a voltage gain less than unity and an output impedance given approximately by

$$Z_{out} \simeq \frac{1}{g_m + g_d}$$

Like the common gate stage discussed previously, by adjusting device periphery we can obtain a constant, real output impedance through 10 GHz with conventional FETs.

By using combinations of the above FET configurations we can make a wide variety of complex circuits with a minimum of reactive elements. Perhaps the simplest low-frequency amplifiers were developed by Hornbuckle et al.[31] and others are shown in Figure 12.31. They consist entirely of four or five FETs, level shifting diodes, an optional resistor to improve the input match, and RF bypass capacitors. The circuit closely resembles a buffered FET logic NOR gate biased into its linear region.

The input FET, F1, is connected in the common source configuration and is

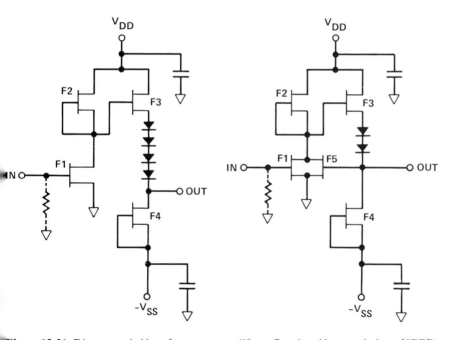

Figure 12.31 Direct coupled low-frequency amplifiers. (Reprinted by permission of IEEE)

biased through a constant current source, F2, which is typically selected to have half the gate periphery of F1 to bias the amplifier in the middle of its operating region. F1 sees an output load consisting of its own and F2's output conductance, and the input impedance of F3, which is high. The voltage gain is then approximated by

$$A = \tfrac{2}{3}(g_m/g_d)$$

Device F3 is connected in the common drain configuration and presents a high input impedance to F1 and has a voltage gain less than unity. Such circuits typically exhibit 12 to 14 dB of gain from 0 to 3 GHz.

The addition of negative feedback with another active device, F5 in Figure 12.31, lowers the gain but increases the bandwidth. The voltage gain of the input common source stage with feedback is given by

$$A = -\frac{g_{m_1} Z_L}{1 + g_{m_2} K Z_L}$$

where g_{m_1} and g_{m_5} are the transconductances of the input and feedback device, respectively, Z_L is the load impedance

$$Z_L = [g_{d_1} + g_{d_2} + g_{d_5} + j\omega(C_{ds_1} + C_{ds_2} + C_{ds_5})]^{-1}$$

and K is the voltage gain of the source follower stage and the voltage dropping diodes.

The gain of these amplifiers falls off rapidly above 3 GHz owing to device capacitance. It is possible to add an inductor between F1 and F3 to peak the gain at a desired frequency by matching the capacitive reactance. Figure 12.32 shows

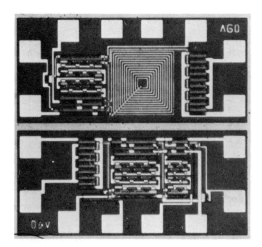

Figure 12.32 Low-frequency amplifiers. (Photo courtesy of Hewlett Packard)

two circuits developed at Hewlett Packard.[36] The top circuit, measuring 300 × 650 μm, is a three-stage direct coupled amplifier and the bottom circuit is a two-stage amplifier including a 13 nH peaking inductor. Note that the single reactive element, while extending the bandwidth from 3 to 5 GHz, is larger than all of the active circuitry.

The noise figure of these amplifiers is poor, 15 to 20 dB, owing to the small periphery of the input devices, typically 20 μm. If a 50 Ω resistor is added at the input to obtain a low VSWR, the noise will degrade a further 3 dB.

The above amplifiers are useful for applications where high gain is needed and the high noise figure can be tolerated. If a low noise figure is required, other circuit designs may be more appropriate.

If we take the expression for voltage gain for a common source FET and add a resistive feedback term in parallel with Y_{gd} we obtain

$$A = \frac{j\omega C_{gd} + Y_f - g_m(1 + j\omega r_g C_f)^{-1}}{j\omega C_{gd} + Y_f + g_d j\omega C_{ds} + Y_L}$$

where Y_f is the admittance of the feedback element. At low frequencies we can write this as

$$A = \frac{Y_f - g_m}{Y_D + Y_f}$$

where

$$Y_D = Y_L + g_d$$

As expected, this degrades the gain, but the input admittance is given by

$$Y_{in} = Y_g + (j\omega C_{gd} + Y_f)(1 - A)$$

which reduces to

$$Y_{in} \simeq \frac{Y_D + g_m}{1 + Y_D/Y_f}$$

at low frequencies. By selecting g_m equal to

$$g_m = Y_0 + Y_D\left(\frac{Y_0}{Y_f} - 1\right)$$

we can achieve a real input admittance of Y_0 over a broad frequency range. A wide variety of such amplifiers have been built using device peripheries from 500 to 1400 μm to obtain sufficient g_m to achieve simultaneously a Y_{in} of 0.02 mho and a reasonable stage gain. Such large devices have much lower noise figures in a 50 Ω system than the 20–40 μm wide FETs used in the previously described amplifiers.

Estreich[32] has given a curve, Figure 12.33, showing noise figure vs FET gate width for various circuit configurations at 1 GHz. The common source FET with resistive feedback has the lowest calculated noise figure of the circuits shown, at the expense of device periphery. A number of firms have demonstrated simple

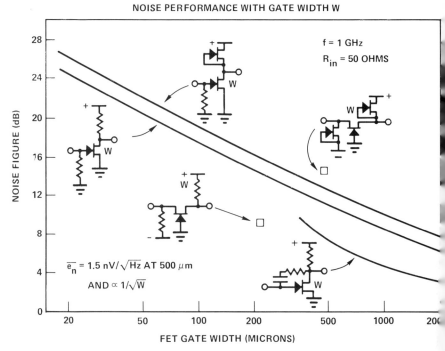

Figure 12.33 Calculated noise figure versus FET gate width. (Reprinted by permission of IEEE)

one- and two-stage amplifiers[33-37] with no reactive tuning elements exhibiting 2–2.5 dB noise figures over wide frequency ranges up to 2 GHz.

Another commonly used amplifier is the 'active matched' circuit shown in Figure 12.34. A common gate FET is used at the input to present a low VSWR, common source device is used to provide gain, and a common drain output is used to present a 50 Ω output impedance. The inductor, L, is used to peak the gain of the amplifier at some desired high frequency. If the first two stages were connected directly together, the gate impedance of the common source stage would heavily load the output of the common gate stage, causing the gain to roll off at high frequencies. The inductor is typically a 5–10 nH spiral inductor or a length of 85–140 Ω transmission line. Rockwell[38] has demonstrated a four-stage amplifier of this type, achieving 8 dB of gain from 0.5 to 9 GHz with a noise figure of 6.8 dB at 8 GHz. The amplifier was formed on a 2.5 × 2.5 × 0.25 mm GaAs chip. The input and output VSWRs were better than 2:1 over the band.

12.6.2 Low-noise amplifiers

A great deal of effort has gone into the development of monolithic X-band (8–12 GHz) low-noise amplifiers. Such circuits are required both for active array

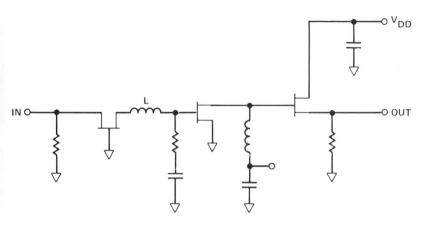

Figure 12.34 Active matched amplifier

nd for the direct broadcast television market. Pengelly and Turner[2] demonstrated the first GaAs monolithic IC containing an FET in 1976, which consisted of a single FET and a lumped element input matching structure. Output matching was achieved with a wire bond to the drain. The amplifier covered the -12 GHz band. Recent single-stage amplifiers at Plessey[39] have demonstrated ide band performance with 6 dB of gain from 6 to 18.5 GHz.

Microwave low-noise amplifiers typically use FETs in the common source onfiguration. This allows good gain and noise figure at the expense of input SWR which is typically 3:1 or poorer in order to achieve proper noise match. If a od input match is necessary along with a low noise figure, either balanced mplifiers or a common gate input circuit must be used. Examples of one- and o-stage single-ended X-band amplifiers[40] are shown in Figure 12.35. The mplifiers consist of 300×0.5 μm FETs, microstrip matching elements, overlay pacitors for RF bypass and d.c. blocking applications, and air bridges to terconnect the FET source pads and to connect the transmission lines to e top plates of the capacitors. The amplifier chips measure $50 \times 50 \times 8$ and $\times 100 \times 8$ mil for the single- and dual-stage circuits, respectively.

To obtain higher microstrip Q and improved performance, a chip thickness of mil was chosen, requiring the use of wire bonding over the edge of the chips to tain grounds. The measured noise figure of the single-stage amplifier is 2.2 dB th an associated gain of 10.5 dB at 12 GHz. At the same frequency, the two- ge amplifiers have achieved 2.5 dB noise figure with an associated gain of .1 dB. This performance is adequate to fulfil a wide range of applications in dar and communication applications, but the amplifiers have an input VSWR 2.5:1 or higher which makes them difficult to use and impossible to cascade. A lanced amplifier can be formed using a pair of amplifiers and two quadrature uplers as shown in Figure 12.36. Monolithic versions of Lange couplers are

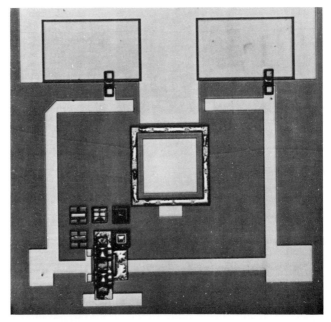

(a)

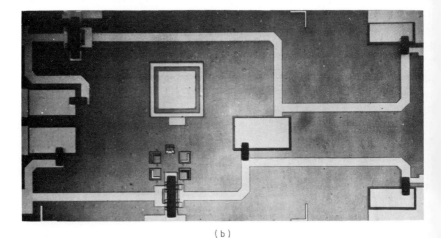

(b)

Figure 12.35 Monolithic LNAs. (Photo courtesy of Hughes Aircraft Company)

easily fabricated on GaAs as shown in Figure 12.37; they have an insertion loss approximately 0.5 dB at 12 GHz and are very broad band. A balanced amplifi could then be formed using a pair of amplifiers and a pair of Lange couplers or single monolithic chip, occupying an area 2.2 times larger than an individu

GaAs microwave monolithic circuits 497

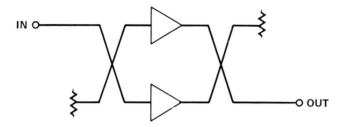

Figure 12.36 Balanced amplifier

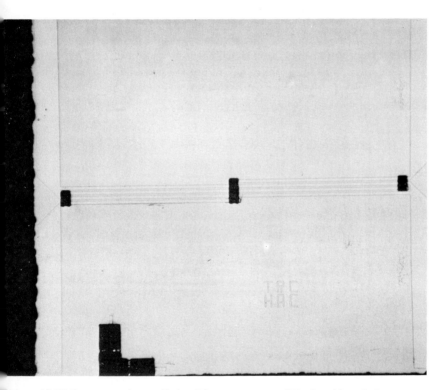

Figure 12.37 Lange coupler on GaAs. (Photo courtesy of Hughes Aircraft Company)

amplifier and with a performance degradation of 0.5 dB in noise figure and 1 dB in gain. Another alternative is to build a low-noise amplifier with a common gate input for which the optimum noise match is close to the conjugate of the input impedance of the device. Lehman et al.[41] have developed a three-stage low-noise amplifier using a common gate FET with a gain of 15 dB, a noise figure of 3.5–4.5 dB and a VSWR of 2:1 over the 9–10 GHz band. The amplifier consists of three 0.5 × 300 μm FETs, microstrip transmission lines, overlay capacitors for RF tuning as well as bypass, and Au–Ge–Ni resistors on a 1.3 × 2.5 × 0.15 mm GaAs chip.

12.6.3 Power amplifiers

The bulk of the work in monolithic power amplifiers has been concentrated in the 9–10 GHz band to satisfy active array radar requirements. The amplifiers are typically designed as single-ended circuits using lumped and distributed passive matching elements to achieve maximum power transfer designs. Power amplifier design can be approached by establishing a resistive load line, extending from the 'knee' of the I–V curve of I_{dss} to the drain-to-gate breakdown voltage at low current as shown in Figure 12.38. At most companies, however, amplifier design proceeds from a thorough large-signal characterization of the FET using load pull techniques to establish optimum load impedance versus frequency. Standard synthesis techniques can then be used to develop wide band constant power amplifiers.

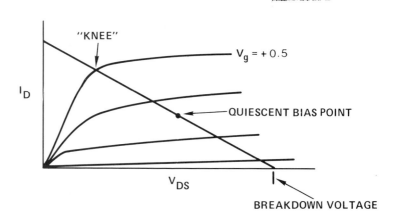

Figure 12.38 Class A power amplifier biasing

One of the first X-band power amplifiers was demonstrated by Pucel et al.[42] in 1979. Shown in Figure 12.39, it consists of four common source FETs, directly combined using microstrip dividers on a 4.8 × 6.3 × 0.1 mm chip. Top surface grounds were achieved using via holes. The amplifier delivered 2.1 W with 3.3 dB of gain at 9.5 GHz. Although good power results were obtained, the problem with this type of design become apparent when you consider that only 2 amplifiers (unyielded) of this size are obtained from one square inch of processed GaAs. This is due primarily to the use of near-quarter-wavelength microstrip

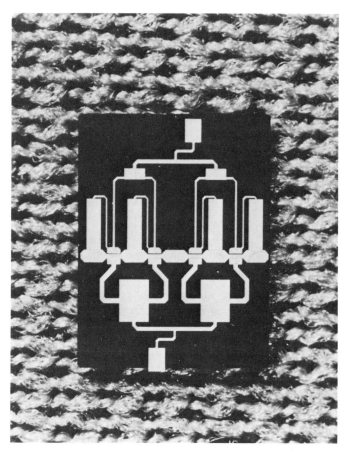

Figure 12.39. Monolithic 2 W power amplifier. (Photo courtesy of Raytheon)

lines as power splitters. An alternative to this design is a semi-lumped approach, using short lengths of high-impedance transmission line and lumped capacitors as matching elements. Figure 12.40 shows a chip containing a three-stage and four-stage power amplifier developed at Texas Instruments.[43] Each amplifier measures 1.0 × 4.0 × 0.1 mm and uses thin-film capacitors and microstrip transmission line for RF matching. Grounds are obtained using tapered microstrip lines from the source to the ground rail at the edge of the trip. This technique can add several ohms of inductive reactance between source and ground which acts as a negative feedback element, but its effect is predictable and repeatable. As indicated by the results shown in Figure 12.41, good performance has been achieved with the four-stage amplifier delivering 1 W of power with 25 dB of gain from 8.7 to 9.3 GHz. More recent amplifiers from TI[44] have used via holes to

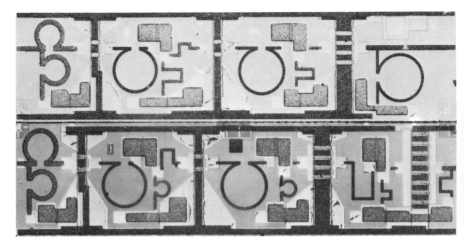

Figure 12.40 Three- and four-stage power amplifiers. (Photo courtesy of Texas Instruments)

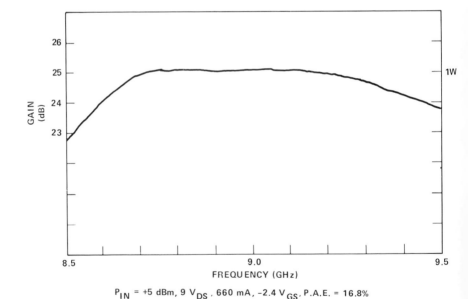

P_{IN} = +5 dBm, 9 V_{DS}, 660 mA, -2.4 V_{GS}, P.A.E. = 16.8%

Figure 12.41 Four-stage monolithic GaAs FET amplifier gain–frequency performance. (Photo courtesy of Texas Instruments)

provide top surface grounds. In general, via holes improve circuit performance, decrease chip area, and decrease the overall circuit yield, with the positive features outweighing the potential yield problem.

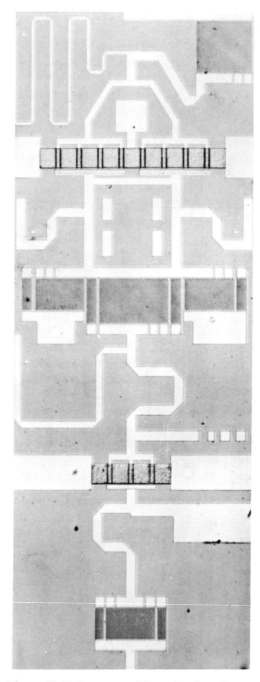

Figure 12.42 Power amplifier using interdigital capacitors. (Photo courtesy of Westinghouse Electric Corporation)

502 Gallium arsenide

Although typical radar bandwidths are 10% or less, there is considerable interest in wider bandwidth power amplifiers. Westinghouse has been pursuing octave bandwidth monolithic power amplifiers since 1978[45] and have published several articles describing 5–10 GHz and 8–12 GHz power amplifiers. The two amplifiers shown in Figures 12.42 and 12.43 present an interesting contrast in design approaches, using interdigital capacitors in one and thin-film capacitors in the other. Both chips use a 2400 μm output device and a first stage of 900 μm in one case and 1200 μm in the other on 2 × 4.75 × 0.1 mm substrates. The active layers are formed by direct implantation into the semi-insulating substrate. The circuit using interdigital capacitors devotes 14% of the chip area to the capacitors and requires four off-chip bypass capacitors wire bonded to the circuit to inject bias. The circuit of Figure 12.43 uses overlay capacitors and via holes to achieve on-chip grounding and bypassing. Although the chip areas are the same, the latter amplifier could have been shrunk considerably in size and the flexibility in choosing circuit topologies is greatly enhanced by not having to design small value, typically under one pF, series capacitors into each matching circuit. This circuit is part of a four-stage power amplifier with a total periphery of 12 mm on a 5.3 × 7 mm chip. The amplifier delivers 3 W of power at 10 GHz.

All of the above amplifiers were single-ended designs and care must be taken when using them in a system. As an example, the impedance seen looking into a single antenna element in an active array radar changes dramatically with scan angle, with VSWRs as bad as 6:1. The amplifiers can be isolated somewhat using circulators or by forming balanced amplifiers. A pair of power amplifiers can be fabricated along with quadrature couplers as described for low-noise amplifiers, providing twice the output power of a single unit, minus coupler losses, at a cost per watt approximately equal to $1/Y$ times that of a single amplifier, where Y is the single amplifier yield.

Texas Instruments has investigated the use of a classic push–pull amplifier[46] as a microwave power amplifier. The basic circuit, shown in Figure 12.44, consists

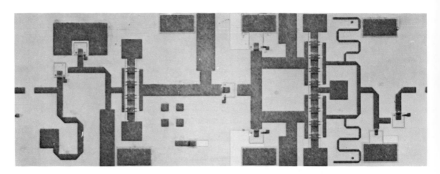

Figure 12.43 Power amplifier using overlay capacitors. (Photo courtesy of Westinghouse Electric Corporation)

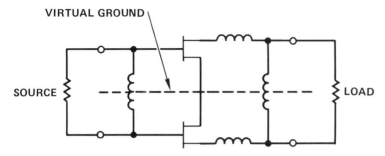

Figure 12.44 Push–pull amplifier

of a pair of FETs driven 180° out of phase with shunt matching connected between the devices rather than to a physical ground plane. A virtual ground then exists between the circuits and the source inductance can be minimized. The circuit that was developed consisted of a two-stage (four FETs) amplifier on a $2 \times 2 \times 0.1$ mm chip. The circuitry was microstrip but no RF connections were needed to the ground plane, with the shunt connections formed using wire bonds between circuits. Using a pair of 1200 μm FETs as output devices, 1 W was achieved with 10 dB gain from 8.5 to 9.5 GHz. Hybrid 180° couplers were used at both ports of the amplifier. The impedance benefits of a classic push–pull amplifier, i.e. a 4:1 impedance transformation, were not obtained due to the presence of the ground plane which was required for heat sinking. In an attempt to further integrate the amplifier, an active 180° phase splitter was developed with a simplified schematic shown in Figure 12.45. This circuit is identical in concept to

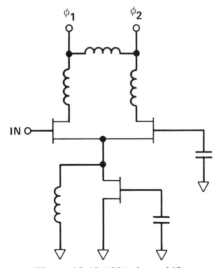

Figure 12.45 180° phase shifter

the low-frequency circuits described earlier with a constant current source biasing a differential pair of FETs, one with an incoming signal applied to the gate, the other with its gate grounded. Although such circuits work well through 3 or 4 GHz, the 9 GHz circuit developed at TI had much higher phase change with frequency than desirable due to device parasitics.

12.6.4 Distributed amplifiers

The amplifiers discussed previously were either direct coupled low-frequency circuits or microwave circuits which used standard filter synthesis techniques to match a device and its associated parasitics to some given impedance over a specified bandwidth. A distributed amplifier divides the active element into a number of small elements and incorporates the parasitics from each into transmission lines connecting all of the inputs and the outputs together as shown in Figure 12.46 using a simplified FET model. This forms a low-pass structure which allows operation of the amplifiers having decades of bandwidth up to frequencies of 30 GHz or higher with current devices.

Distributed amplifiers are well known, having been analysed for the lossless case by Ginzton[47] in 1948 using vacuum tubes and recently by Beyer et al.[48] for FETs. Referring again to Figure 12.46, the input signal flows along the gate line, developing a potential across each device whose phase is proportional to the length along the structure. If the drain transmission line is adjusted to have the same phase velocity as the input, the amplified signals will add constructively, forming a wave increasing in amplitude as it flows to the output port.

For lossless devices, we can write the overall voltage gain A as

$$A = \frac{ng_m}{2} \frac{(Z_{01}Z_{02})^{1/2}}{(1 - f^2/f_c^2)^{1/2}}$$

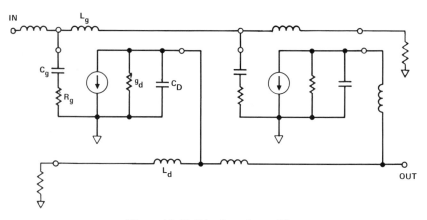

Figure 12.46 Distributed amplifier

GaAs microwave monolithic circuits

where g_m is the transconductance per stage, n is the number of stages, Z_{01} and Z_{02} are the characteristic impedances of the gate and drain transmission lines, f is the frequency, and f_c is the cut-off frequency of the transmission lines given by

$$f_c = \frac{1}{\pi\sqrt{(LC)}}$$

In a realistic amplifier with FETs the loss in the gate and drain is significant and must be accounted for in the design. For the above circuit Beyer et al.[48] have given the voltage gain A as

$$A = \frac{ng_m Z_{02}}{2[1+(\omega/\omega_H)^2]^{1/2}} \frac{\sinh[\frac{1}{2}n(\alpha_d-\alpha_g)]}{\sinh[\frac{1}{2}(\alpha_d-\alpha_g)]} \exp[-\frac{1}{2}n(\alpha_g+\alpha_d)]$$

where ω_H is the gate circuit cut-off frequency, equal to $(R_g C_g)^{-1}$, and α_d and α_g are the attenuation constants per section on the drain and gate line in nepers/section. By setting $dA/dn = 0$ they obtain the optimum number of devices for maximum gain at a frequency

$$n_{opt} = \frac{1}{\alpha_d - \alpha_g} \ln\left|\frac{\alpha_d}{\alpha_g}\right|$$

A practical amplifier providing 11 dB of gain from 2 to 20 GHz has been published by Ayasli et al.[49] The amplifier, shown in Figure 12.47, consists of

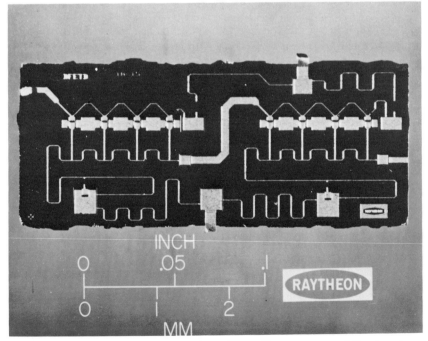

Figure 12.47 Two-stage distributed amplifier. (Photo courtesy of Raytheon)

eight 75 μm FETs connected as two cascaded travelling wave amplifiers of four devices each. The chip size is 1.2 × 5.5 mm and uses microstrip transmission line for reactive matching, overlay capacitors for d.c. blocking and RF bypass, and resistors for bias insertion.

This type of circuit is an excellent candidate for monolithic production, for several reasons. The performance is relatively tolerant of fluctuations in device parameters, which improve the yield and lower the cost of the circuits in production, and the circuit being so wide band will find many different applications, which will increase the number of chips required.

A serious problem with this circuit, however, is its power response. In general, to obtain maximum power from a FET it must be biased such that the maximum voltage swing approaches gate–drain breakdown. If we examine a travelling wave amplifier we find voltage increasing linearly on the output circuit as we approach the load. If the final device sees a voltage V_{max}, the average device sees $V_{max}/2$ and our power is decreased significantly over what could be achieved with a constant drain voltage on each device. This problem can be attacked by altering the characteristic impedance of the drain transmission line as a function of position, but nothing has been published to date on this subject.

12.6.5 Phase shifters

Like power amplifiers, phase shifters are an important component in active array radar systems and a considerable amount of effort has been spent in developing them at frequencies from 1 to 10 GHz. Although a large number of interesting circuits have been published, two designs have been particularly successful. These are the shunt FET switched and the switched amplifier designs.

In the first design we use a FET as a switch as shown in Figure 12.48. When V_{gs} equals zero, the FET is on and, for voltages less than the knee voltage, can be represented by a small resistor whose value is proportional to device periphery. When the device is biased to the pinch-off voltage it can be represented as a capacitance, shunted by the drain conductance. In both cases neither the gate nor

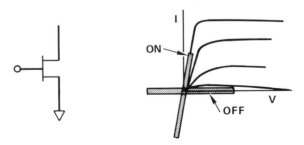

Figure 12.48 FET as switch

GaAs microwave monolithic circuits

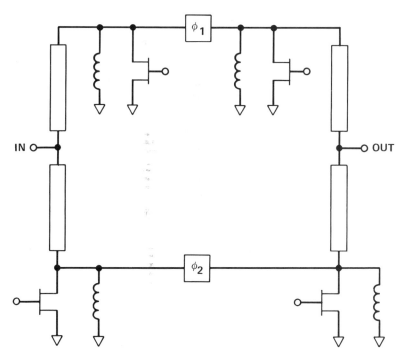

Figure 12.49 Shunt switched one-bit phase shifter

the drain draws any holding current. A one-bit phase shifter can be built using shunt mounted FETs as shown in Figure 12.49. Four FETs are shunt mounted at the end of quarter-wavelength transmission lines which function as impedance inverters. Each device is shunted by an inductor which resonates the drain–source capacitance at the desired centre frequency. These circuits then form a single-pole double throw switch, into which are inserted two circuits providing a differential phase shift. The actual phase shift circuits can be high-pass/low-pass filters which provide a fixed differential phase shift over a frequency range or by different lengths of transmission line which form a constant time delay phase shifter. Figure 12.50 shows a photograph of one-bit phase shifter at 10 GHz. The chip measures 80×100 mil and contains four 1200 μm FETs, over a wavelength of transmission line, and, in the centre of the chip, the two small high-pass and low-pass filters. Although phase shifters of this sort can be made to function well, they have severe inherent, self-evident problems. The chip size is large, especially considering that this is one bit out of the four or five bits normally required, and the 4800 μm of periphery will make this a relatively low production yield circuit. Clever circuit design can shrink the size and periphery somewhat, but alternative approaches are needed.

508　　　　　　　　　　　*Gallium arsenide*

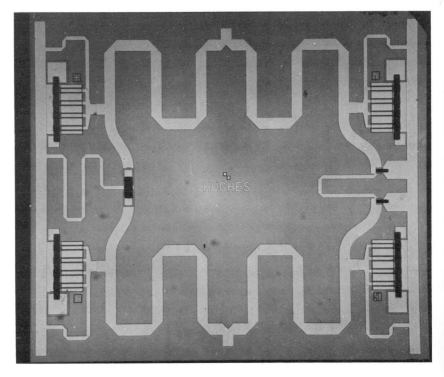

Figure 12.50 One-bit shunt switched phase shifter. (Photo courtesy of Hughes Aircraft Company)

The switched amplifier phase shifter is shown schematically in Figure 12.51. It consists of an input dual-gate dual-drain FET, drain matching circuitry in each of the drain arms, differential phase shift circuitry, and a Wilkinson hybrid to connect the drain circuits together. It is apparent that the signal loses 3 dB by being split between the two input amplifiers and suffers a further 3 dB in the output hybrid. The devices must, therefore, make up 6 dB plus circuit losses in order to provide 0 dB gain for the overall phase shifter. This circuit can be operated in either a digital or analogue mode. As a digital phase shifter, one device is turned on and the other off by appropriately biasing the second gates in each device. This selects either of the two amplifier paths and the differential phase shift elements function as in the previous circuit. In the analogue mode, a 90° phase shift is inserted in one path and both FETs are turned on at the same time. By adjusting second gate bias, we can control the gain of each device and therefore achieve any phase shift between 0 and 90° with a constant gain.

Vorhaus *et al.*[50] have published a digital version of this design shown in Figure 12.52. On a 2.5 × 3.0 × 0.1 mm chip are included the dual-gate dual-drain FET, matching circuitry, overlay capacitors, the isolation resistor for the

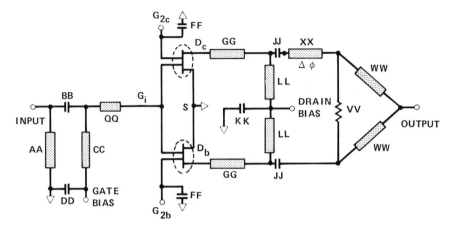

Figure 12.51 Switched amplifier phase shifter. (Photo courtesy of Raytheon)

Wilkinson hybrid, and via holes. Integral beam leads are also formed by etching away the underlying GaAs. This eliminates the need to wire bond directly onto the chip.

12.6.6 Millimetre-wave integrated circuits

While at microwave frequencies the main reasons for using monolithic ICs are cost and large volume demands, monolithic circuits are desirable at millimetre-wave frequencies because of their reproducibility. One mil bondwire has an inductive reactance at 10 GHz of about 1 Ω per mil of length. At 60 GHz the same wire has a reactance of 6 Ω per mil. In a hybrid circuit, the mounting parasitics can become the dominant reactances in the circuit. Forming circuits monolithically helps both to minimize the parasitics and to make them calculable and repeatable from chip to chip.

At high frequencies we can make the following general observations about monolithic ICs:

(a) the Q of inductors and transmission line resonators is proportional to $f^{1/2}$;
(b) the skin depth is proportional to $f^{-1/2}$;
(c) capacitor Q is proportional to $f^{-3/2}$;
(d) area per circuit is proportional to f^{-2};
(e) cost per circuit is proportional to f^{-2}.

At the low millimetre-wave frequencies a single-stage amplifier has been demonstrated at 32 GHz. Shown in Figure 12.53 it consists of a single $75 \times 0.5\,\mu\mathrm{m}$ FET, microstrip matching, and an overlay capacitor for RF bypassing on a

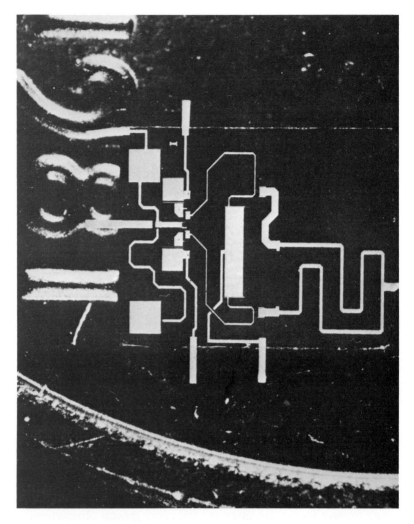

Figure 12.52 Switched amplifier phase bit. (Photo courtesy of Raytheon)

0.3 × 0.4 × 0.1 mm chip. The amplifier had 5 dB gain with a 1 dB bandwidth of 2 GHz. Although the performance is modest by the standards of new quarter-micrometre devices, it is the size which is interesting. On a wafer with two square inches of useful area we would obtain 10 000 of these chips. The yield should also be high considering the small periphery.

Mixers are an important and popular monolithic component at millimetre-wave frequencies and versions have been demonstrated from 30 to 110 GHz. One of the major difficulties with the development of integrated receivers containing

GaAs microwave monolithic circuits 511

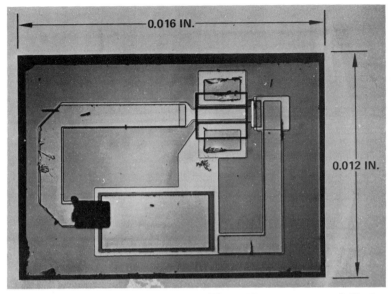

Figure 12.53 Single-stage 32 GHz amplifier. (Photo courtesy of Hughes Aircraft Company)

Schottky barrier diode mixers and FET LOs and IF amplifiers is the basic incompatibility between FET and diode processing. The FET needs an n on semi-insulating layer and the high-quality diodes require n on n$^+$ layers to achieve low loss. Chu et al.[51] have demonstrated an interesting solution to this problem in the development of a 31 GHz mixer–IF amplifier chip. Shown in Figure 12.54 it

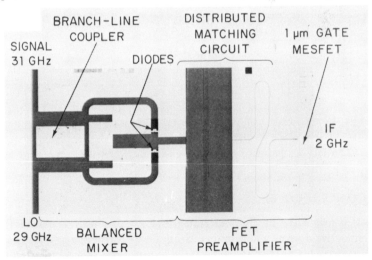

Figure 12.54 31 GHz mixter–IF. (Photo courtesy of MIT Lincoln Laboratory)

512 *Gallium arsenide*

consists of a 5.3 × 2.7 mm chip containing a branch line hybrid, a pair of surface oriented beam lead diodes to form a balanced mixer, a low-pass filter, and a one-stage IF amplifier. The circuit was fabricated by growing an n on buffer layer wafer suitable for FET fabrication. Pits are then selectively etched in the GaAs under the diode locations and n on n^+ layers were grown epitaxially in these locations forming a profile shown in Figure 12.55. The mixer had a conversion loss of 5.5–6.5 dB over a 2 GHz band at 30 GHz and the noise figure of the complete chip varied from 11.5 to 13 dB over the same band.

At frequencies of 50 GHz and higher, the wavelength becomes small enough that it becomes practicable to include not only matching and bias circuitry on monolithic ICs but the transitions to waveguide as well. GaAs integrated circuits have been demonstrated using fin line, quasi-optical, and suspended substrate approaches at frequencies from 50 to 110 GHz. Figure 12.56 shows a simple balanced mixer consisting of a probe that protrudes into a waveguide, a low-pass filter, and a pair of planar Schottky barrier diodes. The conversion loss is 7.6 dB with an LO of 92 GHz and an RF of 75 GHz.

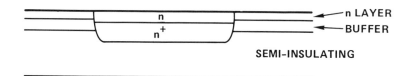

Figure 12.55 Selective epitaxial growth for multiple device types on a common wafer

Figure 12.56 90 GHz balanced mixer. (Photo courtesy of Hughes Aircraft Company)

12.7 CONCLUSIONS

GaAs is a near-ideal medium for integrated circuits with GaAs FETs being excellent active devices and semi-insulating GaAs forming a low-loss and high-resistivity base for passive elements. A wide variety of monolithic circuits have been demonstrated with RF results approaching those of hybrid circuits and specialized circuits have been developed, such as the distributed amplifiers, that take full advantage of the monolithic format and give results which would be difficult to achieve using hybrid techniques. Highly integrated ICs have also been demonstrated including the low-frequency heterodyne generator from Hewlett Packard[30] and the complete TVRO front-end from LEP[52] which includes an FET LO, two-stage preamplifier, and a dual-gate FET mixer on a 1 cm square GaAs chip.

Although the professed goal of many monolithic efforts is the total integration of microwave systems, such as a complete transmit/receive (T/R) module on a single chip, many applications are emerging that use combinations of monolithic and hybrid circuitry components. Raytheon has demonstrated a complete X-band T/R module[53] composed of eleven GaAs monolithic ICs, connected with microstrip on alumina substrates, and the low-frequency generation chip mentioned earlier used hybrid inductors to achieve high Q and small chip area. This combination of monolithic and hybrid circuitry allows an optimal mix of both technologies, using monolithic ICs for the active devices and matching circuitry and hybrid circuits for components that are physically large, such as high-value inductors or power combiners. The use of multichip monolithic circuits rather than total monolithic will also dramatically improve the yield and therefore the cost of systems using them as long as the cost per chip is higher than the cost of interconnecting them.

ACKNOWLEDGEMENTS

The author wishes to acknowledge his many colleagues in the industry who have contributed photographs of their work for inclusion in this chapter. Among them are Drs R. Pucel and Y. Ayasli from the Raytheon Company, Dr J. Degenford from Westinghouse Electric Corporation, Dr A. Chu from MIT Lincoln Laboratory, Mr D. Hornbuckle from Hewlett Packard, Dr W. Wisseman from Texas Instruments, and the members of the Torrance Research Center and the Electron Dynamics Division of Hughes Aircraft Company.

REFERENCES

1. E. Mehal and R. W. Wacker, GaAs integrated microwave circuits, *IEEE Trans. Microwave Theor. Tech.*, **MTT-16**, 451–4 (July 1968).
2. R. S. Pengelly and J. A. Turner, Monolithic broadband GaAs FET amplifiers, *Electron Lett.*, **12**, 251–2 (May 1976).
3. M. V. Schneider, Microstrip lines for microwave integrated circuits, *Bell System Tech. J.*, 1421–4 (May 1969).
4. W. S. Getsinger, Microstrip dispersion model, *IEEE Trans. Microwave Theor. Tech.*, **MTT-17**(8), 415 (August 1969).

5. R. A. Pucel, et al., Losses in microstrip, *IEEE Trans. Microwave Theor. Tech.*, **MTT-16**(6), 342–50 (June 1968).
6. E. O. Hammerstad and F. Berrddal, *Microstrip Handbook*, Electronics Research Laboratory, University of Trondheim, Norway (1975).
7. H. Sobol, Radiation conductance of open-circuit microstrip, *IEEE Trans. Microwave Theor. Tech.*, **MTT-19**(11), 885–7 (November 1971).
8. M. Maeda, An analysis of gap in microstrip transmission lines, *IEEE Trans. Microwave Theor. Tech.*, **MTT-20**(6), 390–5 (June 1972).
9. P. Benedek and P. Silvester, Equivalent capacitances for microstrip gaps and steps, *IEEE Trans. Microwave Theor. Tech.*, **MTT-20**(11), 729–33 (November 1972).
10. P. Silvester and P. Benedek, Microstrip discontinuity capacitances for right-angle bends, T junctions, and crossings, *IEEE Trans. Microwave Theor. Tech.*, **MTT-21**(5), 341–6 (May 1973).
11. C. P. Wen, Coplanar waveguide . . . , *IEEE Trans. Microwave Theor. Tech.*, **MTT-18**(12), 1087–90 (December 1969).
12. W. Hillberg, From approximation to exact relations for characteristic impedances, *IEEE Trans. Microwave Theor. Tech.*, **MTT-17**(5), 259–65 (May 1969).
13. S. B. Cohn, Slot line on a dielectric substrate, *IEEE Trans. Microwave Theor. Tech.*, **MTT-17**(10), 768–78 (October 1969).
14. E. A. Mariani, et al., Slot line characteristics, *IEEE Trans. Microwave Theor. Tech.*, **MTT-17**(12), 1091–6 (December 1969).
15. G. D. Alley, Interdigital capacitors and their application to lumped-element microwave integrated circuits, *IEEE Trans. Microwave Theor. Tech.*, **MTT-18**(12), 1028–33 (December 1970).
16. J. I. Smith, The even and odd mode capacitance parameters for coupled lines in suspended substrate, *1969 G-MTT Symp. Digest*, pp. 324–8 (1969).
17. T. G. Bryant and J. A. Weiss, Parameter of microstrip transmission lines and of coupled pairs of microstrip lines, *IEEE Trans. Microwave Theor. Tech.*, **MTT-16**(12), 1021–7 (December 1968).
18. R. Esfandiari, D. W. Maki, and M. Siracusa, Design of interdigitated capacitors . . . , *IEEE Trans. Microwave Theor. Tech.*, **MTT-31**(1) (January 1983).
19. H. Sobol and M. Caulton, The technology of microwave integrated circuits, *Advances in Microwaves*, vol. 8, New York, Academic Press (1974).
20. F. E. Terman, *Radio Engineer's Handbook*, New York, McGraw-Hill (1943).
21. A. F. Podell, GaAs MICs: expensive, exotic but exciting, *Microwaves*, 56–61 (December 1980).
22. H. E. Bryan, Printed inductors and capacitors, *Tel-Tech*, **14**(12) (December 1955).
23. G. L. Matthaei, L. Young, and E. M. T. Jones, *Microwave Filters, Impedance-Matching Networks and Coupling Structures*, New York, McGraw-Hill (1964).
24. M. Hatzakis, B. J. Canavello, and J. M. Shaw, Single-step optical lift-off process, *IBM J. Res Dev.*, **24**(4), 452–60 (July 1980).
25. W. D. Westwood, *Sputter Deposition and Ion Beam Processing*, American Vacuum Society (1980).
26. M. Durschag and J. Vorhaus, A tantalum-based process for MMIC on-chip thin-film components, *1982 GaAs IC Symposium Digest* (November 1982).
27. A. Chu, et al., A two-stage monolithic IF amplifier utilizing a high dielectric constant capacitor, *1982 Microwave and Millimeter Wave Monolithic Circuits Symposium Digest*, pp. 61–63 (1982).
28. L. A. D'Asaro, J. V. Di Lorenzo, and H. Fukui, Improved performance of GaAs microwave field effect transistors with via connections through the substrate, *IEDM Digest Tech. Papers*, pp. 370–1 (1977).

29. R. Esfandiarfi, D. Maki, H. Yamasaki, M. Feng, and M. Siracusa, Two-stage semi-lumped and distributed GaAs monolithic power amplifiers, *1982 GaAs IC Symposium Digest* (November 1982).
30. R. Van Tuyl, A monolithic GaAs IC for heterodyne generation of RF signals, *IEEE Trans. Electron Devices*, **ED-28**(2), 166–70 (February 1981).
31. D. P. Hornbuckle, GaAs IC direct-coupled amplifiers, *IEEE Int. Microwave Symp. Dig.*, pp. 387–9 (May 1980).
32. D. B. Estreich, A monolithic wide-band GaAs IC amplifier, *IEEE J. Solid State Circuits*, **SC-17**(6), 1166–73 (December 1982).
33. R. L. Van Tuyl, et al., A manufacturing process for analog and digital GaAs ICs, Presented at the *IEEE Gallium Arsenide Integrated Circuit Symposium*, San Diego, CA, October 1981; also in *IEEE Trans. Electron Devices*, **ED-29**, 1031–8 (July 1982).
34. A. K. Gupta, J. A. Higgins, and D. R. Decker, Progress in broadband GaAs monolithic amplifiers, *IEDM Tech. Dig.*, pp. 269–72 (December 1979).
35. R. L. Van Tuyl, A monolithic integrated 4 GHz amplifier, *ISSCC Dig. Tech. Papers*, pp. 72–3 (February 1978).
36. D. P. Hornbuckle and R. L. Van Tuyl, Monolithic GaAs direct-coupled amplifiers, *IEEE Trans. Electron Devices*, **ED-28**, 175–82 (February 1981).
37. K. Honjo, T. Sugiura, and H. Itoh, Ultrabroadband GaAs monolithic amplifier, *Electron. Lett.*, **17**, 927–8 (November 26, 1981).
38. W. C. Petersen, et al., A monolithic GaAs 0.1 to 10 GHz amplifier, *1981 IEEE Int. Microwave Symposium Digest*, pp. 354–5 (1981).
39. R. S. Pengelly, *Microwave Field Effect Transistors*, Research Studies Press, p. 400 (1982).
40. L. C. Liu, D. W. Maki, M. Feng, and M. Siracusa, Single and dual stage monolithic low noise amplifiers, *1982 GaAs IC Symposium Digest*, New Orleans (November 1982).
41. R. E. Lehman, G. Brehm, D. Seymour, and G. Westphal, 10 GHz monolithic low noise amplifier with common gate input, *1982 GaAs IC Symposium Digest*, New Orleans (November 1982).
42. R. A. Pucel, J. L. Vorhaus, P. Ng, and W. Fabian, A monolithic GaAs X-band power amplifier, *IEDM Tech. Dig.*, pp. 266–8 (1979).
43. H. Q. Tserng, H. M. Macksey, and S. R. Nelson, Design, fabrication, and characterization of monolithic microwave GaAs power FET amplifiers, *IEEE Trans. Electron Devices*, **ED-28**, 183–90 (February 1981).
44. H. Q. Tserng, H. M. Macksey, and S. R. Nelson, A four-stage monolithic X-band GaAs FET power amplifier with integral bias networks, *GaAs IC Symposium* (November 1982).
45. J. E. Degenford, R. G. Freitag, D. C. Boire, and M. Cohn, Design considerations for wideband monolithic power amplifiers, *GaAs IC Symp. Res. Abstracts*, Paper No. 22 (1980).
46. V. Sokolov, R. E. Williams, and D. W. Shaw, X-band monolithic push–pull amplifiers, *ISSCC Dig. Tech. Papers*, pp. 118–19 (February 1979).
47. E. Ginzton, et al., *Proc. IRE*, 956–69 (August 1948).
48. J. B. Beyer, S. Prasad, J. Nordman, R. Becker, and G. Hohenwarter, *Wideband Monolithic Microwave Amplifier Study*, Annual Report July 1981–July 1982, N00014-80-C-0923 (1982).
49. Y. Ayasli, D. Reynolds, J. Vorhaus, and L. Hanes, 2–20 GHz GaAs traveling-wave amplifier, *1982 GaAs IC Symposium Digest*, pp. 136–8 (November 1982).
50. J. Vorhaus, R. Pucel, and Y. Tajima, Monolithic dual-gate FET digital phase shifter, *1981 GaAs IC Symposium Digest*, San Diego (October 1981).

51. A. Chu, W. E. Courtney, and R. W. Sudbury, A 31 GHz monolithic GaAs mixer/preamplifier circuit for receiver applications, *IEEE Trans. Electron Devices*, **ED-28**(2), 149–54 (February 1981).
52. P. Harrop, P. Lesartre, and A. Collet, GaAs integrated all FET front-end at 12 GHz, *IEEE GaAs IC Symposium* (November 1980).
53. R. Pucel, *et al.*, A multi-chip GaAs monolithic transmit/receive module for X-band, *1982 IEEE MTT Symposium Digest*, pp. 489–92 (1982).

Gallium Arsenide
Edited by M. J. Howes and D. V. Morgan
© 1985 John Wiley & Sons Ltd

CHAPTER 13

GaAs Digital Integrated Circuit Technology

B. M. WELCH, R. C. EDEN, and F. S. LEE

13.1 INTRODUCTION

During the past 10 years, the superior electronic properties of GaAs over Si has made it the focus of vast material and device research aimed towards ultra-high-speed digital logic applications. This trend has been strongly motivated by the ever increasing demands of modern electronic systems towards increased complexity at very high throughput rates at low power levels. The promise of high-speed, low-power GaAs digital integrated circuits with propagation delays (τ_d) of 100 ps or less should make it possible to attain logic throughput rates well beyond those of any Si integrated circuits presently available or expected in the future. The realization of such potential from GaAs is very difficult, of course, since it necessitates achieving in one technology: (1) ultra-high speed (very low τ_d); (2) low power per gate (P_D); (3) extremely low dynamic switching energies (power × delay products, ($P_D \tau_d$); (4) very high gate densities, VLSI; (5) mature high-yield processing technology; and (6) sophisticated high-speed packing techniques in order to provide practicable commercial implementation of such complex parts.[1-3]

The perceived problem in attaining these requirements is manifested in what are considered classical trade-offs between speed/power and lithographic resolution/yield. For example, in silicon metal–oxide semiconductor (MOS) ICs, the speed may be improved by increasing the supply voltage (V_{DD}) and logic voltage swings in order to increase the average device transconductances or current gain × bandwidth products (f_τ's). Increasing V_{DD}, however, while reducing τ_d, sharply increases gate power dissipation P_D and switching energies $P_D \tau_d$, which would lead to unacceptable power levels in > 10^4 gate VLSI chips.[4] Reducing device geometries to the limits of lithographic resolution is also an approach for minimizing speed × power products and it improves density as well, but this approach quickly arrives at a point of unacceptable yields and high costs resulting

518   *Gallium arsenide*

from extremely complicated processing. An example of pressing Si technology to its limits has been evidenced in sporadic reports of submicrometer Si MOS ring oscillator performance in the range below 100 ps gate delay over the past several years.[5,6] Even with the demonstration of such impressive results, practical implementation of Si IC technology in these speed domains has not and, in all likelihood, will not occur.

The obvious question is why cannot Si integrated circuit performance continue to improve indefinitely? The answer is simply that the methods used to gain higher performance just discussed did not include the most important method for improving performance. That method is to replace silicon with some other semiconductor materials, like GaAs, having superior electronic properties. GaAs, after many years of research and development, has finally reached a level of useful maturity at the same point in time that electronic system requirements are beginning to outstrip present integrated circuit capabilities.[7]

13.2 WHY GaAs IS SUPERIOR FOR HIGH-SPEED DIGITAL ICS

For very high-speed operation, logic gates must have current and power gains greater than their fan-out loadings, i.e. the f_τ and f_{max} (current and power gain × bandwidth products) of the active devices must be very high. Perhaps less obviously, the devices must have a high degree of non-linearity, i.e. they must develop their high transconductance (or f_τ or f_{max}) at control voltages only a small logic voltage swing above threshold. This is necessitated by the need to reduce greatly dynamic switching energies ($P_D \tau_d$) for ultra-high-speed LSI/VLSI. The dynamic switching energy must exceed the stored energy on the switched capacitance C, i.e.

$$P_D \tau_d > \tfrac{1}{2} C (\Delta V_L)^2 \qquad (13.1)$$

where C is the sum of the input capacitances of the fan-out of loading gates plus the parasitic capacitance, and ΔV_L is the logic voltage swing. (More precisely, the $(\Delta V_L)^2$ term is the product of the supply voltage V_{DD} times ΔV_L.) Hence, both the power supply and logic swing voltages must be kept small for low power × delay products, and the devices must not only be fast but must be fast with small logic voltage swings.

A key consequence of this requirement for strong non-linearity is the need for high carrier mobilities and/or very short carrier transit path lengths in the devices. This is easily illustrated for a field-effect transistor (FET) using the Shockley (square law) model,[4] where, for gate voltages V_{gs} near the channel conduction threshold voltage V_p, the saturated drain current I_{ds} is given by

$$I_{ds} = K(V_{gs} - V_p)^2 = \frac{\varepsilon \mu_n W}{2 a L_g}(V_{gs} - V_p)^2 \qquad (13.2)$$

where W is the channel width, L_g the gate length, a the distance between gate and

GaAs digital integrated circuit technology

channel, μ_n the electron mobility, and ε is the dielectric constant of the semiconductor in a metal–semiconductor FET (MESFET) or junction FET (JFET), or of the gate insulator in an insulated gate FET (MOSFET). The transconductance g_m for this Shockley FET model is then

$$g_m = \frac{dI}{dV} = \frac{\varepsilon \mu_n W}{aL_g}(V_{gs} - V_p) \qquad (13.3)$$

and the current gain × bandwidth product in this near-threshold small $V_{gs} - V_p$ region is given by

$$f_\tau = \frac{g_m}{2\pi C_{gs}} = \frac{\mu_n}{2\pi L_g^2}(V_{gs} - V_p) \qquad (13.4)$$

Equation (13.3) illustrates that the strength of the non-linearity, that is, the increase of transconductance with small gate voltages above threshold, increases with μ_n/L_g, while the f_τ expression of equation (13.4), which ignores fringing capacitances, gains an additional $1/L_g$ dependence owing to the assumed capacitance variation. Clearly, to maximize the transconductance at any given $V_{gs} - V_p$ voltage swing, one can reduce the gate length L_g, which is why gate lengths in GaAs FETs are typically 1 μm or shorter. Note, however, that we can obtain an equivalent improvement in near-threshold transconductance by going to a higher channel mobility μ_n. This is, of course, the prime motivation for using GaAs, rather than silicon, in high-speed integrated circuits. Electron mobilities in GaAs of several hundred thousand have been obtained in pure undoped material at 77 K and 8500 cm^2 V^{-1} s^{-1} at room temperature, while with typical MESFET n-channel doping levels ($N_d \sim 10^{17}$ cm^{-3}), electron mobilities in the $\mu_n \sim 4000$ to 5000 cm^2 V^{-1} s^{-1} range are obtained. These room-temperature GaAs electron mobilities are about 5–6 times higher than those for correspondingly doped bulk silicon, and nearly an order of magnitude larger than typical silicon MOS n-channel mobilities under strong inversion conditions.

While the Shockley model clearly predicts that GaAs (having a higher electron mobility) will yield higher performance than Si, the saturated drift velocity model for which the transconductance g_m is given as

$$g_m = \frac{v_s}{L_g} C_{gs} = \frac{\varepsilon v_{sat} W}{a} \qquad (13.5)$$

predicts nearly similar performance levels for both Si and GaAs if one uses the published steady-state bulk saturation velocities (v_{sat}) for GaAs ($v_s \sim 1 \times 10^7$ cm s^{-1})[8] and Si ($v_s \sim 6.5 \times 10^6$ cm s^{-1}).[9] Comparison of GaAs and Si v_{sat} at high electric fields is not consistent with the factor of 5 or more performance improvement indicated by the Shockley (mobility) model.

A more in-depth understanding of the saturation velocity–field characteristics and electronic energy band structure of GaAs can provide insight to why it is not valid to compare GaAs and Si devices at equilibrium saturated field velocities.

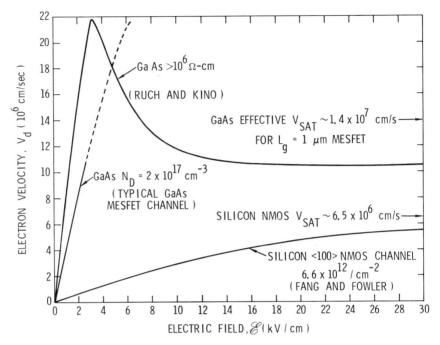

Figure 13.1 Comparison of the equilibrium velocity–field characteristics for electrons in silicon MOS channels and GaAs at 300 K. The high-resistivity GaAs exhibits larger mobilities than typical GaAs MESFET channels (the curve for which is simply extrapolated from the low-field mobility)

Figure 13.1 shows a comparison of the equilibrium velocity–field characteristics for electrons in silicon MOS channels and GaAs at 300 K.

The GaAs steady-state velocity–field data in Figure 13.1 were measured on high-resistivity bulk GaAs by Ruch and Kino.[8] The mobility $\mu_n \sim 8000$ cm^2 V^{-1} s^{-1} for that material is considerably higher than expected for typical MESFET channel dopings ($\mu_n \sim 4000$ cm^2 V^{-1} s^{-1} at $N_D = 2 \times 10^{17}$ cm^{-3}), so an additional GaAs curve extrapolated from this low-field mobility value is also shown.[4] The important feature is that at the $E \sim 10^4$ V cm^{-1} fields in low-power short-channel FETs (for $V_{gs} - V_p \sim 1$ V, $L_g = 1$ µm), the steady-state electron velocities in GaAs are much higher than those in silicon FET channels. In fact, whereas the steady-state velocity–field curve for GaAs shows a saturation velocity of $v_s \sim 1.05 \times 10^7$ cm s^{-1}, analysis based on actual GaAs MESFET data gives an effective saturation velocity of at least $v_s = 1.4 \times 10^7$ cm s^{-1}.[4,10]

There are fairly simple reasons why experimental short-channel GaAs MESFETs perform better than predicted from the steady-state velocity–field characteristics. These reasons derive from the fact that predictions based on 'steady-state' velocity–field curves assume long times, long distances, and voltage drops very large in comparison to the actual internal energy exchange mechanisms of the semiconductor. In GaAs, the drop-off in electron velocity (v_s) for electric fields above $3.1 \times \text{kV cm}^{-1}$ is a reflection of the transfer of electrons from the lowest conduction band minimum, $\Gamma(000)$, to a higher minimum.[11] This point is more easily understood by comparing the electronic energy band structures of GaAs and silicon shown in Figure 13.2. The curves show the energy versus crystal momentum for electrons. The slope (dE/dk) of these curves corresponds to the electron velocity (which has a maximum of $\sim 1 \times 10^8/\text{cm s}^{-1}$ in GaAs), while the reciprocal of the curvature, $(d^2E/dk^2)^{-1}$, corresponds to the effective mass of the electrons. Note that in GaAs, the effective mass is only 6.8% of the free electron mass, while in silicon it is 97% of the free electron mass or 14 times 'heavier' than in the GaAs case. Therefore, it is possible, if applied voltages, electric fields, and times are adequate, for electrons in GaAs with over 0.35–0.4 eV of energy above the minimum of the conduction band to transfer from the lower valley to one of the higher, high-mass, silicon-like valleys (upper valleys). Transfer of electrons between valleys in long devices with high applied voltages leads to the 'negative differential mobility' characteristics (negative slope of the velocity versus electric field strength curve) shown for GaAs in Figure 13.1. Electrons in the higher valleys have high mass and strong inter-valley scattering and so exhibit lower mobility (like conduction electrons in silicon). However, since these minima lie $\Delta E \sim 0.35$ to 0.4 eV above the Γ minimum, it is physically impossible for

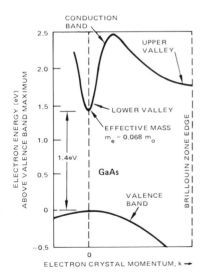
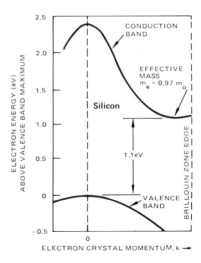

Figure 13.2 Electronic energy band structures of gallium arsenide and silicon, showing the energy (eV) versus crystal momentum k for electrons

electrons to transfer to these low-mobility valleys in low-voltage, short-channel FETs. In fact, even for $V_{gs} - V_p > 0.4$ V operation in very short ($L_g \sim 0.5$ μm) channel GaAs FETs, the electron transit times through the channel are so short (approaching 1 ps) that electron transfer to higher valleys is not completed during transit. These 'velocity overshoot' or 'ballistic' effects are predicted to improve the performance of all short transit path GaAs devices, particularly when the operating voltages are low.

As we have discussed, the f_τ of GaAs FET is interdependent on both the electron mobility, μ_e, and the saturated drift velocity, v_s, in conjunction with the specific electric field and channel length of the device. It should be clear from our discussion that simplistic theoretical predictions of the performance of GaAs and Si FETs are not adequate since they depend solely on the interpretation of the device model used. For instance, the mobility model predicts about 5 times higher g_m for GaAs over Si (for equivalent geometries), while the velocity saturation model implies only a factor of about 1.5 difference in g_m at high electric fields. The mobility models are more appropriate for the low pinch-off (threshold) voltage devices while the velocity saturation model is more applicable at high electrical fields (high pinch-off). However, in practical digital logic applications, FET devices are more nearly between these extremes. Since, the device switches between on and off states, it operates in a mobility controlled domain for a fraction of a cycle and in saturation controlled domain for another fraction of the time.

In view of the obvious inadequacies of predicting the performance of complex devices under dynamic operation using one-dimensional models, it is more appropriate to measure and compare actual working GaAs and Si FET devices. Figure 13.3 shows experimental curves of saturated drain current vs gate voltage above threshold for various GaAs and Si technologies.

All the curves in Figure 13.3 are nearly parabolic at the origin and tend to become linear only at the higher gate voltages, owing to current saturation. Comparing $L_g = 1$ μm MESFETs, both the GaAs and Si devices exhibit approximate square law drain current vs gate voltage characteristics,

$$I_{ds} = K(V_{gs} - V_p)^2$$

following the Shockley model, but with greatly differing K values as would be predicted from a comparison of their transconductances. In this comparison, drain currents differ by about a factor of 5 or 6 (at constant gate voltage) owing to the higher GaAs electron mobility.

As can be observed in Figure 13.3, silicon NMOS devices out-perform Si MESFETs because the gate oxide can be made much thinner (200–300 Å) than the MESFET channel depletion region. (Equation (13.3) indicates the importance of keeping a, the distance between the gate and channel, thin.) However, even though a Si NMOS is quite superior to a Si MESFET, the K value for the $L_g = 0.7$ μm Si NMOS device is over two times smaller than the K value of the

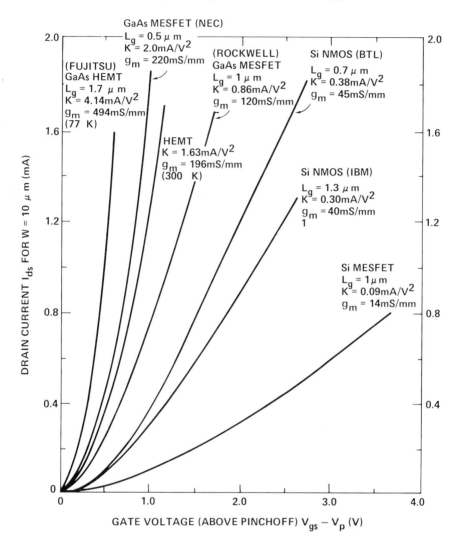

Figure 13.3 Comparison of transconductance g_m and square law K values for various GaAs and Si FET device types

longer $L_g = 1.0\,\mu\text{m}$ GaAs MESFET, over four times smaller than K for the $L_g = 0.5\,\mu\text{m}$ GaAs MESFET, and over ten times smaller for the GaAs HEMT (high electron mobility transistor discussed in Section 13.3.3) at 77 K.

The most outstanding characteristic of GaAs FETs is their ability to develop such high transconductance at relatively low gate voltages. This excellent non-linearity behaviour is probably the most important factor supporting GaAs LSI/VLSI where low dynamic switching energies will be a critical requirement.

Regardless of the countless arguments over theoretical predictions, etc., that are often presented in the literature, it is very clear from the simple analysis of various device performance results presented in Figure 13.3 that GaAs offers significant performance advantages over Si.

13.3 COMPARISON OF GaAs DEVICE APPROACHES FOR IC APPLICATIONS

Gallium arsenide's superior electronic properties (e.g. high electron mobility, high saturation velocity, and semi-insulating substrates) as compared with silicon have placed the main focus of attention in this technology predominantly on high-speed circuit applications. In addition, the ever increasing trend of electronics today towards LSI/VLSI has also created significant pressure towards achieving lower and lower power levels. An increasing interest in GaAs gate arrays demands high current driving capabilities of GaAs logic stages while analogue-to-digital applications require even more varied and different GaAs device and circuit characteristics. This interest in a wide variety of GaAs ICs obviously will require different device characteristics and approaches specifically tailored to the circuit application under consideration. In order fully to understand the directions and strategies currently being pursued in GaAs device technology today, a review of the various available GaAs device approaches and their trade-offs will be very useful. GaAs IC development efforts currently are predominantly directed along the use of the metal–semiconductor field-effect transistor (MESFET) with the exception of the junction field-effect transistor (JFET) and the heterojunction bipolar transistor (HJBT) as opposed to insulated gate FETS (MOSFETs) used in Si ICs. This is because, in spite of considerable efforts to develop satisfactory oxides or dielectrics on GaAs,[12,13] it has proved very difficult to achieve stable dielectric–semiconductor interfaces of low surface state quality on GaAs (please see Chapter 7).

Figure 13.4 illustrates the four main GaAs IC technologies to be discussed; namely planar depletion mode MESFET, enhancement mode FET, high electron mobility transistor, and bipolar heterojunction transistor. These GaAs device types are those which are predominantly used in GaAs IC applications today or are presently under serious development.

In the following, comparisons of the various device technologies will include an assessment of their relative strengths and weaknesses, the state of their relative development and maturity, and a review of their ring oscillator performance results. Integrated circuit performance results will be discussed in Section 13.6.

13.3.1 Depletion mode GaAs devices

The four GaAs device types illustrated in Figure 13.4 can be separated into two very broad basic device families: those fabricated exclusively with ion implan-

GaAs digital integrated circuit technology

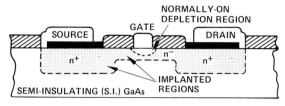

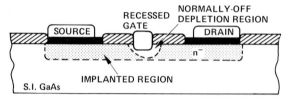

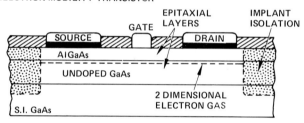

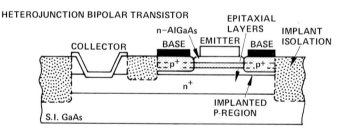

Figure 13.4 GaAs device structures currently in use or under development for digital integrated circuit applications

tation, homojunction depletion, and enhancement FET technologies; and those requiring more sophisticated fabrication approaches using exotic epitaxial layers, heterojunction high electron mobility and bipolar transistors. Both depletion mode MESFET and enhancement mode FET technologies, fabricated simply by ion implanting directly into high-quality semi-insulating GaAs substrates (see Chapters 2 and 5), are significantly more mature than the newly developing heterojunction technologies. Table 13.1 summarizes the fabrication maturity as

Table 13.1 Summary of various GaAs device approaches comparing their relative performance levels, threshold uniformities, fabrication process maturity, and demonstrated circuit complexity

GaAs device type	f_τ (GHz)		Non-linearity; High g_m at low $V_c - V_t$	Threshold voltage uniformity	Fabrication maturity	Circuits demonstrated
Depletion mode MESFET $L_g = 1\ \mu m\ (0.5\ \mu m)$	15 35	typical lab.	Good $V_{gs} - V_p \sim 0.8$ V	very good acceptable V_t control $(\sigma_{V_p} \sim 40\ mV)$	excellent manufacturable mature	various MSI 32 bit adder 8×8 mult. LSI
Enhancement mode MESFET/JFET $L_g = 1\ \mu m\ (0.6\ \mu m)$	12 33	typical lab.	very good $V_{gs} - V_p \sim 0.5$ V	poor/fair very low logic swing $(\sigma_{V_p} \sim 20\ mV)$	good rapidly developing technology	various MSI 1 k RAM 16×16 mult. LSI
Heterojunction E-MESFET $(L_s - 1\ \mu m)$	80	demonstrated	very good $V_{gs} - V_p \sim 0.2$ V	fair/good higher logic swing with AlGaAs Schottky	fair MBE or MOCVD manuf. process required	ring oscillator divide by 2 SSI
Heterojunction bipolar transistor $(W_E \sim 1\ \mu m,$ $W_B = 0.1\ \mu m)$	> 100	projected	excellent (kT limited) $V_{BE} - V_T = 100\ mV$	excellent (energy gap) $(\sigma_{V_p} \sim 1\ mV)$	poor research stage future	simple gate ring oscillator

well as a number of other important elements of the GaAs device technologies under discussion in this section.

Depletion mode MESFET technology[14] has been pursued for GaAs ICs over the past decade and is currently the leader in terms of manufacturability[15] and commercial availability.[16] D-MESFET devices are 'normally on' devices requiring a negative gate potential to deplete (pinch-off) the carriers through the channel region and turn the device off. Numerous circuit approaches have been used successfully with D-MESFETs including buffered FET logic (BFL), Schottky diode FET logic (SDFL), capacitor coupled FET logic (CCFL), and source coupled logic (SCL). Circuits using D-MESFET devices normally require two power supply voltages for operation, along with some kind of voltage level shifting built into the logic gates (see Section 13.4). The two power supply requirement is a disadvantage with respect to increased circuit layout complexity as compared to single supply enhancement mode FET logic.

With respect to fabrication and IC yield, D-MESFETs are the most mature and easiest GaAs device structures to produce. As can be seen in Figure 13.4 the D-MESFET device consists of a shallow (~ 1000–2000 Å deep) n-type implanted channel region doped to about 10^{17} donors/cm^3. Other D-MESFET device features include two ohmic contact source and drain regions typically separated by 3 μm with a $L_g \leqslant 1$ μm long gate strip of Schottky barrier metal located between the source and drain contacts.

Device threshold voltage uniformities (for $V_p = 1$ V) typically exhibit standard deviations (σ_{V_p}) of 60 mV for 2 inch wafers and 100 mV for 3 inch wafers with as low as 34 mV being observed.[15,17] This threshold voltage uniformity is quite acceptable for higher logic swing D-MESFET technology. With D-MESFET logic swings (ΔV_L) of the order of $V_{gs} - V_p \sim 0.8$–1 V, the required uniformity given by

$$\sigma_{V_p} \leqslant \Delta V_L/20 \tag{13.6}$$

must be of the order of 40–50 mV over a circuit chip.[1] Present day D-MESFET standard deviations over MSI/LSI compatible chip areas are typically within 20–25 mV.

Operating characteristics of D-MESFETs are similar to those of a common silicon n-channel JFET, except that the GaAs MESFET transconductance (g_m) is very high (~ 120 mS mm^{-1} for 1 μm gates) and input capacitance is very low, giving a current gain × bandwidth product f_τ of 15–20 GHz.

Performance comparisons of minimally loaded GaAs ring oscillators are summarized in Table 13.2. For D-MESFET ring oscillators using SDFL or BFL logic, propagation delays of $\tau_d \sim 50$ ps are obtained for $L_g = 1$ μm and $W = 10$–20 μm sized gates.[18] SDFL circuits have both good speed and low to moderate power levels while the higher power BFL circuits exhibit the shortest (34 ps) propagation delays.[19] GaAs depletion mode ICs are generally faster than enhancement mode ICs in real logic applications since logic gates like SDFL have

Table 13.2 Comparison of various GaAs depletion and enhancement mode ring oscillator performance results

Company	Technology approach	FET $L_g \times W$ (μm)	Propagation delay (ps)	Power dissipation (mW)	Speed × power product (pJ)
Rockwell	D-MESFET SDFL	1.0 × 10	52	1.2	0.062
Thomson CSF	D-MESFET BFL	0.8 × 20	46	18	0.83
Hughes	D-MESFET BFL	0.5 × 50	34	41	1.4
NTT	E-MESFET DCFL	0.6 × 20	30 (17.5)[a]	1.9 (9.2)[a]	0.057 (0.167)[a]
Bell lab.	E-MESFET DCFL	1.0 × 25	20.9	3.6	0.075
Sony	E-JFET DCFL	1.0 × 10	58	0.64	0.037
Fujitsu	E-MESFET self-aligned	1.5 × 30	50	5.74	0.287
Fujitsu	HEMT E-MESFET	1.1 × 33 1.1 × 33 1.7 × 33	16.8 12.8[a] 17.1[a]	– – 0.96	– – 0.017
Rockwell	HEMT E-MESFET	1.0 × 20	12.2	1.12	0.014

[a] Measured at 77 K.

good fan-out and only experience a small performance loss of about 1.4 times their minimally loaded ring oscillator results. This is not the case for enhancement mode logic (discussed in Section 13.4).

Depletion mode GaAs integrated circuit technology is rapidly maturing as a result of a large amount of research and development over recent years. Present material and fabrication processes can adequately support the logic swing imposed threshold voltage requirements. The present D-MESFET technology will easily support high speeds (75–150 ps) at moderate power levels (1–10 mW/gate). Commercial circuits in the 1–5 GHz range are now beginning to enter the market place.[16] Complexities somewhere under the 10 000 gate level will probably be the upper range of D-MESFET technology as a result of power limitations. As indicated in Table 13.1, D-MESFET technology does not have as high a degree of non-linearity as other GaAs devices, i.e. it does not develop as high a transconductance (g_m) at small logic voltages above threshold as enhancement or bipolar devices.

13.3.2 Enhancement mode GaAs devices

Another device approach being actively pursued is the enhancement mode MESFET. The E-MESFET shown in Figure 13.4 is similar to the D-MESFET in structure, except that the combination of the implanted channel depth and doping concentration is such that the built-in potential of the Schottky barrier completely depletes (pinches off) the channel, so that a positive gate potential must be applied for source–drain conduction to begin (typically $V_p = +0.1$ V in E-MESFETs). Simple direct-coupled FET logic (DCFL) circuits can be used with E-MESFETs, which allows the use of low logic voltage swings and a single low-voltage power supply; this is compatible with large gate complexity and low power dissipations. For these reasons, the E-MESFET approach is considered to have excellent VLSI potential (predominantly from the standpoint of power as indicated in Table 13.1).

Although E-MESFET logic has excellent attributes for achieving VLSI, unfortunately the onset of Schottky gate conduction at $V = 0.7$ or 0.8 V means that the logic swings must be necessarily low (typically ~ 0.5 V), placing severe demands on the uniformity of E-MESFET pinch-off voltages. This enhancement DCFL constraint can be eased slightly by incorporating p–n junction gate FET (JFET) devices (discussed in Section 13.5.3), because the junction barrier height allows forward biasing up to $V \sim 1$ V without excessive conduction.[20]

Maintaining DCFL logic swings at about 0.5 V necessitates achieving pinch-off voltage uniformities of the order of $\sigma_{V_p} \sim 0.025$ V since the standard deviation of V_p must be approximately a factor of 20 below the logic swing. This uniformity has proven exceedingly difficult to achieve on any consistent basis. However, progress in material uniformity and innovative fabrication techniques promise near-term improvement. For example, Fujitsu have demonstrated $\sigma_{V_p} = 0.022$ V E-MESFET uniformities over the central portion of 2 inch wafers.[21]

In addition to the problems of severe uniformity requirements for E-MESFETs, manufacturability is hindered also by more complex device processing requirements than D-MESFETs. The problem with E-MESFETs (or very low pinch-off voltage, $V \sim -0.5$ V, D-MESFETs) is that the substantial depletion region at the surface of GaAs between the source and gate (or gate and drain) tends to pinch off the very lightly doped channel, leading to very high source resistance (R_s) values which degrade the FET transconductance, g_m. One way to avoid high R_s is to recess the gate slightly into the GaAs surface by etching into the thicker channel region as illustrated in Figure 13.4. Although gate recessing techniques have been instrumental in the development of enhancement logic (and to a lesser degree also depletion logic) they have been a key limiting factor in the development towards GaAs enhancement LSI/VLSI. Enhancement GaAs IC fabrication approaches using gate recessing techniques have been traditionally limited to SSI/MSI levels of integration and more recently are being replaced by very sophisticated self-aligned gate E-MESFET device approaches (please see Section 13.5.2).

Various minimally loaded E-MESFET ring oscillator results for E-MESFET and E-JFET device approaches using DCFL logic gates are summarized in Table 13.2. Practical $L_g = 1$ μm E-MESFET or E-JFET ring oscillators[22] using the more traditional non-self-aligned device structures normally obtain propagation delays of $\tau_d \sim 55$–60 ps, while higher performance self-aligned or short-channel

FETs like those reported by Bell[23] and NTT[24] obtain propagation delays of 20 to 30 ps respectively. In general, equivalently sized and structured enhancement mode ring oscillators have yielded slightly slower speed performance than depletion mode devices, but at significantly lower power levels. The new developing self-aligned gate structures also promise increased performance for enhancement technology at somewhat relaxed gate lengths as indicated by Fujitsu's results of $\tau_d = 50$ ps for $L_g = 1.5$ μm sized ring oscillators.[25] One of the highest performance $\tau_d = 20$ ps E-MESFET ring oscillators has been reported by Cornell[26] using self-aligned implanted T-shaped gates with $L_g \sim 0.30$ μm gate lengths. The improved performance obtained using self-aligned gates is largely manifested through a large reduction of source resistance as a result of enhanced doping in the source–gate and drain–gate regions of the devices.

Technology comparisons based on ring oscillator performance can be misleading if other important observations are not made. For instance, enhancement mode ring oscillators have very low parasitic capacitance since they are minimally loaded and are composed of simple DCFL gates. Typical enhancement mode MSI circuits with normal fan-outs of 2 or 3 actually operate with gate delays about 2–3 times the minimally loaded ring oscillator delays owing to the poor fan-out capabilities of DCFL gates as compared to a degradation factor of only 1.3–1.4 for D-MESFET circuits. Therefore, equivalent logic gate speed performance for enhancement technology is generally not as good as depletion mode technology (see Section 13.6.1).

Summarizing, enhancement mode technology is presently undergoing extensive development in order to obtain practical high-yield fabrication methods. Increasingly strong interest in this technology is indicative of the fact that it gives high-speed performance at reduced VLSI compatible power levels (compared to depletion mode logic) with simpler single power supply logic gate structures. Applications have been somewhat limited owing to the relative immaturity of the material and fabrication technologies. Commercial circuits are presently not available. However, the recent proliferation of circuits using enhancement technology indicates that commercial realization is likely within the next few years.

13.3.3 High electron mobility transistor

Newly developing transistor technologies distinguished through their use of heterojunction layer structures provided by sophisticated molecular beam epitaxy (MBE) growth technologies promise even higher performance ($\tau_d = 10$–20 ps) than the traditional FET ion-implanted homojunction device structures just discussed.

One variant of enhancement mode GaAs FET technology offering improved performance over standard E-MESFETs, particularly at low temperature, is called the high electron mobility transistor (shown in Figure 13.4), or HEMT,[27] also referred to as TEGFET[28] for two-dimensional electron gas FET, MODFET[29] for modulation doped field-effect transistor, and SDHT[30] for selectively doped heterojunction transistor. HEMT device structures evolved from the knowledge that GaAs electron mobilities in device channels go up with decreasing temperature (see ring oscillator data in Table 13.2). However, the channel mobilities at 77 K are not significantly higher over 300 K (room temperature) due to electron Rutherford scattering created by ionized donor impurities (typically $N_D \simeq 10^{17}$ cm^{-3}) present in MESFET channels. This observation coupled with the superlattice (alternating layers of GaAs and AlGaAs) material work during the late 1970s,[31] which demonstrated that super high mobility could be realized in semiconductors if electrons could be transferred from a doped AlGaAs layer to an adjacent undoped GaAs layer, led directly to the development of HEMT devices. The concept behind these modulation doped structures is to place donor atoms in a (wide band gap) AlGaAs layer adjacent to an undoped (lower band gap) GaAs channel layer, which then receives the free electrons from the ionized donors by virtue of the electron's inherent affinity to move to the lower band gap region.

HEMT structures in practice are usually fabricated by molecular beam epitaxy (MBE), although metal-organic chemical vapour deposition (MOCVD) is also under development. The most widely used version of this device, similar to that illustrated in Figure 13.4, uses a thick undoped layer of GaAs covered with a thin (~ 700 Å) layer of AlGaAs. The AlGaAs layer under the Schottky gate serves to furnish the electrons for the channel (the undoped GaAs/AlGaAs interface) and also acts as a gate insulator not unlike an n-channel MOSFET (although the gate will begin to conduct at V_{gs} somewhere above 0.9–1.1 V for AlGaAs Schottky).

The principal advantage of the device is that the electron mobility in the channel is higher in the HEMT than in a MESFET because there are no doping ions in the channel to scatter carriers. Electron mobilities of $\mu_e \geqslant 8500$ cm^2 V^{-1} s^{-1} are achieved in HEMTs at room temperature and $\mu_e \geqslant 50\,000$ cm^2 V^{-1} s^{-1} at 77 K are achieved,[32, 33] as compared with typical GaAs MESFET channel mobilities of 4000 to 5000 cm^2 V^{-1} s^{-1} at 300 K. This gives the HEMT devices a fast 'turn-on' characteristic; that is, they develop nearly their full transconductance with gate logic voltage swings only a small amount (~ 200 mV) above threshold. Further, the thin (~ 700 Å AlGaAs) gate insulator region gives these devices comparatively high transconductance values and the effective electron velocities achieved are excellent ($\sim 3 \times 10^7$ cm s^{-1} at 77 K, about twice that of room-temperature MESFETs), so that the current gain × bandwidth products are also better. The highest switching speeds obtained in minimally loaded HEMT ring oscillators, as indicated in Table 13.2, have been reported by Rockwell,[34] with $\tau_d = 12.2$ ps obtained with $L_g = 1$ μm HEMT DCFL inverters at $P_D = 1.1$ mW/gate. Fujitsu, who have generally been credited with popularizing HEMT structures, have reported ring oscillator gate delays in the 12–17 ps range.[35] Initial applications of HEMT structures in D-flip-flop divider architectures are discussed in Section 13.6.1.

Since HEMT devices are a subset of E-MESFET, DCFL application based technology, they naturally enjoy all of the inherent advantages and disadvantages of enhancement technologies. For instance, if we use similar ground rules for threshold uniformity control, threshold voltage standard deviations of about 10 mV must be obtained for HEMT VLSI applications.[1] The superior control offered by MBE is assumed inherently capable of attaining this degree of uniformity and on a limited basis has demonstrated such quality.[31] Of course, making such control repeatable is another matter. Another possible limitation is that HEMT devices may not be able to compete with E-MESFET in radiation hard applications (an important military application area driving GaAs technology development), because low radiation levels generate charge in AlGaAs layers. One advantage of HEMT structures is that the barrier height (ϕ_B) of Schottky barriers on AlGaAs is approximately 1.1 V. Therefore, higher forward voltage logic swings can be used in these structures as opposed to MESFETs.

In comparing HEMT devices with standard homojunction E-MESFETs at similar geometries, as a consequence of their higher mobilities, HEMTs can in principle achieve higher switching speeds ($\sim 30\%$ improvement at room temperature and $\sim 50\%$ improvement at liquid nitrogen temperatures). However, these speed advantages are achieved at far lower logic swings so that the dynamic switching energies are 1–2 orders of magnitude lower. This is, of course, of paramount importance for future HEMT VLSI applications and it is safe to say that this is the key driving force behind research in this technology at present.

13.3.4 GaAs heterojunction bipolar transistor

The realization of very low logic voltage swings requires a high degree of non-linearity (rapid increase in transconductance for small voltage swings ΔV_L above the threshold) combined with extremely uniform threshold voltage devices, naturally attracts the use of device structures in which the threshold voltage is highly insensitive to normal processing variations.

While GaAs MESFET pinch-off voltage is fairly insensitive to horizontal geometry variations (e.g. L_g), V_p is very sensitive to both the thickness (vertical geometry) and doping level in the channel layer. In contrast to the MESFET, an almost ideal device from the standpoint of threshold voltage variations is the bipolar transistor. Threshold uniformities ($\sigma_{V_{BE}}$) of a few millivolts over a circuit or wafer are typically achievable with silicon bipolar transistor technology.

While the inherent threshold uniformity and success that silicon (homojunction) bipolar transistors have enjoyed would seem to make GaAs (homojunction) bipolar transistors also quite attractive, in reality this is not the case for GaAs homojunction (n–p–n) bipolar transistors. This is because of a number of device design compromises centring around the fact that, in order to maintain good emitter injection efficiency, the base doping level N_A must be limited to a fraction of a per cent of the emitter doping. High-speed operation also implies short electron transit times through the base accomplished through reduced base width. Thin base regions (e.g. < 1000 Å) at modest doping levels imply high levels of base resistance (R_B). Therefore, GaAs homojunction bipolar transistors require both high electron mobility, μ_n, for short emitter–collector transit times and high hole mobility, μ_p, for low R_B. The problem lies in the unfortunate fact that the hole mobility ($\mu_p \sim 250 \text{ cm}^2 \text{ V}^{-1} \text{ s}^{-1}$) in GaAs is even lower than the hole mobility ($\mu_p \sim 600 \text{ cm}^2 \text{ V}^{-1} \text{ s}^{-1}$) of Si. Therefore better performance of GaAs homojunction bipolar transistors over Si bipolar is not anticipated.

Fortunately, however, the availability of high-quality AlGaAs/GaAs heterojunction structures makes it possible to fabricate heterojunction bipolar transistors (HJBT) similar to that shown in Figure 13.4, avoiding most of the disadvantages of GaAs homojunction bipolars.[36] In the HJBT device, the use of a wide band gap n-AlGaAs emitter allows the doping level in the p$^+$-GaAs base region to be made very high without degrading the current gain since hole injection from the base into the wider band gap emitter is virtually impossible. The unrestricted use of heavy base doping levels in heterojunction bipolars allows the use of very thin (< 1000 Å) base widths without causing excessive base resistance (providing heavy p-type doping levels of greater than $5 \times 10^{18} \text{ cm}^{-3}$ can be attained). Therefore, extremely high f_t's can be achieved providing, of course, that low emitter contact resistance can also be attained.

Another advantage of HJBT technology is that the threshold voltage is determined strictly by the inherent band gap of the GaAs and AlGaAs, which varies very little in comparison to controlling the channel doping and layer thicknesses of MESFET structures. Also, the inherent high current driving capability of this technology makes it very attractive for gate array logic applications similarly as it does in Si bipolar technology. Fortunately, in HJBTs, it should also be possible to use a wide band gap collector region (along with emitter) to confine the excess stored charge in saturated operation to the very thin base region.[36] This should make it possible for very high-speed VLSI to use simple, very low-power, high-density saturated logic approaches such as I^{2}L.[37,38] Of course, higher speeds would be expected for non-saturated logic approaches (ECL, CSL, etc.), but at somewhat higher power levels and lower gate densities.

The majority of the HJBT development work going on today has been centred around the very difficult material and fabrication requirements.[39] Molecular beam epitaxy has been used exclusively for this work to date and a number of workers are now starting to achieve working devices.[40,41] Early work on AlGaAs/GaAs amplifier applications has been published[42] and recently the first experimental demonstrations of ECL circuits using AlGaAs/GaAs in both ring oscillators and frequency dividers has been reportd by Rockwell.[43] Although only modest performance results (propagation delays of 185 ps for 2 μm design rules and 4.5 GHz in frequency dividers) have been attained to date, these are impressive achievements for such a sophisticated device technology whose idea[44] has only recently come to fruition.

Performance of HJBT devices are ultimately projected to be in the 100–200 GHz range, with gate delays of under 10 ps, and frequency divider performance reaching 20 GHz.[43] In addition, the high current drive capability leading to the low sensitivity of propagation delay to capacitive loading along with the threshold voltage insensitivity of these devices further qualifies GaAs HJBTs as super-high-performance IC candidates.

However, before HJBT technology can become a practical reality, a wide variety of improvements in MBE material, device structures, and fabrication technologies must be achieved. This device is extremely difficult to fabricate (see Figure 13.4) due to its use of complicated MBE heterojunction layer structures in conjunction with ion implantation steps, both for contacting critical p^+ base regions and for device isolation created by implanted damage. Improved performance must be achieved through thinner base regions ($\sim$ 500 Å thick) with carrier concentrations approaching 10^{19} carriers/cm^3 for reduced base resistance. Following the lead of Si, bipolar, self-aligned contact process techniques will likely be employed for reduced emitter strips ($\sim$ 1 μm or less), smaller base contacts, and reduced ohmic contact resistances. As can be observed, GaAs hetrojunction bipolars have all of the lithographic difficulties of the most advanced GaAs or Si device technologies, have even more difficult MBE heterojunction layer requirements than HEMT, coupled with the added complexities of ion implantation and demanding requirements of small low-resistance p- and n-type ohmic contacts.

Improvements in the material and fabrication technology promise to bring HJBTs into the super device class. While the performance pay-off is formidable, the technological complexity of HJBT ICs should extend any practical commercial application out several years.

13.4 LOGIC APPROACHES FOR DIGITAL GaAs ICS

The principal devices used in GaAs digital integrated circuits are the field-effect transistor (FET) and the Schottky barrier diode. Diodes are used both for non-linear logic elements and for voltage level shifting. As a consequence of the very high electron mobility ($\mu_e \sim$ 4500 cm^2 V^{-1} s^{-1}) of GaAs, circuits are normally designed with majority carrier devices using n-type GaAs active layers. The very low hole mobility of GaAs ($\mu_p \sim$ 250 cm^2 V^{-1} s^{-1}) discourages the use of p-channel or homojunction bipolar devices in GaAs. As a result of both the low hole mobility and the unavailability of MOS in GaAs, complementary logic approaches analogous to Si CMOS have not been considered.

GaAs logic approaches, using either enhancement or depletion mode metal–semiconductor Schottky barrier FETs (MESFETs), have received a vast amount of work, although circuits using junction FETs (JFETs) have been used in enhancement mode approaches, while circuits using enhanced mobility heterojunction structures (HEMT, TEGFET, etc.) and heterojunction bipolar devices (HJBT) are now under development and starting to see early application. Since GaAs HJBT technology is in such a very early state of development, bipolar logic circuits will not be discussed in the following. All the logic approaches discussed in this section are based on using either depletion and enhancement mode FET devices and Schottky barrier diodes or a combination of these, as required.

13.4.1 Buffered FET logic (BFL)

The first significant GaAs ICs reported[45] utilized D-MESFETs in a circuit design later called buffered FET logic (BFL).[14] NAND and NOR implementations of BFL logic are illustrated in Figure 13.5. This circuit typically uses $-2.5 < V_p < -1$ V depletion mode MESFETs and hence requires two power supplies. Since it requires a negative gate voltage to turn off an n-channel D-MESFET, while its drain voltage is positive, level shifting must be introduced at some point so that the output logic levels match the input levels. In the buffered FET logic approach, the choice was made to operate with negative logic swings by level shifting the positive drain voltages at the gate output. This is normally accomplished by placing level shifting diodes in the source follower output stage of the gate. The diodes are always forward biased, each one giving a voltage drop of about 0.8 V. In the example shown in Figure 13.5, the three diodes in series accomplished the required approximately 2.5 V level shift.

Placing the level shifting diodes at the output has a drawback. Since the current through the output driver is relatively high, the power dissipation caused by the level shifting diodes is also relatively high. On the other hand, a key attraction of BFL is that the circuit can be easily implemented with a fabrication process paralleling that of a conventional microwave GaAs MESFET process. Originally it was possible to fabricate these early circuits on wafers that had a uniform epitaxially grown layer on the semi-insulating substrate,[45] on which the device structures were isolated from one another by mesa etching.

In the buffered FET logic approach described here, the non-linearity of the FET itself is used to implement logic functions. As shown in Figure 13.5, the positive OR function is performed by FETs in parallel, while the AND function is performed by FETs in series (or equivalently by multiple gate FETs). Higher-level logic functions can also be implemented by combining series and parallel arrangements of FETs.

A large number of circuits have been made using this buffered FET logic circuit approach, starting from simple ring oscillators and proceeding into more complex sequential logic circuits, including complementary clocked master–slave flip-flop stages used in ÷2 and ÷8 binary ripple counter configurations and MSI level logic circuits including a word generator containing 600 active devices

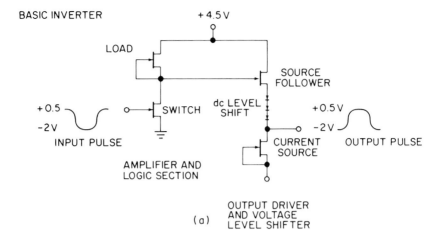

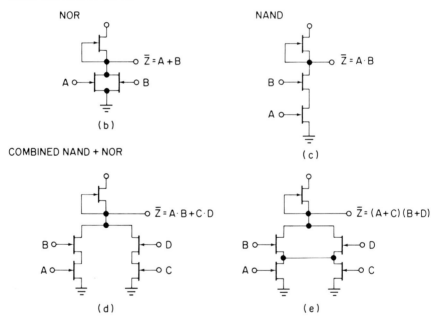

Figure 13.5 Various circuit configurations for buffered FET logic: (a) basic inverter circuit; (b–e) various options for NOR, NAND, and combined NAND + NOR functions

developed by Hewlett Packard[46] and a 32 bit adder containing 2500 devices by NEC[47] (discussed in Section 13.6.1).

Propagation delays of $\tau_d \sim 100$ ps were obtained in ring oscillator measurements on two-level NAND/NOR logic gates fabricated with 1 μm photolithography, performance which translates to 4.5 GHz maximum clocking speeds for frequency dividers.[14] Much simpler chains of $L_g = 0.5$ μm inverters fabricated using E-beam lithography give, as expected, faster ring oscillator speeds, as low as $\tau_d = 34$ ps.[19]

Although, traditionally, buffered FET logic has been considered a relatively high-power (P_D typically more than 4 mW/gate) circuit approach, more recently it has been shown that by reducing the number of level shifting diodes to one, and necessarily by lowering the FET threshold voltage to -0.5 V, power dissipation of the order of a few milliwatts per gate can be achieved.[47] While it is clear that BFL circuits may be designed with lower power than originally thought, application of BFL logic approaches will probably still be limited to MSI levels of integration due to their relatively high power dissipation.

13.4.2 Schottky diode FET Logic (SDFL)

The Schottky diode FET logic (SDFL) approach retains the high speed of the D-MESFET buffered FET logic (BFL) approach, while providing a reduction in power dissipation of up to an order of magnitude.[48] As the name Schottky diode FET logic implies, SDFL utilizes clusters of small, high-performance Schottky diodes to perform the logical positive OR function on groups of inputs, which may then be further processed with the normal FET logic functions (series NANDing, drain dotting, etc.). Figure 13.6 shows SDFL gate circuit diagrams for single-, two- and three-level logic gate configurations.[49] Note that the SDFL gate structure allows virtually unlimited fan-in at the first (positive OR) logic level, but has the same practical restrictions to a fan-in of 2 (or possibly 3) at the second (OR/NAND gate) and third (OR/NAND/WIRED-AND gate) levels if high-speed dynamic performance is to be maintained.

The SDFL circuit approach offers savings, not only in power, but also in circuit area, over previous D-MESFET approaches. The circuit area saving comes about because of the simplicity of the gate design and replacement of (large) FETs with very small (typically 1 μm × 2 μm) Schottky diodes for most logic functions. The fact that the diodes are two-terminal devices also significantly reduces the number of vias and overcrossings required in many circuits as compared to that required with three-terminal devices such as FETs. SDFL NOR gate areas are typically in the 1000 μm² range. The small gate area required for the SDFL gate also implies that the parasitic capacitances associated with these gates tend to be very small. This allows for the attainment of excellent speeds even in gates implemented with low-power ($V_p \sim -1$ V) FETs with small widths. For example, the best speeds obtained with $L_g = 1$ μm, $W = 10$ μm, SDFL fan-in = 2, fan-out = 1 NOR gate ring oscillators is $\tau_d = 52$ ps at $P_D = 1.2$ mW/gate ($P_D\tau_d = 62$ fJ).[18]

It should be noted that although the fan-out is limited to a maximum of 3 (without buffering) for typical SDFL gate designs, the propagation delays are not expected to degrade significantly. The reason for this is that the gate turn-off

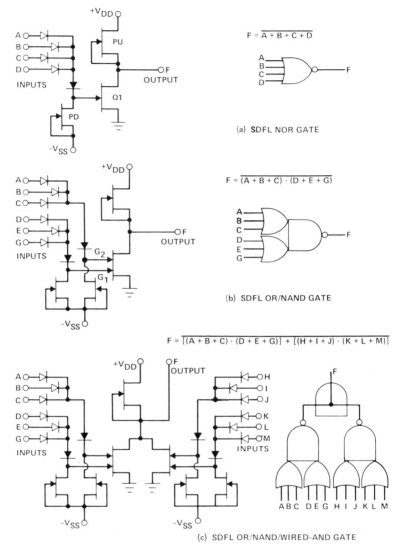

Figure 13.6 Schottky diode FET logic (SDFL) gate configurations: (a) basic SDFL NOR gate; (b) two-level SDFL OR/NAND gate; and (c) three-level SDFL OR/NAND/WIRED-AND gate

current for the switching FET is not provided by the preceding FET drain current, but rather by the current of the pull-down (PD) to $-V_{ss}$, since the previous FET drain is isolated by the switching diode. Hence, in effect, a fan-out of 2 to 3 is built into the SDFL logic gate itself, so that the propagation delay should be relatively weakly dependent on fan-out loading. Heavily loaded gates can be buffered with source follower circuit configurations.

Schottky diode FET logic circuits with D-MESFETs operating at low bias voltages can have power dissipations nearly as low as enhancement mode devices. SDFL circuits have been implemented into complex logic circuits with up to 1000 gate MSI circuit complexities. Demonstration of GaAs LSI (1008 gates, about 6000 active devices) was achieved in 1980 using this SDFL technology approach.[50]

13.4.3 Direct-coupled FET Logic (DCFL)

Enhancement mode devices such as E-MESFET, JFET, or HEMT utilize substantially different fabrication processes as discussed in Section 13.5. However, from the standpoint of circuit implementation, enhancement mode circuit approaches are essentially identical. The simplest circuit approach, direct-coupled FET logic (DCFL), is illustrated for a three-input (positive) NOR gate in Figure 13.7. In this logic approach, the 'normally off' FETs start conducting when their gate voltage becomes positive. A logic '0' corresponds to a voltage near zero. A logic '1' corresponds to a positive voltage capable of fully turning on the 'normally off' FETs, a value usually limited by the onset of gate conduction in the FET, typically of the order of 0.7 to 1.2 V depending on what technology is used (i.e. MESFET, JFET or HEMT).

A significant improvement to the directly coupled logic gate shown in Figure 13.7a is to substitute for the load resistor R_L, an active load current source made with a 'normally on' (depletion mode) FET, with its gate tied to the source. Such a non-linear load sharpens the transfer characteristic and significantly improves the speed and speed × power products of the circuits (by perhaps a factor of 2). The fabrication of the depletion mode active load requires a carrier concentration profile different from that of the enhancement mode devices. Therefore, it requires a multiple localized implantation fabrication technique such as either that used for the Schottky diode FET logic (SDFL), or, in the case of a single active layer, a deep recess must be applied to the enhancement mode device with no gate recess on the depletion mode active load device. Currently, the vast majority of enhancement mode logic circuit designs use depletion mode active loads.

From a static point of view, the fan-out capability of the directly coupled FET logic is excellent since it is only limited by the very low gate leakage currents. However, from a dynamic point of view, the switching speeds are reduced by the gate capacitance loadings by a factor of approximately $1/N$, where N is the number of loading gates, as in silicon MOS. In general, the current through the resistor, R_L, or active load is kept fairly low in DCFL in order to reduce static power and improve noise margin by reducing the 'on' voltage drop of the FET (output 'low' voltage). Consequently, the output rise time under heavy fan-out

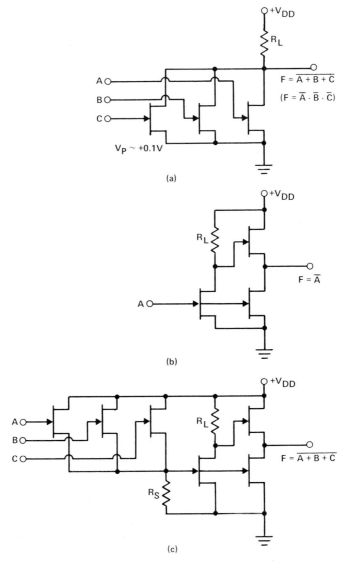

Figure 13.7 Enhancement mode JFET or MESFET circuits: (a) simple direct-coupled FET logic (DCFL) NOR gate with resistor load; (b) pseudo-complementary buffered inverter gate; and (c) combination of source follower logic with the circuit of (b) to give a buffered NOR gate

loading conditions is very poor. This can be greatly improved with the pseudo-complementary output buffer configuration of Figure 13.7b, at very little increase in static power dissipation, but this circuit performs only logic inversion. By combining this inverting buffer with a source follower positive OR input structure as shown in Figure 13.7c, a general multiple input NOR gate can be achieved which has excellent fan-in and fan-out drive capabilities at very modest static power levels.[51]

Enhancement mode logic integrated circuits have been fabricated using E-MESFETs, E-JFETs, and more recently, HEMT devices. The vast majority of these ICs have been designed using E-MESFET devices with typical performance levels on 1 μm ring oscillators of $\tau_d \sim$ 60–80 ps, corresponding to power × delay products of $P_D \tau_d =$ 50 to 100 fJ. The most complex circuits fabricated using enhancement mode DCFL logic is Fujitsu's 16 × 16 parallel multiplier. (Performance of this circuit will be discussed in Section 13.6.)

Although D-MESFET technology has been used in the majority of GaAs IC applications, E-MESFET DCFL based ICs are currently getting increased attention principally due to a greater emphasis on reaching LSI/VLSI complexities where the lower power dissipation, single power supply, and high packing densities of enhancement mode direct-coupled FET logic (DCFL) make it very attractive.

13.4.4 Quasi 'Normally off' Logic

Yield problems arising from poor device threshold voltage control resulting from non-uniform enhancement mode active layers are the most serious limitation preventing broad application of 'normally off' logic. Alternative circuit approaches from the previously discussed DCFL logic gates have focused on designing around the limitations of 'normally off' (self-depleted) layers. Because of the non-linear, approximately square law nature of the FET I_{ds} vs V_{gs} relationship, it is not always necessary to turn off the FET completely (i.e. make V_{gs} more negative than V_p) in order to obtain switching behaviour.

Drain dotting of many FETs as in Figure 13.7a necessitates turning all of the FETs nearly off so that the sum of all of their drain currents is substantially less than the load current (I_L) through R_L. However, if only a single FET switches the load, it is only necessary to reduce its drain current in the 'off' state to a value significantly smaller than I_L, while its 'on' current is well above I_L. This can be achieved in depletion mode MESFETs with reasonably small pinch-off voltages ($V_p \simeq -0.4$ V) with zero or slightly positive gate voltages, so that only a single power supply is required.

A number of circuit approaches for single supply E/D-MESFET logic have been proposed and analysed. Figure 13.8a shows the circuit diagram for an elemental three-input NOR gate in the most promising of these published approaches.[52] This uses source follower logic to obtain the positive OR function, with single diode level shifting and resistor pulldown, R_S, to drive the output inverter FET. The analysis in ref. 52 indicates proper gate operation for MESFET pinch-off voltages in the $-0.4 < V_p < +0.1$ V range, several times the allowable

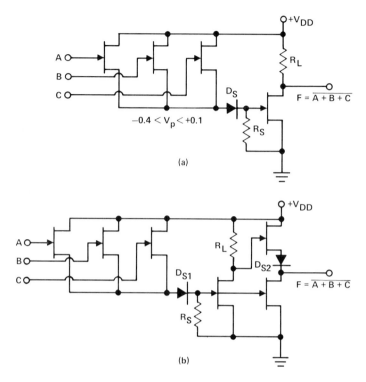

Figure 13.8 Single supply enhancement/depletion MESFET NOR gate circuits: (a) quasi 'normally off' three-input NOR gate; (b) three-input NOR gate with pseudo-complementary buffer

range width for classical E-MESFET logic and much more reasonable in terms of practical fabrication control. The gate output of Figure 13.8a has the same drive problems as that of Figure 13.7a, but this should be improved for heavily loaded gates with the buffer structure of Figure 13.8b. This structure is, of course, very similar (except for the two voltage shifting diodes) to the enhancement circuit of Figure 13.7c.

Development work using the quasi 'normally off' MESFET approach has resulted in circuits operating in the 100–200 ps range with a power level from 0.2 to 2 μW/gate. While these circuits can be operated with a single power supply and can tolerate fabrication non-uniformities of $-0.4\,\text{V} \leqslant V_p < +0.1\,\text{V}$, they do not offer high packing densities ($D = 200$ to 400 gates/mm^2). In the light of the relatively poor packing density, in comparison to DCFL, it is not clear whether going to quasi 'normally off' logic is worth the effort. This is particularly true in the light of the dramatic improvements recently seen in enhancement mode technologies. In any event, useful circuits, such as frequency dividers, have been successfully demonstrated with this technology.[52]

13.4.5 FET Logic gate design variations

The previous sections have discussed the mainstream logic gate approaches presently used in the majority of GaAs digital ICs. For a variety of reasons, numerous other, sometimes unique and novel GaAs logic gates have been proposed. Figure 13.9 illustrates several of the more popular alternative logic gate approaches described in the literature.

Depletion mode Schottky diode FET logic (SDFL), while enjoying larger fan-out (usually 2 or 3) than enhancement mode direct-coupled FET logic (DCFL), still requires some sort of buffering when additional driving capability is required beyond what is normally built into the standard gate configuration. This perhaps

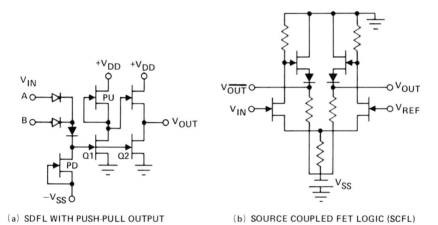

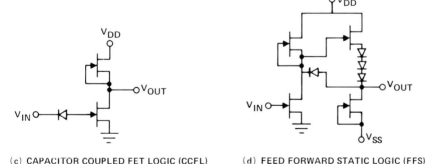

Figure 13.9 Various GaAs logic gates: (a) depletion mode SDFL with push–pull output buffer stage; (b) source-coupled FET logic (SCFL) gate; (c) capacitor-coupled FET logic (CCFL) gate; and (d) feed-forward static (FFS) logic gate (combination of BFL and CCFL)

is most strongly evident in applications requiring large current driving capabilities such as gate arrays. Typically, designers have merely implemented common source follower buffer stages similar to that used in BFL gates in order to gain the additional current driving capability. The problem with this approach is that there is always a finite amount of current present in this buffer stage and this unavoidably increases the power dissipation. An improved implementation of the buffer approach in shown in Figure 13.9a, showing a SDFL gate with a so-called 'push–pull buffer'.[53]

This gate structure contains a source follower plus a switched pulldown FET. The source follower has excellent current supply capability and is relatively insensitive to loading. Because the logic is performed by diodes and the level shifting occurs at the input to the gate, a source follower alone is not enough for SDFL. The pulldown current available for a given gate is dependent on the input states of the gates loading it, but the parasitic capacitance is not. The switched pulldown is added to eliminate this problem. Since it is switched off when the output is high, it does not add more loading or draw power. When turned on it helps to pull down the output node and interconnect lines independent of the input states of the loading gates.

One trade-off associated with SDFL push–pull output is that the pulldown active load must now drive two FET gates. The extra capacitance required of the Q_2 FET gate compromises speed in this logic approach. Propagation delays are typically 25–30% slower in this SDFL circuit configuration. The additional current driving capabilities of the logic gate have been successfully implemented in D-MESFET gate array work at Honeywell.[54]

Another preoccupation of GaAs circuit designers and technologists is concern over the limitations of FET threshold control. This has provided impetus for logic designs like quasi 'normally off' logic (previously discussed) and source-coupled FET logic (SCFL) shown in Figure 13.9b.[55] The basic SCFL inverter consists of a differential amplifier and buffer stages with diode level shifters. The transfer characteristics of the SCFL inverter are nearly independent of the threshold voltage of the switching FETs since the critical level in the transfer characteristic is equal to the externally applied reference voltage V_{ref}. This wide allowable V_p range simplifies processing technology while simultaneously providing large logic swings. In this approach the allowable V_p range is from -0.6 V to $+0.3$ V with logic swings as large as 1.2 V compared to about o.5 V for DCFL. High-speed operation has also been obtained from SCFL with a 4 GHz GaAs binary frequency divider recently being demonstrated using 0.5 μm FET gates.[56] A much more thorough assessment of this technology will be required in the light of the fact that the maximum operating frequency depends strongly on the load resistance of the circuit.

The principal requirements in D-MESFET logic gate designs are the necessity for voltage level shifting between logic stages and maintaining reasonably tight FET threshold control for proper gate operation. Another interesting variation of a GaAs logic gate is shown in Figure 13.9c, called capacitor-coupled FET logic

(CCFL),[57] which uses a capacitively coupled interstage connected to the depletion mode FET, eliminating the requirement for level shifting diodes.

This capacitor coupling is a method of achieving passive, low-power interconnection of logic stages, allowing for large process variation of FET threshold. Circuits in the 1 GHz range have been demonstrated on wafers over which the FET pinch-off voltages varied between -0.5 V and -4.0 V.[58]

In this approach, in order to achieve high transfer efficiency of the gate, the coupling diode must have high capacitance at a reverse bias substantially beyond FET pinch-off. However, in order to accomplish this, typical diodes may occupy as much as 40 % of the total gate inverter area. Another drawback of CCFL is that, in its simplest form, it is a.c. coupled and therefore has a minimum operating frequency of 6 kHz. This feature may limit its applicability, since, accordingly, in most applications, this circuit will require additional initialization and refresh circuitry.

The last GaAs logic gate variation that we will discuss is shown in Figure 13.9d. This circuit is a composite of BFL and CCFL and is called a feed-forward static (FFS) logic gate.[59] In the standard BFL cell, the widths of the voltage shift devices in the driver stage are designed so that their speed compatibility is approximately equal to that of the logic branch. Accordingly, the device widths of the driver stage are usually designed to equal the device widths of the logic stage (the main reason for the relatively high power of BFL). In the FFS cell, the width of the driver stage can be greatly reduced, thereby reducing power dissipation, because the high-frequency signal is transmitted through the capacitance of the reverse biased Schottky diode (in a similar fashion as CCFL). Therefore, the FFS cell maintains the speed of BFL while consuming significantly less power. This circuit concept looks very promising in very high-speed MSI circuit applications where the excessive diode areas required of this approach do not create a problem.

13.5 GaAs IC FABRICATION TECHNOLOGIES

A review of the numerous GaAs IC fabrication technologies presently available (described in the literature) can be divided by both type and relative maturity into several categories. These categories range from the most mature 'now' technology, i.e. planar implanted depletion mode MESFET, demonstrated at the LSI level, to the least mature and most difficult 'future' technology, i.e. heterojunction bipolar transistor, currently only developed at the basic ring oscillator and simple divider level. Between these two extremes are several rapidly developing 'near-term' enhancement mode fabrication technologies, currently receiving the greatest attention throughout the GaAs research and development community. The following will be a review of the main GaAs fabrication technologies, which through their demonstrated results have attained a significant level of maturity and justifiable interest. Heterojunction bipolar transistor technology, owing to its relatively early state of development, will not be reviewed in the context of this section (please see Section 13.3.4).

13.5.1 Conventional planar GaAs IC technology

Numerous GaAs integrated circuit fabrication technologies based on both enhancement mode (E-MESFET) and depletion mode (D-MESFET) field-effect transistors have been under development since the early 1970s. Over the intervening years, the strongest candidate that has emerged, capable of supporting GaAs IC commercialization, is the so-called planar depletion mode MESFET technology.[15] This technology, having received the major development thrust over the past decade, was the earliest (1980) GaAs IC approach to be demonstrated at the LSI level.[50]

Instrumental in leading the Si IC industry forward towards LSI was the early development and availability of planar, ion-implanted, dielectric-passivated fabrication processes. Justifiably, similar developments in GaAs technology[60] were also considered crucial for achieving large-scale integrated circuits. These principles provided the motivation for research efforts at Rockwell, Hewlett Packard, and elsewhere for the development of a planar depletion mode GaAs fabrication technology. Rockwell have generally been credited with the development of planar ion-implanted GaAs fabrication technology, which has subsequently become the industry standard for depletion mode approaches.[61] Typical integrated circuit device structures, shown in Figure 13.10, evolving from these early planar fabrication development efforts, are presently being utilized in the first GaAs IC commercial products at GigaBit Logic.[62] These planar circuits are fabricated by using multiple localized ion implants, providing optimization of individual devices. Owing to the inherent flexibility of this approach, ion implantation is currently being used successfully in all types of depletion mode circuit approaches including BFL, FFS, SDFL, and SDFL with push–pull output (see Section 13.4).

In practice, as illustrated in the upper portion of Figure 13.10, these planar circuits are fabricated by using multiple localized Se and/or Si ion implantation techniques[63] (see Chapter 5) in conjunction with high-quality semi-insulating GaAs substrates[64] (see Chapter 2). Isolation is normally provided by the GaAs semi-insulating (10^6–10^8 Ω cm) substrate, or an alternative approach has been to use local proton, boron or oxygen implants to form high-resistivity damage isolation regions in the GaAs surface.[65, 66] Any number of implantation steps can be incorporated into this fabrication scheme. In the next steps, standard microwave transistor developed AuGe contacts and TiPtAu Schottky barriers (see Chapter 6) and first-level interconnects are fabricated. Devices are normally fabricated using high-resolution direct-step-on-water (DSW) photolithography equipment in conjunction with precise overlay registration techniques. Replication of the circuit patterns is usually a combination of both dry etching and enhanced lift-off processes. The dielectrics used in this planar technology both serve to protect the GaAs surface and are an integral part of the high-yield enhanced lift-off process.[67] Further, after the first-level contacts and intercon-

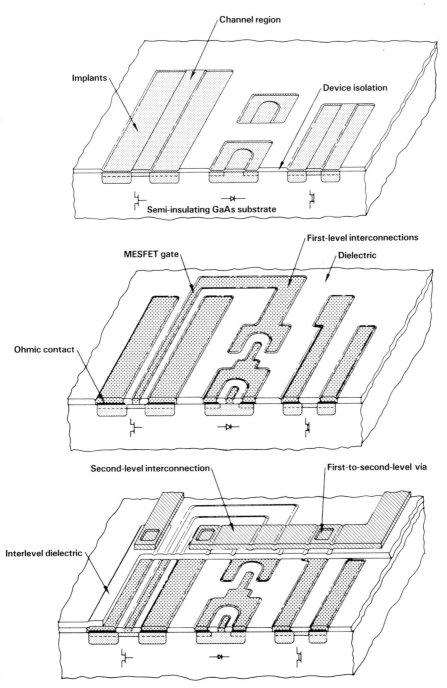

Figure 13.10 Planar ion-implanted D-MESFET fabrication steps showing multiple localized non-implanted layers, gates, Schottky barriers, first-level interconnects, and second-level interconnects

nects are fabricated, a rather planar surface is formed as a result of choosing the thickness of the dielectrics and metals to be similar. This process approach results in a smooth planar surface greatly facilitating the fabrication of complex multilayer interconnect structures.[67] The planarity of this IC structure is illustrated in the lower part of Figure 13.10.

Device control is excellent with this technology since process tolerant depletion mode (usually 1–2 V thresholds) MESFET devices are used. In D-MESFET technology, as a result of using thicker, lower-resistance active layers than those used for 'normally off' E-MESFET applications, channel series resistance is not as serious a concern as it is in enhancement mode technology. Therefore, processes designed to reduce channel resistance, such as deep gate recessing or enhanced doping at the edge of the gate region (self-aligned gates), are not generally required in depletion mode technology. Since the implanted n^+ source/drain regions (see Figure 13.4) do not extend under the gate depletion region, the threshold voltage control of planar D-MESFETs is primarily dependent on the inherent uniformity characteristics of the substrate material. The ease of fabrication associated with simple device structures and straightforward process approaches has allowed D-MESFET technology to mature rapidly in recent times. The rapid advancement and excellent control exhibited of planar D-MESFET devices has provided a strong baseline technology for use in the first commercial GaAs IC applications.

The range of applicability and LSI/VLSI complexity obtainable with GaAs D-MESFET based integrated circuits will ultimately be determined through analysis of the trade-offs between yield, chip area, speed × power product, and gate counts attained. At this time, the ultimate capability and complexity of GaAs D-MESFET ICs are very difficult to assess; however, it is safe to assume that it probably will be beyond our present expectations. Although D-MESFET technology is not likely to satisfy all future GaAs IC applications, it will undoubtedly be the driving technology force that brings GaAs ICs out of the laboratory and into the merchant market for the first time.

13.5.2 Self-aligned Gate Processes

Although GaAs self-aligned gate fabrication techniques are normally thought of in conjunction with enhancement mode FET circuit approaches, a very simple self-aligned gate (SAG) approach has been applied successfully to GaAs depletion mode ICs by both Nippon Electric (NEC)[68] and Philips in France (LEP).[66] The same basic concepts apply to all SAG-FET approaches, namely simply to minimize the source resistance of the FET through (self-alignment) placement of the source–drain regions and/or contacts in as close a proximity as possible to the gate region. The NEC/LEP 'closely spaced electrode' FET scheme shown in Figure 13.11 is accomplished by chemically etching a 0.5–1.0 μm long Al gate by precisely controlling the etch undercut of a 1.5–2.0 μm long resist pattern also used as a lift-off shadow mask for the evaporation of the ohmic contact metal.

GaAs digital integrated circuit technology

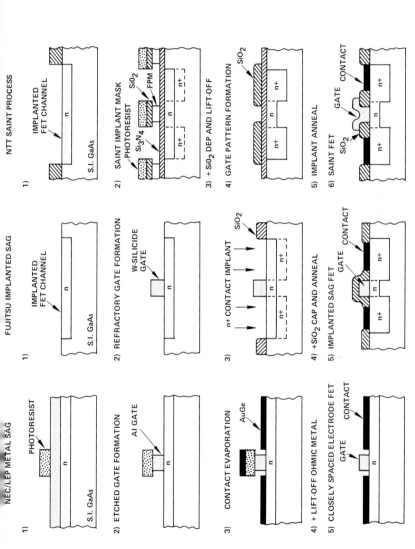

Figure 13.11 Various self-aligned gate (SAG) fabrication technologies; closely spaced electrode FET process, self-aligned tungsten silicide gate process, and self-aligned implanted n-type transistor (SAINT) process

This approach, while used for a number of years for commercial microwave FETs, has not been used extensively for GaAs ICs, primarily due to its limitation to depletion mode applications since recessed gates are not possible with this application.

The main thrust of present-day GaAs LSI/VLSI development efforts is directed towards low-power enhancement mode MESFET IC approaches. Inherent difficulties in the fabrication of E-MESFET structures has made SAG-FET approaches very attractive. This is because the 'normally off' condition of enhancement devices physically requires extremely thin, lightly doped channel regions in order to 'pinch off' the depletion region. These thin active layers are extremely surface sensitive, highly resistive, and difficult to control. The most serious problem is the high series resistance of the channel region, requiring either device structures with gates that are recessed below the surface (e.g. metal gates physically placed below the GaAs surface or use of junction (JFET) p-type buried gates) or additional enhanced doped regions located next to the edges of the gate (e.g. structures similar to self-aligned gates shown in Figure 13.11).

Virtually all of the early enhancement approaches utilized recessed metal gate structures with very limited success.[69, 70] Application beyond SSI/MSI has not proved practicable where etched recessed gate structures were applied. This approach has not been able to provide the $\sigma_{V_p} \sim 20$ mV threshold voltage control required of E-MESFET logic gate circuit approaches.

More recently E-MESFET technologists have directed their energies to more sophisticated device approaches. Borrowing from the well known self-aligned gate (SAG) processes, used in NMOS technology, has resulted in device approaches similar to those illustrated in the centre portion of Figure 13.11. The idea is to use the Schottky gate as a self-aligned mask for implanting the source and drain regions of the FET. Since GaAs n-type layers require post-implant anneals in excess of 800°C, developing a high-temperature compatible Schottky barrier gate metal system became the key technological barrier to be overcome. Fujitsu, among others, have championed this technology quite successfully. Starting with experiments with TiW and then TiW silicide and finally tungsten silicide,[25, 71, 72] researchers at Fujitsu have developed excellent high-temperature Schottky barriers that have provided the basis for the demonstration of 1 k RAMs[21] and 16 × 16 multipliers.[73]

However, even with the successful development of high-temperature Schottky barriers, other formidable problems still remain. When a self-aligned gate FET structure is formed such that the high electron carrier concentration (n^+) regions are placed immediately adjacent to the lightly doped gate region, the presence of the heavier doping concentration can negatively influence the control of the FET threshold voltage. This problem is due, in part, to the high stress of tungsten silicide gates, which enhances the inherent lateral diffusion of the n-type doping under the gate. Both doping and channel length control are serious concerns associated with this device structure, particularly when shorter gate lengths are used.[26] New developments, such as heat pulse annealing for minimizing lateral diffusion,[74] optimization of the vertical n$^+$ doping profiles, and low-stress tungsten silicide films promise improved threshold control of this simple self-aligned gate structure.

Recognition of the device threshold control issues just discussed has led to the development of gate structures similar to that shown in the right portion of Figure 13.11. Laboratories such as NTT,[75] Hitachi,[76] and Toshiba[77] in Japan and HRL[78] and Cornell[26] in the USA are all developing these so-called 'T-bar' gate approaches. The NTT SAINT (for self-aligned implanted N$^+$ transistor) illustrated in Figure 13.11 has the advantage that the heavy n$^+$ doped regions can now automatically be placed, by virtue of the T-bar gate, at a fixed distance (0.1–0.2 μm is typical) away from the critical gate channel depletion regions, thereby minimizing the impact of heavily doped regions on gate threshold control, while still providing low series resistance between the source and drain/gate regions. Without going into explicit details of how the SAINT process is accomplished,[75] it suffices to say that this and other similar process approaches are considerably more complex than normal GaAs IC processing.

The obvious question to consider in regard to these more sophisticated device approaches is whether the increased process complexities required of SAG-FET devices, in order to satisfy the unique constraints of enhancement devices, are LSI/VLSI yield compatible. Since these near-term enhancement technologies are now rapidly emerging, perhaps an answer to this question will be available in the very near future.

13.5.3 FET threshold adjustable processes

GaAs enhancement mode integrated circuits have been largely dominated by self-aligned gate fabrication approaches with the exception of two other approaches, namely, the JFET, pioneered by McDonnell Douglas[79, 80] and Sony[81] and the Pt-buried gate FET approach developed by Toshiba.[82] Both of these approaches, illustrated in Figure 13.12, use virtual recessed gate structures, accomplished by placing the gate below the channel surface region either by using p-type implantation, in the case of JFETs, or by reaction of the Pt gate metal with the GaAs surface in the case of the Pt-buried gate FET. In these two E-FET approaches, the recessed gate structure alleviates the problems normally associated with the high series resistance 'normally off' channel layer. However, more importantly, both technologies are designed for threshold shifting, potentially an even more important requirement for the successful implementation of enhancement mode logic in GaAs ICs.

By carefully monitoring the current levels in the active layers through on-chip I_{ds} measurements of FETs and subsequently adjusting the gate depth during processing, the threshold of the devices on an individual wafer basis can be precisely shifted to the optimal operating level. The ability to implement and

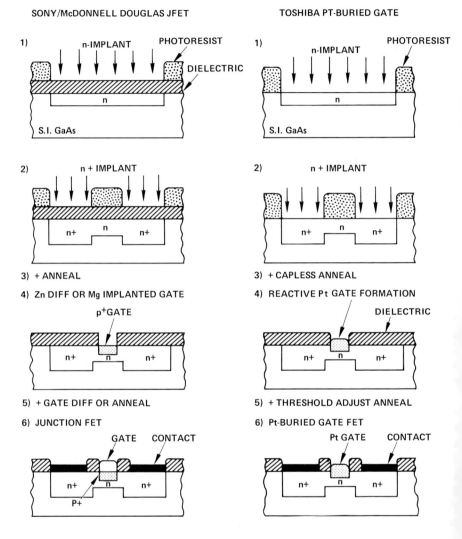

Figure 13.12 Threshold adjustable enhancement mode fabrication technologies: junction field-effect transistor (JFET) process and Pt-burried gate process

control threshold shifting will probably be the most important factor in determining whether GaAs ICs based on enhancement mode devices will become manufacturable. Control of enhancement mode logic depends on both the mean wafer FET threshold (V_p) and the standard deviation (σ_{v_p}) of that threshold. If the average threshold voltage across a wafer can be precisely controlled, then the absolute requirement of threshold voltage variation (σ_{V_p}) over the wafer can be somewhat reduced.

GaAs digital integrated circuit technology

Progress in the threshold control of enhancement mode JFETs has been good using both Zn diffused (Sony) and Mg implanted (McDonnell Douglas) p-type junction gates. In practice, the lightly doped channel layers can be measured prior to the formation of the p junction gate in order to determine what gate depth is optimal for the 'normally off' channel condition. In the case of the implanted gate, an energy adjustment is made[80] and, in the case of the diffused gate, an optimal diffusion time is calculated.[81]

The threshold adjustment offered by the JFET technology has provided a practical means for accommodating the difficult threshold control requirements of enhancement DCFL logic. Using this technology, McDonnell Douglas has satisfactorily fabricated 256 bit SRAM circuits using the implanted p gate approach. Threshold voltage uniformities of the order of 30 mV and RAM yields up to 38% have been claimed on 2 inch diameter wafers.[83] Sony have also reported excellent progress using JFET technology with demonstrated ring oscillator speeds of $\tau_d = 58$ ps at power dissipations of $P_D = 0.64$ mW/gate.[81]

Another quite different and unique FET threshold adjustable technology for E-MESFETs has been developed by Toshiba. In this approach, Toshiba have successfully used a Pt–GaAs reaction to adjust FET device threshold. The observation by Sinha[84] that the barrier height (Φ_B) of Pt–GaAs Schottky barriers degrades above 450°C led to Toshiba's development of the so-called 'Pt-buried gate' process. By reacting Pt metal gates with the GaAs surface at about 400°C in H_2 ambient, a threshold shift of up to 0.35 V can be obtained.[82] Using this technique, Toshiba have reported wafer-to-wafer threshold reproducibility of 40 mV.

This fabrication technique, although very promising, may have some practical restrictions due to the likely possibility of reliability limitations. Current data suggest that this process may not support 'mil spec' reliability standards. Since Pt is extremely reactive with GaAs, it is speculated that it will be difficult to stop the Pt–GaAs reaction completely and that device characteristics may change during typical semiconductor burn-in routines. To date, Toshiba have shown acceptable Pt gate threshold stability under 200°C storage test to 2000 h without bias. Additional reliability verification (particularly under bias) is indicated before this technology will be ready for commercial application.

13.5.4 Heterojunction FET structures

Heterojunction FET integrated circuit approaches offer significant performance advantage over standard homojunction FETs as was discussed in Section 13.3.3. While the advantages of high speed and extremely low power are substantial, a number of difficult technological barriers must be overcome before practical commercial application can be realized. Albeit, numerous development efforts throughout the world promise that this will be a strong future IC technology candidate.[31]

The two most publicized basic enhanced mobility structures are illustrated in

Figure 13.13, namely Fujitsu 'HEMT' structure[27] and the Thompson–CSF 'TEGFET' structure.[28] Both approaches rely on the same basic principle of extremely high mobility in the 'two-dimensional electron gas' interface layer between the AlGaAs and undoped GaAs. These critical epitaxial layers are produced by molecular beam epitaxy (MBE), which in itself is probably the key limiting factor currently preventing early commercial application of this technology. Besides the obvious difficulties of precisely controlling both the doping concentration and layer thickness, the most difficult challenge in growing the AlGaAs is maintaining low ($\leqslant 10^{15}\,\mathrm{cm}^{-3}$) defect and trap concentrations.[31] Extremely high purity is required of these very thin layers in order to minimize

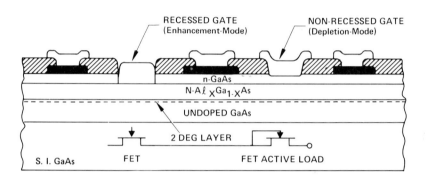

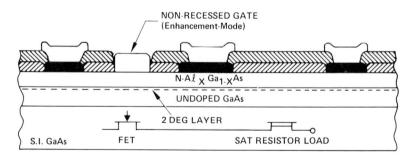

Figure 13.13 FET heterojunction enhancement mode logic gates: high electron mobility transistor (HEMT) inverter with depletion mode load and two-dimensional electron gas field-effect transistor (TEGFET) inverter with saturated resistor load

threshold variations, instabilities in the current–voltage relationship, light sensitivity, and temperature shifts. Continued refinements in the control and reproducibility of MBE technologies and development of metal-organic chemical vapour deposition (MOCVD) growth for this high mobility structure are anticipated.

Fabrication complexity of heterojunction FETs is similar to standard GaAs IC processing although processing is somewhat more difficult as a result of the sensitivities of the extremely thin layered device structure. The most simple process approach, called 'TEGFET', first developed by Thompson[28] and used quite successfully by Rockwell,[34] is shown in the lower portion of Figure 13.13. The approach uses a simple DCFL gate incorporating a saturated resistor load. While the switching FET is in a 'normally off' state under the metal gate channel region, the saturated resistor remains on. This condition exists as a result of the different interface potentials of the AlGaAs Schottky barrier gate (~ 1 V) for the FET and dielectric–GaAs interface potential (estimated to be $\leqslant 0.6$ V). This very simple device structure uses virtually identical processing, with the exception of the active layers, as that used in standard GaAs IC technology. Device isolation is accomplished by using the proton damage isolation method.[65] The most obvious limitations of the TEGFET approach with respect to comparison to the HEMT approach is in making good ohmic contacts directly to the AlGaAs and not buffering the critical AlGaAs surface with another GaAs layer to protect it during processing. With only limited data available from logic circuit applications, it is much to early to assess the strengths and weaknesses of the TEGFET enhanced mobility FET approach over the HEMT approach.

A more sophisticated fabrication approach has been developed by Fujitsu, shown in the upper portion of Figure 13.13. In this 'HEMT' approach, an additional layer of n-GaAs is used at the surface both to facilitate the fabrication of ohmic contacts and to allow the use of a gated FET active load in the DCFL logic gate. This approach requires that the switching FET use a recessed gate in order that the Schottky gate reside directly on the AlGaAs layer in order to obtain the 'normally off' FET condition, whereas the Schottky gate of the active load FET resides on the n-GaAs surface.

Fabrication of this HEMT logic gate structure required the development of selective dry etching of AlGaAs–GaAs.[85] However, early fabrication was accomplished by using wet chemical etching techniques. A planar reactive ion etching technique, using a gas mixture of CCl_2F_2 and helium, was developed, providing a high selectivity ratio (~ 200) between the GaAs and the AlGaAs. In this manner researchers at Fujitsu have been able to etch anisotropically 1 μm long gate structures precisely into the GaAs surface while stopping at the AlGaAs surface. Perhaps one limitation of this approach is in the potential radiation damage that can occur in the AlGaAs channel under the gate as a result of the radiation-induced reactive ion etching of the GaAs material. Little information is available concerning this issue at present. The increased complexity of the HEMT approach has led to a more widespread use of the simpler TEGFET structure. Additional work using both these approaches discussed and others in the literature[29, 30] will be required before all of the various trade-offs can be fully assessed. Ring oscillator performance results are discussed in Section 13.3.3 and divider performance results are presented in Section 13.6.1.

Inasmuch as GaAs heterojunction bipolar transistors (HJBT) are at a very early state of development, it is premature to discuss the ramifications of various fabrication processes. The reader is referred to Section 13.3.4 for a general discussion of HJBT technology.

13.6 PERFORMANCE OF GaAs DIGITAL ICS

The status of GaAs digital integrated circuit technology can best be assessed by reviewing the demonstrated performance results obtained using the various circuit and fabrication approaches discussed throughout this chapter. The comparisons of speed–power performance, while important, should not be viewed without equal consideration to the level of complexity obtainable with any particular technology. Many of the results presented here were obtained with production-compatible processes (e.g. D-MESFET BFL and SDFL technology), while others were obtained from intensively processed, newly developing technologies (e.g. principally E-MESFET DCFL technology). Therefore, an accurate assessment of the GaAs digital IC technology not only must take into consideration performance and complexity, but also must review technology relevant factors such as fabrication complexity, circuit yield, and overall manufacturability.

The main body of this section will focus on the results for the GaAs IC approaches having demonstrated reasonable levels of circuit complexity. In Tables 13.3 and 13.4 a comprehensive summary of the major GaAs IC results are presented. However, before continuing, it would be useful to discuss briefly GaAs ring oscillators inasmuch as these circuits are used quite extensively for assessing integrated circuit performance. A summary of GaAs ring oscillator performance comparisons was presented in Table 13.2.

The ring oscillator circuit provides a convenient and widely used method for evaluation of propagation delay and power dissipation since the ring oscillator provides its own internal signal source. Ring oscillators are composed of a chain of odd number, N, inverters or gates connected back to themselves so as to form a ring in order that the measured frequency of oscillation is related to the intrinsic gate propagation delay by $f = 1/(2N\tau_d)$. Oscillators with gate fan-outs of 1 or 2 are generally used. Note that, in general, GaAs devices provide much higher speed (10–60 ps propagation delays) than Si devices, as expected, owing to their superior electron dynamics and lower substrate parasitics. While Table 13.2 provides the reader with a good perspective of the expected performance differences between various GaAs IC approaches, it should be noted that these ring oscillator results

Table 13.3 Summary of various GaAs IC frequency divider technologies, including circuit architecture, measured toggle frequency, and equivalent propagation delay

GaAs IC technology	Circuit approach	Theor. max. toggle f	Meas'd max. toggle f (GHz)	Power dissipation P_D (mW/gate)	Equiv. $P_D \tau_d$ (pJ)	Equiv. τ_d (ps)
Implemented FET 0.7 μm D-BFL (Thomson CSF)	D-flip-flop T-conn'd ÷ 2 (NOR gates)	$\frac{1}{5\tau_d}$	3	40	2.68	67
1 μm D-SDFL (Rockwell)	D-flip-flop T-conn'd ÷ 4 (NOR-gates)	$\frac{1}{5\tau_d}$	2.6	3	0.231	77
1 μm D-BFL (HRL)	D-flip-flop T-conn'd ÷ 2 (NOR gates)	$\frac{1}{5\tau_d}$	2.2	78	7.1	91
0.6 μm E-DCFL (NTT)	D-flip-flop NOR gate ÷ 8	$\frac{1}{4\tau_d}$	3.8	1.2	0.079	66
1.2 μm DCFL (NEC)	Comp clock NOR gate ÷ 2	$\frac{1}{4\tau_d}$	2.4	3.9	0.39	100
1.0 μm D-BFL (HP)	Comp clock NAND/NOR ÷ 2	$\frac{1}{2\tau_d}$	6.5	–	–	77

Table 13.3 *Continued*

GaAs IC technology	Circuit approach	Theor. max. toggle f	Meas'd max. toggle f (GHz)	Power dissipation P_D (mW/gate)	Equiv. $P_D \tau_d$ (pJ)	Equiv. τ_d (ps)
0.8 μm D-BFL (Thomson CSF)	Comp clock NAND/NOR ÷ 2	$\frac{1}{2\tau_d}$	5.7	60	5.3	88
0.6 μm D-BFL (LEP)	Comp clock (NAND/NOR ÷ 2	$\frac{1}{2\tau_d}$	5.5	40	3.6	91
Heterojunction FET 0.5 μm E-DCFL (Fujitsu)	D-flip-flop ÷ 2	$\frac{1}{5\tau_d}$	5.5 8.9[a]	2.9 2.8[a]	0.1 0.06[a]	36 22[a]
1 μm E-DCFL (Bell)	D-flip-flop ÷ 2	$\frac{1}{5\tau_d}$	3.7 5.9[a]	3.2 5.0[a]	0.17 0.17[a]	54 34[a]
1 μm E-DCFL (Rockwell)	D-flip-flop ÷ 2	$\frac{1}{5\tau_d}$	3.6 5.2[a]	0.46 0.78[a]	0.025 0.03[a]	56 38[a]

[a] Operation at 77 K.

Table 13.4 Summary of various demonstrated GaAs IC logic and static RAM circuits, including technology approach, complexity, and performance results

Company	Logic circuits	Complexity[a]	Technology	Performance
Fujitsu	16 × 16 parallel mult.	3168 gates	E/D-MESFET SAG DCFL	153 ps/gate, 0.3 mW/gate
Rockwell	8 × 8 parallel mult.	1008 gates	D-MESFET	150 ps/gate, 2 mW/gate
NEC	32 bit adder	420 gates	D-MESFET LP/BFL	230 ps/gate, 2.9 mW/gate
Fujitsu	6 × 6 parallel mult.	408 gates	E/D-MESFET SAG DCFL	260 ps/gate, 0.42 mW/gate
Rockwell	5 × 5 parallel mult.	260 gate	D-MESFET SDFL	190 ps/gate, 0.7 mW/gate
Hewlett Packard	word generator	~600 devices	D-MESFET BFL	5 Gbits/s

Company	Static RAMs	Complexity[a]	Technology	Performance
Fujitsu	1 k × 1 bit SRAM	~6852 devices	E/D-MESFET self-aligned	τ_{acc} = 3.6 ns, 68 mW
NTT	1 k × 1 bit SRAM	~7084 devices	E/D-MESFET SAINT	τ_{acc} = 2.0 ns, 459 mW
Toshiba	256 × 4 bit SRAM	—	E/D-MESFET Pt-buried gate	τ_{acc} = 4.0 ns, 160 mW
Mitsubishi	1 k × 1 bit SRAM	—	E/D-MESFET	τ_{acc} = 3.8 ns, 38 mW
Rockwell	256 × 1 bit SRAM	—	D-MESFET	τ_{acc} = 1 ns, 267 mW
McDonnell Douglas	256 × 1 bit SRAM	—	E-JFET	τ_{acc} = 5 ns, 30 mW

[a] Complexity decreases as one descends table

(in terms of propagation delay) cannot be directly translated into propagation delays expected in actual complex logic circuits. The loss in performance translating from GaAs ring oscillators to circuits is not as great as in Si ICs. In typical silicon ICs, NMOS, for example, there is at least a factor for 4–5 difference between small inverter ring oscillator speeds and the speeds seen in actual circuits fabricated with the same technology. About half of this speed loss results from the fan-out loading in the real circuits (as expected for NMOS) and the rest comes from the parasitic capacitances incurred in the larger layout configuration. With semi-insulating GaAs substrates, this latter speed degradation source is greatly reduced, so that there is a much closer correspondence between GaAs ring oscillator speeds and the propagation delays measured in complex circuits. In GaAs, E-MESFET DCFL logic gates have a somewhat larger degradation in speed than D-MESFET approaches as evidenced by comparisons between the ring oscillator speeds of Table 13.2 and the frequency divider performance results of Table 13.3.

13.6.1 Sequential and combinatorial logic circuits

Ring oscillators are useful for monitoring the basic speed–power properties of logic gates. However, in realistic sequential or combinatorial logic, where the majority of the gates are implemented with multiple inputs and outputs, ring oscillators do not always provide an accurate assessment of performance. In order to evaluate the speed–power performance more realistically, the performances of binary ripple frequency dividers made with several GaAs IC technologies are compared in Table 13.3. From this table it is obvious that circuit architectural design has a strong influence on the maximum attainable clock frequency for a given GaAs FET logic. For instance, 0.7–1 μm D-MESFET BFL single-stage ($\div$ 2) dividers yield frequencies in the 2.2–3 GHz range using a $1/5\tau_d$ theoretical maximum toggle frequency while 5.5–6.5 GHz dividers are achieved using a $1/2\tau_d$ logic architecture. Since the average propagation delays of all of these circuits is in the 70–90 ps range, it should be obvious that the ICs with the higher performance are achieved as a direct consequence of using the faster ($1/2\tau_d$) complementary clocked, master–slave flip-flops implemented with NAND/NOR gates.

It can also be seen from Table 13.3 that the propagation delays determined from the dividing frequencies of the depletion mode circuits are reasonably close to those obtained from ring oscillator evaluations (see Table 13.2). This indicates that the speeds obtained with depletion mode logic gate circuits are not greatly reduced by the typical fan-outs of 2 or 3 used in practical IC applications.

However, with enhancement mode GaAs FET logic, this is not the case. The NTT 0.6 μm E-MESFET DCFL implemented divider has a maximum clock frequency of 3.8 GHz corresponding to an equivalent delay of 66 ps,[86] while similar geometry ring oscillators obtained 30 ps delays (see Table 13.2). Still,

GaAs enhancement technology performance does not degrade to as high a degree at typical Si NMOS enhancement circuits.

Very recently, super high-speed GaAs frequency divider performance has been demonstrated with high electron mobility transistor (HEMT) DCFL implemented logic. A divide-by-two counter, using $L_g = 0.5$ μm HEMT, has been successfully operated with a maximum clock frequency of 5.56 GHz at room temperature and a remarkable 8.9 GHz at 77 K.[87] This corresponds to a gate delay time of 36 ps at room temperature and 22 ps at 77 K. Other HEMT results (see Table 13.3) by Rockwell and Bell for 1 μm devices are equally impressive.[34, 88]

Various demonstrated logic circuits well beyond the complexity of simple frequency dividers are summarized by complexity in Table 13.4. Moving up the complexity curve, Hewlett Packard have demonstrated an MSI high-speed GaAs word generator[46] with 600 active devices implemented with D-MESFET BFL logic. The word generator consists of an 8-1 parallel-to-serial data converter, a timing generator, control logic, and ECL interface networks. The circuit generates multiple 8 bit words with dynamic word length control. A maximum data rate of 5 Gbit/s has been successfully demonstrated.

GaAs IC technology provides other very attractive prospects for the implementation of very high speed (combinatorial) data/signal processing components such as: adders, ALUs, multipliers, etc. Progress in this area has also been very encouraging. Development and performance results for a number of circuits including a 4 bit ALU, 32 bit adder, 8 × 8 multiplier, 16 × 16 multiplier, etc., have been reported.[47, 50, 73, 89]

NEC has successfully designed and fabricated a 32 bit adder using D-MESFETs, implemented with a low-power buffered FET logic (LP/BFL) circuit approach.[47] Power dissipation reduction of this BFL approach has been achieved by reducing the number of level shifting diodes to one, consistent with a very low FET threshold voltage (-0.5 V) and equivalently low supply voltages (2 V, -1 V). In this adder, containing about 420 gates, carry look ahead logic operation was utilized for realizing higher speed performance. A maximum addition time of 2.9 ns was reported with power dissipation of 1.2 W. This corresponds to a gate delay time of 230 ps and a power dissipation of 2.9 mW/gate, although the resulting power dissipation was abnormally low for BFL implemented circuits. As expected, the usual compromise of speed–power trade-off was experienced observing the modest gate delays obtained.

In real time digital signal processing systems (for example, in digital filtering or Fourier transform processing), fast multiplication is often an essential function. In many cases, the ultimate speed and bandwidth of a digital system are determined by the multiplication process. Therefore, the successful development of high-speed multipliers are an excellent GaAs logic circuit implementation for demonstration of improved system performance. Multipliers also serve as excellent vehicles for LSI/VLSI circuit development.

In the design of GaAs multipliers, a parallel (array) multiplication architecture is often chosen because of its very modular and repetitive layout. In this approach, only two basic building blocks, a full adder and a half adder, are required to form the multiplier combinatorial array. A straight $N \times N$ parallel multiplier, without any carry look ahead or other more sophisticated architecture (such as Wallace tree), requires $N(N-2)$ full adders and N half adders. To obtain the $2N$ bit product, a maximum of $(N-1)$ sum delays and $(N-1)$ carry delays is needed.[50] While other more complex circuits can reduce the multiply time by reducing carry delays or by recoding multiplier bits to shink the propagation path through the array, the less regular layouts resulting from these approaches significantly increase design time and thus were not considered for early multiplier circuit LSI/VLSI feasibility demonstrations.

Design and fabrication of 3×3 bit, 5×5 bit, and 8×8 bit parallel multipliers have been achieved using D-MESFET, SDFL circuit approaches.[90, 50] The 8×8 bit multiplier shown in Figure 13.14 (initially developed by Rockwell in 1980), consisting of 1008 gates (over 6000 devices) with chip area of 2.25×2.27 mm, was the first GaAs integrated circuit achieving LSI complexity. To date, it still represents the highest integration complexity achieved using depletion mode FET technology.

Figure 13.14 Photograph of the first (1980) GaAs LSI circuit: Rockwell's 8×8 parallel multiplier containing 1008 gates, the largest D-MESFET circuit ever demonstrated

This 8 × 8 bit multiplier uses a NOR gate implementation, forming a 16 bit product of two 8 bit input data bases. Multiplier performance of 5.25 ns with a total power dissipation ranging from 0.61 to 2.2 W was obtained. This performance corresponds to a gate propagation delay of 150 ps and power dissipation of 1–2 mW/gate. In a less complex version of this multiplier (5 × 5 multiplier containing 260 gates) using identical architecture, a 25% wafer yield was reported.[90]

The largest and most complex GaAs logic circuit reported to date is Fujitsu's 16 × 16 bit parallel multiplier, shown in Figure 13.15. This multiplier circuit has been designed using direct-coupled FET logic (DCFL) and fabricated using the newly developed self-aligned gate technology discussed in Section 13.5.2. The 4 × 4 mm chip contains 3168 NOR gates, using both enhancement and depletion FETs. The circuit is fabricated using rather conservative 2.0 μm gate length

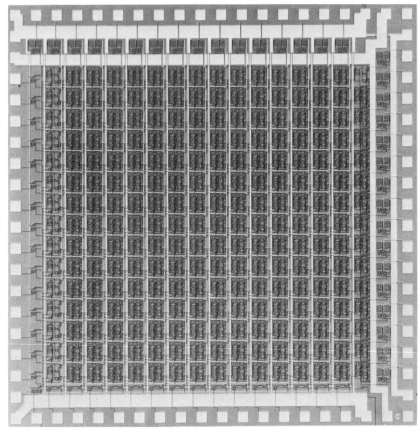

Figure 13.15 Photograph of Fujitsu's 16 × 16 parallel multiplier (1982), the largest (3168 gates) GaAs enhancement mode circuit ever demonstrated

design rules for high yield in conjunction with self-aligned gate FETs for high performance. Performance has been impressive, considering the early stage of development, with gate propagation delay of 153 ps at a power dissipation of 0.3 mW/gate being attained. The multiply time required for a 32 bit product was measured to be 10.5 ns. The 32 bit product is a factor of 2 slower than the 16 bit product (5.25 ns) reported by Rockwell.[50] Yield data for this very complex circuit are not available.

Use of more complex design implementation of multipliers (e.g. a Wallace tree, Booth algorithm, and carry look ahead) is expected to improve greatly the high-speed performance of future multipliers. Nevertheless, these initial multiplier circuits not only demonstrate what a high degree integration GaAs IC can attain, but also present substantial improvement in speed and power performance as compared to the fastest silicon multipliers commercially available.

13.6.2 GaAs random access memory

Memory has long been identified as a key application area for digital GaAs IC technology because of the fast memory access times necessary for very high-speed data requisition. Recently, an increasingly large effort hs been directed towards static RAM (SRAM) development. Interest in GaAs SRAMs, as is the case for other high-speed circuit applications, is centred around the very high current gain × bandwidth products of GaAs devices ($f_\tau = 15$ GHz for $L_g = 1$ μm GaAs MESFET). RAM performance projections of τ_{acc} of 1 ns in 1 kbit or 4 kbit GaAs static RAMs appear very reasonable. Further, since GaAs MESFETs have very high current gain and power gain at d.c., the read access time of GaAs RAMs need not be tied to the static power dissipation in the RAM cells, which means that the fast nanosecond access time should also be attainable at relatively low power dissipation levels.

As summarized in Table 13.4, impressive achievements have been reported in the development of very high-speed GaAs SRAMs (implemented in 1×256 bit, 4×256 bit, and 1×1 kbit versions) by various companies, including Fujitsu, NTT, Toshiba, and Mitsubishi in Japan, and Rockwell and McDonnell Douglas in the USA. Figure 13.16 shows a photograph of Fujitsu's 1×1 kbit static RAM fabricated with the tungsten silicide self-aligned gate technology.[21] Using a 2 μm gate length and 4 μm design rules, this chip measures 3×2 mm and consists of 4652 enhancement FETs and 2200 depletion FETs. The small RAM cell (47 $\times$ 34 μm) used in this work is designed with an enhancement/depletion FET cross-coupled flip-flop circuit comprising two enhancement-switching FETs and two enhancement-transfer FETs.

For this 1 k SRAM, an address access time of 3.6 ns and a minimum write enable pulse width of 1.6 ns were achieved with a power dissipation of only 68 mW. Fully working 1×1 k SRAMs have been obtained recently by Fujitsu with improved 1.3 ns address access times and 300 mW power dissipation. At the

Figure 13.16 Photograph of Fujitsu's 1 × 1 kbit static RAM chip fabricated with self-aligned tungsten silicide gate process

same time, rapid progress is being made towards achieving fully working 4 k SRAMs.[91]

Equally impressive results have been obtained by NTT in Japan towards the development of both 1 k and 4 k static RAMs[92, 93] using the very sophisticated self-aligned gate (SAINT) fabrication technology (discussed in Section 13.5.2). The photographs in Figure 13.17 show both the small (69 × 56 μm) enhancement RAM cell and the 3.32 × 3.32 mm chip used in NTT's 1 k RAM development work. Fully working 1 k RAMs have been attained, however, with very limited yield. Circuit performance at the 1 k level has been demonstrated with access times of 2.0 ns and power levels of 457 mW.[92] Very recent progress towards the development of NTT's 4 k RAM has demonstrated a mininum address access time of 2.8 ns at a power dissipation of 1.2 W.[93] It should be noted that it is not clear from the most recent publications whether any of the 4 kbit static RAM development efforts have actually been able to attain full 4 kbit functionality. The present yield limitations of these enhancement mode FET implemented RAMs can be traced directly to the difficult threshold voltage control problems inherently associated with GaAs enhancement devices (see discussion on enhancement mode FET technology in Section 13.3.2). Further progress in the control and uniformity of the critical 'normally off' GaAs active layers will be required before 1 k and 4 k GaAs SRAMs can become viable commercial products.

Before leaving the subject of static RAMs, two other 256 bit RAM efforts of significant importance should be highlighted. Static RAMS focused on extremely low power are under development by both McDonnell Douglas and Rockwell. In the case of McDonnell, τ_{acc} = 5 ns access times have been achieved at extremely low powers (30 mW), using enhancement mode JFET technology. The impressive circuit yield (up to 38%) seen at the 256 bit level has led to McDonnell's current efforts towards achieving a 1 kbit SRAM.

A static RAM development effort by Rockwell, using a depletion mode MESFET technology, has recently demonstrated 1 ns access times, also at low 267 mW power consumption.[94] Noting that the major efforts in GaAs RAM development are being implemented with enhancement technology (with the exception of Rockwell) supports the consensus of opinion that depletion mode technology, because of power consumption, will not be capable of supporting SRAMs beyond 4 kbit complexities. However, it is highly likely that initial commercial 1 k/4 k RAMs will be introduced using the more yield-tolerant GaAs D-MESFET circuit approach. GigaBit Logic, for instance, have already announced plans for introducing a depletion mode 1 k static RAM during early 1985.

13.6.3 Analogue-to-digital applications

In many real time signal processing systems, the input signals are in analogue form in nature. The data acquisition rate of the system is often limited by the

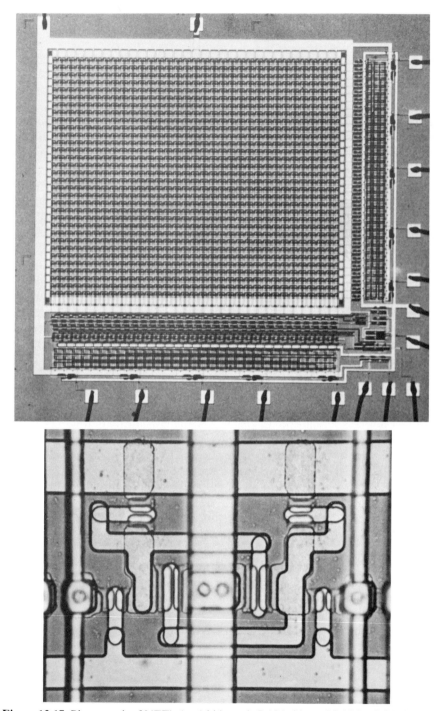

Figure 13.17 Photograph of NTT's 1 × 1 kbit static RAM chip and RAM cell fabricated with the SAINT process

speed performance of the analogue-to-digital converter. Very recently, the development of GaAs A/D, D/A circuits and their components has been reported. A high-speed GaAs strobed comparator[95] has been designed and fabricated using GaAs depletion mode MESFET logic with FET threshold voltages of $V_p = -2$ V. This comparator was designed to be a building block for a 4 bit flash analogue-to-digital converter operating above 1 GHz clock rate. The operation of this comparator has currently been successfully demonstrated up to 1.8 GHz clock rate with a performance suitable for a 4 bit A/D converter. A 4 bit digital-to-analogue converter with 100 ps rise and fall times has also been fabricated using GaAs depletion mode MESFET logic.[96] The output of the D/A converter settles to within 1 % of full scale in 1 ns at rates up to 10^9 samples per second. The implication of these recent GaAs analogue-to-digital development results, using the parallel threshold (flash) design, is that A/D converters operating at 10^9 samples per second rates with 4 to 6 bit resolution may soon be achievable.

Analysis of the various GaAs IC performance results confirms that enhancement mode MESFET circuits have lower power dissipation (but with adequate high-speed performance) when compared to depletion mode MESFET circuits. However, even though a high level of integration of E-MESFET circuits has been demonstrated, circuit yields have been limited due to the tight threshold voltage control required of this technology. Presently, a much broader base of MSI GaAs ICs have been designed and fabricated employing D-MESFET technology. The first commercially available GaAs ICs implemented with D-MESFETs will initially address the high-speed signal processing instrumentation and communication areas, followed shortly thereafter by LSI memory and computer products.[16] These circuits will generally be designed around internal gate delay times of 70–120 ps. Higher performance levels are expected to be achieved with further improvement in GaAs circuit design and processing techniques, and continued developments in super high-speed GaAs device approaches.

ACKNOWLEDGEMENTS

The authors are indebted to the numerous research and development teams throughout the world who, through their hard work and creative efforts, have brought GaAs from the 'material of the future' status to the forefront of a new and exciting high-speed integrated circuit technology. In particular, we would like to acknowledge the early pioneering work of C. A. Liechti, R. L. Van Tuyl, C. Stolte, and R. E. Lee who, at Hewlett Packard, really got the ball rolling in GaAs ICs; the innovation and followthrough of our former team at Rockwell, P. Asbeck, F. E. Eisen, C. Kirkpatrick, C. P. Lee, S. I. Long, and R. Zucca; the persistance of R. Zuleeg at McDonnell; the contributions of P. T. Greiling and C. F. Krumm at Hughes Research Laboratories; and the vision of F. Blum for making commercial GaAs ICs a reality at GigaBit. Also, the GaAs IC community in the USA is indebted for the support and technical guidance provided by R. Reynolds and S. Roosild through the Defense Advanced Research Project Agency. And finally, for the continuing example set by the Japanese in their GaAs development work, which gives us all the added belief and inspiration that this is only a beginning for the GaAs integrated circuit industry.

REFERENCES

1. R. C. Eden, Comparison of GaAs device approaches for ultrahigh-speed VLSI, *Proc. IEEE*, **70**(1), 5–12 (see also pp. 3–4) (Jan. 1982).
2. P. M. Solomon, A comparison of semiconductor devices for high speed logic, *Proc. IEEE*, **70**(5) (May 1982).
3. R. C. Eden and B. M. Welch, Ultrahigh-speed GaAs VLSI: approaches, potential and progress, in *VLSI Electronics: Microstructure Science*, vol. 3, ed. N. Einspruch, New York, Academic Press (1981).
4. R. C. Eden and B. M. Welch, GaAs digital integrated circuits for ultra high speed LSI/VLSI, in *Springer Series in Electrophysics*, vol. 5: *Very Large Scale Integration (VLSI) Fundamentals and Applications*, ed. D. F. Barbe, Berlin, Springer-Verlag, Ch. 5 (19   ).
5. H. Boll, E. Fulest, and R. Johnson, High-speed silicon NMOS circuits, Presented at the *High-Speed Dig. Tech. Conf.*, San Diego, CA, Jan. 13–14 (1981).
6. T. Nakamura, Punchthrough MOSFET for high speed logic, in *ISSCC Tech. Dig.*, San Francisco, CA, pp. 22–23 (Feb. 1978).
7. B. G. Bosch, Gigabit electronics—a review, *Proc. IEEE*, **67**(3) (March 1979).
8. J. G. Ruch and G. S. Kino, Measurements of the velocity–field characteristics of gallium arsenide, *Appl. Phys. Lett.*, **10**, 40 (1967).
9. F. F. Fang and A. B. Fowler, Hot electron effects and saturation velocities in silicon inversion layers, *J. Appl. Phys.*, **41**, 1825–31 (Mar. 15, 1970).
10. H. Morkoc, S. Bandy, R. Sankaran, G. Antypas, and R. Bell, A study of high-speed normally off and normally on $Al_{0.5}Ga_{0.5}As$ heterojunction gate GaAs FETs (HJFET), *IEEE Trans. Electron Devices*, **ED-25**, 619–27 (June 1978).
11. R. C. Eden, J. L. Moll, and W. E. Spicer, Experimental evidence for optical population of the X minima in GaAs, *Phys. Rev. Lett.*, **18**, 597–9 (Apr. 1967).
12. T. Mimura, K. Odani, N. Yokoyama, Y. Nakayama, and M. Fukuta, GaAs microwave MOSFETs, *IEEE Trans. Electron. Devices*, **ED-25**, 573–9 (June 1978).
13. N. Yokoyama, T. Mimura, H. Kusakawa, K. Suyama, and M. Fukuta, Low-power high-speed integrated logic with GaAs MOSFETs, *Jap. J. Appl. Phys.*, **19** (Feb. 1980).
14. R. L. VanTuyl, C. Liechti, R. E. Lee, and E. Gowen, GaAs MESFET logic with 4-GHz clock rate, *IEEE J. Solid-State Circuits*, **SC-12**, 485–96 (Oct. 1977).
15. B. Welch and Y. D. Shen, The manufacturability of GaAs integrated circuits, *Tech. Dig. GaAs IC Symp.*, p. 162 (Nov. 1982).
16. D. MacMillan and T. Gheewala, Learn gallium arsenide basics before applying highspeed ICs, three part series in *EDN Magazine*, March, April, May (1984).
17. R. Zucca, B. M. Welch, C. P. Lee, R. C. Eden, and S. I. Long, Process evaluation test structures and measurement techniques for a planar GaAs digital IC technology, *IEEE Trans. Electron Devices*, **ED-27** (Dec. 1980).
18. S. I. Long, B. M. Welch, R. Zucca, P. Asbeck, C. P. Lee, C. G. Kirkpatrick, F. S. Lee, G. R. Kaelin, and R. C. Eden, High speed GaAs integrated circuits, *Proc. IEEE*, **70**(1), 35–45 (Jan. 1982).

19. P. T. Greiling, C. F. Krumm, F. S. Ozdemir, L. H. Hackett, and R. F. Lohr, Jr, Electron beam fabricated GaAs FETs inverter, *IEEE Trans. Electron Devices*, **ED-25**, 1340 (Nov. 1978).
20. R. Zuleeg, J. K. Notthoff, and K. Lehovec, Femtojoule high-speed planar GaAs E-JFET logic, *IEEE Trans. Electron Devices*, **ED-25**, 628–39 (June 1978).
21. N. Yokoyama, T. Ohnishi, H. Onodera, T. Shinoki, A. Shibatumi, and H. Ishikawa, A GaAs 1 K static RAM using tungsten-silicide gate self-alignment technology, *ISSCC Tech. Dig*, pp. 44–45 (Feb. 1983).
22. Y. Kato, M. Dohsen, J. Kasahara, and K. Taira, Planar normally-off GaAs JFET for high speed logic circuits, *Electron. Lett.*, **17**, 951–2 (1981).
23. P. G. Flahive, S. H. Wemple, R. A. Kiehl, and H. M. Cox, Sub 20 ps DCFL GaAs ring oscillators with self-aligned 1 micron gates, *Proc. GaAs IC Symp.*, pp. 184–6 (Nov. 1982).
24. T. Mizutani, N. Kato, S. Ishida, K. Osafune, and M. Ohmori, GaAs gigabit logic circuits using normally-off MESFETs, *Electronic Lett.*, **16**, 315–16 (April 24, 1980).
25. N. Yokoyama, T. Ohnishi, K. Odani, H. Onodera, and M. Abe, TiW silicide gate technology for self-aligned GaAs MESFET VLSIs, *IEEE IEDM Tech. Dig.*, pp. 80–3 (Dec. 1981).
26. R. A. Sadler and L. F. Eastman, Self-aligned submicron ion-implanted GaAs MESFETS for high speed logic, *41st Annu. Device Res. Conf.*, Burlington, VT, Paper IVA-2 (June 1983).
27. T. Mimura, S. Hiyamizu, T. Fujii, and K. Nanbu, A new field-effect transistor with selectively doped GaAs/m-$Al_xGa_{1-x}As$ heterojunctions, *Jap. J. Appl. Phys.*, **19** (5), 225–7 (May 1980).
28. M. Laviron, D. Delagebeaudeuf, P. Delescluse, P. Etienne, J. Chaplart, and N. T. Linh, Low noise normally on and normally off two-dimensional electron gas field-effect transistors, *Appl. Phys. Lett.*, **40**(6), 530–2 (March 1982).
29. T. J. Drummond, S. L. Su, W. Kopp, R. Fischer, R. E. Thorne, H. Morkoc, K. Lee, and M. S. Shur, High velocity n-on and n-off modulation doped $GaAs/Al_xGa_{1-x}As$ FETs, *IEDM Tech. Dig.*, p. 586 (1982).
30. M. D. Feuer, R. H. Hendel, R. A. Kiehl, J. C. M. Hwang, V. G. Keramidas, C. L. Allyn, and R. Dingle, High-speed low-voltage ring oscillators based on selectively doped heterojunction transistors, *IEEE Electron Devices Lett.*, **EDL-4**(9) (Sept. 1983).
31. H. Morkoc and P. M. Solomon, The HEMT: a superfast transistor, *IEEE Spectrum*, **28** (Feb. 1984).
32. S. Hiyamizu, T. Mimura, and T. Ishikawa, MBE-grown GaAs/n-AlGaAs heterostructures and their application to high electron mobility transistors, *Proc. 13th Conf. on Solid State Devices*, Tokyo, 1981; and *Jap. J. Appl. Phys.*, **21**, Suppl. 21–1, 161–8 (1982).
33. J. C. M. Hwang, J. V. DiLorenzo, P. E. Luscher, and W. S. Knodle, Application of molecular beam epitaxy to III–V microwave and high-speed device fabrication, *Solid-State Technol.*, **25**, 166 (1982).
34. C. P. Lee, D. Hou, S. J. Lee, D. L. Miller, and R. J. Anderson, Ultra high speed digital integrated circuits using GaAs/GaAlAs high electron mobility transistors, *Proc. GaAs IC Symp.*, pp. 162–5 (Oct. 1983).
35. T. Mimura, K. Joshin, S. Hiyamizu, K. Hikosaka, and M. Abe, High electron mobility transistor logic, *Jap. J. Appl. Phys.*, **20**(8), 598–600 (Aug. 1981).
36. H. Kroemer, Heterostructure bipolar transistors and integrated circuits, *Proc. IEEE*, **70**(1), 13–25 (Jan. 1982).
37. H. Yuan, W. V. McLevige, H. D. Shih, and A. S. Hearn, GaAs heterojunction bipolar 1 K gate array, *IEEE ISSCC Tech. Dig.*, pp. 42–3 (Feb. 1984).

38. W. V. McLevige, H. T. Yuan, W. M. Duncan, W. R. Frensley, F. H. Doerbeck, H. Morkoc, and T. J. Drummond, *IEEE Electron Devices Lett.*, **EDL-3,** 43 (1982).
39. For two reviews see: (a) A. Y. Cho and J. R. Arthur, Molecular beam epitaxy, *Prog. Solid State Chem.*, **10,** Pt 3, 157–91 (1975); (b) K. Ploog, Molecular beam epitaxy of III–V compounds, in *Crystals: Growth, Properties and Applications*, ed. H. C. Freyhardt, New York, Springer-Verlag, vol. 3, pp. 73–162 (1980).
40. S. L. Su, O. Tejayadi, T. J. Drummond, R. Fischer, and H. Morkoc, Double heterojunction AlGaAs/GaAs bipolar transistors (DHBJTs) by MBE with a current gain of 1650, *IEEE Electron Devices Lett.*, **EDL-4** (5) (May 1983).
41. P. M. Asbeck, D. L. Miller, R. J. Anderson, and F. H. Eisen, Emittercoupled logic circuits implemented with heterojunction bipolar transistors, *Tech. Dig. GaAs IC Symp.*, pp. 171 (Oct. 1983).
42. J. P. Bailbe, A. Marty, P. H. Hiep, and G. E. Ray, Design and fabrication of high-speed GaAlAs/GaAs heterojunction transistors, *IEEE Trans. Electron Devices*, **ED-27,** 1160–4 (June 1980) (and references cited therein).
43. P. M. Asbeck, D. L. Miller, R. J. Anderson, R. N. Deming, L. O. Hou, C. A. Liechti, and F. H. Eisen, 4.5 GHz frequency dividers using GaAs/GaAlAs heterojunction bipolar transistors, *Tech. Dig. ISSCC*, pp. 50–1 (Feb. 1984).
44. H. Kroemer, Theory of a wide-gap emitter for transistors, *Proc. IRE*, **45**(11), 1535–7 (Nov. 1957).
45. R. L. Van Tuyl and C. A. Liechti, High speed integrated logic with GaAs MESFETs, *IEEE J. Solid State Circuits*, **SC-9**(5) (Oct. 1974).
46. C. Liechti, G. Baldwin, E. Gowen, R. Joly, M. Namjoo, and A. Podell, A GaAs MSI word generator operating at 5 Gbits/s data rate, *IEEE Trans. Microwave Theor. Tech.*, **MTT-30,** 998–1006 (July 1982).
47. R. Yamamoto, A. Higashisaka, S. Asai, T. Tsuji, Y. Takayama, and S. Yano, Depletion-type GaAs MSI 32 b adder, *IEEE Int. Solid-State Circuits Conf. Dig.*, pp. 403 (Feb. 1983).
48. R. C. Eden, B. M. Welch, and R. Zucca, Low power GaAs digital ICs using Schottky diode-FET logic, *1978 Int. Solid State Circuits Conf., Dig. Tech. Papers*, pp. 68–9 (Feb. 1978).
49. R. C. Eden, F. S. Lee, S. I. Long, B. M. Welch, and R. Zucca, Multi-level logic gate implementation in GaAs ICs using Schottky diode-FET logic, *1980 Int. Solid State Circuits Conf., Dig. Tech. Papers* (Feb. 1980).
50. F. S. Lee, G. R. Kaelin, B. M. Welch, R. Zucca, E. Shen, P. Asbeck, C. P. Lee, C. G. Kirkpatrick, S. I. Long, and R. C. Eden, A highspeed LSI GaAs 8 × 8 bit parallel multiplier, *IEEE J. Solid State Circuits*, **SC-17**(4), 638–47 (Aug. 1982).
51. J. K. Notthoff and C. H. Vogelsang, Gate design for DCFL with GaAs E-JFETs, Paper 10 in Research Abstracts of *First Annual Gallium Arsenide Integrated Circuit Symp.*, Lake Tahoe (Sept. 27, 1979).
52. F. Damay-Kavala, G. Nuzillat, and C. Arnodo, Highspeed frequency dividers with quasi-normally-off GaAs MESFETs, *Electron. Lett.*, **17**(25), 968–70 (1981).
53. M. J. Helix, S. A. Jamison, S. A. Hanka, R. P. Vidano, P. Ngo, and C. Chao, Improved logic gate with a push–pull output for GaAs digital ICs, *Proc. GaAs IC Symp.*, pp. 108–11 (Nov. 1982).
54. G. Lee, T. Steeves, M. Schroeder, R. Nelson, T. Vu, B. Hanzal, P. Roberts, M. Helix, S. Jamison, S. Hanka, and J. Brown, A 432-cell SDFL GaAs gate array implementation of a four-bit slice event counter with programmable threshold and time stamp, *Proc. GaAs IC Symp.*, pp. 174–7 (Oct. 1983).
55. S. Katsu, S. Nambu, A. Shimano, and G. Kano, A GaAs monolithic frequency divider using source coupled FET logic, *IEEE Electron. Devices Lett.*, **EDL-3**(8), 197–9 (1982).

56. A. Shimano, S. Katsu, S. Nambu, and G. Kano, A 4 GHz 25 mW GaAs IC using source coupled FET logic, *IEEE Int. Solid State Circuits Conf. Dig.*, p. 42 (Feb. 1983).
57. A. W. Livingstone and P. J. Mellor, Capacitor coupling of GaAs depletion-mode FETs, *IEE Proc.*, pt I, **127**, 297–300 (1980).
58. A. Livingstone, A. Welbourn, and G. Blau, Manufacturing tolerance of capacitor coupled GaAs FET logic circuits, *IEEE Electron. Devices Lett.*, **EDL-3**(10) (Oct. 1982).
59. M. R. Namordi and W. A. White, A low power, static GaAs MESFET logic gate, *Tech. Dig. GaAs IC Symp.*, pp. 21–4 (Nov. 1982).
60. B. M. Welch and R. C. Eden, Planar GaAs integrated circuits fabricated by ion implantation, *1977 IEDM Tech. Dig.*, pp. 205–8 (Dec. 1977).
61. B. M. Welch, Y. D. Shen, R. Zucca, R. C. Eden, and S. I. Long, LSI processing technology for planar GaAs integrated circuits, *IEEE Trans. Electron Devices*, **ED-27** (6) (June 1980).
62. R. C. Eden, A. R. Livingston, and B. M. Welch, Integrated circuits: the case for gallium arsenide, *IEEE Spectrum*, pp. 30–7 (Dec. 1983).
63. F. E. Eisen, B. M. Welch, K. Gamo, T. Inada, H. Mueller, M. A. Nicolet, and J. W. Mayer, Sulfur, selenium and tellurium implantation in GaAs, *Applications of Ion Beams to Materials 1975*, Inst. Phys. Conf. Ser. 28, Bristol and London, Institute of Physics, Chapter 2 (1976).
64. R. N. Thomas, Growth and characterization of large diameter undoped semi-insulating GaAs for direct ion implanted FET technology, *Solid State Electronics*, **24**, 387–99 (1981).
65. D. D'Avanzo, Proton isolation for GaAs integrated circuits, *IEEE Trans. Electron Devices*, **ED-29** (May 1982).
66. M. Berth, M. Cathelin, and G. Durand, Self-aligned planar technology for GaAs integrated circuit, *Proc. 1977 IEDM*, Washington, DC, pp. 201–4 (Dec. 1977).
67. B. M. Welch, Y. D. Shen, and W. P. Fleming, Microstructure pattern replication for advanced planar GaAs LSI/VLSI, *Proc. Microcircuit Engineering 80*, Amsterdam, Sept. 30, Oct. 1–2, p. 489 (1980).
68. T. Furutsuka, T. Tsuji, F. Katano, M. Kanamori, A. Higashisaka, and Y. Takayama, Highspeed E/D GaAs ICs with closely-spaced FET electrodes, *Dig. Tech. Papers, 14th Int. Solid State Devices Conf.* (1982).
69. M. Fukuta, K. Suyama, and H. Kusakawa, Low power GaAs digital integrated circuits with normally off MESFETs, *IEEE Trans. Electron Devices*, **ED-25**, 1340 (Nov. 1978).
70. H. Ishikawa, H. Kusakawa, K. Suyama, and M. Fukuta, Normally-off type GaAs MESFET for low-power high-speed logic circuits, *1977 Int. Solid State Circuits Conf., Dig. Tech. Papers*, pp. 644–5 (Oct. 1977).
71. N. Yokoyama, T. Mimura, M. Fukuta, and H. Ishikawa, A self-aligned source/drain planar device for ultra-high-speed GaAs MESFET VLSIs, *ISSCC Dig. Tech. Papers*, pp. 218–19 (Feb. 1981).
72. Y. Nakayama, K. Suyama, H. Shimizu, S. Yokogawa, and A. Shibatomi, An LSI GaAs DCFL using self-aligned MESFET technology, *GaAs IC Symp. Tech. Dig.*, pp. 6–9 (Nov. 1982).
73. Y. Nakayama, K. Suyama, H. Shimizu, N. Yokoyama, A. Shibatomi, and H. Ishikawa, A GaAs 16 × 16 parallel multiplier using self-alignment technology, *IEEE ISSCC Tech. Dig.*, pp. 48–9 (Feb. 1983).
74. P. M. Asbeck, D. L. Miller, E. J. Babcock, and C. G. Kirkpatrick, Application of thermal pulse annealing to ion-implanted GaAlAs/GaAs heterojunction bipolar transistors, *IEEE Electron Devices Lett.*, **EDL-4** (4) (April 1983).
75. Y. Yamasaki, K. Asai, and K. Kurumada, GaAs LSI-directed MESFETs with self-

aligned implantation for N^+ layer technology (SAINT), *IEEE Trans. Electron Devices*, **ED-29** (11), 1772–7 (Nov. 1982).
76. T. Kohashi, Private Communication.
77. T. Terada, Y. Kitaura, T. Mizoguchi, M. Mochizuki, N. Toyoda, and A. Hojo, Self-aligned Pt-buried gate FET process with surface planarization technique for GaAs LSI, *Tech. Dig. GaAs IC Symp.*, pp. 138–41 (Oct. 1983).
78. R. E. Lee, H. M. Levy, and D. S. Matthews, Material and device analysis of self-aligned gate GaAs ICs, *Tech. Dig. GaAs IC Symp.*, pp. 177–9 (Nov. 1982).
79. G. L. Troeger, A. F. Behle, P. E. Friebertshauser, K. L. Hu, and S. H. Watanabe, Fully ion implanted planar GaAs E-JFET process, *1979 IEDM Tech. Dig.*, pp. 497–500 (Dec. 1979).
80. G. L. Troeger and J. K. Notthoff, A radiation-hard low-power GaAs static RAM using E-JFET DCFL, *GaAs IC Symp. Tech. Dig.*, pp. 78–9 (Oct. 1983).
81. K. Kato, K. Gonoi, K. Taira, J. Kashara, M. Dohsen, M. Arai, and N. Watanabe, High-speed and low-power GaAs JFET IC technology, *4th GaAs IC Symp. Tech. Dig.*, pp. 187–90 (1982).
82. N. Toyoda *et al.*, An application of Pt–GaAs solid phase reaction to GaAs and Related Compounds, *Gallium Arsenide and Related Compounds 1981*, Inst. Phys. Conf. Ser. 63, Bristol and London, Institute of Physics, p. 521 (1982).
83. R. Zuleeg, Private Communication.
84. A. K. Sinha, S. E. Haszko, and T. T. Sheng, *J. Electrochem. Soc.*, **122**, 1714 (1975).
85. K. Hikosaka, T. Mimura, and K. Joshin, Selective dry etching of AlGaAs–GaAs heterojunction, *Jap. J. Appl. Phys.*, **20** (11), 847–50 (1981).
86. T. Mizutani, N. Kato, M. Ida, and M. Ohmori, High-speed enhancement-mode GaAs MESFET logic, *IEEE Trans. Microwave Theor. Tech.*, **MTT-28** (5), 479–83 (1980).
87. K. Nishiuchi, T. Mimura, S. Kuroda, S. Hiyamizo, H. Nishi, and M. Abe, Device characteristics of short channel high electron mobility transistor (HEMT), *Proc. 41st Annu. Device Res. Conf.*, Paper IIA-8 (June 1983).
88. R. A. Kiehl, M. D. Fever, R. H. Hender, J. C. M. Hwang, V. G. Keramidas, C. L. Allyn, and R. Dingle, Selectively doped heterostructure frequency dividers, *IEEE Electron Devices Lett.*, **EDL-4** (10) (Oct. 1983).
89. H. Kusakawa, K. Suyama, S. Okamura, and M. Fukuta, An MSI GaAs integrated circuit: 4-bit arithmetic and logic unit, *Int. Conf. on Solid State Devices*, Tokyo, Japan (1980).
90. S. I. Long, F. S. Lee, R. Zucca, B. M. Welch, and R. C. Eden, MSI high speed low power GaAs ICs using Schottky diode FET logic, *IEEE Trans. Microwave Theor. Tech.*, **MIT-28** (5), 466–72 (May 1980).
91. N. Yokoyama, H. Onodera, T. Shinoki, H. Ohnishi, H. Nishi, and A. Shibatomi, A 3 ns GaAs $4K \times 16$ SRAM, *IEEE ISSCC Tech. Dig.*, pp. 44–5 (Feb. 1984).
92. K. Asai, K. Kurumada, M. Hirayama, and M. Ohmori, 1 Kb static RAM using self-aligned FET technology, IEEE ISSCC Tech. Dig., pp. 46–7 (Feb. 1983).
93. M. Hirayama, M. Ino, Y. Matsuoka, and M. Suzuki, A GaAs 4Kb SRAM with direct coupled FET logic, *IEEE ISSCC Tech. Dig.*, pp. 46–7 (Feb. 1984).
94. S. J. Lee, R. P. Vahrenkamp, G. R. Kaelin, L. D. Hou, R. Zucca, C. P. Lee, and C. G. Kirkpatrick, Ultra-low power, high speed GaAs 256–bit static RAM, *Tech. Dig. GaAs IC Symp.*, pp. 74–7 (Oct. 1983).
95. D. Meignant and M. Binet, A high performance 1.8 GHz strobed comparator for A/D converter, *Tech. Dig. GaAs IC Symp.*, pp. 66–9 (Oct. 1983).
96. G. S. LaRue, A GHz GaAs digital to analog converter, *Tech. Dig. GaAs IC Symp.*, pp. 70–3 (Oct. 1983).

Index

III–V compounds, 1

AES, 106
air bridges, 485
alloyed contacts, 229
amorphous/amorphous junction, 201
amorphous/crystalline junction, 201
anisotrophic etching, 127
annealing, 277
annealing of damage, 163
anodic etching, 143
anodic oxidation, 277
anodization, 283, 471, 483
automatic frequency control, 308
avalanche multiplication, 340

ballistic transport, 398
band gap absorption, 26
bath tub via holes, 485
bond wires, 381
bonding pads, 372
boric oxide, 55, 58
breakdown field, 342
buffer layer, 361, 368
buffered FET logic, 535

carbon in LEC GaAs, 83
carrier accumulation, 382
carrier confinement, 370
carrier removal, 168
chemical etches, 371
chemical etching, 374
chemical polishing, 135
chemical vapour deposition, 278

chirp, 319
   compensation, 320
chloride transport, 99
chromium redistribution, 185
COLDFET, 19
common source, 369
computer simulation, 304
cone angle, 43, 52
contact implant, 377
contact resistivity, 220
coplanar
   impedance, 459
coplanar circuits, 454
correlation impedance, 403
current gain, 367
CVD
   photoenhanced, 278
   plasma-enhanced, 278

damage distribution, 167
DCTL designs, 264
Debye length, 385
density of states, 438
depletion region 340
device line, 305
device performance, 283
diameter control, 59
dielectric growth
   deposited layers, 277
dielectric constant, 455
dielectric growth, 276
dielectric loss, 465
dielectric overlay, 415
differential mobility, 322

diffusion doping, 161
dipole noise, 401
dislocation density, 46
  longitudinal distribution, 51
  radial distribution, 48
dislocation gettering, 82
dislocation networks, 51
distributed feedback concept, 488
DLTS, 274
domain nucleation, 302
doping, 172, 177
double heterostructure, 430
dry etching, 151

effective mass, 9
EL2 defects, 66
electrical activity, 174, 179
electromigration, 204
electron beam alloyed contacts, 247
electron beam annealing, 164
ELED, 434
epitaxial growth, 95
etch rates, 137
etches, 371
evaporation, 278
extrinsic doping, 13

Fermi level, 291
FET, 263
  heterojunction, 553
  low noise microwave, 380
  vertical, 119
FET logic
  buffered, 535
  direct-coupled, 539
  gate design, 543
  Schottky diode, 537
fibre optics, 429
flip-chip, 413
float-off process, 420
free-carrier absorption, 25
free-space wavelength, 441

GaAlAs, 287, 418
GaAs
  avalanche breakdown, 20
  band structure, 7, 521
  barrier height data, 13
  carrier transport, 16
  crystal structure, 3, 5
  device technology, 276

dispersion curves, 7
drift velocity, 17
effective mass, 9, 33
elastic properties, 6
electrical properties, 102
etching, 119
homoepitaxial, 108
intrinsic density, 11
ionization coefficients, 21
mobility, 17–18
optical properties, 23
physical properties, 1
surface properties, 12
v–E characteristic, 300, 520
velocity–field characteristics, 520
vibrational properties, 6
GaAs random access memory, 565
galvanic effect, 135
gate insulation, 283
gate metal resistance, 379
gate metallization, 375
gate recess, 374
gradual channel approximation, 406
ground plane, 457
Gunn domains, 299
Gunn effect, 181
Gunn effect diodes, 204

HEMT, 418, 530
  mutual conductance, 419
heteroepitaxy, 221
heterojunctions, 202
heterolayers, 432
heterostructure bipolar transistor, 532
high temperature contacts, 216
HRLED, 433
hydride process, 99

IGFET, 287
impact ionization, 334, 446
IMPATT, 170, 182, 331
  computer simulations, 349
  cutoff frequency, 348
  double drift, 344
  drift region, 337
  efficiency, 344
  fabrication, 342
  fabrication schedule, 351
  Hi-Lo structures, 347
  Lo-Hi-Lo structures, 347
  model, 331

Index   577

power generated, 341
pulsed, 343
Schotty barrier, 171
single drift, 343
small signal analysis, 334
technology, 350
thermal properties, 345
impedance–field concept, 404
implantation doping, 161, 171
induced current, 340
inductor
  single loop, 468
  spiral, 470
influence fluctuations, 405
injection limiting cathodes, 321
InP, 317
integrated circuits, 185
interdigital capacitor model, 461
interface mixing, 243
intrinsic defects, 62
ion damage, 161
ion implantation, 161, 370
ion induced damage, 163
ion ranges, 162
ion-beam induced etching, 151
ion-crystal interactions, 162
ionization rates, 333
isolation, 372

JFET, 265, 293
Johnson noise, 401
junction temperature, 342

knee voltage, 382, 384

Lange coupler, 497
Langmuir films, 278
LASER, 436
  alloyed contacts, 247
  buried heterostructure, 172, 442
  double heterostructure, 439
  GRIN–SCH, 444
  heterostructure, 436
  homojunction, 436
  longitudinal modes, 441
  optical output, 437
  phased array, 443
  reliability, 445
  stripe, 439
  threshold current, 441, 443
  transient behaviour, 444

transverse mode, 442
laser annealing, 164
LEC growth, 39
LED, 429, 431
  emitting area, 433
liftoff, 282, 373, 483
lithography, 374, 480
logic circuits, 560
logic gates, 369, 420
low barrier contacts, 221
LPE, 96, 98
LSA, 299
luminescence, 28
luminescence spectra, 29
luped element circuits, 310

mask edge trenching, 127
material inhomogeneities, 266
MBE, 95, 105, 288, 356, 378, 429, 449,
     555
  deep levels, 111
  selective area, 111
  system, 107
Melbourn puller, 40
melt stoichiometry, 60
MESFET, 89, 119, 150, 183, 205, 264,
     275, 279, 361, 404
  breakdown, 409
  construction, 362
  d.c. characteristics, 411, 472
  depletion layer profiles, 363
  depletion mode, 524
  dual gate, 418
  enhancement mode, 528
  GaAs, 281
  gain control, 417
  gate oxide, 522
  gradual channel approach, 385
  intrinsic model, 365
  low noise, 400
  mixer, 417
  MMIC, 472
  output power, 408
  photo, 445
  planar, 89
  power, 205, 410
  recessed gate, 525
  reliability, 416
  self-aligned gate, 548
  simple equivalent circuit, 369
  small signal performance, 421

578 Index

structure, 412
switches, 417, 506
thermal resistance, 414
metallization, 195, 368
microstrip, 309, 314, 455, 484
   discontinuities, 458
   impedance, 455
   matching elements, 495
   metallization, 461
   Q, 456
microwave FET, 373
millimetre transistors, 407
MIS, 268
   band diagram, 266
   GaAs, 271–272, 281
   interface, 289
   interfacial effects, 274
   ion contamination, 266
   pinholes, 266
   radiation tolerance, 265
   surface states, 268
   technology, 275, 285
MIS devices, 263
MISFET, 275
   depletion, 264
   GaAs, 275
MISFET medium-power GaAs, 284
MMIC, 453
   active devices, 472
   air bridges, 485
   balanced mixer, 512
   circuit design, 485
   crosstalk, 457
   device isolation, 479
   digital, 517
   distributed amplifiers, 504
   fabrication, 545
   inductor, 468
   interdigital capacitors, 460
   Lange coupler, 497
   low-frequency circuits, 489
   low-noise amplifiers, 494
   lumped elements, 460
   MESFET, 472
   millimetre wave, 509
   overlay capacitors, 465
   parasitics, 462
   phase shifters, 503, 506
   polyimide, 483
   power amplifiers, 498
   power splitters, 499

   processing, 477
   resistors, 471
   Schottky barrier diode, 475
   slot line, 459
   thin film capacitors, 482
   transmission lines, 454
   via hole, 484
MOCVD, 95, 103, 429, 449, 555
MOCVD reactor, 104
modelling, 365
modulation factor, 340
Monte Carlo simulations, 395
Monte Carlo techniques, 300
MOSFET, 282
multilayer contacts, 215
mutual conductance, 369

native oxide, 290
native oxides, 276
necking, 57
negative resistance amplifiers, 323
NMOS technology, 263
noise
   equivalent bandwidth, 402
noise coefficient, 403
noise current, 405
noise figure, 400, 404, 497
noise generator, 403
noise power, 401
noise resistance, 403
non-radiative recombination, 22

ohmic contacts, 195, 219, 394
optical absorption, 24
optical confinement, 431
optical field, 439
optical flux density, 442
optical measurements, 274
optoelectronic devices, 429
optoelectronic repeater, 448
overlay capacitor model, 465
overlay techniques, 414

parasitic resistance, 377, 401, 404
PBT, 420
PECVD, 483
phase equilibrium, 2
phonon absorption, 24
phonon velocity, 459
photoionization, 274
photoresist, 481

Index

PIN diode, 445
pinch off, 364
pinched down 386
planar gunn logic, 144
plasma etching, 152
plasma oscillations, 33
plasma oxidation, 276
plasma oxides, 292
PN diodes, 183
Poisson's equation, 390
polyimide, 483
postgrowth treatments, 280
power added efficiency, 408
power amplifier, 498
preferential etching, 123, 127
processing, 281
projected range, 162, 176
punch through, 419
'purple plague' reaction, 264

quantum efficiency, 434

radiation damage, 161, 166
radiation losses, 315
Ramo–Schockley theorem, 347
range straggling, 162, 176
RBS, 207, 209
reactance compensation, 309
reactive ion etching, 156
Read diode, 335
Read's equation, 336
recessed gate, 525
recombination, 434, 446
reflectivity, 32
refractive index, 34
residual oxide layers, 138
resist, 373
   profiles, 373
RHEED, 106
roundtrip gain, 438

satellite valleys, 397
saturation index, 386
Schottky barrier, 119, 218, 291
Schottky contacts, 195
seed quality, 57
selective etching, 123
selective implantation, 477
self-aligned gate, 548
self-aligned process, 378
semi-insulating layers, 166

semi-insulator behaviour, 60
sheet resistivity, 461, 471
signal-to-noise ratio, 400
silicon nitride, 181
SIMS, 106
sintered contacts, 227
SLED, 434
slot line, 459
solar cells, 288
solid phase reactions, 227
space charge effects, 349
sputtering, 151, 278
steady state oscillation, 305
stoichiometry, 289
surface bonds, 292
surface conversion, 185
surface dielectrics
   GaAs, 147
surface potential, 269
surface preparation, 147, 279
surface states, 196, 269, 286, 290
surface traps, 272

TEA, 321
   reflection, 325
   stabilization, 321
   transducer gain, 324
   transmission, 325
TED, 299
   equivalent circuit, 306
   InP, 317
   structure, 316
   subcritical, 322
   temperature stability, 308
TED thermal analysis, 318
TEGFET, 418, 531
TEO
   bias voltage tuning, 306
   chirp, 319
   electronic tuning, 307
   mechanical tuning, 303
   noise performance, 311
   post-mounted waveguide, 304
   pulsed, 315
   realization, 313
   transmission media, 314
   tuning range, 309
   varactor tuning, 312
   YIG tuning, 307
THLOC, 443
threshold voltage, 91, 301

transducer gain, 324, 325
transferred electron device, 299
transferred electron devices, 181, 204
transit angle, 338
transmission lines, 454
trichloride process, 99
tunnelling, 200
TVRO, 453
two-terminal structures, 268

varactor, 310
v-E characteristic, 366
velocity overshoot, 398

velocity saturation, 379, 405
via hole, 413, 457, 484
  bath tub, 485
  etched, 487
VLSI, 263, 517
VPE, 95, 99

wet chemical etches, 139
work function, 198

XPS, 106

YIG, 307